Plants
for
Environmental Studies

Plants

for

Environmental Studies

Edited by

Wuncheng Wang
Joseph W. Gorsuch
Jane S. Hughes

LEWIS PUBLISHERS

Boca Raton   New York

Acquiring Editor:	Joel Stein
Project Editor:	Joan Moscrop
Cover design:	Denise Craig
PrePress:	Kevin Luong
Manufacturing:	Carol Royal

Library of Congress Cataloging-in-Publication Data

Plants for environmental studies/ edited by
   Wuncheng Wang, Joseph W. Gorsuch, Jane S. Hughes
      p.   cm.
   Includes bibliographical references and index.
   ISBN 1-56670-028-0
   1. Plant indicators. 2. Biological monitoring. 3. Environmental
monitoring. I. Wang, Wuncheng. II. Gorsuch, Joseph W. III. Hughes,
Jane S., 1947– .
QK754.P59 1997
581.7—dc21   96-47075
                                               CIP

© 1997 by CRC Press LLC
Lewis Publishers is an imprint of CRC Press

No claim to original U.S. Government works
International Standard Book Number 1-56670-028-0
Library of Congress Card Number 96-47075
Printed in the United States of America   2 3 4 5 6 7 8 9 0
Printed on acid-free paper

Preface

Environmental issues frequently discussed in today's news media, scientific meetings, and policy decision arenas include global climate change, the greenhouse effect, desertification, and soil erosion. Although different geographical locations may have different combinations and severities of these environmental issues, all of them are related directly or indirectly to the destruction or reduction of plant life on the planet.

Plants, both aquatic and terrestrial, are an essential part of a healthy environment. They produce oxygen and organic carbon essential to animal survival and provide habitat for many species of fish, birds, reptiles, amphibians, insects, and mammals. Plants are the basic component of agriculture, rangeland, forestry, and horticulture. In many ways, the welfare of plants is consistent with that of human beings and the environment.

Mandated by environmental regulations in the United States, many plant species are used as indicators of adverse environmental conditions. For example, phytotoxicity testing is used as part of the ecological risk assessment of industrial and agricultural chemicals, industrial and municipal effluents, food additives, packaging materials, leachates, solid wastes, hazardous sites, and the like. In addition, many reports have shown that plants have important roles in improving wastewater treatment and in remediating hazardous sites, the so-called phytoremediation.

The goal of this book is to present a broad and state-of-the-science overview of plants for environmental studies. The contents encompass vascular plants, in both aquatic and terrestrial environments, and algae. This book is timely because the demand for environmental studies using plant species has become more urgent, as evidenced by the recent literature, workshops, symposia, and consensus methods development.

This book is a joint contribution by authors from academia, governments, research institutions, and industries from the United States, Canada, and Europe. It contains diverse laboratory and *in situ* studies in air, water, wastewater, sediment, and soil. Chapters include environmental impacts on plants (UV and general radiation), plant mutation, statistical analyses of plant data, relationships between plants and water (quantity and quality), algae as ecosystem indicators and laboratory test organisms, use of plant species or communities for ecological risk assessment, and organic and inorganic compound accumulations by plants.

These important subjects constitute only a part of the current knowledge of plant environmental science. Plant studies play prominent roles in many other areas, such as air pollution, soil pollution, wetland ecology, riverine ecology, and aquatic and littoral zone ecology. Ecosystem approaches are being emphasized in scientific and regulatory arenas. Without doubt, plants are an essential component of any comprehensive environmental investigation. It is the hope of the editorial committee that this book will help stimulate a greater interest and shed more light on environmental monitoring and assessment using plant species.

The editorial committee is grateful to all the authors whose combined expertise is the foundation of this book. The committee also acknowledges the contributions of many reviewers whose timely and constructive comments are indispensable. The committee especially wishes to acknowledge Dr. William Lower, formerly of the Cancer Research Center in Columbia, MO and presently serving as a Peace Corps volunteer in Kazakhstan, for his contribution and encouragement; both were essential for the completion of this book.

Wuncheng Wang
Joseph W. Gorsuch
Jane S. Hughes

Editors

Wuncheng (Woodrow) Wang is a hydrologist with the U.S. Geological Survey. Before 1991, he was a principal scientist with the Illinois State Water Survey in Peoria, IL. He has been chairing two Joint Task Groups (*Lemna* and marsh plants) of the Standard Methods Committee since 1987 and chaired the First ASTM Symposium on Use of Plants for Toxicity Assessment in 1989. His interest is the use of plants for water quality assessment.

Joseph (Joe) W. Gorsuch is Director of Silver Issues with Eastman Kodak Company, Rochester, NY. For the 22 years prior to April 1, 1996, he was Group Leader and Senior Environmental Toxicologist at Kodak. He chaired the ASTM E47.11 Subcommittee on Plant Toxicology from 1992 to 1995, chaired the Second ASTM Symposium on Use of Plants for Toxicity Assessment in 1990, and co-chaired the First ASTM Symposium in 1989. He is interested in using plants to evaluate sludge application practices.

Jane S. Hughes is the founder and president of Carolina Ecotox, Inc., a contract environmental toxicology testing laboratory in Durham, NC. She has nearly 20 years of experience conducting and supervising aquatic toxicity testing with a variety of plants and animals to meet diverse regulatory requirements. Her specialty is aquatic plant toxicity testing, and she has chaired methods development activities relating to algae, duckweed, and aquatic macrophytes in ASTM's Committee E-47 on Biological Effects and Environmental Fate. She also served as co-chair and chair for the First and Third ASTM Symposia on Environmental Toxicology and Risk Assessment.

Contributors

Todd A. Anderson
The Institute of Wildlife and
  Environmental Toxicology
Clemson University
Pendleton, South Carolina

Amha Asfaw
Department of Statistics
University of Missouri
Columbia, Missouri

John Cairns, Jr.
Department of Biology
Virginia Polytechnic Institute and
  State University
Blacksburg, Virginia

Cheryl L. Duxbury
Department of Biology
University of Waterloo
Waterloo, Ontario, Canada

Nelson T. Edwards
Environmental Sciences Division
Oak Ridge National Laboratory
Oak Ridge, Tennessee

Mark R. Ellersieck
Department of Statistics
University of Missouri
Columbia, Missouri

Geirid Fiskesjö
Department of Genetics
University of Lund
Lund, Sweden

Joanna Gemel
Cancer Research Center
Columbia, Missouri

Robert W. Gensemer
Department of Biology
Boston University
Boston, Massachusetts

Karen E. Gerhardt
Department of Biology
University of Waterloo
Waterloo, Ontario, Canada

Bruce M. Greenberg
Department of Biology
University of Waterloo
Waterloo, Ontario, Canada

Jerry L. Hatfield
National Soil Tilth Laboratory
Agricultural Research Service
U.S. Department of Agriculture
Ames, Iowa

Brian H. Hill
Environmental Monitoring
  Systems Laboratory
U.S. Environmental Protection
  Agency
Cincinnati, Ohio

Robert W. Holst
Naval Research Laboratory
Washington, D.C.

Anne M. Hoylman
Graduate Program in Ecology
The University of Tennessee
Knoxville, Tennessee

Xiao-Dong Huang
Department of Biological
  Sciences
Wright State University
Dayton, Ohio

Jane S. Hughes
Carolina Ecotox, Inc.
Durham, North Carolina

Lawrence W. Jones
Energy, Environment, and
  Resources Center
The University of Tennessee
Knoxville, Tennessee

Young-Hwa Ju
Department of Crop Sciences
University of Illinois at Urbana-
  Champaign
Urbana, Illinois

Shubender Kapila
Center for Environmental Science
  and Technology
University of Missouri-Rolla
Rolla, Missouri

Lawrence A. Kapustka
ecological planning and
  toxicology, inc.
Corvallis, Oregon

Gary F. Krause
Department of Statistics
University of Missouri
Columbia, Missouri

Michael A. Lewis
National Health and
  Environmental Effects Research
  Laboratory
U.S. Environmental Protection
  Agency
Gulf Breeze, Florida

William R. Lower
Cancer Research Center
Columbia, Missouri

Paul V. McCormick
Everglades Systems Research
  Division
South Florida Water
  Management District
West Palm Beach, Florida

David J. Nagel
Naval Research Laboratory
Washington, D.C.

Niels Nyholm
Department of Environmental
  Science and Engineering
Technical University of Denmark
Lyngby, Denmark

Hans G. Peterson
Water Quality Section
Saskatchewan Research Council
Saskatoon, Saskatchewan,
  Canada

Michael J. Plewa
Department of Crop Sciences
University of Illinois at Urbana-
  Champaign
Urbana, Illinois

Rebecca L. Powell
Monsanto Company
Environmental Sciences Center
St. Louis, Missouri

Ravi K. Puri
Environmental Trace Substances
  Laboratory
University of Missouri-Rolla
Rolla, Missouri

Vivek Puri
Cancer Research Center
Columbia, Missouri

Ye Qiuping
Department of Chemistry
University of Missouri
Columbia, Missouri

Merrilee Ritter
Eastman Kodak Company
Rochester, New York

Otto J. Schwarz
Department of Botany
The University of Tennessee
Knoxville, Tennessee

Kwang-Young Seo
Department of Microbiology
University of Illinois at Urbana-
  Champaign
Urbana, Illinois

Shannon R. Smith
Department of Plant Biology
University of Illinois at Urbana-
  Champaign
Urbana, Illinois

Elizabeth D. Wagner
Department of Crop Sciences
University of Illinois at Urbana-
  Champaign
Urbana, Illinois

Barbara T. Walton
Environmental Sciences Division
Oak Ridge National Laboratory
Oak Ridge, Tennessee

Wuncheng Wang
U.S. Geological Survey
Columbia, South Carolina

Beth Waters-Earhart
Cancer Research Center
Columbia, Missouri

Michael I. Wilson
Department of Biology
University of Waterloo
Waterloo, Ontario, Canada

Contents

chapter one

The effects of ultraviolet-B radiation on higher plants

Bruce M. Greenberg, Michael I. Wilson, Xiao-Dong Huang, Cheryl L. Duxbury, Karen E. Gerhardt, and Robert W. Gensemer

Introduction

The stratospheric ozone layer, which attenuates solar ultraviolet-B (UV-B) radiation (290 to 320 nm), is being depleted by contaminants such as chlorofluorocarbons (Rowland 1989; Blumthaler and Ambach 1990; Frederick 1990; Frederick et al. 1991; Crutzen 1992; McFarland and Kaye 1992; Kerr and McElroy 1993). Increased UV-B has been traced to ozone depletion, and elevated UV-B levels have already been detected (Blumthaler and Ambach 1990; Kerr and McElroy 1993). Furthermore, the U.S. Environmental Protection Agency (EPA) reported that the rate of ozone depletion is proceeding much faster than originally predicted (Pool 1991). Thus, UV-B levels at the surface of the earth will almost certainly continue to increase. One serious concern is that peak chlorofluorocarbon production is occurring now, and it takes more than 20 years for the molecules to reach the stratosphere. Thus, this problem will continue well into the future, with maximal ozone depletion not a reality until well into the 21st century (Greenberg 1993).

Resultant negative impacts of UV-B on biological organisms are inevitable (Tevini et al. 1989; Coohill 1991). Terrestrial plants are especially vulnerable to UV-B due to their requirement for sunlight for photosynthesis. Each 1% decline in ozone is predicted to lead to a 1% diminishment in crop yields (Coohill 1991). The potential molecular sites of UV-B damage in plants are DNA, proteins, membranes, photosynthetic pigments, and phytohormones (Murphy 1983; Greenberg et al. 1989; Tevini et al. 1989; Kochevar 1990; Kramer et al. 1991; Pang and Hays 1991; Chow et al. 1992; Kramer et al. 1992; Quaite et al. 1992; Tevini 1993). In addition, the hazards of many environmental organic contaminants are activated or enhanced by light, and many of these chemicals have strong absorbance bands in the UV-B (Morgan et al. 1977; Cooper and Herr 1987; Newsted and Giesy 1987; Larson and Barenbaum 1988; Huang et al. 1993; Ren et al. 1994). The mechanism of damage to the plant biomolecules is mainly via active oxygen, photooxidation, and free radical reactions (Morgan et al. 1977; Foote 1979; Larson and Barenbaum 1988; Foote 1991; Greenberg et al. 1993a). This is the case whether the UV-B-absorbing species (i.e., the chromophore that initiates the photochemistry) is an endogenous compound (e.g., a protein, DNA or natural product) or an exogenous compound (e.g., a xenobiotic molecule).

The negative effects of increased UV-B on plants are likely to be incremental rather than total devastation. This is because plants have evolved under the selective pressure of ambient UV-B radiation in sunlight and have thus developed adaptive mechanisms (Caldwell 1981; Tevini and Teramura 1989; Tevini et al. 1989). One of the common adaptation processes is alteration in leaf transmittance properties, which results in attenuation of UV-B in the epidermis before it can reach the interior of the leaf (Caldwell 1981; Murali et al. 1988; Tevini et al. 1988, 1991; Day et al. 1992; Cen and Bornman 1993; Wilson and Greenberg 1993a). Biological pathways for the detoxification of active oxygen and free radicals also have elevated levels of activity in some plants following UV-B exposure, and they are a potential protection mechanism (Halliwell 1981; Asada and Takahashi 1987; Murali et al. 1988; Kramer et al. 1991; Strid 1993). DNA repair via photoreactivation and excision repair is another mechanism which can reverse UV-B-induced damage (Sancar and Sancar 1988; Langer and Wellmann 1990; Pang and Hays 1991; Stapleton 1992; Batschauer 1993; Kim and Sancar 1993; Sancar and Tang 1993; Murphy et al. 1993b; Harlow et al. 1994).

UV-B acclimation processes in plants are triggered by UV-B-specific morphogenic photoreceptors, as well as phytochrome and the UV-A/blue light receptor (Pratt and Butler 1970; Galland and Senger 1988; Lecari et al. 1990; Ballaré et al. 1991; Häder and Brodhum 1991; Ensminger and Schäfer 1992; Wilson and Greenberg 1993b). The extent of adaptation of plants is frequently related to UV-B irradiance. Thus, depending on the UV-B level, plants either can be damaged by UV-B (i.e., at higher doses) or can adapt to UV-B (i.e., at lower doses).

In this chapter, we will consider first the nature of UV-B radiation in sunlight, how one can mimic solar UV-B in the laboratory, and the biological effectiveness of UV radiation. Then the detrimental effects of UV-B on plants will be discussed with reference to photoinduced damage to biological systems. Finally, we will examine how plants can ammeliorate UV-B damage and how plants sense UV-B to trigger acclimation mechanisms.

Light sources for the study of UV-B effects

To systematically examine the responses of plants to UV-B, there are many advantages to using controlled laboratory growth conditions. It is possible to simulate solar radiation with respect to the visible to UV-A to UV-B ratio (Adamse and Britz 1992a; Huang et al. 1993; Krizek et al. 1993b; Middleton and Teramura 1993a; Wilson and Greenberg 1993a). As the UV-B content in the source is raised, fundamental processes such as photosynthesis and plant growth are inhibited (Jones and Kok 1966; Bornman 1989; Greenberg et al. 1989; Tevini et al. 1989). However, it has also been suggested that the high visible light backgrounds found in sunlight can diminish the negative effects of UV-B on plants (Warner and Caldwell 1983; Tevini and Teramura 1989; Adamse and Britz 1992b). Therefore, it is important to confirm responses to elevated UV-B observed in the laboratory with field studies.

The solar spectrum

Radiation from the sun with wavelengths greater than 290 nm can reach the surface of the earth (Henderson 1977). Wavelengths less than 290 nm are absorbed by the various gases (O_2, N_2, water, etc.) in the atmosphere and are not of environmental concern. At the surface of the earth, the molar ratio of visible light to UV-A to UV-B is about 100:10:1. The photon fluence rate (or photon flux) of visible light in sunlight is about 2000 μmol m^{-2} s^{-1} on a cloudless day. (This is equivalent to 500 W m^{-2}. Note, micromoles per square meter per second [μmol m^{-2} s^{-1}] is the same as microEinstein per square meter per second [μEinstein m^{-2} s^{-1}]. An Einstein is Avagadro's number [a mole] of photons and was previously used for expressing fluence rates, but it is no longer an SI supported unit.) However, the content of UV-B is highly variable. For example, on a clear day in the summer the UV-B to visible light ratio at latitudes and elevations corresponding to London, England is only about 0.2% of visible light on a photon basis (Figure 1A), while closer to the equator or at higher elevations the UV-B level is much higher: as high as 1.5% of visible light (Caldwell et al. 1980; Caldwell 1981; Gerstle et al. 1986; Blumthaler and Ambach 1990; Bachelet et al. 1991; Frederick et al. 1991). Also, peak levels of UV-B occur at solar noon, and the fraction of UV-B is maximal around the summer solstice and minimal around the winter solstice (Caldwell 1981; Gerstle et al. 1986; Bachelet et al. 1991; Frederick et al. 1991). In designing a laboratory light source that will mimic sunlight, one must take these factors into consideration.

Artificial visible light/UV-B sources

Based on the relative level of UV-B in sunlight (Figure 1A) (Henderson 1977; Reid et al. 1991), light sources with UV-B at about 1% of visible light on a photon basis can be built with fluorescent lamps (Figure 1B). One example is a source containing cool-white fluorescent lamps (visible light) and UV-B fluorescent lamps (FS-20, National Biological, Twinsburg, OH or RPR-3000, Southern New England Ultraviolet Co.) (Adamse and Britz 1992a; Wilson and Greenberg 1993a; Huang et al. 1993). The radiation from the UV-B lamp is filtered through cellulose diacetate (0.08 mm) to remove extraneous UV-C (<290 nm) (Krizek and Koche 1979; Adamse and Britz 1992a; Middleton and Teramura 1993a); the mercury (Hg) gas in all fluorescent lamps emits at 254 nm, but this UV-C radiation is not quantitatively attenuated by the glass and the phosphor in UV-B lamps. UV-C is much more damaging to biological molecules than UV-B and must therefore be completely eliminated. If necessary, the UV-B lamp can be screened with cheesecloth to achieve the desired visible light to UV-B fluence ratio. To mimic loss of the

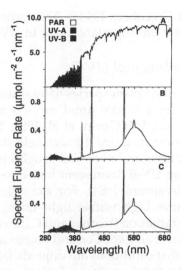

Figure 1 Spectral distribution of sunlight, a visible light plus UV-B source, and a simulated solar radiation (SSR) source. **Panel A:** Sunlight measured on a cloudless day in London, England (September 2, 1991). **Panel B:** Emission spectrum of a visible plus UV-B source filtered through cellulose diacetate. **Panel C:** Emission spectrum of an SSR source filtered through polystyrene. See text for further descriptions of the artificial light sources.

ozone layer, the UV-B level can be raised by removing successive layers of cheesecloth from the UV-B lamp and/or adding UV-B lamps.

Many plants (e.g., *Brassica napus*, rye, soybean, and *Lemna gibba*) can be grown under a visible/UV-B source similar to that described previously (Warner and Caldwell 1983; Mirecki and Teramura 1984; Teramura 1986; Tevini et al. 1991; Huang et al. 1993; Wilson and Greenberg 1993a). If the visible to UV-B ratio is held at 100:1, adverse signs of stress often are not observed. However, the plants will usually demonstrate acclimation responses at this UV-B level (Tevini et al. 1991; Cen and Bornman 1993; Wilson and Greenberg 1993a). If the relative level of UV-B is raised significantly (by 20 to 30%), damage such as inhibition of photosynthesis and leaf wrinkling are generally detected (Ambler et al. 1975; Caldwell 1981; Murali and Teramura 1986; Tevini 1993).

The photon fluence rates (photon flux) used in most laboratory settings are low (<500 μmol m^{-2} s^{-1}). One of the problems sometimes observed with low visible light levels is a lack of adequate acclimation to UV-B, resulting in increased sensitivity to UV-B (see later) (Warner and Caldwell 1983; Mirecki and Teramura 1984; Tevini et al. 1989; Cen and Bornman 1990). However, when the UV-B level is maintained at 1% of visible light, we found that *B. napus* shows strong adaptive responses to UV-B, even under low visible light levels (100 to 200 μmol m^{-2} s^{-1}) (Wilson and Greenberg 1993a). This indicates that the ratio of UV-B to visible light is an important factor in UV-B acclimation.

Sunlight with supplemental UV-B

As stated earlier, for some plant species natural sunlight or at least high fluence rate visible light (>1000 μmol m^{-2} s^{-1}) may be required for optimal acclimation to UV-B (Tevini et al. 1989; Tevini and Teramura 1989). This has been demonstrated through field studies where sunlight (approximately 2000 μmol m^{-2} s^{-1} of visible light) is supplemented with UV-B radiation (using UV-B fluorescent bulbs) to mimic ozone depletion (Mirecki and Teramura 1984). For example, soybean and bean plants seem to require high visible light fluence rates for effective acclimation to UV-B radiation (Warner and Caldwell 1983; Mirecki and Teramura 1984; Tevini et al. 1989; Cen and Bornman 1990). The problem with field studies is that conditions are unpredictable (e.g., visible light levels change due to cloud cover) and plants are often subjected to other forms of stress (e.g., heat). For instance, on a cloudy day, if the supplemental UV-B is not lowered, the ratio of UV-B to visible light will climb. Because some plants may be stressed if the UV-B to visible light ratio rises, this ratio should be meticulously controlled (Caldwell et al. 1983a). Another approach has been to selectively screen out the UV-B present in sunlight and determine how plants perform in full sunlight in the presence or absence of UV-B (Krizek 1978).

Full spectrum simulated solar radiation

Simulating sunlight with respect to both quantity and quality is expensive and technically difficult. However, full solar spectrum artificial lighting at relatively low fluence rates (<400 μmol m^{-2} s^{-1}) is not difficult to obtain and might compensate for the need to have the high fluence rates of full sunlight. For instance, it is likely that the levels of blue light and UV-A are important, as these spectral regions activate DNA repair via photolyase (Langer and Wellmann 1990; Pang and Hays 1991; Adamse and Britz 1992b; Kim et al. 1992; Batschauer 1993; Kim and Sancar 1993). UV-A/blue light also seems to be important for synthesis of protective compounds (e.g., carotenoids) and for stomatal control (Middleton and Teramura 1993b). In addition to combining UV-B with visible light as described earlier, the UV-A spectral region can be readily added to a laboratory light source to better replicate the spectrum of natural sunlight (UV-A fluorescent lamps, black lights, are available from most theatrical lighting suppliers). The level of UV-A in sunlight is relatively constant (about 10% of visible light) and does not vary greatly with latitude, altitude, or season (Henderson 1977; Reid et al. 1991). Also, UV-A will not increase as the ozone layer is depleted.

Light sources that mimic sunlight with respect to the relative amounts of visible light and UV (a visible light to UV-A to UV-B ratio of 100:10:1; Figure 1C) can be assembled with commercially available fluorescent lamps (Huang et al. 1993). One such simulated solar radiation (SSR) source contains four lamps: two cool-white fluorescent lamps, one 350-nm fluorescent lamp, and one 300-nm fluorescent lamp. The radiation from the 300-nm lamp is filtered through cheesecloth to bring the UV-B level down to 1% of visible light. The light is also filtered through cellulose acetate or polystyrene to remove all of the incident UV-C (200 to 290 nm). While the spectral output shown in Figure 1C does not precisely replicate sunlight, the visible light to UV-A to UV-B ratio corresponds to that of terrestrial sunlight in the 290 to 700 nm spectral region from mid-spring to mid-fall in temperate latitudes corresponding to southern Canada and the northern United States (Henderson 1977; Gerstle et al. 1986; Huang et al. 1993; Reid et al. 1991). *B. napus, Spirodela oligorrhiza, L. gibba,* and cucumber have been found to grow well under this source, showing no overt signs of UV-B stress (Huang et al. 1993; Greenberg et al. 1993b). To simulate ozone depletion, the UV-B content of the source can be raised by removing successive layers of cheesecloth from the UV-B lamp.

Biologically effective UV-B

To assess UV-B irradiance, the biologically effective dose is a useful calculation (Caldwell et al. 1986; Madronich 1992; Quaite et al. 1992). This is a weighting factor applied to the UV-B irradiance to account for the wavelength dependence of biologically activating or damaging responses. In general, as the wavelength becomes shorter and the energy increases, increased levels of biological damage are observed. Biologically effective radiation is the integrated cross-sectional overlap between an action spectrum for a biological effect and the spectral output of the light source used to grow the plants. Typical action spectra used for such manipulations are a general plant damage spectrum (Caldwell et al. 1986), inhibition of photosynthesis (Jones and Kok 1966; Bornman et al. 1984), *in vitro* and *in vivo* DNA damage spectra (Setlow 1974; Quaite et al. 1992), and the erythema spectrum (McKinlay and Diffey 1987). Other *in vivo* plant action spectra that can be used include flavonoid synthesis (Beggs and Wellman 1985; Hashimoto et al. 1991), degradation of the D1 protein of photosystem II (PSII) (Greenberg et al. 1989), and curvature of seedlings (Curry et al. 1956). For instance, if biologically effective radiation for plant acclimation to UV-B is needed, an action spectrum for UV-B activation of flavonoid biosynthesis might be used because flavonoids are associated with protection of plants from UV-B. Interestingly, Hashimoto et al. (1991) showed that flavonoid synthesis is inhibited at higher UV-B fluence rates, indicating that the biosynthetic machinery was damaged at higher UV-B levels. The action spectrum for UV-B inhibition of flavonoid biosynthesis was different than the action spectrum for UV-B activation of synthesis. The peak for activation of synthesis was at 290 nm, typical for UV-B photoreceptors associated with morphological and acclimation responses. Conversely, inhibition of flavonoid synthesis increases as the wavelength drops from 295 to 254 nm with a shoulder at 280 nm, indicating damage to proteins. Therefore, these two action spectra can be used to compare biologically effective radiation for UV-B acclimation and UV-B stress.

To determine the content of biologically effective UV radiation, the action spectrum is normalized at a given wavelength (usually at 300 nm). The spectral output of the light source is multiplied by the normalized action spectrum. This weighted spectrum is integrated from 290 to 320 nm to give a biologically effective light dose of UV-B (in joules or micromoles) and then converted to a daily dosage. Typical daily dosages based on the generalized plant damage spectrum are 15, 8.4, and 2.4 kJ m^{-2} for Venezuela, Utah, and Alaska, respectively (Caldwell et al. 1980).

Detrimental effects of UV-B on plants

UV-B radiation triggers numerous responses in plants, including inhibition of photosynthesis, membrane damage, protein damage, DNA damage, delayed maturation, diminished growth, activation of chemical stress, flavonoid synthesis, and leaf thickening (Table 1). Some of

Table 1 Responses of Plants to UV-B Radiation

Acclimation and morphological responses	Damage and injury responses
Altered biomass distribution[a]	Altered gene expression[a]
Altered leaf cell division	Degradation of auxin
Cotyledon curling	Degradation of chlorophyll and carotenoids
Increased DNA repair	Degradation of proteins
Increased flavonoid biosynthesis	Diminished biomass
Increased leaf thickness	Epidermal collapse
Increased number of leaves	Inhibition of growth
Increased number of tillars	Inhibition of photosynthesis
Leaf wrinkling	Increased stomatal conductance
Reduced leaf area	Lower seed yield
Reduced hypocotyl growth	Oxidation of DNA
Reduced shoot height	Peroxidation of lipids
Reduced stomatal density	Pyrimidine dimer formation

[a] Entries are given in alphabetical order.

the effects represent damage (e.g., inhibition of photosynthesis), while others represent acclimation (e.g., flavonoid synthesis). Both laboratory and field studies have shown that elevated UV-B results in plant damage (Jones and Kok 1966; Teramura 1983; Tevini et al. 1989; Tevini and Teramura 1989; Greenberg et al. 1993a). For instance, in a field study performed over 6 years, Teramura and Sullivan (1988) found consistently lower yields for soybean (var. Essex) due to a UV-B level approximating a 25% loss of stratospheric ozone. However, it is unknown specifically which of the negative impacts listed in Table 1 are the most detrimental, at precisely what fluence rates they impact on plants, and to what degree generalizations can be made between different plant species.

Photochemical processes relevant to UV-B effects

UV-B-induced damage to plants depends on a few basic photochemical processes (Foote 1979, 1991; Larson and Barenbaum 1988). Whether the chromophore is an endogenous product or a xenobiotic, there are essentially two ways light can promote damage to a biological system: photomodification and photosensitization (Figure 2). Photomodification, most commonly an oxidation reaction, results in the formation of new compounds that can have altered biological activity (Larson and Baren-

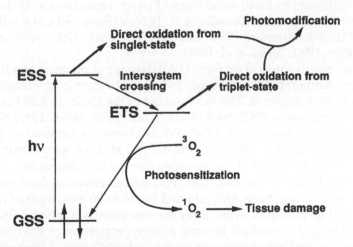

Figure 2 Jablonski diagram for ground and excited states of a chemical. Following absorbance of a photon of energy hv, the molecule is elevated from the ground singlet state (GSS) to an excited singlet state (ESS), from which it can react directly (photomodification), or it can intersystem cross to the excited triplet state (ETS). The triplet-state molecule can also react directly (photomodification) or it can react with ground triplet-state oxygen ($^3O_2$), forming excited singlet-state oxygen ($^1O_2$).

baum 1988; Huang et al. 1993; Greenberg et al. 1993a). A photooxidized biomolecule, such as a protein, will generally have diminished activity (Wolf et al. 1986; Davies 1987).

Photosensitization reactions usually proceed via the formation of highly toxic singlet-state oxygen (Foote 1979, 1991; Krinsky 1979; Larson and Barenbaum 1988). The process begins with the molecule absorbing a photon, which elevates it to an excited singlet state (Figure 2). From there the molecule can be transformed by intersystem crossing to the excited triplet state, where it can react with ground triplet-state oxygen to form excited singlet-state oxygen. Singlet-state oxygen can attack almost any biomolecule to form an organic peroxide. For instance, lipid peroxides greatly inhibit membrane fluidity and function (Krinsky 1979; Thompson 1984; Girotti 1990; Kochevar 1990).

Sensitivity of the photosynthetic apparatus to UV-B

When plants are initially exposed to UV-B, signs of stress are generally observed. The effects on photosynthesis are especially acute (Jones and Kok 1966; Bornman 1989; Tevini et al. 1989). If the UV-B level is low (<1% of visible), many plants ultimately adapt, and control levels of net photosynthetic activity are recovered after continued UV-B exposure (Tevini et al. 1991; Wilson and Greenberg 1993a; Middleton and Teramura 1993b). However, at higher UV-B fluence rates, plants cannot fully acclimate and will show signs of permanent damage to the photosynthetic apparatus (Brandle et al. 1977; Warner and Caldwell 1983; Mirecki and Teramura 1984; Cen and Bornman 1990; Sullivan and Teramura 1990; Tevini et al. 1991).

An effective way to measure UV-B damage of photosystem II (PSII) is room-temperature fluorescence induction (Figure 3; for greater elaboration, see Chapter 8). This is a measure of the ability of PSII to reduce the plastoquinone (PQ) pool (Judy et al. 1990; Miles 1990; Bolhar-Nordenkampf and Oquist 1993). When plants are exposed to photosynthetically active radiation (visible light) at room temperature, PSII emits a small fraction of the absorbed energy as fluorescence. During photosynthetic electron transport, PSII passes electrons from water to PQ. If plants have been dark adapted for 15 to 30 min, most of the PQ pool will be oxidized. When light is subsequently applied, initial fluorescence (F_O) is minimal because a large proportion of the incoming energy is stored as reducing equivalents in the PQ pool. Assuming fully functional PSII, after about 2 s in the light the PQ pool is quantitatively reduced and fluorescence rises two- to fourfold to a maximal level (F_M). This fluorescence induction process is referred to as the Kautsky effect (Kautsky and Hirsch 1934). F_M occurs because the secondary electron acceptor, Q_A, is reduced when the PQ pool is fully reduced, and, therefore, none of the light energy can be stored as chemical energy. This rise in fluorescence due to reduction of the PQ pool is termed F_V where

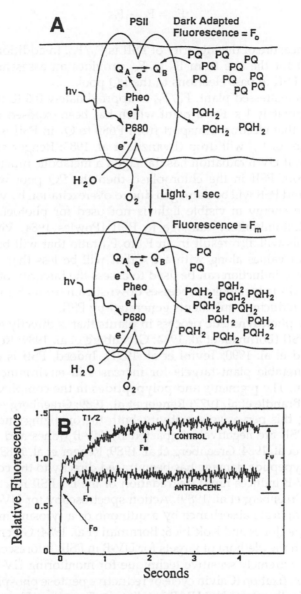

Figure 3 Fluorescence induction from PSII. **Panel A:** Diagram of the PSII reaction center and the PQ pool after dark adaptation and after a short (1 s) light exposure. Minimum fluorescence (F_O) is observed after dark adaptation and maximal fluorescence (F_M) after the short light exposure. If PSII is damaged such that electrons cannot be transferred to Q_A, F_M will be reached immediately. If PSII is damaged such that electrons cannot be transferred to Q_B, then F_O and F_M will be approximately normal, while the time period required to reach F_M will be very short (<1 s). **Panel B:** Typical fluorescence induction curves for control and inhibited PSII. Damage to PSII was induced *in vivo* by incubating *L. gibba* in anthracene (a PAH) under SSR for 24 h.

$$F_V = F_M - F_O$$

A good measure of the integrity of PSII is F_V/F_M. In addition, the half-time $(t_{1/2})$ for the rise from F_O to F_M provides an assessment of the ability of PSII to pass electrons to the PQ pool.

In a nonstressed plant, F_V/F_M is approximately 0.6 to 0.8 and $t_{1/2}$ is approximately 1 s. For a plant which has been exposed to UV-B to an extent that electron transport from P680 to Q_A in PSII is inhibited, F_V/F_M and/or $t_{1/2}$ will drop (Iwanzik et al. 1983; Renger et al. 1989). Moreover, if UV-B radiation has damaged a metabolic function downstream from PSII in the chloroplast, then the PQ pool will remain reduced and PSII will be damaged due to overexcitation by visible light (when the energy in visible light is not used for photochemistry, it causes photoinhibitory damage to PSII [Powles 1984; Virgin et al. 1993]). This will also result in an F_V to F_M ratio that will be less than the control values along with a $t_{1/2}$ that will be less than 1 s. Thus, fluorescence induction can be used to assess the integrity of PSII or to detect blocks in metabolic processes downstream from PSII which consume the reducing equivalents generated by PSII.

A key photosynthetic process in plants that is directly affected by UV-B is PSII (Bornman et al. 1984; Greenberg et al. 1989; Renger et al. 1989; Strid et al. 1990; Tevini et al. 1991). Indeed, PSII is one of the most vulnerable plant targets for increases in environmental UV-B irradiation. The pigments and polypeptides in the complex are labile in UV-B (Brandle et al. 1977; Renger et al. 1989; Greenberg et al. 1989). Reflecting this pervasive sensitivity, both the oxidizing and reducing sides of PSII are negatively impacted by UV-B (Jones and Kok 1966; Bornman et al. 1984; Greenberg et al. 1989; Renger et al. 1989). Furthermore, polypeptide damage has been traced directly to the core of PSII, where UV-B induces rapid degradation of the D1 PSII reaction center protein (Greenberg et al. 1989). Action spectroscopy for UV-B impacts on PSII implicate absorbance by a quinone or a tyrosine in initiating the damage (Jones and Kok 1966; Bornman et al. 1984; Greenberg et al. 1989). With the plethora of targets for UV-B in PSII, fluorescence induction is an extremely sensitive technique for monitoring UV-B stress.

Carbon fixation (Calvin cycle or reductive pentose phosphate cycle) is dramatically affected by UV-B irradiation (Tevini and Teramura 1989; Ziska et al. 1993). In particular, the central enzyme in the carbon fixation pathway, ribulose-1,5-bisphosphate carboxylase/oxygenase (Rubisco), is sensitive to UV-B (Vu et al. 1982, 1984; Strid et al. 1990; Jordan et al. 1992). For instance, in *Pisum*, Rubisco activity is diminished by 90% after 8 d of enhanced UV-B (Strid et al. 1990). Active Rubisco is composed of eight large (54 kDa) and eight small (14 kDa) subunits. Recently, we found that treatment of plants with UV-B results in modification of the large subunit of Rubisco such that it migrates with an apparent molecular weight of 65 kDa in SDS-PAGE (Wilson et al. 1995). Furthermore, after treatment of isolated Rubisco with UV-B, a decrease in fluorescence at 330 nm is observed, indicating photomodification of aromatic residues in the protein (Caldwell 1993). This could represent damage to the protein that renders it inactive; perhaps it is the beginning of a degradation pathway. Indeed, the amount of Rubisco protein drops in the presence of UV-B radiation partially due to enhanced protein degradation (Vu et al. 1982, 1984). Also, the steady-state levels for the mRNA for both subunits of Rubisco decline (Jordan et al. 1992), and *de novo* synthesis of Rubisco is diminished under UV-B irradiation (Jordan et al. 1994).

Damage to the plasma membrane, proteins, and DNA

UV irradiation of plants results in several changes to the plasma membrane, including dissipation of electrical potential, efflux of potassium and bicarbonate, and changes in cellular pH (Murphy 1983; Murphy et al. 1993a). One metabolic activity inhibited by UV-B action is a plasma membrane ATPase (Murphy 1983). An action spectrum for this response shows a peak at 290 nm, which is relevant to UV-B increases due to loss of ozone. Further, inhibition of the ATPase occurs at levels of UV-B ($<$10 μmol m^{-2} s^{-1}) that are already present in the environment. The inactivation mechanism seems to result from a singlet oxygen reaction with a tryptophan residue (Imbrie and Murphy 1984). Microsomal membrane proteins appear to be particularly sensitive to photomodification and may also initiate lipid peroxidation via photosensitization processes (Kochevar 1990; Caldwell 1993).

The effects of UV-B on PSII, Rubisco, and the plasma membrane ATPase indicate that proteins are hypersensitive to oxidation induced by UV-B irradiation. The aromatic amino acids have absorbance maxima around 280 nm, and this absorbance band is still significant from 290 to 300 nm. Thus, UV-B is absorbed by any protein that has a Trp, Tyr, or Phe residue. Absorbance of a photon, especially by a Trp, can lead to a single electron oxidation of the residue and the reduction of oxygen to superoxide (Walrant and Santus 1974; Caldwell 1993), potentially resulting in enzyme inactivation and/or protein aggregation (Walrant and Santus 1974). For instance, Caldwell (1993) demonstrated, using fluorescence spectroscopy, that UV-B treatment of Rubisco resulted in oxidation of a Trp. Further, an action spectrum for inhibition of flavonoid synthesis by high UV-B irradiance indicates protein absorbance is involved in inactivation (Hashimoto et al. 1991). The D1 PSII reaction center protein shows accelerated degradation in UV-B as well, and the peak of the D1 degradation action spectrum is at 300 nm (Greenberg et al. 1989).

Absorbance of UV-B photons by DNA results in damage to the genetic code (Peak and Peak 1986; Stapleton 1992). The modifications are cyclobutane-pyrimidine dimers, 6,4-pyrimidine dimers, and singlet oxygen-mediated oxygenations (Saito et al. 1983; Peak and Peak 1986; McLennan 1987; Pang and Hays 1991; Quaite et al. 1992; Sancar and Sancar 1988). The number of lesions is dependent on the UV-B dose, and the type of lesion that dominates is dependent on wavelength. The most common modifications due to environmental UV-B are cyclobutane-type dimers. Without repair, DNA replication and gene transcription will be disrupted.

Changes in plant morphology

Morphologically, one of the most sensitive areas of a plant to UV-B is the leaf surface (Tevini et al. 1983; Steinmuller and Tevini 1986; Cen and Bornman 1993). This is reflected in the increased stomatal conductance and/or diminished integrity of the epidermal cell layer in plants adapted to UV-B compared to visible-light-grown control plants (Cen and Bornman 1993; Greenberg et al. 1993b). Also, the total number of stomata are diminished after exposure of plants to UV-B (Tevini et al. 1983). Partial collapse of the epidermal layer in response to UV-B has been observed in several plant species (Tevini et al. 1983; Cen and Bornman 1990; Bornman and Vogelmann 1990). Moreover, UV-B can block differentiation of stomatal cells in the upper epidermis (Tevini et al. 1983). This may be the result of one of the major routes of plant adaptation to UV-B, which is attenuation of UV-B by flavonoids in the epidermal layer. Thus, while the mesophyll can be protected from UV-B by screening pigments in epidermal cells, the epidermis itself remains fully exposed to UV-B and subject to damage even following acclimation of the plant. Furthermore, cell division during leaf expansion is inhibited by UV-B and may be the basis for changes in leaf shape (Ambler et al. 1975; Sisson and Caldwell 1976; Dickson and Caldwell 1978; Wilson and Greenberg 1993a). Furthermore, the cytoskeleton of plant cells, which carries out critical functions during cell division and expansion, is disrupted by UV-B (Staxén et al. 1993).

Photoinduced toxicity of xenobiotics

UV-B radiation can increase the hazards of photoactive pollutants in the environment (Cooper and Herr 1987; Larson and Barenbaum 1988). For instance, polycyclic aromatic hydrocarbons (PAHs) absorb strongly in the UV-B and show enhanced phytotoxicity in the presence of actinic radiation (Newsted and Giesy 1987; Schoeny et al. 1988; Huang et al. 1993). One of the major routes is photooxidation of the compounds, which increases the phytotoxicity of the chemicals relative to intact PAHs. In the case of *L. gibba* and *B. napus*, the photooxidized chemicals inhibit growth of leaves and roots (Huang et al. 1993; Greenberg et al. 1993a, 1993b). Chlorosis and inhibition of photosynthesis are common mechanistic manifestations of photoactive pollutants (Duxbury et al. 1993).

The extent that photoactive pollutants can affect plants in the environment is unclear. However, as chemicals move through the environment, they will be exposed to light. This will increase the hazards of some of the chemicals as many common organic pollutants, especially those with aromatic rings, absorb solar UV-B radiation and become photochemically active (Figure 4). Thus, the toxicity of these chemicals has the potential to increase as UV-B increases.

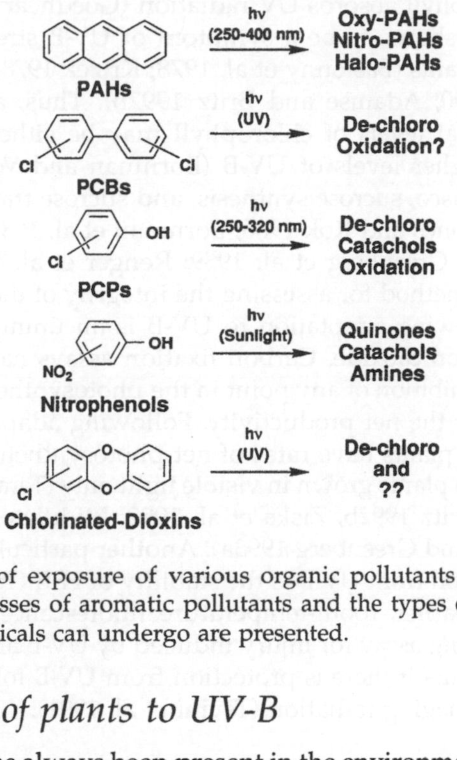

Figure 4 Result of exposure of various organic pollutants to sunlight. The most common classes of aromatic pollutants and the types of photoreactions which these chemicals can undergo are presented.

Acclimation of plants to UV-B

Because UV-B has always been present in the environment, acclimation mechanisms have evolved in plants which diminish the damaging effects of the radiation (Caldwell 1981; Caldwell et al. 1989; Tevini and Teramura 1989; Tevini et al. 1989; Barnes et al. 1990; Ensminger 1993). This is especially true for plants from regions of naturally high levels of UV-B irradiation. For instance, UV-B fluence rate gradients exist in the environment due to changes in latitude and/or elevation. Studies of plant species distributed along these natural UV-B gradients have revealed that many plants from lower latitudes or higher elevations, where UV-B is greater, have more pronounced adaptive mechanisms than plants from higher latitudes and/or lower elevations (Robberecht et al. 1980; Caldwell 1981; Larson et al. 1990; Sullivan et al. 1992). Even within a given plant species, some cultivars are more tolerant of UV-B than others (Biggs et al. 1981; Murali and Teramura 1986; Murali et al. 1988; Sullivan et al. 1992; Middleton and Teramura 1993b).

Measures of protection

One way to assess if plants are adapted to growth in UV-B and thus are protected from higher levels of UV-B is to assay chlorophyll content and photosynthetic activity during or after growth in UV-B (Caldwell et al. 1982; Tevini et al. 1988; Barnes et al. 1990; Wilson and Greenberg 1993a). Chlorophyll absorbs UV radiation (Goedheer 1966), and chlorosis has been shown to be a symptom of UV-B stress, especially in UV-sensitive plants (Basiouny et al. 1978; Krizek 1978; Teramura 1983; Strid et al. 1990; Adamse and Britz 1992b). Thus, a plant that can maintain normal levels of chlorophyll may be either tolerant of, or adapted to, higher levels of UV-B (Bornman and Vogelmann 1990). Also, PSII, Rubisco, sucrose synthesis, and sucrose transport are sensitive to UV-B (Jones and Kok 1966; Bornman et al. 1984; Vu et al. 1984; Bornman 1989; Greenberg et al. 1989; Renger et al. 1989; Strid et al. 1990). A good method for assessing the integrity of the photosynthetic apparatus following adaptation to UV-B is an unimpaired ability to assimilate carbon dioxide. Carbon fixation assays have the ability to determine if inhibition of any point in the photosynthetic chain is great enough to alter the net productivity. Following adaptation to growth in UV-B, many plants have rates of net photosynthetic carbon assimilation similar to plants grown in visible light only (Teramura et al. 1990; Adamse and Britz 1992b; Ziska et al. 1992; Middleton and Teramura 1993b; Wilson and Greenberg 1993a). Another particularly useful measure of protection from UV-B is the stability of PSII (Tevini et al. 1991). As discussed earlier, room-temperature fluorescence induction from PSII is a revealing assay for injury induced by UV-B and has thus been used to determine if there is protection from UV-B following acclimation to this damaging radiation (Tevini et al. 1991).

Alteration of leaf transmittance properties

Leaf morphology

A potential mechanism that would allow the photosynthetic apparatus to remain fully viable under UV-B insult is alterations in leaf transmittance properties (Figure 5). The result of this process is to attenuate UV-B radiation before it reaches a target in the mesophyll (Caldwell 1981; Tevini et al. 1989). Leaf thickening has been suggested as a UV-B screening mechanism because it would, for example, increase the path length for photons (Bornman and Vogelmann 1990). Fiberoptic and light microscopic analyses revealed that *B. napus* grown in UV-B plus visible light had modestly thicker leaves than control plants

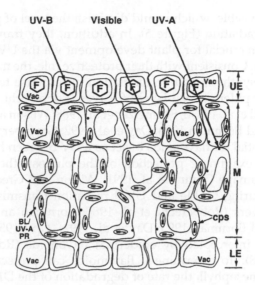

Figure 5 Transmittance of solar radiation through leaf tissue. A drawing of a cross section of a leaf is shown. UV-B is absorbed by flavonoids (F) in vacuoles (Vac) of the upper epidermis (UE). Visible light and UV-A radiation pass through the upper epidermis unimpeded by flavonoids. Visible light is absorbed by photosynthetic pigments in chloroplasts (cps) in the mesophyll (M). UV-A radiation activates the blue light/UV-A photoreceptor (BL/UV-A PR). The lower epidermis (LE) is also shown, illustrating the absence of UV-B protecting flavonoids.

(Cen and Bornman 1993). Similar results were obtained by measuring leaf density thickness (fresh weight per unit area) (Wilson and Greenberg 1993a).

Changes in leaf shape

Changes in leaf shape have also been suggested as possible UV-B protection mechanisms (Barnes et al. 1990). This could diminish the leaf cross-sectional area exposed to UV-B. For example, it was found that *B. napus* cotyledons curl upward in response to UV-B (Wilson and Greenberg 1993b). It has been shown also that leaves of many UV-B-acclimated plants have smaller areas, wrinkled morphologies, and waxy buildups, all of which could lower UV-B exposure (Steinmuller and Tevini 1986; Cen and Bornman 1993; Middleton and Teramura 1993b; Wilson and Greenberg 1993a).

Epidermal flavonoids

UV-B screening pigments, usually phenylpropanoids (flavonoids), are synthesized in response to UV-B in many plant species (Robberecht et al. 1980; Strack et al. 1982; Harborne 1988; Tevini et al. 1989; Bornman and Vogelmann 1990; Day et al. 1992; Li et al. 1993). They are an efficient screen because they absorb strongly from 280 to 340 nm, but do not absorb in the visible, which would diminish the level of photosynthetically active radiation (Figure 5). In addition, they transmit UV-A, a spectral region crucial for plant development via the UV-A/blue light photoreceptor. Consistent with their protective role, the majority of new flavonoid biosynthesis and accumulation in response to UV-B occurs in the upper epidermal cell layer of leaves (Robberecht and Caldwell 1978; Caldwell et al. 1983b; Schmelzer et al. 1988; Wilson and Greenberg 1993a; Cen and Bornman 1993; Day et al. 1992; Greenberg et al. 1993b).

The potential for flavonoids to act as a UV-B screen has been demonstrated by examining protection of the photosynthetic apparatus from UV-B damage (Tevini et al. 1991; Wilson and Greenberg 1993a) and by measuring visible and UV-B radiation transmission through epidermal layers (Robberecht et al. 1980; Bornman and Vogelmann 1990; Cen and Bornman 1993; Day et al. 1992). Up to 95% of incident UV-B can be intercepted by epidermal flavonoids (Robberecht and Caldwell 1978). To determine if flavonoids can protect a molecular target in the mesophyll, the rate of degradation of the D1 PSII reaction center protein in UV-B can be assayed (Mattoo et al. 1984; Greenberg et al. 1989; Wilson and Greenberg 1993a). The rate of degradation of this protein is proportional to the fluence of UV-B photons reaching the mesophyll (Greenberg et al. 1989). In the case of *B. napus*, UV-B-induced degradation of the D1 PSII reaction center protein was 50% slower in UV-B-adapted plants than in control plants (Wilson and Greenberg 1993a). Notably, there was a concomitant twofold increase in leaf flavonoid levels in the adapted plants relative to the control plants. However, when visible light was used to promote degradation, the rate was the same in adapted and control plants, indicating that only UV-B photons were intercepted by the flavonoids. Similar results were achieved by Tevini et al. (1991), where protection of PSII was measured by fluorescence induction. Thus, flavonoids in the epidermal layer can effectively and specifically attenuate UV-B.

UV-B induction of flavonoid biosynthesis is regulated at the gene level under the control of an unidentified UV-B photoreceptor (Kruezaler et al. 1983; Beggs et al. 1986; Dangl et al. 1987; Schmelzer et al. 1988; Hahlbrock and Scheel 1989; Kubasek et al. 1992). The best-characterized gene for an enzyme in the flavonoid biosynthetic pathway is the gene for chalcone synthase (CHS), which is present as a multigene family (Koes et al. 1989; Wingender et al. 1989). CHS catalyzes the committal step in flavonoid biosynthesis (Hahlbrock 1981; Harborne 1988). In addition to UV-B activation, the flavonoid pathway, and in particular the CHS gene, is activated by a variety of other environmental stimuli, e.g., fungal elicitors and pathogen attack (Koes et al. 1989; Wingender et al. 1989, 1990; Lozoya et al. 1991; Kubasek et al. 1992). For plant species examined to date, the majority of CHS gene transcription in response to UV-B occurs in epidermal cells, where the flavonoids are ultimately localized (Hrazdina et al. 1982; Beerhues et al. 1988; Jahnen and Hahlbrock 1988; Schmelzer et al. 1988; Cen and Bornman 1993; Jahne et al. 1993). However, this pattern may differ in other species (Day et al. 1992). Analysis of the CHS gene promoter revealed DNA segments 5' to the transcription start site (*cis*-elements) that are required for UV-B induction of gene expression (Schulze-Lefert et al. 1989; Staiger et al. 1989; Block et al. 1990; van der Meer 1990). Also, proteins (*trans*-acting factors) that bind to CHS promoter regions have been characterized (Staiger et al. 1990; Weishaar et al. 1991). However, little else is known about the signal transduction pathway for activation of the CHS gene following UV-B exposure. Other genes in the flavonoid pathway that are known to be activated by UV-B include phenylalanine ammonia lyase, 4-coumarate:CoA ligase, chalcone flavone isomerase, and dihydroflavonol-4-reductase (Fritzmeier et al. 1983; Douglas et al. 1987, 1991; van Tunen et al. 1988; Beld et al. 1989; Blyden et al. 1991).

DNA repair

DNA repair appears to be an important UV-B protective mechanism. DNA damaged by UV-B radiation can be restored by three mechanisms: photoreactivation, excision repair, and recombination (Sancar and Sancar 1988; Stapleton 1992). It should be noted that DNA repair involves a response to damage from UV-B, not the prevention of damage. Nonetheless, an increased ability to repair DNA is an important aspect of adaptation of UV-B radiation. UV-B-induced lesions in DNA are known to be diminished in the presence of UV-A and blue light, a process which occurs in plants, animals, and microbes (Sancar and Sancar 1988; Pang and Hays 1991). The enzyme that carries out this reaction, photolyase, reverses cyclobutane-type pyrimidine dimers (one of the major UV-B-induced DNA lesions [see Changes in Plant Morphology section]). UV-A and/or blue light is required for this process because the enzyme uses light energy of these wavelengths to drive the repair reaction (McLennan 1987; Pang and Hays 1991). Recently, a DNA photolyase from mustard was found to be highly homologous to yeast and bacterial photolyases (Batschauer 1993). In addition, the transcription of the photolyase gene is under light control, although UV-B is not absolutely required for activation of transcription.

DNA damage other than cyclobutane-type dimers between pyrimidines can only be repaired by excision or recombination (Stapleton 1992). Excision repair involves nicking DNA near the lesion, removal of several bases from one DNA strand around the lesion, and resynthesis with DNA polymerase (McLennan 1987; Sancar and Sancar 1988). An endonuclease from carrot that nicks the DNA to start the excision repair process has been reported (McLennan and Eastwood 1987), although it is not yet known if this mechanism is important for repair of UV-B-induced lesions. Interestingly, a number of *Arabidopsis* mutants putatively deficient in excision repair of 6,4-pyrimidine dimers and/or photoreactivation are more sensitive to UV-B than the wild-type plants (Britt et al. 1993; Harlow et al. 1994). Recombinational repair has not been reported in plants (Stapleton 1992), so it is unknown if this is a possible repair mechanism for UV-B damage to plant DNA.

Free radical and active oxygen detoxification

Another route for protection against UV-B stress is scavenging of active oxygen and other radical species (Asada and Takahashi 1987). Upon absorbance of a photon by a molecule, there are two possible routes for photosensitized oxidation (Foote 1979, 1991). Type I photosensitization results in peroxide production, while type II photosensitization leads to generation of singlet oxygen (Figure 2) (Kochevar 1990; Salmon et al. 1990; Caldwell 1993). DNA photoproducts and tryptophan radicals can be generated by UV-B radiation. In the case of UV-B-mediated PAH toxicity to plants, singlet oxygen is considered to be a primary factor in damage (Greenberg et al. 1993a).

Carotenoids are good quenchers of oxygen radicals and can protect against chlorophyll photobleaching (Foote et al. 1970). The concentration of carotenoids was reported to rise in response to UV-B in soybean, and these increases were found to correlate with protection of net photosynthesis from UV-B (Middleton and Teramura 1993b). Also, flavonoids have free radical scavenging activity (Husain et al. 1987; Takahama 1988; Huguet et al. 1990). This property of flavonoids could be especially important for protecting epidermal cells from UV-B. Additionally, singlet oxygen is rapidly reduced to superoxide, which can be detoxified by superoxide dismutase/glutathione reductase/ascorbate peroxidase systems in plants (Halliwell 1981; Murali et al. 1988; Bowler et al. 1991; Asada 1992; Krizek et al. 1993). The relative levels of these enzymes change in some plant species following UV-B adaptation. For instance, Strid (1993) demonstrated that glutathione reductase gene expression in *Pisum sativum* is induced by UV-B, although surprisingly superoxide dismutase gene expression seemed to drop concomitantly. Thus, active oxygen scavenging capacity apparently can increase following adaptation to UV-B radiation, although the precise mechanism remains unclear.

UV-B photomorphogenic photoreceptors

The ability of plants to acclimate to UV-B depends on their capacity to detect elevated levels of damaging UV-B radiation. It is clear that many plants can specifically respond to UV-B (Caldwell 1981; Steinmetz and Wellmann 1986; Caldwell et al. 1989; Tevini and Teramura 1989; Barnes et al. 1990; Ballaré et al. 1991; Sullivan et al. 1992; Tevini 1993). Accordingly, there must be a photoreceptor(s) that detects the UV-B photons and triggers the necessary changes in development. Three general classes of morphogenic photoreceptors have been identified in higher plants: phytochrome, a UV-A/blue light photoreceptor(s), and a UV-B photoreceptor(s) (see references in Attridge 1990).

Considerably more is known about phytochrome and the UV-A/blue light photoreceptor(s) than the UV-B photoreceptor(s) (Attridge 1990; Briggs and Short 1991; Eilfeld and Haupt 1991; Sage 1992; Ahmed and Cashmore 1993). Much of the information about phytochrome and UV-A/blue light detection was derived first from studies on specific, readily quantifiable, morphogenic responses, followed by characterization of the photoreceptors at the molecular level (Mohr 1986; Baskin and Iino 1987; Vierstra and Quail 1986; Briggs and Short 1991; Eilfeld and Haupt 1991; Warpeha et al. 1992). The amount of light required to activate phytochrome or the UV-A/blue light photoreceptor is on the order of 10^{-3} to 10 μmol m^{-2} total fluence. Responses that have been used to characterize phytochrome include induction of seed germination, flowering, enzyme activity, leaf expansion, and gene expression (Attridge 1990; Eilfeld and Haupt 1991; Sage 1992). Processes triggered by UV-A/blue light include phototropism and regulation of growth and development (Gaba and Black 1987; Liscum et al. 1992; Ahmed and Cashmore 1993; Palmer et al. 1993).

Generally, the total fluence of photons required to trigger UV-B photomorphogenic responses are 10 to 10^3 μmol m^{-2}, considerably higher than the threshold for many responses mediated by phytochrome and UV-A/blue light (Beggs et al. 1986; Steinmetz and Wellmann 1986; Baskin and Iino 1987; Takeda and Abe 1992; Wilson and Greenberg 1993b). As discussed previously, UV-B-induced photomorphogenic responses include growth inhibition, changes in leaf shape, increases in expression of specific genes, and synthesis of flavonoids and anthocyanins (Curry et al. 1956; Beggs et al. 1986; Steinmetz and Wellmann 1986; Caldwell et al. 1989; Takeuchi et al. 1989; Barnes et al. 1990; Ballaré et al. 1991; Hashimoto et al. 1991; Sullivan et al. 1992; Takeda and Abe 1992; Li et al. 1993; Wilson and Greenberg 1993b). These photomorphogenic processes appear to be developmental in nature, potentially leading to acclimation. They do not appear to be side effects of nonspecific cellular damage that erroneously appear as a morphogenic process.

A few possibilities have been suggested as UV-B photoreceptor(s). It could be a protein–pigment complex associated with the plasma membrane or a UV-B-sensitive protein such as a K$^+$-ATPase (Popescu et al. 1989; Ensminger and Schäfer 1992; Murphy et al. 1993a). Other possibilities are aromatic metabolites such as flavins and/or pterins free or bound to proteins (Galland and Senger 1988; Häder and Brodhun 1991; Ensminger and Schäfer 1992). This is because the action spectra for many responses have peaks at about 290 nm, an absorbance maxima corresponding to flavins or flavoproteins (Ghisla et al. 1974; Galland and Senger 1988). Braun and Tevini (1993) reported that a

UV-B-induced *trans* to *cis* isomerization of cinnamic acid leads to activation of phenylalanine ammonia lyase (PAL) activity and thus increased synthesis of flavonoids. In addition, *trans*-cinnamic acid represses expression of genes for PAL (Loake et al. 1991). This would make *trans*-cinnamic acid a potential UV-B photoreceptor (Yamamoto and Towers 1985; Mavandad et al. 1990; Loake et al. 1991). Clearly, determination of the actual photoreceptor species will require extensive research to determine the classes of UV-B photoreceptors and why there are similar action spectra for different UV-B responses (Yamamura et al. 1977; Hashimoto et al. 1991; Yatsuhashi et al. 1982; Wellman 1983; Beggs and Wellman 1985).

One of the major problems in identifying the UV-B photoreceptor(s) is the lack of responses which are both photomorphogenic in nature and UV-B specific. Many of the UV-B-induced developmental processes mentioned earlier are not readily quantifiable and/or have major interactions with other photoreceptors, greatly limiting specific investigation of the UV-B detection system(s) (Beggs et al. 1986; Mohr 1986; Gaba and Black 1987; Hashimoto et al. 1991). For instance, there seems to be an interaction between the UV-B photoreceptor and phytochrome for flavonoid synthesis (Beggs et al. 1986). Rapid and readily quantifiable morphological assays could begin to alleviate the shortfall in information on the UV-B photoreceptor. Along these lines, we have discovered a response to UV-B, upward curling of *B. napus* L. cotyledons, that may be useful for probing the mechanism of UV-B photoreception (Wilson and Greenberg 1993b). Cotyledon curling occurs when *B. napus* is germinated in visible light plus UV-B radiation at a ratio of 100:1. In addition, curling can be induced in 4-d-old visible-light-grown seedlings with a 60 min pulse of UV-B. However, pulses of red light, blue light, far-red light, and UV-A did not induce curling, indicating UV-B specificity. The degree of curling showed a log-linear dependence on UV-B fluence (from 3.6 to 24 mmol m^{-2}) and reciprocity with respect to length of exposure and fluence rate. These data indicate that curling is photomorphogenic in nature and may be triggered by a single photoreceptor species, both of which are important properties for examining a photoreceptor. It will be interesting to determine if cotyledon curling is triggered by the same photoreceptor(s) that are involved in synthesis of flavonoids.

Summary

As depletion of the stratospheric ozone layer continues, the biosphere will likely be exposed to higher levels of UV-B radiation (290 to 320 nm). For plants, damage from UV-B can occur to several cellular components such as the photosynthetic apparatus, proteins, and DNA. Thus, assessments of how plants develop in response to UV-B and if some of the resultant processes confer protection to the plants are crucial areas of research. Interestingly, many plants exhibit responses to UV-B that confer protection from the damaging radiation. These include UV-B screening pigments, free radical scavenging, and DNA repair. For instance, when the UV-B to visible light ratio is maintained at a level close to that found in solar radiation (approximately 1:100), one finds that many plants grow without signs of overt UV-B stress. Under these conditions, altered leaf morphology, increased leaf thickness, synthesis of flavonoids, and appearance of DNA photolyase occur, which are considered to be adaptive mechanisms. However, if the UV-B level is raised significantly beyond 1% of visible light, photosynthesis is inhibited and net damage to DNA is observed. Also, many environmental organic pollutants are photoactive in the UV region of the spectrum and could have greater impacts on plants as the UV-B content of solar radiation increases. Thus, it is clear that UV-B can have major impacts on plants. At current UV-B levels, most (if not all) plants acclimate. As the UV-B to visible light ratio increases, the ability to adapt diminishes and the amount of damage rises. Thus, a major objective of future research on UV-B effects on plants will be to elucidate the most effective UV-B adaptation mechanisms and define which plants carry these responses naturally.

Acknowledgments

We wish to thank C. Marwood, B. McConkey, V. Gaba, D. McCormac, M. Edelman, A.K. Mattoo, D.G. Dixon, and E.B. Dumbroff for many fruitful discussions. We are grateful to Dr. D. Lean for use of his spectroradiometer. This work was supported by research and strategic grants from NSERC to BMG.

References

Adamse P, Britz SJ (1992a) Spectral quality of fluorescent UV-B sources during long term use. *Photochem Photobiol* 56:641–644.

Adamse P, Britz SJ (1992b) Amelioration of UV-B damage under high irradiance. I. Role of photosynthesis. *Photochem Photobiol* 56:645–650.

Ahmed M, Cashmore AR (1993) HY4 gene of *Arabidopsis thaliana* encodes a protein with characteristics of a blue-light receptor. *Nature* 366:162–166.

Ambler JE, Krizek DT, Semeniuck P (1975) Influence of UV-B radiation on early seedling growth and translocation of ^{65}Zn from cotyledons in cotton. *Plant Physiol* 34:177–181.

Asada K (1992) Ascorbate peroxidase — a hydrogen peroxide-scavenging enzyme in plants. *Physiol Plant* 85: 235–241.

Asada K, Takahashi M (1987) Production and scavenging of active oxygen in photosynthesis. In *Photoinhibition*, Kyle DJ, Osmond CJ, Arntzen CJ, Eds., Elsevier, New York, pp. 227–287.

Attridge TH (1990) *Light and Plant Responses,* Edward Arnold, New York.

Bachelet D, Barnes PW, Brown D, Brown M (1991) Latitudinal and seasonal variation in calculated UV-B irradiance for rice growing regions of Asia. *Photochem Photobiol* 54:411–422.

Ballaré CL, Barnes PW, Kendrick RE (1991) Photomorphogenic effects of UV-B radiation on hypocotyl elongation in wild-type and stable-phytochrome-deficient mutant seedlings of cucumber. *Physiol Plant* 83:652–658.

Barnes PW, Flint SD, Caldwell MM (1990) Morphological responses of crop and weed species of different growth forms to ultraviolet-B radiation. *Am J Bot* 77:1354–1360.

Baskin TI, Iino M (1987) An action spectrum in the blue and ultraviolet for phototropism in alfalfa. *Photochem Photobiol* 46:127–136.

Basiouny FM, Van TK, Biggs RH (1978) Some morphological and biochemical characteristics of C3 and C4 plants irradiated with UV-B. *Physiol Plant* 42:29–32.

Batschauer A (1993) A plant gene for photolyase: an enzyme catalyzing the repair of UV-light-induced DNA damage. *Plant J* 4:705–709.

Beerhues L, Robenek H, Weirmann R (1988) Chalcone synthases from spinach (*Spinacia oleracea* L.). II. Immunofluorescence and immunogold localization. *Planta* 173:544–553.

Beggs CJ, Wellmann E (1985) Analysis of light controlled anthocyanin formation in coleoptiles of *Zea mays* L.: the role of UV-B, blue, red and far red light. *Photochem Photobiol* 41:481–486.

Beggs CJ, Wellmann E, Grisebach H (1986) Photocontrol of flavonoid biosynthesis. In *Photomorphogenesis in Plants,* Kendrick RE, Kronenberg GHM, Eds., Martinus Nijhoff, Dordrecht, Holland, pp. 467–499.

Beld M, Martin C, Huits H, Stuitje AR, Gerats AGM (1989) Flavonoid synthesis in *Petunia hybrida*: partial characterization of dihydroflavonol-4-reductase genes. *Plant Mol Biol* 13:491–502.

Biggs RH, Kossuth SV, Teramura AH (1981) Response of 19 cultivars of soybeans to ultraviolet-B irradiance. *Physiol Plant* 53:19–26.

Block CJ, Dangl J, Hahlbrock K, Schulze-Lefert P (1990) Functional borders, genetic fine structure and distance requirements of cis-elements mediating light responsiveness of the parsley chalcone synthase promoter. *Proc Natl Acad Sci USA* 87:5387–5391.

Blumthaler M, Ambach W (1990). Indication of increasing solar ultraviolet-B radiation flux in alpine regions. *Science* 248:206–208.

Blyden ER, Doerner PW, Lamb CJ, Dixon RA (1991) Sequence analysis of a chalcone isomerase cDNA of *Phaseolus vulgaris* L. *Plant Mol Biol* 16:167–169.

Bolhar-Nordenkampf HR, Oquist G (1993) Chlorophyll fluorescence as a tool in photosynthesis research. In *Photosynthesis and Production in a Changing Environment: A Field and Laboratory Manual,* Hall DO, Scurlock, JMO, Bolhar-Nordenkampf, HR, Leegood RC, Long SP, Eds., Chapman and Hall, London, pp. 193–206.

Bornman JF (1989) New trends in photobiology: target sites of UV-B radiation in photosynthesis of higher plants. *J Photochem Photobiol* 4:145–158.

Bornman JF, Vogelmann TC (1990) Effects of UV-B radiation on leaf optical properties measured with fibre optics. *J Exp Bot* 42:547–554.

Bornman JF, Bjorn LO, Akerlund HE (1984) Action spectra for photoinhibition by ultraviolet radiation of photosystem II activity in spinach thylakoids. *Photobiochem Photobiophys* 8:305–313.

Bowler C, Slooten L, Vandenbranden S, DeRycke R, Botterman J, Sybesma C, Van Montagu M, Inze D (1991) Manganese superoxide dismutase can reduce cellular damage mediated by oxygen radicals in transgenic plants. *EMBO J* 10:1723–1732.

Brandle JR, Campbell WF, Sisson WB, Caldwell MM (1977) Net photosynthesis, electron transport capacity, and ultrastructure of *Pisum sativum* L. exposed to ultraviolet-B radiation. *Plant Physiol* 60:165–169.

Braun J, Tevini M (1993) Regulation of UV-protective pigment synthesis in the epidermal layer of rye seedlings. *Photochem Photobiol* 57:318–323.

Briggs WR, Short TW (1991) The transduction of signals in plants: responses to blue light. In *Phytochrome Properties and Biological Action*, NATO ASI Series H, Cell Biology, Vol. 50, Thomas B, Johnson CB, Eds., Springer-Verlag, Berlin, pp. 287–301.

Britt AB, Chen JJ, Wykoff D, Mitchell D (1993) A UV-sensitive mutant of *Arabidopsis* defective in the repair of pyrimidine-pyrimidinone(6-4) dimers. *Science* 261:1571–1574.

Caldwell CR (1993) Ultraviolet-induced photodegradation of cucumber (*Cucumis sativus* L.) microsomal and soluble protein tryptophanyl residues in vitro. *Plant Physiol* 101:947–953.

Caldwell MM (1981) Plant response to solar ultraviolet radiation. In *Encyclopedia of Plant Physiology*, Vol. 12A, Lange OL, Nobel PS, Osmond CB, Ziegler H, Eds., Springer-Verlag, Berlin, pp. 169–197.

Caldwell MM, Robberecht R, Billings WD (1980) A steep latitudinal gradient of solar ultraviolet-B radiation in the arctic-alpine life zone. *Ecology* 61:600–611.

Caldwell MM, Robberecht R, Nowak RS, Billings WD (1982) Differential photosynthetic inhibition by ultraviolet radiation from the arctic-alpine life zone. *Arctic Alpine Res* 14:195–202.

Caldwell MM, Gold WG, Harris G, Ashurst CW (1983a) A modulated lamp system for solar UV-B (280–320 nm) supplementation studies in the field. *Photochem Photobiol* 37:479–485.

Caldwell MM, Robberecht R, Flint SD (1983b) Internal filters: prospects for UV-acclimation in higher plants. *Physiol Plant* 58:445–450.

Caldwell MM, Camp LB, Warner CW, Flint SD (1986) Action spectra and their key role in assessing biological consequences of solar UV-B radiation change. In *Stratospheric Ozone Reduction, Solar Ultraviolet Radiation and Plant Life*, NATO ASI, Vol. G8, Worrest RC, Caldwell MM, Eds., Springer-Verlag, Berlin, pp. 87–111.

Caldwell MM, Teramura AH, Tevini M (1989) The changing solar ultraviolet climate and the ecological consequences for higher plants. *Trends Ecol Evol* 4:363–367.

Cen Y-P, Bornman J (1990) The response of bean plants to UV-B radiation under different irradiances of background visible light. *J Exp Bot* 41:1489–1495.

Cen Y-P, Bornman J (1993) The effect of exposure to enhanced UV-B radiation on the penetration of monochromatic and polychromatic UV-B radiation in the leaves of *Brassica napus. Physiol Plant* 87:249–255.

Chow WS, Strid A, Anderson JM (1992) Short term treatment of pea plants with supplementary ultraviolet-B radiation: recovery time-course of some photosynthetic functions and components. In *Research in Photosynthesis*, Vol. 4, Murata N, Ed., Kluwer Academic Publishers, Dordrecht, Holland, pp. 361–364.

Coohill TP (1991) Action spectra again? *Photochem Photobiol* 54:859–870.

Cooper WJ, Herr FL (1987) Introduction and overview. In *Photochemistry of Aquatic Environmental Systems*, Heitz JR, Zika WJ, Eds., ACS Symposium Series 339, American Chemical Society, Washington, D.C., pp. 1–8.

Crutzen PJ (1992) Ultraviolet on the increase. *Nature* 356:104–105.

Curry GM, Thimann KV, Ray PM (1956) The base curvature response of *Avena* seedlings to the ultraviolet. *Physiol Plant* 9:429–440.

Dangl JD, Hauffe KD, Lippardt S, Hahlbrock K, Scheel D (1987) Parsley protoplasts retain differential responsiveness to UV light and fungal elicitor. *EMBO J* 9:2551–2556.

Davies KJA (1987) Protein damage and degradation by oxygen radicals. I. General apsects. *J Biol Chem* 262:9895–9901.

Day TA, Vogelmann TC, DeLucia EH (1992) Are some plant life forms more effective than others in screening out ultraviolet-B radiation? *Oceologia* 92:513–519.

Dickson JA, Caldwell MM (1978) Leaf development of *Rumex patientia* L. (Polygonceae) exposed to UV irradiation (280–320 nm). *Am J Bot* 65:857–863.

Douglas CJ, Hoffman H, Schulz W, Hahlbrock K (1987). Structure and elicitor or UV-light stimulated expression of two 4-coumarate:CoA ligase genes in parsley. *EMBO J* 6:1189–1195.

Douglas CJ, Hauffe KD, Ites-Morales ME, Ellard M, Paszkowski U, Hahlbrock K, Dangl JL (1991). Exonic sequences are required for elicitor and light activation of a plant defense gene, but promotor sequences are sufficient for tissue specific expression. *EMBO J* 10:1767–1775.

Duxbury CL, Marwood CA, Huang X-D, Ren L, Dixon DG, Greenberg BM (1993) Photoinduced assimilation and toxicity of polycyclic aromatic hydrocarbons by duckweed. In *Environment and Energy Investment: A Key to Economic Renewal*, Proceedings, Technology Transfer Conference, Ontario Ministry of the Environment and Energy, Toronto, Ontario, POSTC22.

Eilfeld P, Haupt W (1991) Phytochrome. In *Photoreceptor Evolution and Function*, Holmes MG, Ed., Academic Press, New York, pp. 203–239.

Ensminger PA (1993) Control of development in plants and fungi by far-UV radiation. *Physiol Plant* 88:501-508.

Ensminger PA, Schäfer E (1992) Blue and ultraviolet-B light photoreceptors in parsley cells. *Photochem Photobiol* 55:437-447.

Foote CS (1979) Mechanisms of photooxidation. In *Singlet Oxygen*, Wasserman HH, Murray RW, Eds., Academic Press, New York, pp. 135–146.

Foote CS (1991) Definition of type I and type II photosensitized oxidation. *Photochem Photobiol* 54:659.

Foote CS, Chang YC, Denny RW (1970) Chemistry of singlet oxygen. X. Carotenoid quenching parallels biological protection. *J Am Chem Soc* 92:5216–5218.

Frederick JE (1990) Trends in atmospheric ozone and ultraviolet radiation: mechanisms and observations for the northern hemisphere. *Photochem Photobiol* 51:757–763.

Frederick JE, Weatherhead EC, Hayward EK (1991) Long-term variations in ultraviolet sunlight reaching the biosphere: calculations for the past three decades. *Photochem Photobiol* 54:781–788.

Fritzemeier K-H, Rolfs C-H, Pfau J, Kindl H (1983) Action of ultraviolet-C on stilbene formation in callus of *Arachis hypogaea*. *Planta* 159:25–29.

Gaba V, Black M (1987) Photoreceptor interaction in plant photomorphogenesis: the limits of experimental techniques and their interpretations. *Photochem Photobiol* 45:151–156.

Galland P, Senger H (1988) The role of pterins in the photoreception and metabolism of plants. *Photochem Photobiol* 48:811–820.

Gerstle SAW, Zardecki A, Wiser HL (1986) A new UV-B handbook, volume 1. In *Stratospheric Ozone Reduction, Solar Ultraviolet Radiation and Plant Life*, NATO ASI, Vol. G8, Worrest RC, Caldwell MM, Eds., Springer-Verlag, Berlin, pp. 63–74.

Ghisla S, Massey V, Lhoste J-M, Mayhew SG (1974) Fluorescence and optical characteristics of reduced flavines and flavoproteins. *Biochemistry* 13:589–597.

Girotti AW (1990) Photodynamic lipid peroxidation in biological systems. *Photochem Photobiol* 51:497–509.

Goedheer JC (1966) Visible absorption and fluorescence of chlorophyll and its aggregates in solution. In *The Chlorophylls*, Vernon LP, Seely GR, Eds., Academic Press, New York, pp. 147–184.

Greenberg BM (1993) Acclimation of plants to current UV-B levels. In*Workshop on the Effects of Increased UV-B Radiation*. Environment Canada, Toronto, Ontario, Canada, pp. 3–4.

Greenberg BM, Gaba V, Canaani O, Malkin S, Mattoo AK, Edelman M (1989) Separate photosensitizers mediate degradation of the 32-kDa photosystem II reaction center protein in the visible and UV spectral regions. *Proc Natl Acad Sci USA* 86:6617–6620.

Greenberg BM, Huang X-D, Dixon DG, Ren L, McConkey BJ, Duxbury CL (1993a) Structure activity relationships for the photoinduced toxicity of polycyclic aromatic hydrocarbons to plants — a preliminary model. In *Environmental Toxicology and Risk Assessment*, Vol. 2, Gorsuch JW, Dwyer FJ, Ingersoll CG, La Point TW, Eds., ASTM STP 1216, American Society for Testing and Materials, Philadelphia, pp. 369–378.

Greenberg BM, Wilson MI, Gaba V, Ren L, Huang X-D (1993b) Responses of *Brassica napus* (oilseed rape) to ultraviolet-B. *Life Sci Adv-Plant Physiol* 12:167–176.

Häder D-P, Brodhun B (1991) Effects of ultraviolet radiation on the photoreceptor proteins and pigments in the paraflagellar body of the flagellate, *Euglena gracilis*. *J Plant Physiol* 137:641–646.

Hahlbrock K (1981) Flavonoids. In *The Biochemistry of Plants, Vol. 7, Secondary Plant Products*, Conn EE, Ed., Academic Press, New York, pp. 425–456.

Hahlbrock K, Scheel D (1989) Physiology and molecular biology of phenylpropanoid metabolism. *Annu Rev Plant Physiol Mol Biol* 40:347–369.

Halliwell B (1981) Toxic effects of oxygen on plant tissues. In *Chloroplast Metabolism: The Structure and Function of Chloroplasts in Green Leaf Cells*. Clarendon Press, Oxford, pp. 179-205.

Harborne JB (1988) The flavonoids: recent advances. In *Plant Pigments*, Goodwin TW, Ed., Academic Press, New York, pp. 299–343.

Harlow GR, Jenkins ME, Pittalwala TS, Mount DW (1994) Isollation of uvh1, an *Arabidopsis* mutant hypersensitive to ultraviolet light and ionizing radiation. *Plant Cell* 6:227–235.

Hashimoto T, Shichijo C, Yatsuhashi H (1991) Ultraviolet action spectra for the induction and inhibition of anthocyanin synthesis in broom sorghum seedlings. *J Photochem Photobiol* 11:353–363.

Henderson ST (1977) *Daylight and Its Spectrum*, Adam Hilger, Ltd., Bristol, England.

Hrazdina G, Marx GA, Hoch HC (1982) Distribution of secondary plant metabolites and their biosynthetic enzymes in pea (*Pisum sativum* L.) leaves. *Plant Physiol* 70:745–748.

Huang X-D, Dixon DG, Greenberg BM (1993) Impacts of UV radiation and photomodification on the toxicity of PAHs to the higher plant *Lemna gibba* (Duckweed). *Environ Toxicol Chem* 12:1067–1077.

Huguet AI, Manez S, Alcaraz MJ (1990) Superoxide scavenging properties of flavonoids in a non-enzymic system. *Z Naturforsch* 45c:19–24.

Husain SR, Cillard J, Cillard P (1987) Hydroxyl radical scavenging activity of flavonoids. *Phytochemistry* 26:2489–2491.

Imbrie CW, Murphy TM (1984) Mechanism of photoinactivation of plant plasma membrane ATPase. *Photochem Photobiol* 40:243–248.

Iwanzik W, Tevini M, Dohnt G, Voss M, Weiss W, Graber P, Renger G (1983) Action of UV-B radiation on photosynthetic primary reactions in spinach chloroplasts. *Physiol Plant* 58:401–407.

Jahne A, Frtizen C, Weissenbock G (1993) Chalcone synthase and flavonoid products in primary-leaf tissues of rye and maize. *Planta* 189:39–46.

Jahnen W, Hahlbrock K (1988) Differential regulation and tissue-specific distribution of enzymes of phenylpropanoid pathways in developing parsley seedlings. *Planta* 173:453–458.

Jones LW, Kok B (1966) Photoinhibition of chloroplast reactions II. Multiple effects. *Plant Physiol* 41:1037–1043.

Jordan BR, He J, Chow WS, Anderson JM (1992) Changes in mRNA levels and polypeptide subunits of ribulose 1,5-bisphosphate carboxylase in response to supplementary ultraviolet-B radiation. *Plant Cell Environ* 15:91–98.

Jordan BR, James PE, Strid A, Anthony RG (1994) The effect of UV-B radiation on gene expression and pigment composition in etiolated and green leaf tissue: UV-B induced changes in gene expression are gene specific and dependent upon tissue development. *Plant Cell Environ* 17:45–54.

Judy BM, Lower WR, Miles CD, Thomas MW, Krause GF (1990) Chlorophyll fluorescence of a higher plant as an assay for toxicity assessment of soil and water. In *Plants for Toxicity Assessment*, Wang WW, Gorsuch JW, Lower WR, Eds., ASTM STP 1091, American Society for Testing and Materials, Philadelphia, pp. 308–318.

Kautsky H, Hirsch A (1934) Chlorophyllfluoreszenz und kohlensaureassimilation. Das fluoreszenzverhalten gruner pflanzen. *Biochem Zeitschrift* 274:423–434.

Kerr JB, McElroy CT (1993) Evidence for large upward trends of ultraviolet-B radiation linked to ozone depletion. *Science* 262:1032–1034.

Kim S-T, Sancar A (1993) Photochemistry, photophysics and mechanism of pyrimidine dimer repair by DNA photolyase. *Photochem Photobiol* 57:895–904.

Kim S-T, Feng Y, Sancar A (1992) The third chromophore of DNA photolyase: Trp-277 of *Escherichia coli* DNA photolyase repairs thymidine dimers by direct electron transfer. *Proc Natl Acad Sci USA* 89:900–904.

Kochevar IE (1990) UV-induced protein alterations and lipid oxidation in erythrocyte membranes. *Photochem Photobiol* 52:795–800.

Koes RE, Spelt CE, Mol JNM (1989) The chalcone synthase multigene family of *Petunia hybrida* (V30): differential, light regulated expression during flower development and UV light induction. *Plant Mol Biol* 12:213–225.

Kramer GF, Norman HA, Krizek DT, Mirecki RM (1991) Influence of UV-B radiation on polyamines, lipid peroxidation and membrane lipids in cucumber. *J Phytochem* 30:2101–2108.

Kramer GF, Krizek DT, Mirecki RM (1992) Influence of UV-B radiation and spectral quality on UV-B-induced polyamine accumulation in soybean. *Phytochemistry* 31:1119–1125.

Krinsky NI (1979) Biological roles of singlet oxygen. In *Singlet Oxygen*, Wasserman HH, Murray RW, Eds, Academic Press, New York, pp. 597–641.

Krizek DT (1978) Differential sensitivity of two cultivars of cucumber (*Cucumis sativus* L.) to increased UV-B irradiance. I. Dose-response studies. In *Final Report on Biological and Climatic Effects Research*, USDA-EPA, U.S. Environmental Protection Agency, Washington, D.C., 33 pp.

Krizek DT, Koche J (1979) Use of regression analysis to estimate UV spectral irradiance from broad band radiometer readings under FS-40 fluorescent sunlamps filtered with cellulose acetate. *Photochem Photobiol* 30:483–489.

Krizek DT, Kramer GF, Upadhyaya A, Mirecki RM (1993a) UV-B response of cucumber seedlings grown under metal halide and high pressure sodium lamp/deluxe lamps. *Physiol Plant* 88:350–358.

Krizek DT, Mirecki RM, Britz SJ, Harris WG, Thimijan RW (1993b) Use of microwave powered lamps as a new high intensity lighting source in plant growth chambers: spectral characteristics. *Plant Physiol* 102(suppl.):806.

Kruezaler F, Ragg H, Fautz E, Kuhn DN, Hahlbrock K (1983) UV induction of chalcone synthase mRNA in cell suspension cultures of *Petroselinium hortense. Proc Natl Acad Sci USA* 80:2591–2593.

Kubasek WL, Shirley BW, McKillop A, Goodman HM, Briggs W, Ausebel, F (1992) Regulation of biosynthesis gene in germinating *Arabidopsis* seedlings. *Plant Cell* 4:1229–1236.

Langer B, Wellmann E (1990) Phytochrome induction of photoreactivating enzyme in *Phaseolus vulgaris* L. seedlings. *Photochem Photobiol* 52:861–863.

Larson RA, Barenbaum MR (1988) Environmental phototoxicity. *Environ Sci Technol* 22:354–360.

Larson RA, Garrison WJ, Carlson RW (1990) Differential responses of alpine and non-alpine *Aquilegia* species to increased ultraviolet-B radiation. *Plant Cell Environ* 13:983–987.

Lecari B, Sodi F, diPaola ML (1990) Photomorphogenic responses to UV radiation: involvement of phytochrome and UV photoreceptors in the control of hypocotyl elongation in *Lycopersicon esculentum*. *Physiol Plant* 79:668–672.

Li J, Ou-lee T-M, Raba R, Amundson RG, Last RA (1993) *Arabidopsis* flavonoid mutants are hypersensitive to UV-B irradiation. *Plant Cell* 5:171–179.

Liscum E, Young JC, Poff KL, Hangarter RP (1992) Genetic separation of phototropism and blue light inhibition of stem elongation. *Plant Physiol* 100: 267–271.

Loake GJ, Choudhary AD, Harrison MJ, Mavandad M, Lamb CJ, Dixon RA (1991) Phenylpropanoid pathway intermediates regulate transient expression of a chalcone synthase gene promoter. *Plant Cell* 3:829–840.

Lozoya E, Block A, Lois R, Hahlbrock K, Scheel D (1991) Transcriptional repression of light-induced flavonoid synthesis by elicitor treatment of cultured parsley cells. *Plant J* 1:227–234.

Madronich S (1992) Implications of recent total atmospheric ozone measurements for biologically active ultraviolet radiation reaching the earth's surface. *Geophys Res Lett* 19:37–40.

Mattoo AK, Hoffman-Falk H, Marder JB, Edelman M (1984) Regulation of protein metabolism: coupling of photosynthetic electron transport to in vivo degradation of the rapidly metabolized 32-kilodalton protein of the chloroplast membranes. *Proc Natl Acad Sci USA* 81:1380–1384.

Mavandad M, Edwards R, Liang X, Lamb CJ, Dixon RA (1990) Effects of trans-cinnamic acid on expression of the bean phenylalanine ammonia-lyase gene family. *Plant Physiol* 94:671–680.

McFarland M, Kaye J (1992) Chlorofluorocarbons and ozone. *Photochem Photobiol* 55:911–929.

McKinlay AF, Diffey BL (1987) A reference action spectrum for ultraviolet-induced erythema in human skin. In *Human Exposure to Ultraviolet Radiation: Risks and Regulations*, Passchler WR, Bosnajokovic BFM, Eds., Elsevier, Amsterdam, pp. 83–87.

McLennan AG (1987) DNA damage, repair, and mutagenesis. In *DNA Replication in Plants*, Bryant JA, Dunham VL, Eds., CRC Press, Boca Raton, FL, pp. 135–186.

McLennan AG, Eastwood AC (1987) An endonuclease activity from suspension cultures of *Daucus carota* which acts upon pyrimidine dimers. *Plant Sci* 46:151–157.

Middleton EM, Teramura AH (1993a) Potential problems in the use of cellulose diacetate and mylar filters in UV-B radiation studies. *Photochem Photobiol* 57:744–751.

Middleton EM, Teramura AH (1993b) The role of flavonol glycosides and carotenoids in protecting soybean from ultraviolet-B damage. *Plant Physiol* 103:741–752.

Miles D (1990) The role of chlorophyll fluorescence as a bioassay for assessment of toxicity in plants. In *Plants for Toxicity Assessment,* Wang WW, Gorsuch JW, Lower WR, Eds., ASTM STP 1091, American Society for Testing and Materials, Philadelphia, pp. 297–307.

Mirecki RM, Teramura AH (1984) Effects of ultraviolet-B radiation on soybean. V. The dependence of plant sensitivity on the photosynthetic photon flux density during and after leaf expansion. *Plant Physiol* 74:475–480.

Mohr H (1986) Coaction between pigment systems. In *Photomorphogenesis in Plants*, Kendrick RE, Kronenberg GHM, Eds., Martinus Nijhoff, Dordrecht, Holland, pp. 547–564.

Morgan DD, Warshawsky D, Atkinson T (1977) The relationship between carcinogenic activities of polycyclic aromatic hydrocarbons and their singlet, triplet, and singlet-triplet splitting energies and phosphorescence lifetimes. *Photochem Photobiol* 25:31–38.

Murali NS, Teramura AH (1986) Intraspecific differences in *Cucumis sativus* sensitivity to ultraviolet-B radiation. *Environ Exp Bot* 26:89–95.

Murali NS, Teramura AH, Randall SK (1988) Response differences between two soybean cultivars with contrasting UV-B radiation sensitivities. *Photochem Photobiol* 48:653–657.

Murphy TM (1983) Membranes as targets of ultraviolet radiation. *Physiol Plant* 58:381–388.

Murphy TM, Qian YC, Auh CK, Verhoeven C (1993a) UV-induced events at plant plasma membranes. In *Frontiers of Photobiology,* Shima A, Ed., Elsevier Science Publishers, Amsterdam, pp. 555–560.

Murphy TM, Martin CP, Kami J (1993b) Endonuclease activity from tobacco nuclei specific for ultraviolet radiation-damaged DNA. *Physiol Plant* 87:417–425.

Newsted JL, Giesy JP (1987) Predictive models for photoinduced acute toxicity of polycyclic aromatic hydrocarbons to *Daphnia magna*, Strauss (Cladocera, Crustacea). *Environ Toxicol Chem* 6:445–461.

Palmer JM, Short TW, Briggs WR (1993) Correlation of blue light-induced phosphorylation to phototropism in *Zea mays* L. *Plant Physiol* 102:1219–1225.

Pang Q, Hays JB (1991) UV-B inducible and temperature-sensitive photoreactivation of cyclobutane pyrimidine dimers in *Arabidopsis thaliana*. *Plant Physiol* 95:536-543.

Peak MJ, Peak JG (1986) Molecular biology of UVA. In *The Biological Effects of UVA Irradiation*, Urbach F, Gange RW, Eds., Praeger, Westport, CT, pp. 42–52.

Pool R (1991) Ozone loss worse than expected. *Nature* 350:451.

Popescu T, Roessler A, Fukshansky L (1989) A novel effect in *Phycomyces* phototropism: positive bending and compensation spectrum in far UV. *Plant Physiol* 91:1586–1593.

Powles SB (1984) Photoinhibition of photosynthesis induced by visible light. *Annu Rev Plant Physiol* 35:14–44.

Pratt LH, Butler WL (1970) Phytochrome conversion by ultraviolet light. *Photochem Photobiol* 11:503–509.

Quaite FE, Sutherland BM, Sutherland JC (1992) Action spectrum for DNA damage in alfalfa lowers predicted impact of ozone depletion. *Nature* 358:576–578.

Reid DM, Beallard FD, Pharis RP (1991) Environmental cues in plant growth and development. In *Plant Physiology: A Treatise, Vol. X: Growth and Development*, Bidwell RGS, Ed., Academic Press, San Diego, CA, pp. 65–181.

Ren L, Huang X-D, McConkey BJ, Dixon DG, Greenberg BM (1994) Photoinduced toxicity of three polycyclic aromatic hydrocarbons (fluoranthene, pyrene and naphthalene) to the duckweed *Lemna gibba*. *Ecotoxicol Environ Saf* 28:160–171.

Renger G, Volker M, Eckert HJ, Fromme R, Hohm-Veit S, Graber P (1989) On the mechanism of photosystem II deterioration by UV-B irradiation. *Photochem Photobiol* 49:97–105.

Robberecht R, Caldwell MM (1978) Leaf epidermal transmittance of ultraviolet radiation and its implications for plant sensitivity to ultraviolet-radiation induced injury. *Oecologia* 32:277–287.

Robberecht R, Caldwell MM, Billings WD (1980) Leaf ultraviolet optical properties along a latitudinal gradient in the arctic-alpine life zone. *Ecology* 61:612–619.

Rowland FS (1989) Chlorofluorocarbons and the depletion of stratospheric ozone. *Am Sci* 77:36–45.

Sage L (1992) *Pigment of the Imagination*, Academic Press, New York, 562 pp.

Saito I, Sugiyama H, Matsura T (1983) Photochemical reactions of nucleic acids and their constituents of photobiological relevance. *Photochem Photobiol* 38:735–743.

Salmon S, Maziere JC, Santus R, Moliere P, Bouchemal N (1990) UV-B induced photoperoxidation of lipids of human low and high density lipoproteins. A possible role of tryptophan residues. *Photochem Photobiol* 52:541–545.

Sancar A, Sancar GB (1988) DNA repair enzymes. *Annu Rev Biochem* 57:29–67.

Sancar A, Tang M-S (1993) Nucleotide excision repair. *Photochem Photobiol* 57:905–921.

Schmelzer E, Jahnen W, Hahlbrock K (1988) In situ localization of light-induced chalcone synthase mRNA, chalcone synthase, and flavonoid end products in the epidermal cells of parsley leaves. *Proc Natl Acad Sci USA* 85:2989–2993.

Schoeny R, Cody T, Warshawsky D, Radike M (1988) Metabolism of mutagenic polycyclic aromatic hydrocarbons by photosynthetic algal species. *Mutat Res* 197:289–302.

Schulze-Lefert P, Becker-Andre M, Schulz W, Hahlbrock K, Dangle JL (1989) Functional architecture of the light-responsive chalcone synthase promoter from parsley. *Plant Cell* 1:707–714.

Setlow RB (1974) The wavelength in sunlight effective in producing cancer: a theoretical analysis. *Proc Natl Acad Sci USA* 71:3363–3366.

Sisson WB, Caldwell MM (1976) Photosynthesis, dark respiration and growth of *Rumex patientia* L. exposed to ultraviolet radiation (288 to 313 nm) simulating or reduced atmospheric ozone column. *Plant Physiol* 58:563–568.

Staiger D, Kaulen H, Schell J (1989) A CACGTG motif of the *Antirrhinum majus* chalcone synthase promoter is recognized by an evolutionarily conserved nuclear protein. *Proc Natl Acad Sci USA* 89:6930–6934.

Staiger D, Kaulen H, Schell J (1990) A nuclear factor recognizing a positive regulatory upstream element of the *Antirrhinum majus* chalcone synthase promoter. *Plant Physiol* 93:1347–1353.

Stapleton AE (1992) Ultraviolet radiation and plants: burning questions. *Plant Cell* 4:1353–1358.

Staxén I, Bergounioux C, Bornman JF (1993) Effect of ultraviolet radiation on cell division and microtubule organization in *Petunia hybrida* protoplasts. *Protoplasma* 173:70–76.

Steinmetz V, Wellmann E (1986) The role of solar UV-B in growth regulation of cress (*Lepidium sativum* L.) seedlings. *Photochem Photobiol* 43:189–193.

Steinmuller D, Tevini M (1986) Action of ultraviolet radiation (UV-B) upon cuticular waxes in some crop plants. *Planta* 164:557–564.

Strack D, Meurer B, Weissenböck G (1982) Tissue-specific kinetics of flavonoid accumulation in primary leaves of rye (*Secale cereale* L.). *Z Pflanzenphysiol* 108:131–141.

Strid A (1993) Alteration in expression of defence genes in *Pisum sativum* after exposure to supplementary ultraviolet-B radiation. *Plant Cell Physiol* 34:949–953.

Strid A, Chow WS, Anderson JM (1990) Effects of supplementary ultraviolet-B radiation on photosynthesis in *Pisum sativum*. *Biochim Biophys Acta* 1020:260–268.

Sullivan JH, Teramura AH (1990) Field study of the interaction between solar ultraviolet-B radiation and drought on photosynthesis and growth in soybean. *Plant Physiol* 92:141–146.

Sullivan JH, Teramura AH, Ziska LH (1992) Variation in UV-B sensitivity in plants from a 3,000-m elevational gradient in Hawaii. *Am J Bot* 79:737–743.

Takahama U (1988) Hydrogen peroxide-dependent oxidation of flavonoids and hydroxycinnamic acid derivatives in epidermal and guard cells of *Tradescantia virginiana* L. *Plant Cell Physiol* 29:475–481.

Takeda J, Abe S (1992) Light induced synthesis of anthocyanins in carrot cells in suspension. IV. The action spectrum. *Photochem Photobiol* 56:69–74.

Takeuchi Y, Akizuki M, Shimizu H, Kondo N, Sugahara K (1989) Effect of UV-B (290-320 nm) irradiation on growth and metabolism of cucumber cotyledons. *Physiol Plant* 76:425–430.

Teramura AH (1983) Effects of ultraviolet-B radiation on the growth and yield of crop plants. *Physiol Plant* 58:415–427.

Teramura AH (1986) Overview of our current state of knowledge of UV effects on plants. In *Effects of Changes in Stratospheric Ozone and Global Climate*, Vol. 1, Titus JG, Ed., U.S. Environmental Protection Agency, Washington, D.C., pp. 165–173.

Teramura AH, Sullivan JW (1988) Effects of ultraviolet B radiation in soybean yield and seed quality. *Environ Pollut* 53:416–469.

Teramura AH, Sullivan JW, Ziska LH (1990) Interaction of elevated ultraviolet-B radiation and CO_2 on productivity and photosynthetic characteristics in wheat, rice and soybean. *Plant Physiol* 94:470–475.

Tevini M (1993) Effects of enhanced UV-B radiation on terrestrial plants. In *UV-B Radiation and Ozone Depletion: Effects on Humans, Animals, Plants, Microorganisms, and Materials*, Tevini M, Ed., Lewis Publishers, Boca Raton, FL, pp. 125–153.

Tevini M, Teramura AH (1989) UV-B effects on terrestrial plants. *Photochem Photobiol* 50:479–487.

Tevini M, Thoma U, Iwanzik W (1983) Effects of UV-B radiation on plants during mild water stress II. Effects on growth, protein and flavonoid content. *Z Pflanzenhysiol* 109:435–448.

Tevini M, Grusemann P, Fieser G (1988) Assessment of UV-B stress by chlorophyll fluorescence analysis. In *Applications of Chlorophyll Fluorescence*, Lichtenthaler HK, Ed., Kluwer Academic, New York, pp. 229–238.

Tevini M, Teramura AH, Kulandaivelu G, Caldwell MM, Bjorn LO (1989) Terrestrial plants. In *Environmental Effects Panel Report Pursuant to Article 6 of the Montreal Protocol on Substances that Deplete the Ozone Layer*, van der Luen JC, Tevini M, Worrest RC, Eds., United Nations Environment Programme, United Nations, New York, pp. 25–37.

Tevini M, Braun J, Fieser G (1991) The protective function of the epidermal layer of rye seedlings against ultraviolet-B radiation. *Photochem Photobiol* 53:329–333.

Thompson JE (1984) Physical changes in the membranes of senescing and environmentally stressed plant tissues. In *Physiology of Membrane Fluidity*, Vol. 2, Shinitzky M, Ed., CRC Press, Boca Raton, FL, pp. 85–108.

van der Meer IM, Spelt CE, Mol JNM, Stuitje AR (1990) Promoter analysis of the chalcone synthase (chsA) gene of *Petunia hybrida*: a 67 bp promoter region directs flower-specific expression. *Plant Mol Biol* 15:95–109.

van Tunen AJ, Koes RE, Spelt CE, van der Krol AR, Stuitje AR, Mol JNM (1988) Cloning of the two chalcone flavanone isomerase genes from *Petunia hybrida*: coordinate, light-regulated and differential expression of flavonoid genes. *EMBO J* 7:1257–1263.

Vierstra RD, Quail PH (1986) Phytochrome: the protein. In *Photomorphogenesis in Plants*, Kendrick RE, Kronenberg GHM, Eds., Martinus Nijhoff, Dordrecht, Holland, pp. 158–165.

Virgin I, Andersson B, Aro E-M (1993) Photoinhibition of photosystem II. Inactivation, protein damage and turnover. *Biochim Biophys Acta* 1143:113–134.

Vu CV, Allen LH Jr., Garrard LA (1982) Effects of supplemental UV-B radiation on primary photosynthetic carboxylating enzymes and soluble proteins in leaves of C3 and C4 crop plants. *Physiol Plant* 55:11–16.

Vu CV, Allen LH Jr., Garrard LA (1984) Effects of enhanced UV-B radiation (280–320nm) on ribulose-1,5-bisphosphate carboxylase in pea and soybean. *Environ Exp Bot* 24:131–143.

Walrant P, Santus R (1974) Ultraviolet and N-formyl-kynurenine-sensitized photoinactivation of bovine carbonic anhydrase: an internal photodynamic effect. *Photochem Photobiol* 20:455–460.

Warner CW, Caldwell MM (1983) Influence of photon flux density in the 400–700 nm waveband on inhibition of photosynthesis by UV-B (280–320 nm) irradiation in soybean leaves: separation of indirect and immediate effects. *Photochem Photobiol* 38:341–346.

Warpeha KMF, Kaufman LS, Briggs WR (1992) A flavoprotein may mediate the blue light-activated binding of guanosine 5'-triphosphate to isolated plasma membranes of *Pisum sativum* L. *Photochem Photobiol* 55:595–603.

Weisshaar B, Armstrong GA, Block A, da Costa e Silva O, Hahlbrock K (1991) Light-inducible and constitutively expressed DNA binding proteins recognizing a plant promoter element with functional relevance in light responsiveness. *EMBO J* 10:1777–1786.

Wellman E (1983) UV radiation in photomorphogenesis. In *Encyclopedia of Plant Physiology, New Series, Vol. 16B: Photomorphogenesis,* Mohr H, Shropshire W, Eds., Springer-Verlag, Heidelberg, pp. 745–756.

Wilson MI, Greenberg BM (1993a) Protection of the D1 photosystem II reaction center protein from degradation in ultraviolet radiation following adaptation of *Brassica napus* L. to growth in ultraviolet-B. *Photochem Photobiol* 57:556–563.

Wilson MI, Greenberg BM (1993b) Specificity and photomorphogenic nature of ultraviolet-B induced cotyledon curling in *Brassica napus* L. *Plant Physiol* 102:671–677.

Wilson MI, Ghosh S, Gerhardt, KE, Holland N, Babu TS, Edelman M, Dumbroff EB, Greenberg BM (1995) In vivo photomodification of ribulose-1,5-bisphosphate carboxylase/oxygenase holoenzyme by ultraviolet-B radiation: formation of a 66 kDa variant of the large susunit. *Plant Physiol* 109:221–229.

Wingender R, Rohrig H, Horicke C, Wing D, Schell J (1989) Differential regulation of soybean chalcone synthase genes in plant defense, symbiosis and upon environmental stimuli. *Mol Gen Genet* 218:315–322.

Wingender R, Rohrig H, Horicke C, Schell J (1990) cis-Regulatory elements involved in ultraviolet light regulation and plant defense. *Plant Cell* 2:1019–1026.

Wolf SP, Garner A, Dean RT (1986) Free radicals, lipids and protein degradation. *TIBS* 11:27–31.

Yamamoto E, Towers GHN (1985) Cell wall bound ferulic acid in barley seedlings during development and its photoisomerization. *J Plant Physiol* 117:441–449.

Yamamura S, Kumagai T, Oda Y (1977) An action spectrum for photoinduced conidiation in *Helminthosporium oryzae. Plant Cell Physiol* 18:1163–1166.

Yatsuhashi H, Hashimoto T, Shimizu S (1982) Ultraviolet action spectrum for anthocyanin formation in broom sorghum first internodes. *Plant Physiol* 70:735–741.

Ziska LH, Teramura AH, Sullivan JH (1992) Physiological sensitivity of plants along an elevational gradient to UV-B radiation. *Am J Bot* 79:863–871.

Ziska LH, Teramura AH, Sullivan JH, McCoy A (1993) Influence of ultraviolet-B (UV-B) radiation on photosynthetic and growth characteristics in field-grown cassava (*Manihot esculentum* Crantz). *Plant Cell Environ* 16:73–79.

chapter two

Radiation effects on plants*

Robert W. Holst and David J. Nagel

* This is a "work of the U.S. Government" (prepared by an officer or employee of the U.S. Government as a part of official duties), and therefore, is not subject to U.S. copyright.

Introduction

The study of radiation effects is tremendously complex because of the many types of radiations, widely different nonliving targets, and diverse responses. In fact, the study of radiation effects spans what are essentially three distinct fields. Radiation *physics* deals with primary and secondary interaction mechanisms which are active on fast (usually subnanosecond) time scales. Targets of primary interest include homogeneous gases, liquid and solids, and structured materials, notably active electronic and optical devices. Radiation *chemistry* involves molecular changes which occur up to seconds after the initial interaction. Pure liquids and solutions get the most attention. Radiation *biology* concerns temporary or permanent changes during the lifetime of exposed individuals and over periods spanning many generations. Objects studied range from cellular components to single-celled organisms to complex plants and animals.

This chapter discusses the interaction of radiation and plants and the effects which result. The organization of this chapter can be represented schematically in a three-dimensional "space," as in Figure 1. Radiation (type, energy, intensity, etc.) is represented conceptually as one "axis." The individual plant (species/variety, age, condition, etc.) or community of plants (diversity, density, etc.) being irradiated makes up the second "axis." The effects due to specific radiations on particular individuals or groups of plants, ranging from low-level physical and chemical effects of little consequence to prompt death from high-level radiations, constitutes the third "axis."

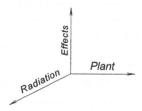

Figure 1   Representation of factors relevant to radiation effects on plants.

In studying or exploiting radiation effects on plants, three groups of actions are required: (1) the plants must be chosen, and the experimental arrangement must be designed; (2) the amount and character of absorbed radiations have to be ascertained from a combination of measurements and calculations; and (3) the effects of interest must be quantified and correlated with the absorbed radiation characteristics.

The next section provides an overview of radiation types and sources, with emphasis on those employed for plant studies. It includes attention to photon, electron, neutron, and ion radiations of diverse energies and intensities. Penetration lengths for electromagnetic radiations are graphed since they are basic to the design and interpretation of most experiments on the radiation response of plants. Then various terminology use in the radiation study and methods for estimation of absorbed doses are briefly discussed.

The third section provides a review of radiation effects on plants, primarily those caused by γ-irradiation, with some reference and comparison to work that has been done using X-ray and neutron radiations. The effects of these radiations on plants will be addressed in the following order: plant communities, individual plants, physiological response, and genetics.

Radiation effects on plants are ordinarily thought of as deleterious, but they are often useful, either in scientific studies or to generate desirable changes. The fourth section provides a brief synopsis of the uses of γ-irradiation in plant studies to induce effects of interest.

The final section contains a brief summary of what is generally known from the hundreds of studies of radiation effects on plants. Many research opportunities remain, some of them quite broad and fundamental. Examples of needed studies are also presented and discussed in this last section.

This chapter is written with the expectation that most readers will have more background in botany than radiation sciences. Therefore, some basic aspects of radiation interactions are included in the text. Additional general information on radiation and its use in the biological and chemical sciences are included in the Bibliography. Those technical papers used as the basis for the plant study section are given in the References section.

Radiation sources and characteristics

Radiations that impact plants can come from either natural sources, i.e., cosmic or terrestrial, or manmade sources, either in the laboratory or the environment, for example, nuclear test facilities and nuclear accidents. Description of a radiation field incident on a plant depends on the type of radiations and can be complex. Most important is the spectrum that indicates the specific or continuous energies and the intensities at each energy. Spatial and temporal nonuniformities can be important. Factors such as the polarization of an electromagnetic field are usually ignored. Although most of this chapter deals with γ-irradiation and its effects on plants, we will describe the general characteristics of photon, electron, neutron, and ion sources which have been or could be employed to irradiate plants.

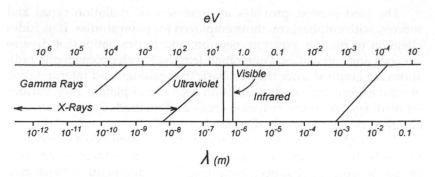

Figure 2 Electromagnetic spectrum that may affect plant growth and development.

Radiation sources

Photons

The electromagnetic (EM) spectrum, ranging from near infrared through ultraviolet (UV) and X-rays to γ-rays (Figure 2), has been employed in the past for plant radiation studies. Radiations with energies above about 1 eV, which can excite electrons or cause atoms to move, are of interest here. Very-low-energy radiations, which heat molecules or excite electrons but do not ionize atoms or cause atomic displacements, are excluded. One example is UV-induced effects which are reviewed in Chapter 1. Numerous practical and developmental sources are available in each spectral region. In the UV region, the most convenient source is a mercury vapor lamp. Its spectrum consists of five strong lines in the 300- to 600-nm range. X-rays are most conveniently available from X-ray machines. Fixed anodes, or rotating anodes operated at about ten times higher electrical and X-ray power, are widely available. High voltage X-ray tubes emit radiation extending into the in the γ-ray region. Radioactive sources can emit X-rays, but they are preferentially γ-ray sources. Their simplicity and intensity are attractive features, but they require significant shielding to protect personnel. Cobalt 60 (^{60}Co) is the most widely employed γ-ray source. It emits two dominant lines at 1.17 and 1.33 MeV. Orbiting electrons in storage rings produce synchrotron radiation which spans all regions of interest for plant studies. It is brighter than other radiation sources in and above the X-ray region. However, synchrotron radiation beams generally have only small (<1 cm) diameters, which are insufficient to irradiate entire plants. They are sufficient for localized studies such as high-dose effects on leaves.

The interaction lengths of photons in materials depend on the chemical composition and density of the absorber. For incident intensity [I(o)], the intensity of EM radiation, typically measured in photons per square centimeter, at a depth x in a material is given by

$$I(x) = I(o)e^{-\mu_m \rho x} \tag{1}$$

where μ_m is the energy-dependent mass absorption coefficient and equals μ/ρ, the linear absorption coefficient divided by the density. For a homogeneous material composed of i elemental components, the absorption coefficient is given by

$$\mu_m = \overline{z_i}\left(\frac{\mu}{\rho}\right)_i \rho_i \tag{2}$$

Absorption coefficients are readily available for relevant energies and all elements. The absorption length is defined as the thickness x_e for which

$$\frac{I(x)}{I(o)} = \frac{1}{e} \tag{3}$$

That is, $x_e = 1/\mu_m\rho$. It is a convenient measure of the ability of EM radiation to penetrate a material. Figure 3 is a plot of absorption length in a material similar to the water component and composition of plant material. It shows that energies below a few kiloelectronvolts are superficially absorbed. Photons above a few kiloelectronvolts are sufficient to uniformly dose thin leaves, while photon energies above a few tens of kiloelectronvolts will penetrate stems 1 cm in diameter.

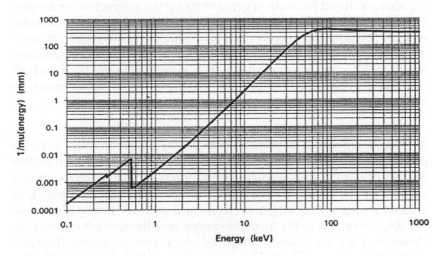

Figure 3 Energy spectrum of γ-irradiation penetrating a hypothetical plant cell composed of 90% water and 10% solid material.

Electrons

Electron beams can be conveniently produced from hot or cold filament sources. They can be accelerated to low (<100 keV) energies in tabletop (~1 m) electrostatic devices, to intermediate (<100 MeV) energies in laboratory-scale (~10 m) accelerators, and to high (>1 GeV) energies in large (~ 1 km) facilities. Low-energy electron beams can be defocused or scanned to irradiate objects the size of plants. However, electrons with energies less than 100 keV penetrate air less than 1 cm. Higher energy beams which can be brought through a thin vacuum window into the atmosphere generally have small (< 1 cm) diameters and are not easily scanned. Because of these factors (limited range of low energies and small irradiation areas at high energies), electrons have been used rarely for study of radiation effects on plants.

Neutrons

Neutrons are neutral and massive in contrast to electrons, so their interactions with samples are significantly different. Both photons and electrons produce ionization but little atomic displacement. For neutrons, the situation is generally reversed; that is, they predominantly produce displacement damage. Fluxes of neutrons are available from nuclear reactors or from targets impacted by ion beams from accelerators. Depending upon the configuration of a reactor source, the neutrons can be either fast fission neutrons with energies from 10 keV to over 10 MeV or thermal neutrons with energies less than 0.1 eV. Fast fission neutrons are the primary neutron type produced by nuclear reactors. Thermal neutrons emerge from reactor moderators where fast fission neutrons have been slowed without being captured. Neutrons are also produced in neutron generators by the interaction of accelerated charged particles with specific target materials of low molecular weight. The generators create neutrons either by positive-ion bombardment, for example, (p,n) or (d,n) reactions, or by electron bombardment, which are effectively (γ,n) reactions. These generators can produce both fast and thermal neutrons and are often the source of neutrons for biological studies.

Ions

Ion beams can consist of particles which vary widely in mass, from the proton (1 amu) to uranium (235 to 238 amu) and beyond. The transient charges of fast (i.e., relativistic) ions in targets can approach the atomic number of the projectile. Therefore, like photons and electrons, ions interact strongly with target electrons and produce heavy ionization. For ions, over 99% of the incident energy usually goes into ionization. However, they also cause numerous and extensive atomic displacements. The ranges of ions depend mostly on the ion species, the ion beam characteristics (ion beam energy), and the target characteristics (composition and structure). Ion beams, like electron beams, which have sufficient energy for air exposures, usually have cross sections less than 1 cm in diameter. Hence, they are not employed for whole-plant exposures.

Terminology

The terminology and units of measure employed in the study of radiation effects are numerous and somewhat arcane due to their historical origins. However, the primary terms are few in number and can be easily comprehended when viewed in physical terms, such as the number of particles and the energy carried or deposited by a beam of radiation.

Whatever the source, the radiation which proceeds toward and falls upon the plants under study is most clearly defined in terms of quanta per area (square centimeters usually) normal to the beam, either integrated over the exposure or per unit of time (usually the second). The integral value is called the particle (actually quantum) "fluence," and the value per second is termed the "fluence rate" or "flux." This is true for photons, whether in the UV, X-ray, or γ-ray regions of the spectrum, as well as for particles, including electrons, neutrons, ions, or atoms. Additional factors may be used to describe the radiation, for example, polarization in the case of photons. However, these are rarely important in radiation effects studies.

Historically, a unit of *exposure* is used to describe the intensity of photons or charged particles incident upon the object under study. Its unit, the *roentgen*, is defined in terms of the amount of charge (either electrons or ions) produced in air by the passage of the radiation. This characterization occurs because it is relatively easy to measure the liberated charge with an electroscope or the more modern electrometer. Readings in roentgens are still available from many instruments, and allowable exposure rates in health physics are commonly given in units of milliroentgens per hour. Hence, this unit is still used and valuable. However, quantitative interpretation of radiation effects requires conversion of measured roentgen values to quanta per square centimeter per second, which is straightforward only for beams with one quantum and energy, e.g., electrons of 200 keV.

If a beam of particles and/or photons with mixed quanta or energies is employed, as is often the case, then a full measurement of that radiation flux requires the use of calibrated spectrometers appropriate to the type of radiation and its energy range in order to obtain the spectral output for each radiation type. In this case, *spectral units* are employed, namely quanta per square centimeter per second per bandwidth, where the spectral slice (bandwidth) can be a portion of the wavelength or energy spectrum of the radiation. Since absorbed energy is central to the cause and interpretation of radiation effects, use of energy spectral units is preferred. For example, appropriate units for

X-ray or γ-ray beams are photons per square centimeter per second per 1 keV of bandwidth of the photon energy spectrum. The *kiloelectronvolt* (keV) is a common unit of spectral energy for ionizing radiations. In some instances, it is preferable to convert the number of quanta into the equivalent energy carried by the beam through the use of the energy per quantum. This is convenient since a fraction of the energy carried by the beam of radiation is deposited in the target (i.e., plant) under study, producing the various effects of interest.

The *dose* is the amount of radiant energy absorbed per unit mass in the material of interest. The modern unit of dose is the *gray* (Gy), which equals 1 J kg^{-1}. The historical unit is the *rad*, which is 1/100 of a gray. The *dose rate* is merely the dose per unit time (usually seconds). Although the dose and dose rate are absolutely central to the quantitative study of the effects of any type of radiation in any target, their determination is complex, as reviewed in the next subsection. In fact, even the definition of the dose is subject to detailed discussion. While its definition is clear, the complexities of energy deposition and non-uniformities in the target are both important. First, different radiations produce different amounts of ionization, atomic displacement, heating, etc. in the same target. Second, most targets, including all plants, are nonuniform on spatial scales, ranging from the molecular to the whole organism, so that a given radiation fluence will deposit different amounts of energy in various tissues and in the diverse components of cells. The point is that a given dose (absorbed energy/mass) integrates over all the different forms in which the energy resides in both transient and final forms, as well as over all the different compositional and structural forms in the target. These complexities have forced a consideration of levels of complexity beyond the simple dose.

Calculation of dose from a known incident spectral fluence or flux requires knowledge of the cross sections or interaction (i.e., penetration) lengths for the particular form of radiation and the composition and structure of the target. Such calculations are complex, but a simple measure of the strength of the radiation–target interaction is available. The *linear energy transfer* (LET) describes the rate at which a particle loses energy to the medium, commonly measured in kiloelectronvolts per micron of length along the path of radiation. The LET gives the initial linear density of energy deposition without regard to the form of that energy, for example, ionization or kinetic energy of particles excited in the target by the passage of the radiation. The *kerma* (also with units of joules per kilogram) is defined to describe the transient kinetic energy of all charged particles (electrons and ions) released within the target material due to absorption of some or all of the incident radiation. Similarly, the *radiation chemical yield* is defined to describe the level of chemical changes induced by radiation absorption, as distinct from heating or the escape of some of the energy by radiation. Radiation chemical yield is given in units of moles of product created per joule deposited and is denoted by G.

One other term in use in radiation biology deserves mention in this brief summary. For the same absorbed dose, different radiations or the same radiation of different energies have different effects on the individual plant or its progeny. This is described in terms of the *relative biological effectiveness* (RBE) of the various radiations and energies. Determination of the quantitative RBE for different radiations and energies and for different targets, ranging from cells to whole plants, is complex and beyond description here. A description of how biologically effective radiation is used in UV-B effects on plants is provided in Chapter 1.

Dosimetry

Whatever the radiation employed for study of effects on plants, advances in understanding almost always require quantitative measurement of cause and effect. Insights can be gained only rarely by merely noting the relative effects of different types or energies of radiation. In some cases, useful correlations between the quantity of *incident* radiation and the observed effects will suffice. However, in most radiation effects studies, it is necessary to determine quantitatively the amount of radiant energy which is *absorbed* in the entire plant or the part under study. Further, it is highly desirable to know the character of that absorbed energy. For example, is it entirely due to electron excitation and ionization, or does some of the deposited energy also lead to atomic displacements and broken molecular bonds? Dosimetry is the determination of the amount and character of absorbed energy. The terms and units used to quantitatively measure the radiation incident upon and absorbed by plants under study were reviewed in the previous subsection.

While the goals of dosimetry are simple to state, their attainment is relatively complex. Accurate dosimetry requires a combination of careful measurements and calculations. There are two ways to arrive at the needed information on the dose deposited in the region of interest in a plant. First, the incident spectrum is quantified, and tabulated absorption coefficients are employed to obtain the dose. Second, a dosimeter is used to obtain a measure of the deposition, which is then corrected to take into account differences between the composition and structure of the dosimeter and the plant. In both cases, the apparatus used in front of the source, usually at the place of the plant, requires calibration. Each of these two approaches will be described in the following subsections.

Source dosimetry

The first is the preferred approach to dose determination, in principle. It begins with experimental measurements and concludes with a detailed calculation of the deposition. The initial goal is to obtain a quantitative measure of the radiation traveling from the source to the position of the plant to be exposed. The measurements must be able to relate quantitatively the spectrum, that is, the intensity as a function of the photon or particle energy, for the radiation of interest. Quantitative spectroscopy of energetic photons, neutrons, electrons, or ions requires the use of specialized instrumentation tailored to the type of radiation and the energy range which will expose the plant samples. Such instrumentation is not readily available in all cases. Therefore, sometimes expensive spectroscopic instruments must be custom built to perform the needed measurements. No matter where the instrumentation is obtained, it requires calibration in order to convert whatever is recorded into the quantitative spectrum. Such calibrations require the employment of sources of known characteristics. Calibration sources are commonly too weak to be employed for the exposure of plants to the desired doses in reasonable times, but this is not always the case. When a known or otherwise calibrated radiation source is adequate for exposure of plants to the needed levels, then the source measurement (the first experimental step) is avoided.

If the source of radiation to be used has already been characterized to some degree, for example, the shape of its spectrum determined, then the situation is significantly easier. In that case, one or a few absolute measurements at specific energies can be used to produce a quantitative spectrum over the entire spectral region of interest by providing the scale factor needed to place the known spectral shape on an absolute basis. Some photon sources, notably radioactive γ-ray sources, emit spectra which are well known; in these cases, it is necessary only to use a calibrated γ-ray detector to obtain the scale factor in order to have the absolute intensity level. In the case of fission sources of neutrons, the emitted spectral shape is sometimes known adequately from the literature, and use of a calibrated neutron detector will give the scale factor. Electron and ion beam energies are usually monochromatic and can be determined, assuming proper functioning of the sources, from the voltage settings on the accelerators producing the beams. Beam currents can also be read from instruments, such as ammeters, built into the source of electrons or ions. In addition, the current can be measured by running the beam into a properly designed Faraday cup and reading the current with an electrometer.

Once the absolute fluence of quanta, usually expressed in quanta per square centimeter per second for monochromatic (that is, monoenergetic) spectra, is known at the position in which the plants will be exposed, then calculations are employed to obtain the dose, that is, the amount and character of the deposited energy. This procedure depends on the availability and accuracy of the interaction (absorption and scattering) cross sections for (1) the specific radiation type and energy and (2) the materials making up the plant samples. Also, codes must be available to take the absolute spectrum, the sample, and the geometry as input for computation of the dose as a function of position in the plant with adequate numerical accuracy. Proper account has to be taken of both electronic excitations and ionizations and of the displacement and migration of atoms. The ultimate disposition of the incident energy, for example, in heat or chemical changes, should result from the calculations of the absorbed dose. The codes, which produce the absorbed dose, can be based on different approximations, ranging from relatively simple analytical calculations, through transport theories of intermediate complexity, to Monte Carlo calculations which are conceptually simple but computationally intensive.

Incident dosimetry

The second approach to dose determination also involves the use of sequential experimental and calculational steps. It is simpler in practice, but is not as thorough as the first approach, nor does it yield as much detailed information on the absorbed energy. In this approach, dosimeters replace the plants. They can be anything from small and simple instruments, which electronically present the dose reading, to small samples, which require further steps to yield their dosage. An example of the latter is a thermoluminescent dosimeter (TLD), which usually is a few-millimeter-sized piece of a salt such as lithium fluoride. After exposure, the TLD is put in a small oven, where the temperature can be ramped to 200 to 300°C, depending on the TLD material. Doing so releases electrons that were excited and then trapped at defects during the radiation exposure. The result is emitted light (luminescence) which is recorded to provide the dose quantity through use of a calibration curve quantitatively relating the amount of light to the absorbed dose.

The simplicity and diversity of commercial dosimeters make them attractive for quantification of the dose a plant receives. However, the material of the dosimeters is never the same in composition or structure as that of the plant of interest. Hence, the absorption and distribution of energy in the dosimeter is not the same as that of the plant. This requires the application of some corrections to the dose recorded in the dosimeter in order to obtain an estimate of the dose in a given volume within the plant. Such corrections are not readily available, so radiation effects on plants are sometimes gauged relative to the dose read by a specified type of dosimeter with composition close to that of plants. This approach is practical, but not rigorous.

Quantitative determination of dose for any radiation effects study can be laborious and time consuming. In general, studies of the effects of radiations on plants employ routine approaches to dosimetry and concentrate mostly on the response of the plants. That is, dose determination in most studies of radiation effects on plants generally have not been as thorough as studies of radiation effects on animals, especially cell cultures.

Plants and effects

Effects on plants caused by various types of radiation have been studied using plant communities; individual plants; plant organs or tissues; and whole cells, organelles, or chemicals within plants. The effects of radiation on plants will be addressed in this chapter in the following order: plant community, individual plant, physiological, and genetic. The order is based on large-scale or long-term biological sequences where effects such as irradiation on plant communities have often been observed first. The individual plants are then studied more closely at the various levels of their structure, organs, cells, and cell organelles for anatomical, physiological, and hereditary changes to that plant or on succeeding generations. We do not review the basics of botany or of the various sciences that are related to each of these levels, e.g., ecology, morphology, physiology, cytology, and genetics, but only mention how radiation may affect the plant at the level(s) of complexity as tested by the researchers cited.

Plant community studies

Since the early 1950s, the effects of radiation on natural plant communities in the United States (Woodwell 1963; Monk 1966; Fraley and Whicker 1973a; Vollmer and Bamberg 1975; Rudolph 1971; Kaaz et al. 1971), France (Fabries et al. 1972; Saas et al. 1975), Spain (Gomez-Campo and Casas-Builla 1972; Hernandez-Bermejo 1977), and Canada (Dugle and Mayoh 1984) have been studied using γ-irradiation sources. The predominant γ-irradiation source is ^{60}Co, which provides controlled long-term exposure of a chronic nature (generally <400 R/d). One of the first sites to be established was at Brookhaven National Laboratory in Upton, NY (Woodwell and Sparrow 1963). In all of these community studies, the general effects associated with plant survival and reproduction were noted.

In old-field plant communities, studied at a variety of sites, the general conclusion is that the species composition and diversity of the community changes proportionally to the distance from the source. Woodwell and Oosting (1965) noted that there was generally a 50% reduction in coefficient of community (a measure of species similarity within vs. among samples along an environmental gradient), with a threefold increase in radiation exposure rate, and that there was a species-specific threshold. The general threshold in the oak-pine forest at Brookhaven was approximately 50 R/d; most species were affected in some manner at 320 R/d. However, *Erigeron canadensis* only began to show detrimental effects at 640 R/d, illustrating species differences in the ability to resist radioactive exposure. Older plants were found to be relatively more radioresistant. Similar results were noted by Monk (1966) in a longleaf pine field study, where a 50% reduction in species similarity and diversity occurred at 530 and 750 R/d, respectively. In the near-climax forest study at Brookhaven, Woodwell and Sparrow (1963), in predicting the effects of chronic γ-irradiation, observed that enhanced physiological stresses characteristic of natural ecosystems affect results. Woody species seemed generally more sensitive to ionizing radiation than herbaceous species. The greater sensitivity of woody (or ligneous) plants over herbaceous plants was noted in a study using Mediterranean species (Fabries et al. 1972). Also, typical Mediterranean species were found to be more resistant than sub-Mediterranean species.

Changes in species dominance and diversity has also been found in grass communities. In short-grass plains vegetation given irradiation at chronic exposure rates of 0.01 to 650 R/h, the dominant species below 3.3 R/h was *Bouteloua gracilis* (Fraley and Whicker 1973a; Fraley 1987). *Lepidium densiflorum* became the dominant species at exposures between 4 and 12 to 28 R/h. *Gaura coccina* and *Carex heliophila* were found to be the most radioresistant. With 12 to 15 R/h exposure rates, a 50% effect on species diversity and the coefficient of community was experienced. In a second study utilizing acute doses (8 R to 517 kR) during a growing season, Fraley and Whicker (1973b) found short-grass plains to be very resistant. The summer vegetative growth was more resistant in both coefficient of community and diversity than the period of seedling growth or growth during the late autumn. Community changes occurred within constant species diversity values and constant coefficients of community, illustrating the ineffectiveness of these values to fully quantify specific species changes within mass changes (Fraley 1987).

Desert shrubs have also been studied using *in situ* γ-irradiation. *Krameria parvifolia* was given 10 years of chronic radiation (0.1 to 10 R/d). Fruit and leaf reductions were found to occur at 6 R/d; above 7 R/d, 16% of the plants of this highly adapted and successful desert shrub died (Vollmer and Bamberg 1975). In a field experiment in the Northern Mojave Desert, *Ephedra nevadensis* failed to flower and exhibited reduced vegetative growth with a chronic exposure totalling 4 to 10 kR (Kaaz et al. 1971). For boreal shrub communities, some relationship was observed between increased plant height and reduced survival, as would be expected (Dugle and Mayoh 1984).

Regeneration of irradiated communities followed the normal succession of species dominance for that area except where perennials with underground meristems persisted and then became the dominant species (Fraley 1987). In a boreal forest study in Canada, strawberry was the first species to reinvade an affected area due primarily to its ability to propagate vegetatively, not because of any unusual radioresistance (Sheppard et al. 1992). Also, the recovering communities are different from the original, with selective elimination of sensitive plants (McCormick and Platt 1962). In the regeneration of the Brookhaven oak-pine forest understory, the sedge, *C. pensylvanica*, dominated in the 125 to 250 R/d area with a *Rubus* succession over a 5-year period. The distribution of *Rumex acetosella* changed, suggesting possible genetic alteration of radiosensitivity (Flaccus et al. 1974). In a controlled field study in Spain, the recovery of natural grasslands was retarded and the community structure was simplified in both biomass and degree of complexity (Hernandez-Bermejo 1977). The selection process favored weeds and other herbs where irradiation occurred in the younger, successional stages; older, irradiated communities tended to have their disturbances buffered. Low levels of irradiation were also found to have some stimulatory effect on growth.

Not only were the plants in the various communities affected by the γ-irradiation, but so was the soil. The most profound effect was the decrease in complex soil organic matter content with increased radiation (Armentano et al. 1975). This was attributed to the increased decay of leaves and subjacent litter together with release of leachable organic compounds and minerals (Saas et al. 1975). Phosphate ions became more abundant with increasing radiation, having been released from the complex organic and mineral matter. With less organic matter in the upper layers of soil, the storage capacity of water in irradiated decay material is decreased, thereby enhancing the leaching of organic and mineral compounds.

The soil algal community of the Brookhaven oak-pine community underwent principle changes at less than 1000 R/d (Franz and Woodwell 1973). The coefficient of community and percent of similarity dropped to 50% at ~2250 R/d, where eukaryotic forms were replaced with prokaryotic forms (at radiation levels from <1000 to 2000 R/d). The algal community was found to be more resistant than higher plants and exhibited a resistance similar to lichens of this oak-pine forest.

Studies using low, changing exposure rates have been performed to note effects simulating exposure to fallout. In a unique field study, Chandorkar and Clark (1986) planted *Pinus strobus* and *P. sylvestris* seedlings at a radioactive waste disposal area where they were subjected to a low-level, continuous γ-irradiation of 10.15 mR/h. Even at this low exposure rate, photosynthesis was suppressed by 16 to 19%, respiration by 14 to 23%, and ethanol soluble sugar content by 14 to 25%. General growth was depressed with a reduction of stem elongation and crowding of needles. Lateral branch growth was stimulated, due in part to the presence of free auxin (indole acetic acid, IAA). Using a 36-h decreasing exposure rate source to simulate fallout decay, Bottino and Sparrow (1971) found that yield is more sensitive as an endpoint than survival and that reproductive survival is more sensitive than somatic (plant) survival.

Plant species studies

By far, the most work that has been done in studying the effects of irradiation on plants has been performed using individual plant species under controlled conditions. There is wide variability between plants in their tolerance of exposure to γ-irradiation and absorbed dose. Table 1 provides a summary of γ-irradiation dose- or exposure-effect levels as they relate to germination, whole-plant growth, gross organ development, and/or yield. When performing an analysis of irradiation effects to economically important plants (i.e., crops), it is generally considered that for comparative purposes a 10% effect on lethality represents about a 50% reduction in yield (Sparrow et al. 1970; Bottino and Sparrow 1971; Dugle and Mayoh 1984).

General effects

In many field and laboratory studies, the general radiation sensitivity to various species and genera have been compared. Gymnosperms have been found to be usually more sensitive to acute irradiation (LD_{50}s of 0.4 to 1.2 kR) than angiosperms (LD_{50}s of 0.4 to 17.5 kR) (Sparrow et al. 1968b). In predictions based upon analysis of data from numerous studies (Sparrow et al. 1971), cereal crops are among the more sensitive, with 50% yield reductions with doses ranging from 1 to 4 kR (chronic exposure simulating fallout). Rice, on the other hand, is much more resistant. Broadleaf and root crops exhibited a wide range of resistance (1 to 12 kR and 1 to 16 kR, respectively). In reassessments regarding γ-irradiation susceptibility, plants have been determined to be generally more sensitive than when they were first assessed (Woodwell 1963).

There are differences within plant genera in their sensitivity to the effects of γ-irradiation. *Lycopersicum esculentum* is more resistant than *L. pimpinellifolium* to a vegetative stage chronic exposure approaching 7.5 R/d (Nettancourt and Contant 1966). Donini et al. (1964) determined that bread wheats are more resistant than durum wheats when the seeds received chronic irradiation. Shamsi and Sofajy (1980) noted differences between cultivars of broadbean, Killion and Constantin (1972) observed cultivar differences in corn, and Dutta et al. (1986) found differences in groundnut cultivars.

For many of these studies, either total dose (or dose) or dose rates are provided. In order to determine the relationship between dose and dose rate, Killion et al. (1971) observed that, in damage to soybeans, dose and dose rate are generally related factors at low doses (2.5 kR) incurred with low rates (12.5 R/min). When the total dose and dose rates are increased to 5 kR and 25 R/min, respectively, they become

Table 1   A Summary of γ-Irradiation Dose/Exposure Relationships Affecting Yield, Whole-Plant Growth, or Organelle Growth

Plant	Dose level[a]	Affected part[b]	Ref.
Gymnosperms			
Pinus rigida and *P. strobus*	6 R/d over 7 years	Trees killed	Bostrack and Sparrow 1970
	1–2 R/d	Increased tree height and needle length and weight	
P. monophylla (Pinyon Pine)	1.3–8 kR	Immediate growth inhibition	Branderburg et al. 1962
	200–500 R	Terminal buds died	
	100 R	No needles	
	80–100 R	Needles number reduced	
	>15 R	Stem tip elongation inhibited	
P. resinosa	195 R	GR_{50} for seedlings	Rudolph 1971
P. banksiana	16 kR	LD_{50} seed germination	Rudolph 1979
	12.6 kR	6 month survival	
	11.3 kR	10 year survival	
White Pine	685 R	LD_{50} seed germination and survival	Clark et al. 1968
Jack Pine	6310 R		
Abies balsamea	110 Gy	LD_{50} mature trees	Dugle 1986
	1.5 mGy/h	LR_{50} mature trees	
Picea mariana	380 R	GR_{50} for seedlings	Rudolph 1971
P. glauce	28 R/20 h/d	LD_{50} for shoots	Cecich and Miksche 1970
	2 R/20 h/d	LD_{50} for needle primordia	
P. abies	2.33 kR	LD_{50} growth	Rudolph and Miksche 1970

Species	Dose	Effect	Reference
P. glauce	3.29 kR	(seed irradiated)	
Thuja occidentalis	3.93 kR		
Pinus sylvestris	3.92 kR		
Larix laricina	4.56 kR		
Picea mariana	5.19 kR		
P. resinosa	6.02 kR		
Pinus contorta	6.87 kR		
P. banksiana	11.35 kR		Davies 1973
Angiosperms			
Potato	4 kR	YD_{50}	
Sugar beet	4 kR	YD_{50}	
Broad bean	0.1–0.25 kR	YD_{50} in flowering plants	
	0.25–0.5 kR	YD_{50} at early vegetative stages	
Clover	8 kR	YD_{50}	
Perennial ryegrass	16 kR	YD_{50}	
Meadow fescue	16 kR	YD_{50}	
Rumex hydrolapathum	10.0 kR	LD_{50} vegetative stage	Sparrow et al. 1968a
R. crispus	11.5 kR	LD_{50} vegetative stage	
R. sanguineus	13.5 kR	LD_{50} vegetative stage	
Wheat (spring sown)	2 kR	Death (exposure before tillering)	Davies 1968
Avena sp.	1 kR	(GR_{50}) 2 mm coleoptile	Miura et al. 1974
	300 kR	(GR_{50}) 14 mm coleoptile	
Nigella sativa	20–30 kR	LD_{30} germination	Datta et al. 1986
Wheat	0.5–1.25 kR at 1–2 leaf stage	Stimulated height, tillering, ear number, grain yield	Iqbal 1980

Table 1 (continued)

Plant	Dose level[a]	Affected part[b]	Ref.
Wheat	0.5 to 1.25 kR at ear emergence and anathesis	Reduced height and yield	Killion and Constantin 1971
	1.5 kR at 5 R/min	YD_{100}	
	1.6 kR at 40 R/min	LD_{100}	
	1.6 kR at 5 R/min 1–4 leaf stage	No effect	
	1.5 kR at 20 R/min to winter dormant	No effect	Bottino and Sparrow 1971
Wheat	1.73 kR (fallout decay)	YD_{50} (seed)	
	2.26 kR (constant exposure)	YD_{50} (seed)	
Oat	1.42 kR (fallout decay)	YD_{50} (seed)	
	1.51 kR (constant exposure)	YD_{50} (seed)	Siddiqui et al. 1976
Sugarcane	1.1–1.8 kR (setts)	LD_{50} for seedling	
Solanum chacoense	2–3 kR	GR_{50} plant survival	
	6–9 kR	26% recovery of parental stock tubers	McCrory and Grun 1969
Soybean	2.8 kR	GR_{50} height and survival	Killion and Constantin 1974
	17.8 R/min	YD_{50} seed and vegetative	
Soybean	2 kR	LD_{50} (plant height and shoot and root dry wt.)	Witherspoon and Corney 1970
Broadbean	750 R	Stimulatory to growth and yield	Shamsi and Sofajy 1980
		LD_{100}	
Lucerne	>3000 R		
	45–50 kR	YD_{50} field	Fautrier 1976

Species	Dose	Effect	Reference
Lupinus albus	16 R/h for 8 weeks	50% reduction in flowering	Cordero 1982
Tomato	2.75 R/h	No fruits	
	500–1000 R	Increased fruit wt.	Sidrak and Suess 1972
Flax	100–200 R/d	Seed yield increase	Bari 1971
	400–1000 R/d	Increased plant height	
Peach	15 kR @ 12 R/min or 3.7 kR @ 800 R/min	LD_{50} for seedling survival	Lapins et al. 1969
Prunus avium	2.75 kR	LD_{50} for accessory buds	Katagari and Lapins 1974
	4.5 kR	LD_{50} for main buds	
Cucumis melo	3 kGy	LD_{50} pollen tube growth	Cuny and Roudot 1991
Acer spicatum	8.7 mGy/h	LD_{50} (field, 6 years)	Dugle and Mayoh 1984
Alnus rugosa	10.4 mGy/h	LD_{50} (field, 6 years)	
Cornus rugosa	10.4 mGy/h	LD_{50} (field, 6 years)	
Linnaea borealis	3.3 mGy/h	LD_{50} (field, 6 years)	
P. pensylvanica	23 mGy/h	LD_{50} (field, 6 years)	
Ribes americanum	11.8 mGy/h	LD_{50} (field, 6 years)	
Rosa spp.	20 mGy/h	LD_{50} (field, 6 years)	
Salix discolor	2.1 mGy/h	LD_{50} (field, 6 years)	
Vaccinium angustifolium	12.6 mGy/h	LD_{50} (field, 6 years)	
Viburnum lentago	3.8 mGy/h	LD_{50} (field, 6 years)	
Bryophyta			
Pelia epiphylla	1–10 kR	LD_{50}	Woollon and Davies 1981
Marchantica polymorpha	1–10 kR	LD_{50}	
Lucularia cruciata	1–10 kR	LD_{50}	
Hookeria lucens	10–25 kR	LD_{50}	
Polytrichum commune	10–25 kR	LD_{50}	

Table 1 (continued)

Plant	Dose level[a]	Affected part[b]	Ref.
Sphagnum cuspidatum	10–25 kR	LD_{50}	
Thuidium tamariscinum	50–100 kR	LD_{50}	
Funaria hygrometrica	>100 kR	LD_{50}	
Fissidens adianthoides	>100 kR	LD_{50}	
Plagiothecium denticulatum	>100 kR	LD_{50}	
Fungi			
Aspergillus flavus	25–200 kR	Young mycelia	Mohyuddin et al. 1972
Lichen fungi	5–600 kR	GR_{50}	Gannutz 1972
	20–750 kR	GR_{90}	
Free-living fungi	60–220 kR	GR_{50}	
	245–470 kR	GR_{90}	
Algae			
Chlorella pyrenoidosa	24 kR	GR_{50}	Gannutz 1972
Trebouxia sp. (free living)	40 kR	GR_{50}	
Trebouxia sp. (lichen algae)	2–150 kR	GR_{50}	
Chlorella	23 kR @ 8 kR/min	LD_{90} acute exposure	Posner and Sparrow 1964
	15 kR @ 5 kR/d	LD_{90} colony formation	
Chlamydomonas	10 kR @ 8 kR/min	LD_{90} acute exposure	
	90 kR @ 30 kR/d	LD_{90} colony formation	
Sirogonium melanosporum	3 kR	Cell division reduced 13%	Wells and Hoshaw 1980
	15 kR	Lethal within 3 months	
S. sticticum	15 kR	Giant cells w/double-sized nuclei	

Species	Dose[a]	Effect[b]	Reference
Pediastrum boryanum	50 Gy	Increase in colony density	Slavinski and Hillson 1984
	500 Gy	Lethal	
Anacystis nidulans	20 kR	GR_{50}	Asato 1971

[a] Units of radiation dose: 1 Rad (R) = 1 cGray (cGy) or 1 Gray (Gy) = 100 Rad (R).

[b] For dose/exposure level: GR_{50} = 50% growth reduction; GR_{90} = 90% growth reduction; LD_{50} = 50% lethal dose; LD_{90} = 90% lethal dose; YD_{50} = 50% yield reduction; LR_{50} = dose rate for 50% growth reduction.

independent, with total dose being the deciding factor of damage sustained. The generality and basis of this behavior are unknown.

Growth stages

The ability of a plant to resist exposure to radiation is dependent upon the stage of growth during irradiation and the plant parts or organs that receive the irradiation (Kawai and Inoshita 1965). Davies (1970) found that cereals (barley, wheat, and oats) exhibited a lower likelihood of survival to irradiation during the vegetative stage. Bajaj et al. (1970) noted similar results for beans. Killion and Constantin (1972) observed that chronic radiation during corn vegetative and gametogenic stages was more damaging to yield and growth than similar radiation during the embryogenic stage. Pfahler (1970) determined that the reproductive viability of corn plants was greatly reduced (male pollen transmission by 25%, female fertility by 90%) when the seeds received doses of 40 kR.

Numerous studies have been performed on the effects of γ-radiation on seed germination and seedling development using seeds that have been irradiated. Seed irradiation has been studied for many years because of the possible use of radiation to enhance storage, without spoiling, of viable seeds for food or as crop seed. The criteria for assessing seed viability for these types of studies are percent of seed germination and crop yield, that is, seed production. Results of some of these studies are presented in Table 1.

The oxygen and water content of the seed has a direct bearing on the radiosensitivity of seeds. Constantin et al. (1970) found that the less storage oxygen there is in a seed, the greater the protection from (or less damage as caused by) irradiation. Constantin et al. (1970) and Conger and Caribia (1972) determined that the water content of rice seed is also crucial with respect to radiosensitivity to γ-irradiation. That is, radiation sensitivity decreased with increasing water content from 2 to 13%, but then increased with increasing water content to 20%.

γ-Irradiation has also been found to stimulate growth and development when applied at low doses and rates during early seedling growth. Generally, the doses are on the order of 500 to 1250 R, and/or dose rates are less than 1000 R/d for enhanced vegetative growth and less than 200 R/d to increase fruiting and yield (Abifarin and Rutger 1982; Bari 1971; Bostrack and Sparrow 1970; Iqbal 1980; Shamsi and Sofajy 1980; Sidrak and Suess 1972; Slavinski and Hillson 1984). Austrian and Scots pine plants required up to 10 kR for growth stimulation (Vidakovic 1967).

Plants that employ vegetative propagation in the form of bulbs, such as onions, have also been studied with respect to storage effectiveness. Thomas et al. (1975) demonstrated that onion, if irradiated (6 to 9 kR) within 2 weeks of harvest, would not sprout, but would be stimulated to sprout if irradiated 3 or more weeks after harvest. The quality of the bulb in all cases was reduced with discoloration of the growth center.

Plant organs

When studying irradiation effects of plant parts, a general rule is the more the growing parts (those parts undergoing rapid mitotic or meiotic cell division, e.g., meristems, reproductive cells, cambium) are exposed to the radiation, the more at risk is the plant, or, more specifically, those parts being produced (meristematic organs [Cordero and Gunckel 1982a; Graham and Generoso 1971; Branderburg et al. 1962; Fabries et al. 1972; Lapins and Hough 1970; Chauchan and Singh 1975; Mednik and Usmanov 1985], gametogenic area [Cordero and Gunckel 1982b; Gastel and Nettancourt 1975; Killion and Constantin 1971; Killion et al. 1971]). Retamal et al. (1990) went so far as to show that even the location of the seed in the pod can have an effect on its radioresistance. For example, basal seeds of peanut had 2.38 times greater resistance than the apical seeds.

Generally, cells other than meristems or primordia stop growing with increased radiation dosage (especially that of γ-irradiation). Cordero and Gunckel (1982b) and Dutta et al. (1986) observed that the flower parts were more sensitive than the vegetative parts. Some detrimental growth effects of the vegetative plant can be attributed to the breakdown of the vascular system. Branderburg et al. (1962) noted some vascular strand anomalies in *P. monophylla*, while Cordero and Gunckel (1982a,b) found altered vascular anatomy in *Lupinus albus*, each at low doses (<3 kR).

Lower plant forms

Simpler plants such as algae, fungi, and symbiotic lichens exhibit a wide range of resistance to γ-irradiation. Kraus (1969) studied the effects of acute γ-irradiation doses on the cyanobacterium (formerly blue-green algae) and found no systematic behavior (morphologically or taxonomically) to the range of resistance. These organisms though are a great deal more resistant than terrestrial higher plants, with LD_{90}s ranging from below 400 kR to greater than 1200 kR. *Anacystis nidulans* recovered faster from γ-irradiation effects when exposed to continuous white light than when kept in continuous darkness or exposed only to red light (Asato 1971). The survival of two green algae (*Chlorella* and *Chlamydomonas*) occurred at a lower total dose level for acute exposure than for the chronic exposure (Posner and Sparrow 1964). When cells of another green alga, *Sirogonium sticticum*, were irradiated at a dose of 15 kR during cell division, giant cells with double-sized nuclei (with no nuclear divisions) and microcells (with no nuclei) were observed (Wells and Hoshaw 1980).

For fungi, doses of 50 to 100 kR are needed to effect a dry-weight reduction in the mycelia of *Penicillium expansum*, along with temporary increases in RNA and DNA levels for less than 4 d (Chou et al. 1971). The slime mold *Physarum polycephalum*, at doses of 30 to 225 kR, under-

goes a coalescence of asynchronous microplasmodia into a multinu-
clear macroplasmodia which is more radiosensitive (Dmitriev et al.
1980). The maximum sensitivity to irradiation occurred during mitosis
or at early synthetic (S) stage of interphase. The nongerminated conidia
of toxigenic strain of *Aspergillus flavus* are more resistant to γ-irradiation
treatments than the nontoxigenic strain (*A. flavus oryzae*) (Padwal-
Desai et al. 1976a). Susceptibility to radiation injury increased only for
a short period, followed by development of resistance for either strain
(Padwal-Desai et al. 1976b). The germination of *A. flavus* and *A. ochraceus* sclerotia, fungi that attack stored grains, was reduced by increased
γ-irradiation (1.0 or 1.5 kGy) and reduced CO_2 (<40%) (Menasherov et
al. 1992).

Lichens are somewhat unique due to their symbiotic relationship
of fungi and algae. When studied as separate components and compared to free-living forms of their respective components, the free-
living algae were more sensitive (LD_{50} of 8 to 30 kR) than the algal
component (LD_{50}s of 10 to 50 kR) (Gannutz 1972). The fungal component was the least sensitive (12 to 148 kR) and comparable to free-living
fungi. Erbisch and Kalosis (1973) noted similar results, but with greater
irradiation (1000 and 2000 kR) to *Cladonia sylvatica* and its phycobiont.
The podetia of the fungi was damaged somewhat at 2000 kR, while the
phycobiont suffered repairable damage at only 100 kR.

Comparative radiation types

Some comparison studies using different radiation types to induce
effects in plants have been done. Killion and Constantin (1974) studied
the effects of separate and combined β- and γ-irradiation on soybeans.
γ-Irradiation was only a little more effective (1.2 times) than β-radiation
at a constant dose for reducing lateral growth and vegetative and seed
yield. While survival and height were dependent on total dose, they
were independent of dose rate and β- to γ-irradiation ratio. In a study
of differential and combined effects of β-, γ-, and fast neutron irradiation
on soybean seedling growth and survivability, Witherspoon and Corney (1970) determined that γ- and β-irradiation are about equal in
effectiveness, while fission neutrons were 6 to 15 times more robust.
Roth (1978) found that, with respect to RBE in *Vicia faba*, 14.8 MeV
neutrons are 7.6 times more effective than 1.25 MeV ^{60}Co γ-rays. The
effects of 18 MeV electrons are small, indicating the need for energetic
and massive particles to cause physiological or genetic changes in
plants which are manifested in vegetative growth, reproductive ability,
or yield. Nakai et al. (1980) found that thermal neutrons effects are
not affected by seed moisture and post-irradiation storage effects like
γ-irradiation. In general, there is a paucity of data on the effects of
different radiation on plants due to the complexity of using diverse
sources and performing accurate dosimetry.

Physiological studies

Radiation-induced changes in the biochemicals and processes within plants constitute an important class of effects. Depending upon the cell, the absorbed dose or exposure, and the total dose or exposure of irradiation received, various physiological processes can be disrupted or altered by these biochemical changes. Single processes and whole metabolic pathways are affected by changes in the types and quantities of metabolites.

Metabolite alterations

One primary process altered by radiation is the breakdown of the complex carbohydrates to hexoses and trioses. Irradiated *Fagus silvatica* wood yielded several types of sugars not found in control plants, suggesting a possible random mechanism of depolymerization (Bonotto and Milic 1969). Kurobane et al. (1979) found germinating barley seed to exhibit enhanced leaching of sugars and phosphates from the seed when irradiated with four doses of 500 kR. Dubery et al. (1988) identified several stress-induced metabolites, some with antifungal activity, in the pitted regions of γ-irradiated *Citrus* peel. In soft fruit cells irradiated and observed histochemically, the polysaccharide or functional molecule itself was not affected directly (Foa et al. 1980). Rather, the regulatory mechanism was altered.

Proteins can fragment during irradiation. A peanut storage protein, arachin, extracted from cotyledons, breaks down with acute irradiations of 0.5 and 1 MR (Huystee 1971). Generally, γ-irradiation has been shown to affect protein quality and quantity in plants and cultures. O'Hara (1969) noted an increase in the various protein fractions in wheat. Chronic irradiation (53.5 Gy/year) of *Synechococcus lividus* in culture causes changes in protein content, which are sensitive to different phases of synthesis and catabolism (ammonia excretion) (Conter 1987). Pai and Gaur (1983), using mitochondria isolated from γ-irradiated (250 and 500 kR) kidney bean hypocotyl segments, noted that reduced protein content and its qualitative alteration were directly related to enhanced protease activity.

The amino acid content of seeds and seedlings fluctuate with irradiation. Amounts of free amino acids increase with increasing irradiation doses (to 15 kR) in corn seedling roots, while they drop in the shoots. Bound amino acids, specifically lysine and valine, also increase in concentration (in the corn seedling roots) at lower levels of irradiation (Iqbal et al. 1974). In γ-irradiated wheat seeds (20 to 200 kR), the quantity of endogenous amino acids was low, while the specific activity of leucine was high when compared to the controls. There was also decreased incorporation of leucine in nascent α-amylase in irradiated seeds attributable to its impaired *de novo* synthesis (Ananthaswamy et al. 1971).

The dose to produce enhanced quantities or unmasking of anthocyanins and other flavonoids, which are indicators of degradation in other plant systems, varies with each species. When buckwheat seeds were irradiated (up to 12 kR), formation of some flavonoids was stimulated while others was suppressed, and the content varied both with the length of storage and subsequent location in the seedlings (Margna and Vainjarv 1976). Anthocyanins are not degraded during irradiation (up to 16 kR exposure) (Sparrow et al. 1968a). Similarly, carotenoid–protein complexes (98% β-carotene) from fresh mangoes are stable to γ-irradiation (up to 750 kR exposures) when contained in water rather than in petroleum ether (Ramakrishnan and Francis 1980).

Oxidative stress mechanisms

γ–Irradiation has been shown to induce oxidative stress (Edreva et al. 1989; Singh and Kesavan 1992). One form of oxidative stress due to γ-irradiation is free radicals (hydroxyl), which, along with a greatly enhanced presence of oxygen, have been shown to be the primary causal agents of damage in plant tissues and cells (Constantin et al. 1970; Donaldson et al. 1979a,b). Exposure of a plant to low levels of γ-irradiation (0.02 Gy over 7 d) induces the formation of free radicals (Kuzin et al. 1986). Unprotected gibberellic acid (GA) in barley seedlings in the [γ]-lactone ring is reduced by these radicals (Sideris et al. 1971). When barley seeds were soaked in water with various concentrations of oxygen (1.6 to 12.5%) and then irradiated (up to 50 kR), the increased presence of oxygen allowed greater damage, as manifested in the seedling. By using γ-irradiation and inhibitors of the enzymatic ($NADPH+Fe^{+3}$-induced) lipid peroxidation, Pai and Gaur (1988) determined the participation of superoxide, hydrogen peroxide, and hydroxyl radicals as effective oxygen species that initiate lipid peroxidation in mitochondria with subsequent cellular damage. Singh and Kesavan (1992) have developed a hypothesis concerning the roles of catalase and superoxide dismutase (SOD) as radioprotection agents from hydroxyl and superoxide in the irradiated cells. Catalase, in addition to converting hydrogen peroxide to oxygen and water, scavenges the radiation-induced trapped electrons as well as hydroxyl ions in dry seeds, preventing them from forming superoxide anions. Where superoxide anions are formed, SOD converts them into hydrogen peroxide and oxygen, which are much less damaging to cells than superoxide ions.

Enzymes

Enzymes are affected (stimulated or degraded) by γ-irradiation. With an acute dose of 200 kR to citrus fruits, phenylalanine ammonia-lyase (PAL) activity in the peel was at a maximum 1 d after irradiation, resulting in enhanced levels of phenolics. It then decreased, especially in those fruits which would exhibit no subsequent peel damage (Riov et al. 1968). In a later study by Riov (1975), the elevated levels of phenolics were only found in damaged cells of the outer peel (flavedo), not undamaged cells. The induction of PAL by light in excised potato parenchymatous tissue, or by γ-irradiation in excised bud tissue, can be abolished completely by submersion of the tissue in water, an oxygen-limiting process (Shirsat and Nair 1976).

Enzymes employed in growth and development of plants are affected differentially by γ-irradiation. In corn seeds, irradiation with 50 kR produced an increase in most arylesterases of the first internode after 5 d and a decrease in pH 7.5 esterases of the shoot after 7 d (Endo 1967). Satyanarayan and Nair (1986) noted the enhancement of the 4-aminobutyrate shunt in potatoes irradiated with 100 Gy. The enzymes that are used in growth of pollen tubes in onion and beet are adversely affected by lower doses of irradiation (<100 kR) as compared to doses affecting those enzymes of general respiration (>100 kR) (Georgieva and Atanassov 1986). In corn seed, sweet potato root tissue, barley seed, *Lilium ragale* pollen tubes, and tobacco leaves, irradiation (up to approximately 100 kR) indirectly stimulates the development or expression of peroxidase (Endo 1967; Ogawa and Uritani 1971; Balachandran and Kesavan 1978; Georgieva 1987; Edreva et al. 1989; respectively). Ethylene generation in plants is independent of peroxidase development (Ogawa and Uritani 1971; Strydom et al. 1991). Like peroxidase, ribonuclease activity is enhanced by irradiation (10 kR) (as noted by an increase in enzyme production), with a corresponding decrease in RNA levels in the petiole of coleus leaves (Gordon and Buess 1972). ATPase and mevalonate kinase of GA syntheses are impaired in germinating irradiated wheat seeds (Machaiah et al. 1976). The irradiation reduces the efficiency of GA interconversion, especially from less active to more active. This effect is related to disturbance of the protein-synthesizing system.

In a study of the substructures of the peroxidase enzyme in *Saintpaulia*, thermal neutrons were used to cause permanent changes in its enzyme activity. This radiation (250 and 1000 R) permanently altered the peroxidase isozymes, as noted by the increased numbers of isozymes and their different electrophoretic mobility (Kelly and Lineberger 1981). This illustrates how neutrons, and possibly other forms of radiation, fragment enzymes, destroying their integrity and activity.

Growth regulators

γ-Irradiation has generally been found to decrease the amount and thereby the effectiveness of endogenous growth regulators. In studies of the effects of low levels of continuous γ-irradiation (10.15 mR/h) on vascular cambium activity in Scotch pine (*Pinus sylvestris*), there was a notable decrease in stem elongation (21%) with reduced endogenous IAA levels (Chandorkar and Dengler 1987). The mitotic activity of the cambium was stimulated with notable increases in the supply of nutrients and increased ratios of GA and cytokinins over auxins. Gordon et al. (1971) had also found reduced levels of auxin (along with RNA and protein) in coleus leaf petioles when they had received acute doses of irradiation (up to 10 kR). Irradiation (10 kR) accelerates the degradation of the IAA synthesis enzyme by activating the proteolytic activity of hydrogen peroxide-dependent IAA oxidase, with associated decreases in tryptophan, a substrate for IAA synthesis (Ussuf and Nair 1974). However, the proteolytic activity of peanut cotyledon cells is greatly suppressed with large doses of irradiation (0.5 and 1 MR) (Huystee 1971).

When these growth regulators are applied exogenously, the injurious effect(s) is minimized or eliminated. GA (GA_3 2 × 10^{-5} M) caused normal germination in wheat despite irradiation (Ananthaswamy et al. 1971). Similar results have been found for rice seedlings using GA_3, IAA, or indole butyric acid (IBA) (El-Aishy et al. 1976); in sorghum treated with cysteine (Reddy and Smith 1978); and in wheat with the antioxidants hydroquinone (0.01%) (Babaev et al. 1990) and 5-methyl-resorcinol (0.01%) (Babaev and Morozova 1991). Auxins (IAA at 3 × 10^{-4} M) inhibited post-irradiation breakdown of DNA in blue-green alga, *Anacystis nidulans* (Dmitriev and Grodzinsky 1974), while 10 ppm 2,4-D (an auxin-like substance) restricts the injurious effect of γ-irradiation (up to 30 kR) in safflower apical meristems (Chauchan 1976).

Organelle integrity

A number of studies have shown direct correlations between cell structural component degradation and increased degradation enzyme presence or activity due to γ-irradiation. *Penicillium expansum*, exposed to acute doses of irradiation (50 to 100 kR), experienced a decrease in mycelial dry weight, which could be directly correlated with increases in polygalacturonase and tissue macerating enzymes, specifically pectolytic enzymes (Chou et al. 1970a,b). The photosynthetic apparatus of corn (*Zea mays*), where seeds had been exposed to γ-radiation at 500 R, was unaffected, as were the quantities of chlorophylls *a* and *b* (Ahmad et al. 1976). In irradiated, young, white pine seedlings, as the exposure dose increases between 230 and 7500 R, the photosystem itself became adversely affected (Ursino et al. 1974). Ursino et al. (1977) determined that the reduced photosynthetic rates (observed in 11.25 kR irradiated soybean leaves) were not due to stomatal restrictions regardless of oxygen levels (1 and 21%).

The ultrastructure of chloroplasts and mitochondria are reasonably stable to γ-irradiation. More than 1 MR of γ-irradiation is needed to produce a damaging effect on the chloroplast ultrastructure of wheat leaf cells (Walne and Haber 1968). The physiological deterioration of the chloroplast structure in the dark was also retarded with 1 MR of irradiation. The energy transfer process from the photoreaction center to electron traps is preserved at 1 MR, but not at 2 MR. Pai and Gaur

(1988), in studying enzyme inhibition in mitochondria isolated from bean hypocotyls, noted uncoupling and structural disintegration with between 1 and 5 kGy of γ-irradiation.

Mineral and metabolite transport within a cell relies heavily on the permeability and integrity of the cell membranes. In studying membrane permeability, Pai and Gaur (1987) and Hayashi et al. (1992) found that, with 1 to 5 kGy of γ-irradiation, membranes such as those of the mitochondria will leak electrolytes. Where Rb$^+$ was used as an indicator for K$^+$ uptake in *Chlorella*, two mechanisms were found (Paschinger and Vanicek 1974). At 3×10^{-5} *M* Rb$^+$, 60 kR of γ-irradiation reduced Rb$^+$ uptake at 30°C, whereas it increased at 10°C. At 3×10^{-3} *M* Rb$^+$, its uptake was stimulated by irradiation. The K/K exchange system is stimulated, but the K$^+$ net transport is injured. The increase in K efflux after irradiation is not attributed entirely to enhanced membrane permeability. Dong et al. (1994) have characterized the effects of γ-irradiation on the plasma membrane of cultured apple cells. H$^+$-ATPase activity was lost immediately with doses as low as 0.5 kGy and was found to be irreversible. This loss of activity was associated with increases in passive ion effluxes and decreased cell capacity to regulate external pH. In membrane permeability studies, *Chelidonium majus* latex vacuoles, exposed to less than 252 Gy, tended not to release fatty acids or alkaloids (Sato et al. 1992). In both irradiated potato tubers and kidney bean hypocotyl, individual phospholipids in membranes increased or decreased, respectively, in content, e.g., linolenic acid contents increased with accompanying decrease of linoleic acid content (Hayashi et al. 1992). The effect of γ-irradiation on cell integrity can be characterized by losses in uncontrolled ion effluxes with mixed changes in phospholipid content.

Genetic studies

The mutation of genes, that is, the alteration of the genetic message for growth and reproduction, can be caused by the use of radiation. Numerous mutation studies using radiation have been performed for the purposes of permanently altering the genetic composition of the tested plants. These alternations result in changes to the morphology, anatomy, and/or productivity in subsequent generations of the plant. Through selection processes, desirable plant characteristics are selected and cultivated. Such work is discussed in the Irradiation as a Tool in Plant Research section. The relative effectiveness of different radiations, correlations of radiation resistance with other factors, and damage repair are reviewed in the subsection on genetics.

Gastel and Nettancourt (1974) determined that chronic irradiation is less effective than acute irradiation for inducing self-compatibility mutations. Acute γ-irradiation is comparable to X-rays for such work. This latter point was verified by Sanda-Kamigawara et al. (1991). Das

(1978), on the other hand, determined that γ-rays can be more lethal than X-rays, at the same absorbed dose, to growth, pollen activity, and second generation mutant frequency in *Pennisetum typhoides*. Contant et al. (1971) determined that pollen is not the most efficient stage during ontogenesis of the tomato for production of γ-ray-induced effects. Pollinated zygotes gave higher mutation frequencies than irradiated gametes. However, seed irradiation gave the highest efficiency, i.e., the highest incidence of mutation frequencies. Because genetic mutations are stochastic, there is no threshold for occurrence, and the severity of the resulting morphologic or physiologic changes is unrelated to the dose (Alpen 1990).

Interphase chromosome volume (ICV) has been measured in cells of plants that have undergone irradiation in an effort to measure the susceptibility or resistance of the plant to irradiation. The most predictive work on plant survivability to irradiation has centered on equating survivability of higher plants to the size of the ICV, with a larger volume correlated with greater sensitivity (Sparrow and Woodwell 1962; Yamakawa and Sparrow 1966; Sparrow et al. 1971). Kaaz et al. (1971) noted this relationship in the Mojave Desert shrub, where *Ephedra nevadensis*, having a larger ICV, was more sensitive compared to other shrubs. Conger (1976) verified Miksche and Rudolph's (1968) findings that there was no consistent relationship between radiation response and nuclear volume within ecotypes or cultivars of a species. Woollon and Davies (1981) extended the findings of an inverse correlation between radiation sensitivity dose response and ICV (nuclear volume/haploid chromosome number) to include bryophytes. ICV and nuclear volume can used as a general measure of plant resistance to irradiation.

A second relationship to radioresistance was also noted by Sparrow et al. (1961), where a direct relationship existed between it and polyploidy. Ichikawa (1972) verified this in a study using *Tradescantia* (triploid clone) stamen hairs. The higher radioresistance of higher polyploidies was explained by their smaller chromosome volumes rather than genetic redundancy. However, meiotic abnormalities induced by γ-irradiation are greater for tetraploid then diploid species of *Phyalis* (Gupta and Roy 1985). Therefore, no general relationship can be made between polyploidy and radioresistance.

Mutant-induced changes and the ability of the cell to repair its DNAs are among the more studied aspects of radiation genetics. Netrawali and Nair (1986) noted the necessity of supernatant protein or whole chloroplasts of light-grown cells to repair dark-grown *Euglena* nuclear and spheroplast DNA. Oxygen is required for the fast rejoining processes in chromosomes of *Nigella damascena* (Gilot-Delhalle et al. 1973). For barley, the capacity to repair induced lesions is directly related to their γ-irradiation sensitivity (Inoue et al. 1980).

Irradiation as a tool in plant research

It is at the genetic level that most research has been done with respect to using radiation as a tool. γ-Irradiation (and X-ray radiation) has been used as a tool to induce mutagenic changes in flowering plants in order to develop new varieties with different flower colors. Buiatti et al. (1965) found that color changes in *Gladiolus* flowers were a function of variety with irradiations less than 25 kR, as did Gupta and Samata (1967) for *Cosmos bipinnatus*. Also, Kawai and Inoshita (1965) used γ-irradiation to determine that differentiation of egg and pollen initial cells had already initiated at the stage of ear primordia formation in rice. Most of this type of genetic work has been with the use of large acute doses of irradiation (500 to 2500 Gy) in order to cause a mutation without greatly affecting other cellular processes (Haber 1968). In more recent genetic work, lower doses of γ-irradiation (~100 Gy) have been used to fragment the chromosomes of a donor plant so that desirable genetic traits may be transferred to other plant protoplasts and thereby become a viable part of the recipient plant (Dudits et al. 1987; Kumar and Cocking 1987).

Some plants have been altered genetically to provide better resistance to one or more environmental insults such as insects or drought. γ-Irradiation (10 to 40 kR) alteration of the gene composition of the seed embryo improved mung beans by selection in the second generation, M_2 (Rajput 1974). The same was found with rice (Virk et al. 1978). Resistance to the rust, *Melampsora lini*, in flax was enhanced by irradiating the flax seed (Srinivasachar and Seetharam 1971). However, sorghum resistance to witchweed is not increased with increased doses of irradiation, although the incidence of infestation declines because the irradiation reduced the sorghum root growth (Bebawi 1984).

In the early 1980s, a specialized type of mutation work, protoplast fusion using γ-irradiated donor protoplasts, became available as a means of transferring desirable traits between plant species and genera. Dudits et al. (1987), working with information concerning irradiation-induced genomic fragmentation, produced a viable fusion system. Resistant traits in carrot were infused into tobacco with resulting regeneration of fertile tobacco plants. Others did similar work using asymmetric somatic hybrids, for example, *Lycopersicon esculentum* and irradiated *L. peruvianum* (Wijbrandi et al. 1990) tomatoes and potatoes (McCabe et al. 1993), ryegrass cultivars (Creemers-Molenaar et al. 1992), and tobacco and *Petunia* (Hinnisdaels et al. 1991). Usually, the recipient species is viable with little or no dramatic protoplast DNA fragmentation as Derks et al.(1992) were able to determine using *L. peruvianum* to achieve inter-intraspecific chloroplast transfer.

In the area of plant taxonomy, scientists have been noting differences in physiological processes and using various types of instrumentation to distinguish between plant species, races, and ecotypes. McCormick and Rushing (1964) used γ-irradiation as a means to identify various races of *Sedum pulchellum* by noting the delay of seed germination as a function of irradiation dose and polyploidy.

Because γ-irradiation causes disruptions in the genetic system that manifest themselves as altered physiological processes, it can be used to stop, alter, or shunt these processes, thereby identifying them. Pai and Gaur (1988), in studying enzyme inhibition, used γ-irradiation to study oxidation of succinate and NADPH in bean hypocotyls. Metabolic shunts for accumulated amino acids were studied by Satyanarayan and Nair (1986). Machaiah and Vakil (1979) found that by using a combination of irradiation and gibberellins (GA₃), three isoenzymes were noted in the developing seedlings operating in two systems for the synthesis of functional α-amylase molecules. Radiosensitivity itself can be used to measure oxygen concentration in root tip meristem tissue (Donaldson et al. 1979b). Radiation is used as a tool to study membrane permeability. Pai and Gaur (1987) and Hayashi et al. (1992) found that with 1 to 5 kGy of γ-irradiation, membranes such as those of the mitochondria will leak electrolytes but not fatty acids (Sato et al. 1992).

In general, plant growth and differentiation can also be studied using γ-irradiation and other radiations (X-ray and neutron). The number of leaf initials has been determined through the use of directed radiation at meristematic tissues (Haber 1968).

Haber (1968) also lists other uses of ionizing radiation as tools in plant research. These include determining cross sections, thickness, shape, and internal structures; somatic mutations in studying cell lineages in plant development; gas movement in plants; more extensive studies using mitosis-inhibited cells; and senescence studies.

Summary

In this chapter we have surveyed the effects caused by various types of radiation on parts and activities of the plant communities, individual plants, plant organs, and cell(s). Of the forms of radiation available for plant study, γ-radiation has proven to be the radiation of choice because of its availability and controllable dose in an energy range that can be either beneficial or deleterious. Normally observed is the effect on the whole plant, i.e., its health and reproductive capability. The central question is whether radiation directly affects the chemical structure of the cellular building blocks or does it bring about its effect through the process of forming those structures, i.e., through enzymatic changes (presence and/or rate changes).

γ-Radiation can have a wide range of effects on plant growth and development, ranging from stimulatory at low levels (~5 to 10 Gy), through increasingly harmful effects to vegetative growth, to pronounced decreases in reproductive effectiveness and viable yields at higher levels of radiation (generally >10 Gy to over 10 kGy). The degree of effects on the various plant species is a function of species, age, plant morphology, physiological stresses, and genetics. Large chromosome size and polyploidy can increase the level of deleterious effect. The amount of protection afforded by the structure of the plant is also a factor, as is the presence of certain chemicals such as growth regulators or antioxidants.

Most structural components of the cells and organs (complex carbohydrates and proteins) are not directly affected by γ-radiation, especially at doses less than 10 kGy. The plant organs and growth stages that are most susceptible to radiation are the meristematic tissues and the gametogenic areas, those parts that undergo rapid DNA division and/or replication. DNA is very vulnerable to outside influences during these phases of growth. The DNA altered by the radiation then directs the production of altered enzymes or their nonproduction. In either case, degradation enzymes, and other suppressed enzymes, can function in a less-hindered manner, causing changes in metabolic pathways, which in turn show their effects in altered plant growth and development. The effects are only noted at the whole-plant level if a significant number of cells within an organ are affected. The number of significant cells may be only one if it is the "mother" cell for a meristem or gamete or many cells if an older organ is irradiated.

Research opportunities available in plant science using γ-irradiation, or other types of radiation, are primarily in the areas of (1) gene division to allow genetic exchanges in asymmetrical somatic organelles and (2) physiological disruptions as a means to map various metabolic processes in the cells. Controlled field studies are also needed to better assess the possible long-term effects to the plant communities that are near manmade radiation sources (radiation waste dumps, nuclear accident sites, and weapons test sites). Other studies are necessary to evaluate the effects of radiation on individual plants or plant products that may be used as food sources in space flights.

As we learn more about the genome of various organisms and how modification of specific genes or nucleic acids can lead to desired plant types or undesirable results, controlled use of the various radiations could come into greater play in this area of plant science.

Bibliography

Alpen EL (1990) *Radiation Biophysics*, Prentice Hall, Englewood Cliffs, NJ.
Kiefer J (1990) *Biological Radiation Effects*, Springer-Verlag, Berlin, 444 pp.
Spinks JWT, Woods RJ (1990) *An Introduction to Radiation Chemistry*, Wiley-Interscience, New York, 574 pp.

Tabata Y, Ito Y, Tagawa S (1991) *CRC Handbook of Radiation Chemistry,* CRC Press, Boca Raton, FL, 937 pp.
Travis EL (1989) *Primer of Medical Radiobiology,* 2nd ed., Year Book Medical Publ. Inc., Chicago, 302 pp.
Tubiana M, Dutreix J, Wambersie A (1990) *Introduction to Radiobiology,* Taylor and Francis Inc., Bristol, PA, 371 pp.
Wang CH, Willis DL, Loveland WD (1975) *Radiotracer Methodology in the Biological, Environmental and Physical Sciences,* Prentice-Hall, Englewood Cliffs, NJ.

References

Abifarin AO, Rutger JN (1982) Effect of low gamma radiation exposures on rice seedling development (*Oryza sativa*). *Environ Exp Bot* 22:285–291.
Ahmad MB, Ashour NI, El-Basyouni SZ, Sayed AM (1976) Response of the photosynthetic apparatus of corn (*Zea mays*) to presowing seed treatment with gamma rays and ammonium molybdate. *Environ Exp Bot* 16:217–222.
Ananthaswamy HN, Vakil UK, Sreenivasan A (1971) Biochemical and physiological changes in gamma-irradiated wheat during germination. *Radiat Bot* 11:1–12.
Armentano TV, Holt BR, Bottino PJ (1975) Soil nutrient content of old-field and agricultural ecosystems exposed to chronic gamma irradiation. *Radiat Bot* 15:329–336.
Asato Y (1971) Photorecovery of gamma irradiated cultures of blue-green alga, *Anacystic nidulans. Radiat Bot* 11:313–316.
Babaev MS, Morozova IS (1991) Antimutagenic activity of 5-methylresorcinol in spontaneous and radiation mutagenesis of wheat. *Cytol Genet* 25:28–30.
Babaev MS, Zoz NN, Serebryanyi AM, Morozova IS (1990) Effect of hydroquinone on wheat plants under field conditions. *Cytol Genet* 24:44–47.
Bajaj YPS, Saettler AW, Adams MW (1970) Gamma irradiation studies on seeds, seedlings and callus tissue cultures of *Phaseolus vulgaris* L. *Radiat Bot* 10:119–124.
Balachandran R, Kesavan PC (1978) Effect of caffeine on peroxidase activity and gamma-ray-induced oxic and anoxic damage in *Hordeum vulgare. Environ Exp Bot* 18:99–104.
Bari G (1971) Effects of chronic and acute irradiation on morphological characters and seed yield in flax. *Radiat Bot* 11:293–302.
Bebawi FF (1984) Effects of gamma irradiation on *Sorghum bicolor–Striga hermonthica* relations. *Environ Exp Bot* 24:123–129.
Bonotto S, Milic G (1969) Formation De Sucres Dan Le Bois De *Fagus silvatica* L Irradie Par Les Rayons Gamma Du Cobalt-60. *Radiat Bot* 9:375–378.
Bostrack JM, Sparrow AH (1970) The radiosensitivity of gymnosperms. II. On the nature of radiation injury and cause of death of *Pinus rigida* and *P. strobus* after chronic gamma radiation. *Radiat Bot* 10:131–143.
Bottino PJ, Sparrow AH (1971) Comparison of the effects of simulated fallout decay and constant exposure-rate gamma-ray treatments on the survival and yield of wheat and oats. *Radiat Bot* 11:405–410.

Branderburg MK, Mills HL, Richard WH, Shields LM (1962) Effects of acute gamma radiation on growth and morphology in *Pinus monophylla* Torr. and Frem. (Pinyon Pine). *Radiat Bot* 2:251–263.

Buiatti M, Ragazzini R, Tognoni F (1965) Effects of gamma radiation on *Gladiolus*. *Radiat Bot* 5:97–98.

Cecich RA, Miksche JP (1970) The response of white spruce (*Picea glauce* (Moench) Voss) shoot apices to exposure of chronic gamma radiation. *Radiat Bot* 10:457–467.

Chandorkar KR, Clark GM (1986) Physiological and morphological responses of *Pinus strobus* L. and *Pinus sylvestris* L. seedlings subjected to low-level continuous gamma irradiation at a radioactive waste disposal area. *Environ Exp Bot* 26:259–270.

Chandorkar KR, Dengler NG (1987) Effect of low level continuous gamma irradiation on vascular cambium activity in Scotch pine *Pinus sylvestris* L. *Environ Exp Bot* 27:165–175.

Chauchan YS (1976) Morphological studies in safflower (*Carthamus tinctorius* L.) with special reference to the effect of 2,4-D and gamma rays. II. Cellular responses. *Environ Exp Bot* 16:235–240.

Chauchan YS, Singh RP (1975) Morphological studies in safflower (*Carthamus tinctorius* Linn.) with special reference to the effect of 2,4-D and gamma rays. I. Vegetative shoot apex. *Radiat Bot* 15:68–77.

Chou TW, Singh B, Salunkhe DK, Campbell WF (1970a) Effects of gamma radiation on *Penicillium expansum* L. I. Some factors influencing the sensitivity of the fungus. *Radiat Bot* 10:511–516.

Chou TW, Singh B, Salunkhe DK, Campbell WF (1970b) Effects of gamma radiation on *Penicillium expansum* L. II. Some enzymatic changes in the fungus. *Radiat Bot* 10:517–520.

Chou TW, Salunkhe DK, Singh B (1971) Effects of gamma radiation on *Penicillium expansum* L. III. On nucleic acid metabolism. *Radiat Bot* 11:329–334.

Clark GM, Sweeney WP, Bunting WR, Baker DG (1968) Germination and survival of conifers following chronic gamma irradiation of seed. *Radiat Bot* 8:59–66.

Conger BV (1976) Response of inbred and hybrid maize seed to gamma radiation and fission neutrons and its relationship to nuclear volume. *Environ Exp Bot* 16:15–170.

Conger BV, Caribia JV (1972) Modification of the effectiveness of fission neutrons versus ⁶⁰Co gamma radiation in barley seeds by oxygen and seed water content. *Radiat Bot* 12:411–420.

Constantin MJ, Conger BV, Osborne TS (1970) Effects of modifying factors on the response of rice seeds to gamma-rays and fission neutrons. *Radiat Bot* 10:539–549.

Contant RB, Devreaux M, Ecochard RM, Monti RM, DeNettancourt D, Mugnozza GTS, Verkerk K (1971) Radiogenetic effects of gamma- and fast neutron irradiation on different ontogenetic stages of the tomato. *Radiat Bot* 11:119–136.

Conter A (1987) Changes in protein content during growth phases under chronic gamma irradiation at a low dose-rate in *Synechococcus lividus* in culture. *Environ Exp Bot* 27:85–90.

Cordero RE (1982) The effects of acute and chronic gamma irradiation on *Lupinus albus* L. III. Chronic effects. *Environ Exp Bot* 22:359–372.

Cordero RE, Gunckel JE (1982a) The effects of acute and chronic gamma irradiation on *Lupinus albus* L. I. Effects of acute irradiation on the vegetative shoot apex and general morphology. *Environ Exp Bot* 22:105–126.

Cordero RE, Gunckel JE (1982b) The effects of acute and chronic gamma irradiation on *Lupinus albus* L. II. Effects of acute irradiation on floral development. *Environ Exp Bot* 22:127–137.

Creemers-Molenaar J, Hall RD, Krens FA (1992) Asymmetric protoplast fusion aimed at intraspecific transfer of cytoplasmic male sterility (CMS) in *Lolium perenne* L. *Theor Appl Genet* 84:763–770.

Cuny F, Roudot AC (1991) In vitro pollen growth and melon pollen germination (*Cucumis melo* L.) after gamma irradiation. *Environ Exp Bot* 31:277–283.

Das LDV (1978) Effects of radiations on the R_1 and R_2 progenies of *Pennisetum typhoides* S & H. *Environ Exp Bot* 18:121–125.

Datta AK, Biswas AK, Sen S (1986) Gamma radiation sensitivity in *Nigela sativa* L. *Cytologia* 51:609–615.

Davies CR (1968) Effects of gamma irradiation on growth and yield of agricultural crops. I. Spring sown wheat. *Radiat Bot* 8:17–30.

Davies CR (1970) Effects of gamma irradiation on growth and yield of agricultural crops. II. Spring sown barley and other cereals. *Radiat Bot* 10:19–27.

Davies CR (1973) Effects of gamma irradiation on growth and yield of agricultural crops. III. Root crops, legumes and grasses. *Radiat Bot* 13:127–136.

Derks FHM, Hall RD, Colijn-Hooymans CM (1992) Effect of gamma-irradiation on protoplast viability and chloroplast DNA damage in *Lycopersicon peruvianum* with respect to donor-recipient protoplast fusion. *Environ Exp Bot* 32:255–264.

Dmitriev AP, Grodzinsky DM (1974) Influence of some radiosensitizers on gamma-ray-induced degradation of DNA in the blue-green alga, *Anacystis nidulans*. *Radiat Bot* 14:11–15.

Dmitriev AP, Gushcha NI, Grodzinsky DM (1980) Radiosensitivity of the slime mould *Physarum polycephalum* to gamma irradiation. *Environ Exp Bot* 20:247–250.

Donaldson E, Nilan RA, Konzak CF (1979a) Interaction of oxygen, radiation exposure and seed water content on [gamma]-irradiated barley seeds. *Environ Exp Bot* 19:153–164.

Donaldson E, Nilan RA, Konzak CF (1979b) Minimum gamma-radiation exposure and oxygen concentration to produce post-irradiation oxygen-enhancement of damage in barley seeds. *Environ Exp Bot* 19:165–173.

Dong C-Z, Montillet J-L, Triantaphylidès C (1994) Effects of gamma irradiation on the plasma membrane of suspension-cultured apple cells. Rapid irreversible inhibition of H^+-ATPase activity. *Physiol Plant* 90:307–312.

Donini B, Mugnozza GTS, D'Amato R (1964) Effects of chronic gamma irradiation in durum and bread wheats. *Radiat Bot* 4:387–393.

Dubery IA, Holzapfel CW, Kruger GJ, Schabort JC, Van Dyk M (1988) Characterization of a gamma-radiation-induced antifungal stress metabolite in citrus peel. *Phytochemistry* 27:2769–2772.

Dudits D, Maroy E, Praznovszky T, Olah A, Gyorgyey J, Cella R (1987) Transfer of resistance traits from carrot into tobacco by asymmetric somatic hybridization: regeneration of fertile plants. *Proc Natl Acad Sci USA* 84:8434–8438.

Dugle JR, (1986) Growth and morphology in balsam fir: effects of gamma radiation. *Can J Bot* 64:1484–1492.

Dugle JR, Mayoh KR (1984) Responses of 56 naturally-growing shrub taxa to chronic gamma irradiation. *Environ Exp Bot* 24:267–276.

Dutta M, Bandyopadhyay A, Arunachalam V, Kaul SL, Frasad MVR (1986). Radiation induced pollen sterility and enhanced outcrossing in groundnut (*Arachis hypogaea* L). *Environ Exp Bot* 26:25–29.

Edreva AM, Georgieva ID, Cholakova NI (1989) Pathogenic and non-pathogenic stress effects on peroxidases in leaves of tobacco. *Environ Exp Bot* 29:365–377.

El-Aishy SM, Abd-Alla SA, El-Keredy MS (1976) Effects of growth substances on rice seedlings grown from seeds irradiated with gamma rays. *Environ Exp Bot* 16:69–75.

Endo T (1967) Comparison of the effects of gamma-rays and maleic hydrazide on enzyme systems of maize seed. *Radiat Bot* 7:35–40.

Erbisch FH, Kalosis JJ (1973) Initial observations of the effects of gamma radiation on oxygen consumption, ^{32}P uptake and phycobiont of *Cladonia sylvatica* (L.) Hoffm. *Radiat Bot* 13:361–367.

Fabries M, Grauby A, Trochain JL (1972) Study of a Mediterranean type phytocenose subjected to chronic gamma radiation. *Radiat Bot* 12:125–135.

Fautrier AG (1976) The influence of gamma irradiation on dry seeds of *Lucerne*, cv Wairau. I. Observations on the M-1 generation. *Environ Exp Bot* 16:77–81.

Flaccus E, Armentano TV, Archer M (1974) Effects of chronic gamma radiation on the composition of the herb community of an oak-pine forest. *Radiat Bot* 14:263–271.

Foa E, Jona R, Vallania R (1980) Histochemical effects of gamma radiation on soft fruit cell walls. *Environ Exp Bot* 20:47–54.

Fraley L Jr. (1987) Response of shortgrass plains vegetation to gamma radiation. III. Nine years of chronic irradiation. *Environ Exp Bot* 27:193–201.

Fraley L Jr., Whicker FW (1973a) Response of shortgrass plains vegetation to gamma radiation. I. Chronic irradiation. *Radiat Bot* 13:331–341.

Fraley L Jr., Whicker FW (1973b) Response of shortgrass plains vegetation to gamma radiation. II. Short-term seasonal irradiation. *Radiat Bot* 13:343–353.

Franz EH, Woodwell GM (1973) Effects of chronic gamma irradiation on the soil algal community of an oak-pine forest. *Radiat Bot* 13:323–329.

Gannutz TP (1972) Effects of gamma radiation on lichens. I. Acute gamma radiation on lichen algae and fungi. *Radiat Bot* 12:331–338.

Gastel AJG van, Nettancourt D de (1974) The effects of different mutagens on self-incompatibility in *Nicotiana alta* Link and Otto. I. Chronic gamma irradiation. *Radiat Bot* 14:43–50.

Gastel AJG van, Nettancourt D de (1975) The sensitivity of the pollen and stylar component of the self-incompatibility reaction to chronic gamma irradiation. *Radiat Bot* 15:445–447.

Georgieva ID (1987) Cytochemical investigation of pollen and pollen tubes after gamma irradiation. II. Effect of the irradiation on quinone formation. *Phytomorphology* 37:159–163.

Georgieva ID, Atanassov AI (1986) Cytochemical investigation of pollen and pollen tubes after gamma-irradiation: effect of the irradiation on some dehydrogenases and hydrolases. *Phytomorphology* 36:337–346.

Gilot-Delhalle J, Thakare R, Moutschen J (1973) Fast rejoining processes in *Nigella damascena* chromosomes revealed by fractionated ^{60}Co [gamma]-ray exposures. *Radiat Bot* 13:229–242.

Gomez-Campo C, Casas-Builla M (1972) Some effect of ionizing radiation on plant phyllotaxis. *Radiat Bot* 12:165–172.

Gordon SA, Buess EM (1972) Correlative inhibition of root emergence in the gamma irradiated coleus leaf. II. Abscopal increase of ribonuclease activity. *Radiat Bot* 12:361–364.

Gordon SA, Gaur BK, Woodstock L (1971) Correlative inhibition of root emergence in the gamma-irradiated coleus leaf. I. Premorphogenic levels of auxin, RNA and metabolites in the petiole base. *Radiat Bot* 11:453–461.

Graham ET, Generoso EE (1971) Cytoplasmic injury in the shoot apex of maize after gamma-irradiation. *Radiat Bot* 11:73–74.

Gupta MN, Samata Y (1967) The relationship between developmental stages of flower-buds and somatic mutations induced by acute X- and chronic gamma-irradiation in *Cosmos bipinnatus*. *Radiat Bot* 7:225–240.

Gupta SK, Roy SK (1985) Comparison of meiotic abnormalities induced by gamma-rays between a diploid and a tetraploid species of *Physalis*. *Cytologia* 50:167–175.

Haber AH (1968) Ionizing radiations as research tools. *Annu Rev Plant Physiol* 19:463–489.

Hayashi T, Todoriki S, Nagao A (1992) Effect of gamma-irradiation on the membrane permeability and lipid composition of potato tubers. *Environ Exp Bot* 32:265–271.

Hernandez-Bermejo JE (1977) Effets de la Radiation Gamma Aigue sur des Phytocenoses (De Prairies et Planctoniques). *Environ Exp Bot* 17:87–97.

Hinnisdaels S, Bariller L, Mouras A, Sidorov V, Del-Favero J, Vauskens J, Negrutiu I, Jacobs M (1991) Highly asymmetric intergeneric nuclear hybrids between *Nicotiana* and *Petunia*: evidence for recombinogenic and translocation events in somatic hybrid plants after "gamma"-fusion. *Theor Appl Genet* 82:609–614.

Huystee RB van (1971) Fragmentation of a storage protein, arachin, by gamma radiation. *Radiat Bot* 11:287–292.

Ichikawa S (1972) Radiosensitivity of a triploid clone of *Tradescantia* determined in its stamen hairs. *Radiat Bot* 12:179–189.

Inoue M, Ito R, Tabata T, Hasegawa H (1980) Varietal differences in the repair of gamma-radiation-induced lesions in barley. *Environ Exp Bot* 20:161–168.

Iqbal J (1980) Effects of acute gamma irradiation developmental stages and cultivar differences on growth and yield of wheat and sorghum plants. *Environ Exp Bot* 20:219–232.

Iqbal J, Kutacek M, Jiracek V (1974) Effects of acute gamma irradiation on the concentration of amino acids and protein-nitrogen in *Zea mays*. *Radiat Bot* 14:165–172.

Kaaz HW, Wallace A, Romney RM (1971) Effect of a chronic exposure to gamma radiation on the shrub *Ephedra nevadensis* in the Northern Mojave Desert. *Radiat Bot* 11:33–37.

Katagiri K, Lapins KO (1974) Development of gamma-irradiated accessory buds of sweet cherry, *Prunus avium* L. *Radiat Bot* 14:173–178.

Kawai T, Inoshita T (1965) Effects of gamma-ray irradiation on growing rice plants. I. Irradiations of four main development stages. *Radiat Bot* 5:233–255.

Kelly JW, Lineberger RD (1981) Thermal neutron induced changes in *Saintpaulia*. *Environ Exp Bot* 21:95–102.

Killion DD, Constantin MJ (1971) Acute gamma irradiation of the wheat plant: effects of exposure, exposure rate, and developmental stage on survival, height, and grain yield. *Radiat Bot* 11:367–373.

Killion DD, Constantin MJ (1972) Gamma irradiation of corn plants: effects of exposure, exposure rate, and developmental stage on survival, height, and grain yield of two cultivars. *Radiat Bot* 12:159–164.

Killion DD, Constantin MJ (1974) Effects of separate and combined beta and gamma irradiation the soybean plant. *Radiat Bot* 14:91–99.

Killion DD, Constantin MJ, Siemer EG (1971) Acute gamma irradiation of the soybean plant: effects of exposure, exposure rate, and developmental stage on growth and yield. *Radiat Bot* 11:225–232.

Kraus MP (1969) Resistance of blue-green algae to ^{60}Co gamma radiation. *Radiat Bot* 9:481–489.

Kumar A, Cocking EC (1987) Protoplast fusion: a novel approach to organelle genetics in higher plants. *Am J Bot* 74:1289–1303.

Kurobane I, Yamaguchi H, Sander C, Milan RA (1979) The effects of gamma irradiation on the leaching of reducing sugars, inorganic phosphate and enzymes from barley seed during germination in water. *Environ Exp Bot* 19:41–47.

Kuzin AM, Vagabova ME, Vilenchik MM, Gogvadze VG (1986) Stimulation of plant growth by exposure to low level gamma-radiation and magnetic field, and their possible mechanism of action. *Environ Exp Bot* 26:163–167.

Lapins KO, Hough LF (1970) Effects of gamma rays on apple and peach leaf buds at different stages of development. II. Injury to apical and axillary meristems and regeneration of shoot apices. *Radiat Bot* 10:59–68.

Lapins KO, Bailey CH, Hough LF (1969) Effects of gamma rays on apple and peach leaf buds at different stages of development. I. Survival, growth and mutation frequencies. *Radiat Bot* 9:379–389.

Machaiah JP, Vakil UK, Sreenivasan A (1976) The effect of gamma irradiation on biosynthesis of gibberellins in germinating wheat. *Environ Exp Bot* 16:131–140.

Machaiah JP, Vakil UK (1979) The effect of gamma irradiation on the formation of alpha-amylase isoenzymes in germinating wheat. *Environ Exp Bot* 19:337–348.

Margna U, Vainjarv T (1976) Irradiation effects upon flavonoid accumulation in buckwheat seedlings. *Environ Exp Bot* 16:201–208.

McCabe PF, Dunbar LJ, Guri A, Sink KC (1993) T-DNA-tagged chromosome 12 in donor *Lycopersicon esculentum* x *L. pennellii* is retained in asymmetric somatic hybrids with recipient *Solanum lycopersicoides*. *Theor Appl Genet* 86:377–382.

McCormick JF, Platt RB (1962) Effects of ionizing radiation on a natural plant community. *Radiat Bot* 2:161–188.

McCormick JF, Rushing WM (1964) Differential radiation sensitivities of races of *Sedum pulchellum* Michx. *Radiat Bot* 4:247–251.

McCrory GJ, Grun P (1969) Effects of gamma irradiation on radiation sensitivity in the diploid potato species *Solanum chacoense*. *Radiat Bot* 9:407–413.

Mednik IG, Usmanov PD (1985) Influence of gamma-rays on the number of initial cells in *Arabidopsis thaliana*. *Arabidopsis Inform Serv* 22:65–70.

Menasherov M, Paster N, Nitzan R (1992) Effects of physical preservation methods on sclerotial germination in *Aspergillus flavus* and *Aspergillus ochraceus* in stored grain. *Can J Bot* 70:1206–1210.

Miksche JP, Rudolph TD (1968) Use of nuclear variables to investigate radio-sensitivity of gymnosperm seed. *Radiat Bot* 8:187–192.

Miura K, Hashimoto T, Yamaguchi H (1974) Effect of gamma-irradiation on cell elongation and auxin level in *Avena* coleoptiles. *Radiat Bot* 14:207–215.

Mohyuddin M, Osman N, Skoropad WP (1972) Inactivation of conidiospores and mycelia of *Aspergillus flavus* by [gamma]-radiation. *Radiat Bot* 12:427–431.

Monk CD (1966) Effects of short-term gamma irradiation on an old field. *Radiat Bot* 6:329–335.

Nakai H, Saito M, Yamagata H (1980) RBE of thermal neutrons for M_1 damage in rice. *Environ Exp Bot* 20:191–200.

Netrawali MS, Nair KAS (1986) Enhancement of DNA repair in gamma-irradiated spheroplasts and nuclei of dark-grown *Euglena* cells by the supernatant of light-grown *Euglena* cells or their chloroplasts. Proc Symp Cellular Control Mechanisms: Bhabha Atomic Research Centre, Trombay, Bombay, Jan 6–8, 1982, pp. 19–32.

Nettancourt D de, Contant RB (1966) Comparative study of the effects of chronic gamma irradiation on *Lycopersicum esculentum* Mill and *L. pimpinellifolium* Dunal. *Radiat Bot* 6:545–556.

O'Hara CE (1969) The effects of gamma irradiation on the dehydrogenase activities and on the proteins of irish grown wheats. *Radiat Bot* 9:33–36.

Ogawa M, Uritani I (1971) Metabolic changes in tissues prepared from irradiated sweet potato roots with regard to ethylene effects. *Radiat Bot* 11:67–71.

Padwal-Desai SR, Ghanekar AS, Sreenivasan A (1976a) Studies of *Aspergillus flavus*. I. Factors influencing radiation resistance of non-germinating conidia. *Environ Exp Bot* 16:45–51.

Padwal-Desai SR, Ghanekar AS, Sreenivasan A (1976b) Studies of *Aspergillus flavus*. II. Responses of germinating conidia to single and combined treatments of gamma radiation and heat. *Environ Exp Bot* 16:53–57.

Pai KU, Gaur BK (1983) Quantitative and qualitative changes in mitochondrial protein isolated from gamma-irradiated kidney bean hypocotyl segments. *Environ Exp Bot* 23:143–148.

Pai KU, Gaur BK (1987) Gamma-ray-induced biochemical changes in the outer mitochondrial membrane from kidney bean hypocotyls. *Environ Exp Bot* 27:297–304.

Pai KU, Gaur BK (1988) Enzymatic lipid peroxidation and its prevention by succinate in mitochondria isolated from gamma-irradiated bean hypocotyls. *Environ Exp Bot* 28:259–265.

Paschinger H, Vanicek T (1974) Effects of gamma irradiation on the two mechanisms of Rb (K) uptake by *Chlorella*. *Radiat Bot* 14:301–307.

Pfahler PL (1970) Reproductive characteristics of *Zea mays* L. plants produced from gamma-irradiated kernels. *Radiat Bot* 10:329–335.

Posner HB, Sparrow AH (1964) Survival of *Chlorella* and *Chlamydomonas* after acute and chronic gamma radiation. *Radiat Bot* 4:253–257.

Rajput, MA (1974) Increased variability in the M_2 of gamma-irradiated mung beans (*Phaseolus aureus* Roxb). *Radiat Bot* 14:85–89.

Ramakrishnan TV, Francis FJ (1980) Carotenoid-protein complexes and their stability towards oxygen and radiation. *Environ Exp Bot* 20:1–6.

Reddy CS, Smith JD (1978) Effects of delayed post treatment of gamma-irradiation seed with cysteine on the growth of *Sorghum bicolor* seedlings. *Environ Exp Bot* 18:241–243.

Retamal N, Lopez-Vences M, Duran JM (1990) Effect of gamma-irradiation on germination and leakage of electrolytes from peanut (*Arachis hipogaea* L.) seeds. *Environ Exp Bot* 30:1–7.

Riov J (1975) Histochemical evidence for the relationship between peel damage and the accumulation of phenolic compounds in gamma-irradiated citrus fruit. *Radiat Bot* 15:257–260.

Riov J, Monselise SP, Kahan RS (1968) Effect of gamma radiation on phenylalanine ammonia-lyase activity and accumulation of phenolic compounds in citrus fruit peel. *Radiat Bot* 8:463–466.

Roth J (1978) Wirkungen Verschiedener Strahlenarten auf das Langenwachstum der Wurzelspitze von *Vicia faba*. *Environ Exp Bot* 18:177–183.

Rudolph TD (1971) Gymnosperm seedling sensitivity to gamma radiation: its relation to seed radiosensitivity and nuclear variables. *Radiat Bot* 11:45–51.

Rudolph TD (1979) Effects of gamma irradiation of *Pinus banksiana* Lamb. seed as expressed by M_1 trees over a 10-year period. *Environ Exp Bot* 19:85–91.

Rudolph TD, Miksche JP (1970) The relative sensitivity of the soaked seeds of nine gymnosperm species to gamma radiation. *Radiat Bot* 10:401–409.

Saas A, Bovard P, Grauby A (1975) The effect of chronic gamma irradiation on decay of oak (*Quercus pubescens* Willd.) and dogwood (*Cornus mas* L.) leaves and subjacent litter. *Radiat Bot* 15:141–151.

Sanda-Kamigawara M, Ichikawa S, Watanabe K (1991) Spontaneous, radiation- and EMS-induced somatic pink mutation frequencies in the stamen hairs and petals of a diploid clone of *Tradescantia*, KU 27. *Environ Exp Bot* 31:413–421.

Sato M, Watanabe M, Hiraoka A (1992) Gamma-irradiation damage to latex vacuole membranes of *Chelidonium majus*. *Phytochemistry* 31:2585–2590.

Satyanaraya V, Nair PM (1986) Enhanced operation of 4-aminobutyrate shunt in gamma-irradiated potato tubers. *Phytochemistry* 25:1801–1805.

Shamsi SRA, Sofajy SA (1980) Effects of low doses of gamma radiation on the growth and yield of two cultivars of broad bean. *Environ Exp Bot* 20:87–94.

Sheppard SC, Ross HA, Hawkins JL (1992) Reciprocal transplant study of clones of strawberry proliferating in an irradiation field: morphometrics. *Environ Exp Bot* 32:383–389.

Shirsat SG, Nair PM (1976) The requirement of oxygen for the induction of phenylalanine ammonia lyase in potatoes by light or gamma irradiation. *Environ Exp Bot* 16:59–63.

Siddiqui SH, Mujeeb KA, Keerio GR (1976) Gamma irradiation effects on sugarcane (*Saccharum* sp) clone Co-547. *Environ Exp Bot* 16:65–68.

Sideris EG, Nawar MM, Nilan RA (1971) Effects of gamma radiation on gibberellic acid solutions and gibberellin-like substances in barley seedlings. *Radiat Bot* 11:209–214.

Sidrak GH, Suess A (1972) Effects of low doses of gamma radiation on the growth and yield of two varieties of tomato. *Radiat Bot* 13:309–314.

Singh SP, Kesavan PC (1992) Post-irradiation modification of oxygen-, nitrogen- and nitrous oxide-mediated damage in dry barley seeds by catalase and super oxidedismutase: influence of post-hydration temperature. *Environ Exp Bot* 32:329–342.

Slavinski AJ Jr., Hillson CJ (1984) Effects of gamma radiation on the growth and form of *Pediastrum boryanum*. *Environ Exp Bot* 24:247–252.

Sparrow AH, Woodwell GM (1962) Prediction of the sensitivity of plants to chronic gamma irradiation. *Radiat Bot* 2:9–26.

Sparrow AH, Cuany RL, Miksche JP, Schairer LA (1961) Some factors affecting the responses of plants to acute and chronic radiation exposures. *Radiat Bot* 1:10–34.

Sparrow AH, Furuya M, Schwemmer SS (1968a) Effects of x- and gamma radiation on anthocyanin content in leaves of *Rumex* and other plant genera. *Radiat Bot* 8:7–16.

Sparrow AH, Rogers AF, Schwemmer SS (1968b) Radiosensitivity studies with woody plants. I. Acute gamma irradiation survival data for 28 species and predictions for 190 species. *Radiat Bot* 8:149–186.

Sparrow AH, Schwemmer SS, Klug EE, Puglielli L (1970) Changes in survival in response to long-term (8 year) chronic gamma irradiation. *Science* 169:1082–1084.

Sparrow AH, Schwemmer SS, Bottino PPJ (1971) The effects of external gamma radiation from radioactive fallout on plants with special reference to crop production. *Radiat Bot* 11:85–118.

Srinivasachar D, Seetharam A (1971) Induction of resistance to rust, *Melampsora Lini*, in flax, by gamma radiation. *Radiat Bot* 11:143–145.

Strydom GJ, Staden J van, Smith MT (1991) The effect of gamma radiation on the ultrastructure of the peel of banana fruits. *Environ Exp Bot* 31:43–49.

Thomas P, Srirangarajan AN, Limaye SP (1975) Studies on sprout inhibition of onions by gamma irradiation. I. Influence of time interval between harvest and irradiation, radiation dose and environmental conditions on sprouting. *Radiat Bot* 15:215–222.

Ursino DJ, Moss A, Stimac J (1974) Changes in the rates of apparent photosynthesis in 21% and 1% oxygen and of dark respiration following a single exposure of three-year-old *Pinus strobus* L. plants to gamma radiation. *Radiat Bot* 14:117–125.

Ursino DJ, Schefski H, McCabe J (1977) Radiation-induced changes in rates of photosynthetic CO_2 uptake in soybean plants. *Environ Exp Bot* 17:27–34.

Ussuf KK, Nair PM (1974) Effect of gamma irradiation on the indole acetic acid synthesizing system and its significance in sprout inhibition of potatoes. *Radiat Bot* 14:251–256.

Vidakovic M (1967) Growth of Austrian and Scots Pine plants after gamma-irradiation of pollen. *Radiat Bot* 7:529–542.

Virk DS, Saini SS, Gupta VP (1978) Gamma radiation induced polygenic variation in pure-breeding and segregating genotypes of wheat and rice. *Environ Exp Bot* 18:185–191.

Vollmer AT, Bamberg SA (1975) Response of the desert shrub *Krameria parvifolia* after ten years of chronic gamma irradiation. *Radiat Bot* 15:405–409.

Walne PL, Haber AH (1968) Actions of acute gamma radiation on excised wheat leaf tissue. II. Immediate lethal dose effects of chloroplast ultrastructure and subsequent retardation of normal disintegration in darkness. *Radiat Bot* 8:399–406.

Wells CV, Hoshaw RW (1980) Gamma irradiation effects on two species of the green alga, *Sirogonium*, with different chromosome types. *Environ Exp Bot* 20:39–45.

Wijbrandi J, Posthuma A, Kok JM, Rijken R, Vos JGM, Koornneef M (1990) Asymmetric somatic hybrids between *Lycopersicon esculentum* and irradiated *Lycopersicon peruvianum*. *Theor Appl Genet* 80:305–312.

Witherspoon JP, Corney AK (1970) Differential and combined effects of beta, gamma and fast neutron irradiation of soybean seedlings. *Radiat Bot* 10:429–435.

Woodwell GM (1963) Design of the Brookhaven experiment on the effects of ionizing radiation on a terrestrial ecosystem. *Radiat Bot* 3:125–133.

Woodwell GM, Oosting JK (1965) Effects of chronic gamma irradiation on the development of old field plant communities. *Radiat Bot* 5:205–222.

Woodwell GM, Sparrow AH (1963) Predicted and observed effects of chronic gamma radiation on a near-climax forest ecosystem. *Radiat Bot* 3:231–237.

Woollon FBM, Davies CR (1981) The response of bryophytes to ionizing radiation. *Environ Exp Bot* 21:89–93.

Yamakawa K, Sparrow AH (1966) The correlation of interphase chromosome volume with pollen abortion induced by chronic gamma irradiation. *Radiat Bot* 6:21–38.

chapter three

Plant–water interactions

Jerry L. Hatfield

Introduction

Water is an essential component of all plant life. Without fresh water there would be no plant or animal life on earth. However, water is often taken for granted and assumed to be adequate and available to supply the needs of plant growth and development. To illustrate the importance of water, all one has to do is drive across the United States from the East Coast to the West Coast during the summer to realize the role that water plays in plant growth and food production. Desert shrubs and cactus of the arid west are adapted to survival with a minimum amount of water, while even the slightest delay in rainfall in the temperate and subtropical areas of the east will cause a major impact on plant growth and development.

Plant–environment interactions are an integral part of the everyday process that plants undergo through the course of their lifecycle. If we are to improve our comprehension of the response of plants to water, we need to continue to develop a basic understanding of the response of plants to different environmental conditions. Plant–water interactions are a topic on which volumes have been written. To discuss every nuance of our current understanding would require a monograph devoted to the subject of water in plants. The intent of this chapter is to provide an overview of the plant–water interactions in the context of how plants can be used for environmental studies.

Meidner (1983) stated that understanding plant–water relations requires an understanding of plant anatomy; properties of the plant water transport system; the physics of the internal and external plant environment, particularly water vapor diffusion; and the application of control theory to the link between cell water status and metabolism. Within this chapter the focus will be on the link between the environment and plant water response. This chapter will not discuss methods for evaluating plant water status, but rather the function of water within the plant.

Water deficits that arise within the plant are a result of atmospheric conditions, soil water availability, ability of the plant to extract water from a sufficient volume of soil, and transport of water through the plant. Thus, a water deficit could occur within a plant because of excessive atmospheric demand as typified by a high rate of advection with hot, dry air moving across a field; a lack of soil water to meet the atmospheric demand; a limited root system that prevents uptake of an adequate water supply; or a physiological limitation within the root and vascular system of the plant. The first three factors are the most common in occurrence; however, the physiological limitations are the least quantified.

Role of water in plants

The role of water in the physiological function of plants has been reviewed by Slatyer (1967), Hsiao (1973), and Boyer (1985). Together, these publications provide the foundation of our current understanding of plant response to water. An actively growing agricultural plant contains 80 to 90% water, while the percent of water within a tree may be considerably less. A lush field of growing corn represents a volume of standing water. The change in the water status within a leaf, root, stem, or reproductive area of the plant reflects a complex balance among the water available within the soil, the evaporative demand of the atmosphere, and the ability of the plant to transport water from the soil to the leaf. Water has such an important role that even a slight reduction in the water content can impact both growth and physiological function. However, the plant has an active role in its response to water status at both the root and shoot portions of the plant. In this section, the role of water in plants will be discussed to show how photosynthesis, respiration, and growth are affected by the water status of the plant.

Photosynthesis

Photosynthesis represents the mechanism for the initial metabolism of the plant through the accumulation of free CO_2 in the atmosphere into sugars that are further used by the plant in the process of growth and development. There are many complex metabolic reactions that occur from the time that CO_2 is captured until the starch compounds are stored in the harvested grain or other reproductive organs. If we examine photosynthesis as a flux of CO_2 from the air into the leaf, a simple flux gradient equation can be used to describe the principal factors. The rate of photosynthesis (Ps) is a function of the gradient between the internal CO_2 concentration (C_i) within the leaf and the ambient air (C_a) and the resistance terms that can be represented by the internal leaf resistance (R_{mi}), stomatal resistance (R_s), and atmospheric resistance (r_a) across the leaf boundary (Equation 1).

$$Ps = [C_a] - [C_i]/(r_{mi} + r_s + r_a) \tag{1}$$

Resistance terms in Equation 1 can be expressed as conductance or 1/resistance. Many current mathematical models use resistance as an electrical analog for leaves and canopies. From a physiological viewpoint, conductance provides a better description of plant response to water deficits; however, to assist in the comparison among models, the resistance approach will be used in this chapter. This model would apply to a single leaf suspended in the air and represents three important factors that change in response to water status of the leaf. All three resistance terms vary with water status and are expressed as a function of leaf water potential (Figure 1). Equation 1 also can be easily expanded to represent a canopy of leaves by changing the resistance terms to represent the canopy resistance.

Leaf water potential is a measure of the energy of the free water within the leaf cells. Leaf water potential values decrease as water stress or water deficits become more severe and are commonly used as a direct measure of leaf water status. The internal mesophyll resistance, r_{mi}, is affected by changes in the free water within the leaf. Molz and Ferreir (1982) developed a mathematical model of water movement with a plant cell and across plant tissue. Water flow within a plant is dependent upon the gradients of water potential. Therefore, if we assume that as the water potential of the leaf declines there is a linear increase in leaf internal resistance, there would be a decline in photo-

Plants for environmental studies

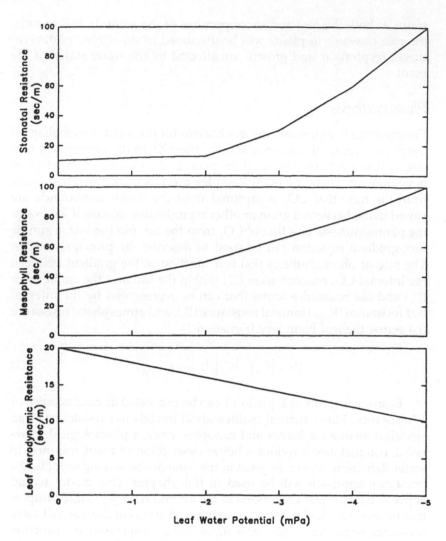

Figure 1   Generalized response functions of stomatal resistance, mesophyll resistance, and leaf aerodynamic resistance to leaf water potential.

synthetic rate. Likewise, if stomatal resistance increases as the leaf water potential declines, photosynthesis would also decline. These relationships are fairly straightforward since the change in water status is directly linked to changes in the factors that influence both the mesophyll and stomatal resistance. Boyer (1970) showed that photosynthesis of corn (*Zea mays* L.) and soybean (*Glycine max* L. Merr.) declined as the water potential decreased due to stress caused by a reduction in available water to the plant. These experiments revealed that species varied in their response to available water. There are general reactions by plants to changing water potentials; however, for specific varieties, the responses to changes in water availability may have to be experimentally determined.

Stomatal resistance

Stomatal resistance within Equation 1 is the most variable of the resistance terms and the most responsive to changes in leaf water potential. There are, however, seasonal and diurnal changes in the factors that influence the response of stomatal resistance to changes in water availability. A typical response curve of stomatal resistance, mesophyll resistance, and leaf aerodynamic resistance to leaf water potential is shown in Figure 1. Stomatal resistance is the most responsive to decreases in water potential, followed by mesophyll resistance and then leaf aerodynamic resistance. The shape of the curves describing these relationships for the three parameters is different, suggesting that there are different mechanisms being invoked for each of the photosynthetic resistances. Hall et al. (1976) summarized that stomatal response and function are a balance between control of water use and development of plant water deficits while maintaining adequate levels of photosynthesis and evaporative cooling. The conflict between these two functions lead to a balance in response to the changing environment of a plant.

Stewart and Dwyer (1983) proposed a steady-state model to describe the stomatal response to plant water deficit. The model was based on the interaction between stomatal resistance and leaf water potential. In their model, stomatal resistance remained constant until some threshold value of leaf water potential was reached and then resistance increased rapidly. They found that the relationship between resistance and leaf water potential was affected by differences in the elastic properties of guard cell walls and walls of the surrounding epidermal cells. The guard cell from the opening of the stomata and the changes of the water content of these cells cause a change in the stomatal opening. In their model, the accumulation of solutes within cells reduces the threshold values of leaf water potential. However, on a daily basis, the accumulation of osmotically active photosynthates can raise the threshold values of leaf water potential (Stewart and Dwyer 1983).

Stomatal resistance may change throughout the season independent of the effect of water deficits on the plant. Evenson and Rose (1976) found that in cotton (*Gossypium hirsutum* L.) stomatal resistance doubled during the season. A change in the stomatal density on the upper leaf surface and the stomatal silt width and length accounted for 70% of the change in stomatal resistance (Evenson and Rose 1976). Spence et al. (1986) reported that guard cells in *Vicia faba* plants grown under water deficit were considerably smaller than those from plants not subjected to water deficits. They found that the smaller pores could open under lower guard cell pressures. The changes in the mechanical aspects of a stomatal guard cell under water stress provide an improved understanding of some of the adaptations that a plant undergoes during water deficits.

Muchow (1985) examined the stomatal resistance of six legumes: soybean, green gram (*Vigna radiata* cvs Berken and CES-1D-21), black gram (*V. mungo* cv Regur), cowpea (*V. unguiculata* cv Red Caloona), lablab bean (*Lablab purpureus* cv Highworth), and pigeon pea (*Cajanus cajan* cvs Royes and insensitive ICP 7179). He found that the stomatal resistance fluctuated over time in a manner unrelated to soil water availability even though the measurements were taken around solar noon. There were differences among the legumes in their response to water deficits imposed throughout the season.

Resistances of canopies (aggregate of leaves) have been determined for a number of crops. van Bavel (1967) reported that canopy resistance for alfalfa (*Medicago sativa* L.) increased rapidly with decreasing soil water potential because the plant was no longer able to transpire at the potential rate. Sziecz and Long (1969) showed that the canopy resistance of several different species responded similarly to decreasing soil water availability. They found that for alfalfa the canopy resistance value under no soil water deficit was 25 s/m. Hatfield (1985) concluded that for wheat the canopy resistance without a soil water deficit was 20 s/m and increased 0.4 s/m for each 1% of decrease in soil water availability. He also reported that canopy resistance was not affected until 20% of the available soil water was removed from the soil volume occupied by wheat roots. Canopy resistance can provide an integrative term which relates leaf area of the crop, leaf age, soil water availability, and solar radiation for a given species (Monteith et al. 1965).

Leaf aerodynamic resistance

Changes in leaf aerodynamic resistance, r_a, are more subtle in their response to available water. There are several factors that influence leaf aerodynamic resistance: leaf shape and thickness, temperature of the leaf, air temperature, leaf orientation in the wind flow, and windspeed. Factors that affect the leaf aerodynamic resistance to water vapor transport are affected after the stomatal resistance begins to increase. Hatfield and Burke (1991) showed that leaf size and plant water availability affects leaf temperature and the diurnal changes in leaf temperature relative to plant water availability. By comparing three different species in the same environment, they demonstrated a difference in leaf temperature of cool- and warm-season plants in response to available water. In their study, the warm-season cucumber (*Cucumis sativa* L.) and bell pepper (*Capsium frutescens* L.) responded differently than the cool-season summer cotton (*G. hirsutum* L.). Midday stomatal resistances, with an optimum water availability, for single leaves were 17.5 s/m for cotton, 64.7 s/m for cucumber, and 25.0 s/m for bell pepper, while the leaf resistance for all three species was between 10 and 17.5 s/m (Hatfield and Burke 1991).

Changes in leaf resistance are more of a protection mechanism against water deficits than a first indicator of water status. For example, a loss of leaf turgor, i.e., wilting, will change the leaf resistance through orientation to the wind flow across the leaf, but also decrease the radiation load upon the leaf. Comparisons between stomatal and leaf resistances show that throughout the daylight hours, stomatal resistances range from 40 to 100 s/m, while leaf resistances vary between 10 to 20 s/m (Hatfield and Burke 1991). All three resistance values in Equation 1 vary during the day. Changes in leaf water status affect mesophyll and stomatal resistances more than leaf aerodynamic resistance. Hence, stomatal and mesophyll resistance terms are more strongly coupled with water availability than the aerodynamic properties of the leaf.

Water loss from a leaf (transpiration) lowers the temperature of the leaf, just as the skin temperature of a human cools as water evaporates during the process of perspiration. The latent heat flux is the energy consumed in the process of converting water from a liquid to a vapor. One of the simplest forms of calculating the water evaporation flux from an object is shown in Equation 2.

$$LE = R_n - \rho C_p (T_s - T_a)/r_{ah} \tag{2}$$

In Equation 2, LE represents the latent heat of evaporation from a leaf (joules per square meter per second), R_n the net radiation balance of the leaf (joules per square meter per second), ρ the density of air (kilograms per cubic meter), C_p the heat capacity of air (joules per kilogram per °C), T_s the surface temperature of the leaf (°C), T_a the temperature of the surrounding air (°C), and r_{ah} the aerodynamic resistance for sensible heat transfer. Equation 2 reveals that the temperature of a leaf transpiring rapidly will be lower than the surrounding air. This same principle is used in a wet bulb psychrometer, where the temperature of the wet bulb thermometer with the evaporating surface is cooler than that of the dry bulb temperature without evaporation. This simple concept is used to measure the relative humidity of air. In a similar way, a leaf with a rapid rate of water loss will have a temperature below that of air temperature. The r_{ah} term is the same as in Equation 1 and represents the interaction of the shape of the object and the windspeed on the energy exchange rate between a leaf and the surrounding air. Leaf shape and morphological changes that occur during water stress minimize the leaf–air temperature differences.

Leaf evaporation rates

As a plant undergoes water deficits, leaf temperature increases relative to air temperature. Under optimum leaf water supply and a midday solar radiation level, the leaf may be 5 to 10°C below air temperature. However, as water supply to the leaf decreases, leaf temperature may increase to 5 to 10°C above ambient temperatures. During water deficits, the plant undergoes adaptations to reduce the negative impact of these high leaf temperatures. An increase in stomatal resistance and the concurrent reduction in the rate of water loss from the leaf will cause the leaf temperature to increase because there is a reduced availability of water for the evaporation process. Leaf and stomatal resistance interact during water deficits to influence leaf temperature. Jackson et al. (1981) developed a quantitative method of using leaf or foliage temperature as a measure of stress experienced by the plant during the course of a day or season. Burke and Hatfield (1987) found that for cotton there was a relationship between foliage temperature and the potential activity of glutathione reductase. These studies suggest that the physiological changes in plants experiencing high foliage temperatures during a water deficit are reflected in enzyme activity levels.

As plants undergo water deficits, there is a change in the leaf morphology. One of the primary factors in the energy exchange between a leaf and the surrounding air is the leaf shape, termed r_{ah} in Equation 2. When we examine the response of a grass leaf to water deficits, we find that as the severity of the stress increases, the leaf begins to roll or assume a cylindrical shape. Begg (1980) showed that leaf rolling in sorghum reduced the effective leaf area by 68%. Leaf rolling is related to leaf water potential. In grass plants, leaf rolling not only reduces the effective leaf area for the interception of net radiation, but also provides a more efficient shape for the transfer of sensible heat away from the leaf (Leyton 1975). Thus, the capacity to form cylindrical leaf shapes is a valuable adaptation to water stress. Broad-leafed plants, on the other hand, wilt to reduce the interception of radiation (Leyton 1975). In broad-leafed plants, leaf wilting decreases the leaf aerodynamic resistance because the leaves move more in the air stream because of the reduced turgor pressure (Parlange and Heichel 1971). Leaf morphological changes as a result of water deficits are reversible. They occur during periods of stress and revert to normal leaf appearances once the stress is relieved.

Water use rates by plants

The rate of water use or transpiration by the plant is an extension of leaf evaporation rates. It is a function of the available energy or net radiation, water vapor pressure deficit of the atmosphere, windspeed, and the soil water available to the plant. Three of the terms represent the energy available for the evaporation of water, the potential gradient of water from the leaf or canopy into the atmosphere, and the rate at which saturated air will be transported away from the leaf (Equation 2). Often, the most controlling factor on water use rates is the availability of the soil water supply to the plant. There are numerous mathematical equations that have been derived to estimate the rate of crop water use, and the variation among these equations is due to the time scale of the measurements, the available meteorological and plant data, and the sophistication of the model. These equations have been summarized by Hatfield (1990).

The rate of water loss by plants and the onset of stress is determined by the interaction of the soil water availability and the prevailing meteorological conditions. Denmead and Shaw (1960) showed that corn plants would be able to maintain growth under low soil water conditions when there was a low atmospheric demand or low evapotranspiration rate. Conversely, with high atmospheric demands and high soil water availability, a stress condition could result if the plant transport system was not able to extract water at a sufficient rate from the soil. Boyer (1985) summarized the available information on water transport within plants and the interaction with plant growth and concluded that we need to improve our understanding of the water potential process within plants and the linkage between water loss by evaporation and water uptake for cell expansion. Westgate and Boyer (1984) concluded that transpiration and growth compete for the available water within the plant. The water use rate is dependent upon the resistances within the plant and water potential gradient. Growth is dependent upon the water potential within the plant.

Water use rates by crops determine the rate at which the soil water content within the root volume may become limiting. Slatyer (1967) provided one of the classic examples of how leaf water potential increased each day as the soil water availability decreased and an example of the diurnal trend in water status within the leaf. These principles remain as much in force today as when they were first reported. Leaf water status is an important part of the physiological reaction of the plant. Water use rates and the soil water supply determine how quickly the plant may be subjected to stress conditions.

Water use by plants has been related to estimations of the water requirements for optimal plant growth. In a classic study, de Wit (1958) showed that the amount of water transpired by a crop throughout the growing season was related to the dry matter production. He showed that the relationship was linear and that the slope of the line was dependent upon the variety and species. Howell and Musick (1984) extended the original concept of de Wit to relate the water use differences among species to radiation interception by the crop, vapor pressure deficit of the atmosphere, and daily potential water use by the crop. They were able to show that there was considerable difference in water use among four species that included cotton (*G. hirsutum* L.), grain sorghum (*Sorghum bicolor* L. Moench.), sugar beet (*Beta vulgaris* L.), and winter wheat (*Triticum aestivum* L.). However, within a species there are large differences among years that prevent a standardized comparison among species for the amount of water used. A similar analysis was conducted by Hanks (1983) to show that water requirements for crop production could be estimated from meteorological data and expected crop yield. Unfortunately, there are no exact numbers on water use by a particular plant species or variety because of the dependence upon the meteorological conditions and the availability of soil water. Doorenbos and Pruitt (1975) have developed guidelines that are currently available to estimate the water requirements of different crops throughout the world. These methods are routinely used for estimated crop water requirements.

In many areas of the world, there is concern about the water use within riparian zones along streams and, in particular, the water use by phreatophytes along stream channels. There is an abundant supply of water for a phreatophyte species because these plants have their root system within the saturated soil layer along streams and rivers. The water use rate, however, could be quite large because these species would have a nonlimiting amount of available water. It would be expected that along streams the water use and thus the water relations would be greatly altered. Phreatophytes are dependent upon the groundwater as the source of moisture. Nichols (1993) estimated that greasewood (*Sarcobatus vermiculatis*) could use 24 cm of groundwater annually. There is little quantitative evidence for the effect of a groundwater source on the water relations of phreatophytes. Foster and Smith (1991) found that the stomatal conductance patterns of three high elevation phreatophytes, cottonwood (*Populus angusifolia* James) and willows (*Salix monticola* Nutt. and *S. exigua* Bebb), increased during the day. They attributed this increase to a large soil-to-leaf hydraulic gradient, access to water tables, and low atmospheric pressures. They concluded that high stomatal conductances were beneficial to these plants and allowed for the attainment of maximum photosynthetic rates. Young et al. (1985) also studied the water relations of three phreatophytes, *S. planifolia*, *S. wolfii*, and *Betula occidentalis*, and concluded that stomatal conductance and water use rates were influenced by solar irradiance, dew on leaves, minimum air temperatures, and phenology. Busch et al. (1992) used stable isotopes to determine the patterns of water use and the source of water for phreatophytes. They concluded that the phreatophytes, *Populus* and *Salix*, extract water dominantly from the groundwater or saturated soil, while *Tamarix* used water from unsaturated alluvial soils. These studies show that there is a variation in the water use patterns among phreatophytes and that the water relations and response are different than nonphreatophytes. There is a need for detailed research to determine the impact of riparian vegetation on the water balance and stream hydrology and the effect of the stream conditions on the water use rates of these vegetation areas.

Leaf photosynthesis rates

Photosynthetic processes respond to leaf water status, and responses vary among species. Since water enters the plant via the roots, differences in rooting depth, root elongation, and root characteristics among species and cultivated varieties within a species result in variable responses to water deficits. For example, a plant with a more extensive root system would be able to extract water more efficiently from a soil profile than one with limited root development. This effect is often seen when root extension is reduced by soil compaction, feeding insects, or root diseases. Water transport through the root system is a function of the soil water availability and the potential gradient from the soil to the leaf. Any change in water availability within the soil, ability of the root to explore the maximum amount of soil volume, and ability of the root to transport water effectively will impact the photosynthetic efficiency of the leaf (Boyer 1985). Reduction in soil available water decreases photosynthesis (Boyer 1970).

The number of times a plant is subjected to water deficits also affects the photosynthetic response to water deficits. Bennett and Sullivan (1981) showed that subjecting grain sorghum (*Sorghum bicolor* L. Moench.) plants to osmotically induced water deficits by adding polyethylene glycol to hydroponically grown plants invoked a different response depending upon the stress history of the plant. They found that when sorghum plants were preconditioned with a water stress of –0.6 MPa, imposed at early vegetative growth (10 to 18 d after emergence) for 8 d, and then subjected to another stress at 37 d after emergence, the previously stressed plants maintained a higher net photosynthetic rate. The different response could not be explained by changes in the plant that would lead to avoidance of water deficits. They noted that the initial stress caused a reduction in the number of green leaves, green leaf area, plant height, total dry matter, and shoot to root ratio. The hardening effect of prior stress on the response of plants to a subsequent water deficit also has been reported for sugarcane (*Saccharum officinarum* L.) (Ashton 1956), beans (*Phaseolus acutifolius* L.) (Kinbacher et al. 1967), corn (Boyer and McPherson 1975), and wheat (*Triticum aestivium* L.) (Todd and Webster 1965).

The nature of the photosynthetic response to water deficits is dependent upon the species and the variety. The primary factors associated with this response are due to changes in the internal resistances and response of the turgor of the guard cells upon the stomatal resistance. Relationships among these parameters have been detailed in the articles of Hsiao (1973) and Boyer (1970). Levitt (1980) summarized different species' reactions to water stress on the photosynthetic process.

Leaf respiration rate

Respiration is a fundamental process within plants that provides the energy to convert photosynthetic products (sugars) into starches, proteins, and other metabolic products needed for growth. Respiration consumes O_2 in the conversion of sugars and releases CO_2 into the atmosphere around the leaf as a product of the metabolic process. Respiration response to water status is different than photosynthesis. Brix (1962) found that respiration rate in tomato (*Lycopsersicon esculentum* L.) declined linearly with decreasing leaf water potentials, while in loblolly pine (*Pinus taeda* L.) respiration rate increased after an initial decline. Levitt (1980) suggested several reasons for the respiratory response to water deficits in plants. The primary reason for changes in respiration is linked to changes in the cell water status and is related to either increased concentrations of solutes within the cell or decreased diffusion rates across the cell wall as affected by the internal water potential of the cell.

The response of respiration to water deficits is a complicated process. During the initial stages of water loss from plant tissue, there is an increase in respiration rate. As water deficits become more severe and the water status within the plant decreases, respiration rate will decrease. These responses are linked to both the changes in gaseous diffusion across the cell walls and the availability of substrate for use in the respiration process. Hanson and Hitz (1982) outlined the metabolic responses of plants to water deficits and showed the linkages between substrate availability and respiration. These responses to water deficits are important in determining how different species will react to changing water status. Respiration can be used as a measure of plant reaction to water deficits. However, the response function will vary with species, history to prior stress, and stage of plant development.

Growth and development

Plant response to water deficits throughout a season is a balance between the growth, morphological, and physiological processes within the plant. Turner and Kramer (1980) compiled the available information on adaptation to water and high temperature stress and concluded that water stress within plants and plant–water interactions are a complex balance between the morphological adaptations and the physiological reactions to water deficits. We need to be aware of the balance between these two processes and reactions as we develop studies or techniques that quantify plant response to water. The most noticeable effects of water deficits on plants are on growth and development. Hsiao et al. (1976a) discussed the interaction among water stress, growth, and osmotic adjustment and ranked the following sensitivities to water stress. Cell expansive growth is the process most sensitive to water deficits. Growth inhibition is often seen before any other physiological response (Hsiao 1973). An example of this response was shown by Bennett and Sullivan (1981), who found that mild water deficits (−0.6 MPa) in the early stages of plant growth decreased the plant size, leaf number, and the shoot to root ratio. The magnitude of the water deficits sufficient to affect growth would be typical of those a plant might experience during a relatively mild drought period. Growth could be limited during midday when the water loss could exceed water uptake from the soil and a water deficit occurs within the plant. Expansive growth occurs at night when the water status is more favorable within the plant. Hsiao further proposed that stomatal resistance in mesophytes could be considered as a moderately sensitive parameter.

The interactions among water deficits, growth, and the yield of crop plants are a complicated set of interactions dependent upon the plant species, genetic variation within the species, stage of growth, and severity of the water deficit (Hsiao et al. 1976b). A comprehensive feedback model can be developed to show that deficits sufficient to cause reductions in leaf growth would lead to reduced leaf area development (Hsiao et al. 1976b). Reductions in leaf area development would lead to reduced interception of radiation and, consequently, less photosynthate produced for growth. Continued reduction in leaf expansion would then lead to less cell expansion and reduced numbers of leaves on a plant. These are both negative feedbacks between water deficits and plant growth. A positive feedback from slower cell expansion is solute accumulation by the cells. More solutes per cell causes an increase in the osmotic potential of the leaf. In turn, this could lead to increased water uptake and restoration of leaf turgor potential. The end result is a continuation of expansion growth. These examples illustrate the complexities of the interactions among water deficits and plant growth.

Salter and Goode (1967) summarized the available literature on the effects of water deficits on annual, biennial, and perennial crops at different growth stages. This extensive review of over 1250 different references serves as an excellent resource for individuals wanting to compare the water deficit effects on different crops. A study by Stegman (1982) illustrates the effect of timing on the yield of crops. Three different stress levels (as defined by the degree of soil water depletion and leaf water potential in the midafternoon) produced different results in corn. When stress was imposed at the reproductive stage when the seeds were fertilized or during the grain filling, there was a maximum impact on yield. Stegman (1982) reported yields of 62% of the control when stress was imposed during the reproductive stage and 69% when stress was imposed during the grain filling stage. These results support the concepts proposed by Hsiao and Acevedo (1974) and Hsiao et al. (1976a) regarding the maximum effect of water deficits on growth and yield.

Species differ in their reaction to water stress. In a comprehensive study comparing grain sorghum and soybean in Australia, it was found that grain sorghum was more resistant to soil water deficits and that C_4 plants would be better adapted than C_3 plants to water deficits. Constable and Hearn (1978) found that grain sorghum produced more dry matter than soybean, even though sorghum had less leaf area. The seasonal patterns of growth and development they observed suggested that there was a translocation of stored assimilates from the stem into the grain during the grain filling period. Soil water extraction patterns were different between two soybean cultivars and the sorghum crop. One variety of soybean was able to extract more water below the 0.8-m depth, while the other cultivar was able to maintain a deeper root system and was able to remove water from a 1.2-m depth (Burch et al. 1978). Sorghum with a fibrous root system was able to extract more water from the upper soil profile and was able to effectively use the water from the upper regions of the soil profile where there is more available soil water. The sorghum crop would be able to take advantage of the rewetting of the upper soil profile by small rainfall events more easily than the taprooted soybean crop. They also found that single leaf responses were not indicative of a canopy. Single leaf data are collected under similar environmental conditions because of the leaf position at the upper part of the canopy, while the canopy represents an integration of a wide range of conditions.

Growth response to water deficits differs as a function of growth stage, degree and duration of water deficits, and cultivars within a species. Photosynthesis and transpiration measured on single leaves during the reproductive growth stage showed that, as water deficits increase, there is an increase in the early morning and late afternoon photosynthesis (Rawson et al. 1978). Midday photosynthesis rates were greatly affected by soil water deficits because of stomatal closure. They also noted that the plants rapidly recovered from soil water deficits, and photosynthetic rates recovered quickly from stressed conditions. In a companion study, Turner et al. (1978) found that stomatal closure occurred in two soybean cultivars at −1.5 MPa. The differences in yield, water extraction, and growth between the two cultivars of soybean in this study were attributed to the different phenological patterns of the two cultivars. Turner et al. (1978) found little evidence in their study for osmotic adjustment in field-grown plants as proposed by Hsiao (1973). Turner et al. (1978) found the relationships among leaf water potential, osmotic potential, turgor, and relative leaf water content did not change during the season with soil water extraction. These results indicate the complexities of the interactions between the physiological reactions of plants and the soil water deficits that plants experience throughout a growing season.

Neyshabouri and Hatfield (1986) compared semideterminate and indeterminate soybean and found no differences between the two cultivars in their leaf water potential, stomatal conductance, and net photosynthesis response to soil water deficit. They found that the semideterminate soybean responded more favorably to water deficits during the reproductive stage because of the lack of competing sinks for photosynthate. In water-limiting environments, the semideterminate cultivar would have an advantage during the reproductive growth stage. The semideterminate cultivar retained more pods and hence more seed than the indeterminate cultivar at all irrigation levels (Neyshabouri and Hatfield 1986).

Muchow (1985) observed that there are a variety of reasons that the crop growth differs among species relative to water deficits. For six legumes the reaction to water deficits were postponement of dehydration through control of water loss, no adjustment in osmotic potential, maintenance of a high water potential, tolerance to dehydration through partially open stomata, a decreasing water potential over time, and a moderate osmotic adjustment coupled with partial stomatal opening and maintenance of a high water potential. The highest yields were from plants exhibiting one or more from each of these physiological responses. This information led Muchow to conclude that there is no single physiological mechanism that explains the observed responses to water deficits in a semiarid environment.

Reproductive growth

Water deficits during the flowering, seed set, and seed filling periods of plant growth can cause major impacts on the harvestable yield. There are different reactions than during the expansive growth stages. Water deficits during reproduction often decrease seed size. Reductions in seed size are often thought to be the result of a shortened seed filling period rather than being due to the rate of seed growth (Meckel et al. 1984). Westgate and Thomson-Grant (1989a) found that the water status of a developing soybean seed was not affected by short-term water deficits severe enough to inhibit the metabolic activity of the plant. Their results would suggest that the water status within the seed could be maintained at a favorable status at the expense of the parent plant. Westgate et al. (1989) found that both source activity and sink activity were altered by the change in water deficits and that these respond in a way to maintain the movement of assimilates to the developing seed. These changes permit the seed filling process to continue, although for a reduced period of time. In a similar study on corn, Schussler and Westgate (1991a) found that water deficits during flowering and early kernel growth reduced the yield potential by decreasing the number of kernels per ear. They attributed this reduction to an inhibition of photosynthesis during the early kernel growth. Since changes in photosynthesis are one of the first physiological responses to water deficits, any change in water status would impact kernel development. In corn plants, water deficits at pollination decrease the number of kernels, and part of kernel loss can be attributed to developmental failure after fertilization. In a detailed study on ovary development, Schussler and Westgate (1991b) found that kernel set was dependent upon the rate of movement of assimilates into the ovaries. However, they also noted that the complete reduction in kernel set from water-deficient plants was more than could be attributed to reduced assimilate supply to the kernel. In a companion study, Westgate and Thomson-Grant (1989b) found that reproductive development in corn is sensitive to water deficits during anthesis and sensitivity decreases as reproductive stages progress.

In a study conducted at five locations across the North American Great Plains on winter wheat, Major et al. (1988a,b) reported the effects of soil water deficits and nitrogen interactions on growth and yield. There was an interaction between the nitrogen level and soil water deficit. High levels of nitrogen fertilizer increased the number of kernels per spike and spike per unit area; however, this sink was too high relative to supply because the weight per seed was the lowest at the high nitrogen level. Seed yield could only be maintained under conditions with an optimal soil water supply. Throughout the Great Plains, the soil water deficit develops gradually during the growth of the wheat crop and is a condition to which the plant must adjust. Grain growth rates from the rain-fed plants were 62%, and dry matter growth rates were 70% of the irrigated plants (Major et al. 1988b). Soil water and nitrogen interactions were found to be consistent among locations and within the common variety and locally adapted variety that was evaluated at each location.

The effect of water deficits on grain filling can be summarized by the research of Westgate and Boyer (1986). They found that during grain filling the water potential within the grain is associated with large osmotic forces and prior to physiological maturity the grain water potential is controlled by internal metabolic events. The high solute concentrations between the grain and the vascular supply by the plant are responsible for the continued movement of assimilates to the grain even under water deficit conditions. It is expected that similar reactions would occur in all plants that produce seed. Therefore, annual crops would have adapted physiological mechanisms that would ensure the production of viable grain at the expense of the parent plant. This adaptation would be necessary to ensure the survival of the species under less than optimal conditions.

Water surplus effects on plants

Most of the emphasis in plant–water interactions is on the topic of water deficits. The majority of research has been directed toward understanding how plants cope with water deficits rather than water excess. The recent occurrence of floods in the Midwest raised a number of questions about the effect of excess water on plants. Levitt (1980) proposed that there are primary and secondary effects of excess water on plants. The primary effect is due to excess water potential. Secondary effects from excess water are due to the effects of water on gaseous diffusion in the soil and stress caused by the leaching of nutrients from the root zone. The literature on the primary effect of water surplus on plants is somewhat limited. Ricard et al. (1994) presented a review on plant metabolism response to low oxygen and showed that the response may be less variable among plant species than originally thought.

Oxygen depletion in the root zone

Plants in normal growing conditions with water contents that range from field capacity to lower limits of extractable soil water are subjected to oxygen concentrations typical of free air. Respiration within the root is necessary for proper plant growth. When the soil environment changes from aerobic to anaerobic as the oxygen and other gases are forced from the soil by water, the physiological processes within the plant are affected. Wignarajah et al. (1976) showed that in barley (*Hordeum vulgare* L.) and rice (*Oryza sativa* L.) flooding reduced both the shoot and root growth. Corn exhibited a 40% reduction in dry matter production after being subjected to a 20-d period of flooding (Hoehler et al. 1976). It appears that reduction in the soil oxygen concentration due to flooding may be responsible for the large dry matter decreases observed in flooded fields. Bornstein et al. (1984) found growth effects in alfalfa subjected to different water table levels within the soil. The oxygen diffusion rate was influenced by the length of time the soil was subjected to flooding. Inadequate drainage within clay soils would cause a reduction in growth compared to soils that drained rapidly (Bornstein et al. 1984).

Effect of water surplus on physiological processes

Stress due to water excess within the root zone can cause both physiological and growth responses. The changes that occur within plants are similar to those of water deficits, with growth being the most sensitive parameter followed by photosynthesis and other metabolic changes.

Waterlogging of plants can occur when the soil is inundated or the water table rises so that a portion of the root zone is saturated. Reicosky et al. (1985) constructed a water table gradient unit to investigate the effects of flooding on cotton growth. They could vary the water table at depths of 0.1 to 0.66 m below the surface. Plants with more than 55% of their root system below the water table began to exhibit reduced leaf growth 3 to 4 d after the flooding commenced (Reicosky et al. 1985). Visible symptoms of wilting and decreased leaf water potential were not observed until 7 to 8 d after flooding. A similar pattern has been seen for reductions in leaf growth due to water deficits. When the plants were subjected to a second period of flooding, there were no symptoms of wilting other than reduced leaf growth. They attributed the inhibition of leaf growth to a decrease in the nutrient status of the plant caused by the excessive water in the root zone. Meyer et al. (1987) found that a single exposure to waterlogging did not cause a yield reduction comparable to that which occurred when waterlogging events persisted. Measured soil oxygen levels decreased rapidly as the soil was flooded because the water displaced air, and root zone respiration consumed the limited oxygen in the root zone. The rate of oxygen consumption was almost three times greater in the high-nitrogen treatment than in the low-nitrogen treatment. They also found that the number of roots was increased almost two times in the high-nitrogen treatment compared to the low-nitrogen treatment. In response to a single flooding event, leaf growth decreased by 28% and apparent photosynthesis by 16% (Meyer et al. 1987). The reduction in photosynthetic activity was noticeable 2 d after flooding, while the other symptoms were not detectable until 6 d after flooding. Hocking et al. (1987) reported that waterlogging impaired the uptake of most nutrients in young cotton plants, but had less effect when the plants were older and exposed to additional flooding events. They found that nitrogen, phosphorus, and potassium uptake was impaired in waterlogged cotton plants. The reductions in uptake were noticed within 2 d of the flooding event.

Excess water can impair growth through a direct physiological effect. The impact of these changes in water status need to be addressed to understand the differing species' response to excess water.

Summary

Plant growth and response to water deficits are a result of the atmospheric demand on the plant and the available soil water. If the deficit within the plant is short-lived, e.g., a daily response during the midday, then the plant will continue to grow and function normally. If, however, the water deficits become increasingly severe as the soil water supply becomes more limited and the plant is unable to extract water from the soil profile, then both physiological and morphological changes occur. Cell expansion is the most sensitive process to water deficits. Growth reductions can lead to changes in the shoot to root ratio that would favor increased water uptake relative to leaf area for transpiration. Photosynthetic rates decrease as water deficits increase; however, these changes are more sensitive than respiration rate. There are a number of intricate feedbacks and interrelationships that exist within a plant as water deficits are imposed. Plant response to environmental conditions is a continual integration of those feedbacks.

Plants are excellent indicators of environmental conditions. However, one cannot select a single response to observe without understanding the causal mechanisms. Water is one of the more variable parameters in the environment, and the response of a plant to changing water deficits or excess water varies among species and stages of growth within a species. Nevertheless, plant response to water can be an effective measure of environmental changes if we learn to interpret the signals and quantify the response.

Plants respond to environmental conditions in many ways. There is no standard response curve and quantitative value that one can ascribe for all plants or even cultivated varieties within a species. To understand fully how a plant responds to a water deficit or surplus, the general concepts given within this chapter must be applied to a particular set of conditions. However, the concepts are applicable across a broad range of conditions. Plant–water interactions will continue to be a topic of interest and study because of the variation in the water across the earth and the growing need to produce plants for human and animal consumption.

References

Ashton FM (1956) Effects of a series of cycles of alternating low and high soil water contents on the rate of apparent photosynthesis in sugar cane. *Plant Physiol* 31:266–274.

Begg JE (1980) Morphological adaptations of leaves to water stress. In *Adaptation of Plants to Water and High Temperature Stress*, Turner NC, Kramer PJ, Eds., Wiley-Interscience, New York, pp. 33–42.

Bennett JM, Sullivan CY (1981) Effects of water stress preconditioning on net photosynthetic rate of grain sorghum. *Photosynthetica* 15:330–337.

Bornstein J, Benoit GR, Scott FR, Hepler PR, Hedstrom WE (1984) Alfalfa growth and soil oxygen diffusion as influenced by depth to water table. *Soil Sci Soc Am J* 48:1165–1169.

Boyer JS (1970) Differing sensitivity of photosynthesis to low leaf water potentials in corn and soybean. *Plant Physiol* 46:236–239.

Boyer JS (1985) Water transport. *Annu Rev Plant Physiol* 36:473–516.

Boyer JS, McPherson HG (1975) Physiology of water deficits in cereal crops. *Adv Agron* 27:1–23.

Brix H (1962) The effect of water stress on the rates of photosynthesis and respiration in tomato plants and loblolly pine seedlings. *Physiol Plant* 15:10–20.

Burch GJ, Smith RCG, Mason WK (1978) Agronomic and physiological responses of soybean and sorghum crops to water deficits. II. Crop evaporation, soil water depletion and root distribution. *Aust J Plant Physiol* 5:169–177.

Burke JJ, Hatfield JL (1987) Plant morphological and biochemical responses to field water deficits. III. Effect of foliage temperature on the potential activity of glutathione reductase. *Plant Physiol* 85:100–103.

Busch DE, Ingraham NL, Smith SD (1992) Water uptake in woody riparian phreatophytes of the Southwestern United States: a stable isotope study. *Ecol Appl* 2:450–459.

Constable GA, Hearn AB (1978) Agronomic and physiological responses of soybean and sorghum crops to water deficits. I. Growth, development, and yield. *Aust J Plant Physiol* 5:159–167.

Denmead OT, Shaw RH (1960) Availability of soil water to plants as affected by soil moisture content and meteorological conditions. *Agron J* 54:385–389.

de Wit CT (1958) Transpiration and crop yields. *Versl Landbouwkd Onderzock* 64.6:1–88.

Doorenbos J, Pruitt WO (1975) Guidelines for Predicting Crop Water Requirements. FAO Irrigation and Drainage Paper No. 24, FAO, Rome, 144 pp.

Evenson JP, Rose CW (1976) Seasonal variation in stomatal resistance in cotton. *Agric Meteorol* 17:381–386.

Foster JR, Smith WK (1991) Stomatal conductance patterns and environment in high elevation phreatophytes of Wyoming. *Can J Bot* 69:647–655.

Hall AE, Schulze ED, Lange OL (1976) Current perspectives of steady-state stomatal responses to environment. In *Water and Plant Life*, Lange OL, Kappen L, Schulze ED, Eds., Springer-Verlag, New York, pp. 169–188.

Hanks RJ (1983) Yield and water-use relationships: an overview. In *Limitations to Efficient Water Use in Crop Production*, Taylor HM, Jordan WR, Sinclair TR, Eds., American Society of Agronomy, Crop Science Society of America, and Soil Science Society of America, Madison, WI, pp. 393–411.

Hanson AD, Hitz WD (1982) Metabolic responses of mesophytes to plant water deficits. *Annu Rev Plant Physiol* 33:163–203.

Hatfield JL (1985) Wheat canopy resistance determined by energy balance techniques. *Agron J* 77:279–283.

Hatfield JL (1990) Methods of estimating evapotranspiration. In *Irrigation of Agricultural Crops*, Stewart BA, Nielsen DR, Eds., American Society of Agronomy, Madison, WI, Agronomy Monograph No. 30. pp. 435–474.

Hatfield JL, Burke JJ (1991) Energy exchange and leaf temperature behavior of three plant species. *Environ Exp Bot* 31:295–302.

Hocking PJ, Reicosky DC, Meyer WS (1987) Effects of intermittent waterlogging on the mineral nutrition of cotton. *Plant Soil* 101:211–221.

Hoehler T, Grothus R, Schaub H, Egle K (1976) Influence of oxygen on dry matter production and daily changes in CO_2 uptake: 2. Growth, net photosynthesis, and transpiration of *Amaranthus paniculatus* and *Zea mays* grown in 4% oxygen compared to normal air. *Photosynthetica* 10:59–70.

Howell TA, Musick JT (1984) Relationship of dry matter production of field crops to water consumption. In *Crop Water Requirements,* Perrier A, Riou C, Eds., INRA, Versailles, France, pp. 247–269.

Hsiao TC (1973) Plant responses to water stresses. *Annu Rev Plant Physiol* 24:519–570.

Hsiao TC, Acevedo E (1974) Plant response to water deficits, water-use efficiency, and drought resistance. *Agric Meteorol* 14:59–84.

Hsiao TC, Acevedo E, Fereres E, Henderson DW (1976a) Water stress, growth, and osmotic adjustment. *Phil Trans R Soc London R* 273:479–500.

Hsiao TC, Fereres E, Acevedo E, Henderson DW (1976b) Water stress and the dynamics of growth and yield of crop plants. In *Water and Plant Life,* Lange OL, Kappen L, Schulze ED, Eds., Springer-Verlag, New York, pp. 281–305.

Jackson RD, Idso SB, Reginato RJ, Pinter PJ Jr. (1981) Canopy temperature as a crop water stress indicator. *Water Resour Res* 17:1133–1138.

Kinbacher EJ, Sullivan CY, Eastin JD (1967) Thermal stability of malic dehydrogenase from heat hardened Phaseolus acutifolius 'Tepary Buff'. *Crop Sci* 7:148–151.

Levitt J (1980) *Responses of Plants to Environmental Stresses. Volume II. Water, Radiation, Salt, and Other Stresses,* Academic Press, New York, 607 pp.

Leyton L (1975) *Fluid Behavior in Biological Systems,* Clarendon Press, Oxford, UK, 235 pp.

Major DJ, Blad BL, Bauer A, Hatfield JL, Hubbard KG, Kanemasu ET, Reginato RJ (1988a) Winter wheat yield response to water and nitrogen in the North American Great Plains. *Agric For Meteorol* 44:141–149.

Major DJ, Blad BL, Bauer A, Hatfield JL, Hubbard KG, Kanemasu ET, Reginato RJ (1988b) Seasonal patterns of winter wheat phytomass as affected by water and nitrogen on the North American Great Plains. *Agric For Meteorol* 44:151–157.

Meckel L, Egli DB, Phillips RE, Radcliffe D, Leggett JE (1984) Effect of moisture stress on seed growth in soybean. *Agron J* 76:647–650.

Meidner H (1983) Our understanding of plant water relations. *J Exp Bot* 34:1606–1618.

Meyer WS, Reicosky DC, Barrs HD, Smith RCG (1987) Physiological responses of cotton to a single waterlogging at high and low N levels. *Plant Soil* 102:161–170.

Molz FJ, Ferreir JM (1982) Mathematical treatment of water movement in plant and cell tissue: a review. *Plant Cell Environ* 5:191–206.

Monteith JL, Szeicz G, Waggoner PE (1965) The measurement and control of stomatal resistance in the field. *J Appl Ecol* 2:345–355.

Muchow RC (1985) Stomatal behavior in grain legumes grown under different soil water regimes in a semi-arid tropical environment. *Field Crops Res* 11:291–307.

Neyshabouri MR, Hatfield JL (1986) Soil water deficit effects in semi-determinate and indeterminate soybean growth and yield. *Field Crops Res* 15:73–84.

Nichols WD (1993) Estimating discharge of shallow ground water by transpiration from greasewood in the Northern Great Basin. *Water Resour Res* 29:2771–2778.

Parlange JY, Heichel GH (1971) Boundary layer resistance and temperature distribution on still and flapping leaves. I. Theory and laboratory experiments. *Plant Physiol* 46:437–442.

Rawson HM, Turner NC, Begg JE (1978) Agronomic and physiological responses of soybean and sorghum crops to water deficits. IV. Photosynthesis, transpiration and water use efficiency of leaves. *Aust J Plant Physiol* 5:195–209.

Reicosky DC, Meyer WS, Schaefer NL, Sides RD (1985) Cotton response to short-term waterlogging imposed with a water-table gradient facility. *Agric Water Manage* 10:127–143.

Ricard B, Couee I, Raymond P, Saglio PH, Saint-Ges V, Pradet A (1994) Plant metabolism under hypoxia and anoxia. *Plant Physiol Biochem* 332:1–10.

Salter PJ, Goode JE (1967) *Crop Responses to Water at Different Stages of Growth*, Research Review No. 2, Commonwealth Agricultural Bureaux, Farnham Royal, Bucks, England, 246 pp.

Schussler JR, Westgate ME (1991a) Maize kernel set at low water potential. I. Sensitivity to reduced assimilates during early kernel growth. *Crop Sci* 31:1189–1195.

Schussler JR, Westgate ME (1991b) Maize kernel set at low water potential. II. Sensitivity to reduced assimilates at pollination. *Crop Sci* 31:1196–1203.

Slatyer RO (1967) *Plant-Water Relationships*, Academic Press, New York.

Spence RD, Wu H, Sharpe PJH, Clark KG (1986) Water stress effects on guard cell anatomy and the mechanical advantage of the epidermal cells. *Plant Cell Environ* 9:197–202.

Stegman EC (1982) Corn grain yield as influenced by timing of evapotranspiration deficits. *Irrig Sci* 3:75–87.

Stewart DW, Dwyer LM (1983) Stomatal response to plant water deficits. *J Theor Biol* 104:655–666.

Szeicz G, Long IF (1969) Surface resistance of crop canopies. *Water Resour Res* 5:622–633.

Todd GW, Webster DL (1965) Effect of repeated drought periods on photosynthesis and survival in cereal seedlings. *Agron J* 57:399–403.

Turner NC, Kramer PJ, Eds. (1980) *Adaptation of Plants to Water and High Temperature Stress*, Wiley-Interscience, New York, 482 pp.

Turner NC, Begg JE, Rawson HM, English SD, Hearn AB (1978) Agronomic and physiological responses of soybean and sorghum crops to water deficits. III. Components of leaf water potential, leaf conductance, $^{14}CO_2$ photosynthesis, and adaptation to water deficits. *Aust J Plant Physiol* 5:179–194.

van Bavel CHM (1967) Changes in canopy resistance to water loss from alfalfa induced by soil water depletion. *Agric Meteorol* 4:165–176.

Westgate ME, Boyer JS (1984) Transpiration- and growth-induced water potentials in maize. *Plant Physiol* 74:882–889.

Westgate ME, Boyer JS (1986) Water status of the developing grain of maize. *Agron J* 78:714–719.

Westgate ME, Thomson-Grant DL (1989a) Effect of water deficits on seed development in soybean. I. Tissue water status. *Plant Physiol* 91:975–979.

Westgate ME, Thomson-Grant DL (1989b) Response of the reproductive tissues to water deficits at anthesis and mid-grain fill. *Plant Physiol* 91:862–867.

Westgate ME, Schussler JR, Reicosky DC, Brenner ML (1989) Effect of water deficits on seed development in soybean. II. Conservation of seed growth rate. *Plant Physiol* 91:980–985.

Wignarajah K, Greenway H, John CD (1976) Effect of water-logging on growth and activity of alcohol-dehydrogenase in barley and rice. *New Phytol* 77:585–592.

Young DR, Burke IC, Knight DH (1985) Water relations of high-elevation phreatophytes in Wyoming. *Am Midl Nat* 114:384–392.

chapter four

Plant activation of environmental agents: the utility of the plant cell/microbe coincubation assay

Michael J. Plewa, Kwang-Young Seo, Young-Hwa Ju, Shannon R. Smith, and Elizabeth D. Wagner

Introduction

Public health concerns about the presence of environmental toxins are usually limited to the direct exposure of humans to xenobiotics. Terrestrial and aquatic plants are likewise exposed and serve as a sink for such chemicals. The solar energy flow and the food chain of the biosphere begins with plants. The global plant biomass is approximately 8.3×10^{14} kg dry weight — a value that represents over 90% of the total mass of the living biota (Woodwell et al. 1978). Plants can metabolize many xenobiotics by oxidation, transformation, and conjugation and can compartmentalize these products in their tissues (Lamoureux and Rusness 1986). Approximately 6.3 billion pounds of pesticides are used worldwide each year (Briggs 1992). The environmental impact of plant metabolism of xenobiotics and the sink function of the global plant biomass is unknown, but it may play an important role in environmental toxicology (Plewa and Gentile 1982; Sandermann 1992).

Interest in the genetic effects of toxic plant metabolites of xenobiotics has grown since the demonstration of *in vivo* and *in vitro* plant activation (Plewa and Wagner 1993). Plant activation is the process by which a promutagen is activated into a mutagen by a plant system (Plewa 1978). A promutagen is a chemical that is not mutagenic in itself, but can be biologically transformed into a mutagen. Depending upon the developmental stage of an individual, a mutagen can exert teratogenic effects, precipitate coronary disease, produce mutations involving germinal cells, or cause mutations in somatic cells that may lead to cancer (Sorsa 1980). Plant systems have been widely employed in classical and environmental mutagenesis (Nilan 1978). With the realization of the immense diversity of xenobiotics to which plants are intentionally and unintentionally exposed, the capability of plants to bioconcentrate environmental agents and activate promutagens into stable, toxic metabolites becomes a significant concern. For a partial list of plant-activated agents see Table 1.

Aromatic amines (arylamines) are classical promutagens and procarcinogens and have been used to resolve questions involving mutagenic activation, cancer induction, and human polymorphic sensitivity to environmental agents (Weisburger 1988). These agents are substrates for mammalian activation by the cytochrome P-450-mediated pathway (Guengerich et al. 1988), as well as by cellular peroxidases such as prostaglandin H synthase (Eling et al. 1988). We employ arylamine promutagens as a class of model compounds to study the mechanisms of plant activation (Figure 1) (Seo et al. 1993; Plewa et al. 1993, 1995). The purpose of this chapter is to review data obtained from the plant cell/microbe coincubation assay, to elucidate the mechanisms and discuss a working model for the plant activation of arylamines, and to present data on the molecular effects of these activated arylamines.

The plant cell/microbe coincubation assay

This assay was developed in our laboratory and employs living plant cells in suspension culture as the activating system and specific microbial strains as the genetic indicator organism (Plewa et al. 1983, 1988; Wagner et al. 1994). In general, plant and microbial cells are coincubated in a medium with a promutagen. The activation of the promutagen is detected by counting revertant colonies after plating the microbe on

Table 1 List of Chemical Promutagens Activated by Plant Systems

Studies on Intact Plants or Plant Cells with Genetic Endpoints

N–Acetylaminofluorene		2-Aminofluorene
Aflatoxins	Chlordane	Trp-P-2
Aniline	1,2-Dibromoethane	Glu-P-1
m-, *o*-, *p*-Aminophenol	Heptachlor	Benzidine
Benzo[α]pyrene	Maleic hydrazide	Ethanol
Furylfuramide	*s*-Triazines	N-Nitrosamines
Pyrrolizidine alkaloids		Phenol
		Sodium azide

Studies Using Plants, Tissues, or Cells with External Genetic or Biochemical
Endpoints

2-Aminofluorene		4-Aminobiphenyl
Benzidine	Alachlor	Benzo[α]pyrene
Cyclophosphamide	Chlordane	2,4-Diaminotoluene
Ethanol	Heptachlor	2-Naphthylamine
4-Nitro-*o*-	Propachlor	Sodium azide
phenylenediamine	*s*-Triazines	*m*-, *o*-Phenylenediamine

Studies on Plant Cell-Free Systems

N-Acetylaminofluorene		Aflatoxin B₁
Aniline	Atrazine	2-Aminofluorene
Benzo[α]pyrene	Captan	Cyclophosphamide
7,12-Dimethyl-	Chlordane	Ethanol
benz[α]anthracene	1,2-Dibromoethane	Pyrenes
4-Nitro-*o*-	Diquat	
phenylenediamine	Maleic hydrazide	
m-, *o*-Phenylenediamine	Niclofen	
Sodium azide	Pentac	
	Ziram	

Note: For references see Plewa and Wagner (1993).

selective media; the viability of the plant and microbial cells may be monitored, as well as other components of the assay.

More specifically, the assay is conducted as follows. Long-term plant cell suspension cultures of tobacco (*Nicotiana tabacum*), the non-photosynthetic cell line TX1, were maintained in MX medium, a modified liquid culture medium developed by Murashige and Skoog (1962). Continuous cultures of this cell line were maintained by aseptically transferring 3 g fresh weight of cells from a 7-d culture into 100 ml of MX medium. All cultures were grown at 28°C with shaking (150 rpm) under dark conditions. For the plant cell/microbe coincubation assay, a TX1 cell culture was grown at 28°C to early stationary phase. The plant cells were harvested by centrifugation at 138 ×*g* for 3 min at 4°C. The supernatant was removed, and the cells were suspended in MX medium without plant growth hormone (MX⁻). The cells were centri-

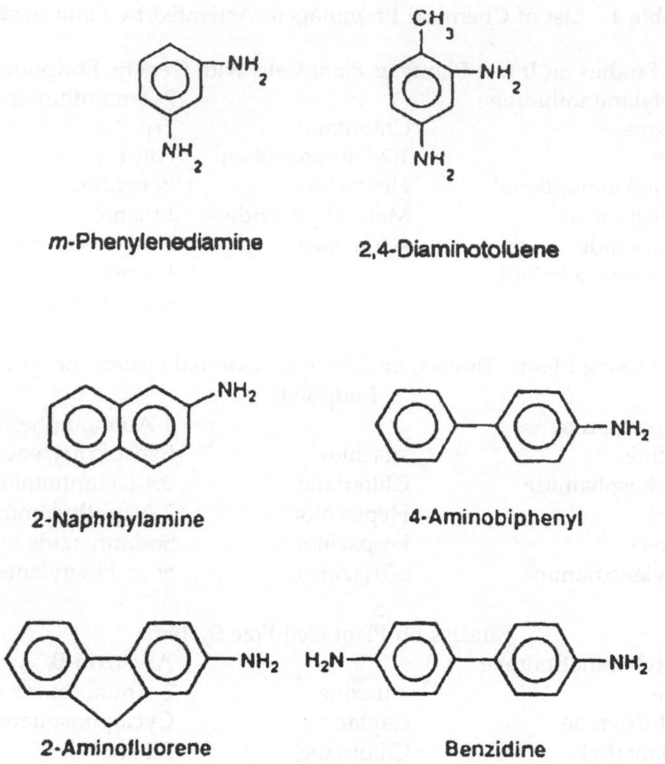
NH2
NH2
m-Phenylenediamine
CH3
NH2
NH2
2,4-Diaminotoluene
NH2
2-Naphthylamine
NH2
4-Aminobiphenyl
NH2
2-Aminofluorene
H2N
NH2
Benzidine

Figure 1 Arylamine promutagens that are plant activated by cultured tobacco cells.

fuged as before, the supernatant was removed, and the cells were suspended in MX⁻ medium in a volume approximately one half the original volume. The fresh weight of the plant cells was determined by placing a 5.0-ml sample of suspension on Miracloth filter fabric in a Büchner funnel with vacuum filtration until water droplets no longer formed from the funnel. The pad of cells was placed in a tared weighing vessel and weighed. Appropriate volumes of the plant cell suspension and MX⁻ medium were combined to yield a final concentration of 100 mg/ml. The suspension was stored on ice (≤30 min) until its use in the plant cell/microbe coincubation assay.

An overnight culture of *Salmonella typhimurium* was grown from a single colony isolate in 100 ml of Luria broth at 37°C with shaking. The bacterial suspension was centrifuged at 4000 ×*g* for 5 min at 4°C. The supernatant was decanted, and the bacterial pellet was suspended in 100 m*M* potassium phosphate buffer, pH 7.4. This suspension was centrifuged as before, the supernatant was decanted, and the pellet was suspended in 10 ml of 100 m*M* potassium phosphate buffer, pH 7.4. The optical density of the cell suspension was determined spectrophotometrically from a sample prepared with 0.05 ml of the bacterial suspension and 4.95 ml of buffer. The optical density of a buffer blank (5 ml) and the bacterial sample were measured at 660 nm. The titer (in colony-forming units) was calculated according to the following formula: cell titer = $(OD_{660\ nm})$ $(6.67 \times 10^8$ CFU/ml) (5.0/0.05 ml). Appropriate amounts of the bacterial suspension and potassium phosphate buffer were combined to yield a final concentration of 1×10^{10} cells/ml, and this cell suspension was placed on ice.

In the coincubation assay, each reaction mixture consisted of 4.5 ml of the plant cell suspension in MX⁻ medium (100 mg/ml), 0.5 ml of the bacterial suspension $(5 \times 10^9$ cells), and a known amount of the promutagen in microliter amounts. Concurrent negative controls consisted of plant and bacterial cells alone, heat-killed plant cells plus bacteria and the promutagen, and both buffer and solvent controls. These components were incubated at 28°C for 1 h with shaking at 150 rpm under dark conditions. After the treatment time, the reaction tubes were placed on ice. Triplicate 0.5-ml aliquots ($\sim 5 \times 10^8$ bacteria) were removed and added to molten top agar supplemented with 550 μM histidine and biotin. The top agar was poured onto Vogel Bonner (VB) minimal medium plates, incubated for 72 h at 37°C, and revertant his^+ colonies were scored (Maron and Ames 1983). Examples of the plant activation of a monocyclic, bicyclic, and polycyclic arylamine are presented in Figure 2.

In additional studies, the remainder of the reaction mixture was used to determine the viability of the plant and bacterial cells. One volume of cold 250 mM sodium citrate buffer, pH 7, was added to each reaction tube, and 0.5 ml of this suspension was removed and mixed with 2 ml of MX⁻ medium. The viability of the TX1 cells was immediately determined using the phenosafranin dye exclusion method. A 0.1% phenosafranin solution (1 mg/ml) was prepared with MX⁻ medium and made immediately prior to use. One drop of the diluted plant cell suspension was mixed with one drop of the stain on a microscope slide. Living cells with intact cell walls exclude the dye and appear clear; dead cells stain a pink to red color. A minimum of 100 cells was counted from each treatment group. The heat-killed TX1 cells were also analyzed to verify proper staining.

The viability of the bacterial cells was determined by adding 1 ml of the cold reaction mixture to 1 ml of cold 100 mM phosphate buffer, pH 7.4. After a dilution series in phosphate buffer was conducted, a specific volume was added to each of three molten LB top agar tubes and poured upon LB plates. After incubation at 37°C for 24 h, the viable bacterial cells were counted.

The viability of the plant and bacterial cells in the coincubation assay is exceedingly important. Figure 3A illustrates a direct, linear relationship between the activation of *m*-phenylenediamine and TX1

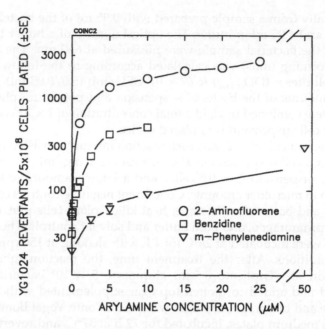

Figure 2 Concentration-response data for the plant activation of a monocyclic arylamine (*m*-phenylenediamine), a bicyclic arylamine (benzidine), and a polycyclic arylamine (2-aminofluorene) employing the plant cell/microbe coincubation assay.

cell viability. However, a decrease in the number of viable bacterial cells plated is less important (Figure 3B).

Inhibitors to identify biochemical pathways in plant activation

The use of enzyme inhibitors for specific pathways has aided the biochemical characterization of the metabolism of xenobiotics in mammalian and plant species. With inhibition studies care must be taken to ensure that the biological integrity of the assay is not compromised. In many studies employing *S. typhimurium* strains, a decrease in the number of revertants with increasing "inhibitor" concentration is equated with true inhibition. This same decline would be observed under other circumstances. The analysis of the viability of the plant cells and microbial cells is very important to discriminate between decreases in mutation induction due to toxicity in the microbial cells, toxicity in the activating system from the agent alone, toxicity in the activating system from a synergic interaction between the agent and the mutagen, and declines due to a true repression of mutagen activation. An example of this was observed with the effect of metyrapone on the plant activation of *m*-phenylenediamine. Using *S. typhimurium* strain TA98, the

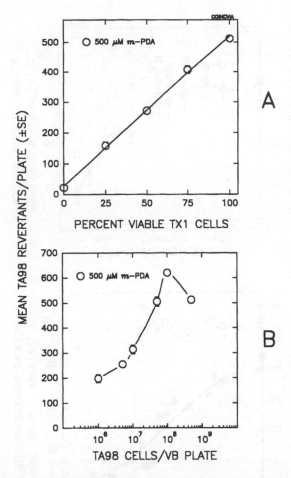

Figure 3   The effect of TX1 cell death on the induction of *S. typhimurium* strain TA98 revertants induced by the tobacco cell activation of 500 µM *m*-phenylenediamine (A). The effect of TA98 cell death on the induction of TA98 revertants caused by the plant activation of *m*-phenylenediamine (B).

activation of *m*-phenylenediamine was diminished by approximately 70% by 15 mM metyrapone (Figure 4A). However, this concentration of metyrapone was toxic to the TX1 cells, thus reducing the activation of *m*-phenylenediamine with the net effect of reducing the number of TA98 revertants per plate. An example of the repression of the plant activation of *m*-phenylenediamine by diethyldithiocarbamate is presented in Figure 4B. With 10 mM diethyldithiocarbamate, there is approximately a 95% reduction in the mutagenic potency without toxicity to the plant cells. This represents a true repression of activation by a peroxidase inhibitor. These data illustrate the necessity of monitoring all of the components when studying the inhibition of activation or mutagenesis.

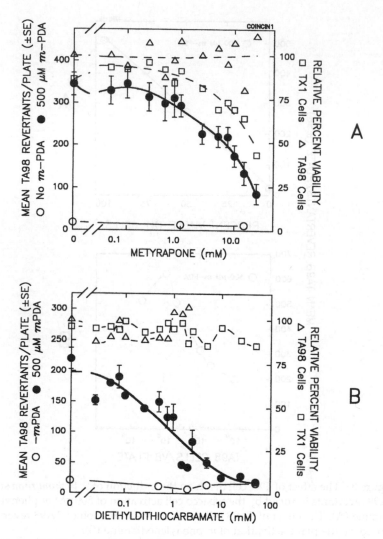

Figure 4 (A) The repression of the tobacco cell activation of 500 μM *m*-phe-nylenediamine by metyrapone. The concentration range for the repression curve is presented along the abscissa in a linear-log scale in a range of 0 and 50 μM to 25 mM (●). The concentration-response curve for the direct-acting mutagenicity of metyrapone was evaluated in *S. typhimurium* TA98 cells and was conducted in a range of 0 and 1 to 15 mM (○). The toxicity of metyrapone was evaluated in TA98 cells ( ) or TX1 cells (□) in a concentration range of 0 and 50 μM to 25 mM. (B) The repression of the tobacco cell activation of 500 μM *m*-phenylenediamine by diethyldithiocarbamate. The concentration range for the repression curve is presented along the abscissa in a linear-log scale in a range of 0 and 25 μM to 50 mM (●). The concentration-response curve for the direct-acting mutagenicity of diethyldithiocarbamate was evaluated in *S. typhimurium* TA98 cells and was conducted in a range of 0 and 500 μM to 50 mM (○). The toxicity of diethyldithiocarbamate was evaluated in TA98 cells ( ) or TX1 cells (□) in a concentration range of 0 and 25 μM to 50 mM.

Seven inhibitors — diethyldithiocarbamate, metyrapone, 7,8-ben-zoflavone, potassium cyanide, (+)-catechin, methimazole, and acetami-nophen — were incorporated in the plant cell/microbe coincubation assay (Wagner et al. 1989, 1990). These chemicals have been used to study metabolism in animal systems (Harvison et al. 1988; Kawanishi et al. 1985; Poulsen et al. 1974; Steele et al. 1985; Ullrich et al. 1973) as well as in plants (Dennis and Kennedy 1986; Gichner and Veleminsky 1984; Gichner et al. 1988; Kuwahara et al. 1988; Nelson et al. 1981). Using these inhibitors, plant cell peroxidases were found to be involved in the activation of the promutagenic aromatic amines, 2-aminofluorene and *m*-phenylenediamine by tobacco cells (Wagner et al. 1989, 1990). Figure 5 illustrates the relationship between the concentration of dieth-yldithiocarbamate, the repression of the activation of 2-aminofluorene or *m*-phenylenediamine, and the inhibition of the activity of tobacco cell peroxidase (Wagner et al. 1989; Plewa et al. 1991). We recently confirmed that the plant activation of arylamines was repressed as a function of peroxidase inhibition in intact *Tradescantia* plants (Gichner et al. 1994).

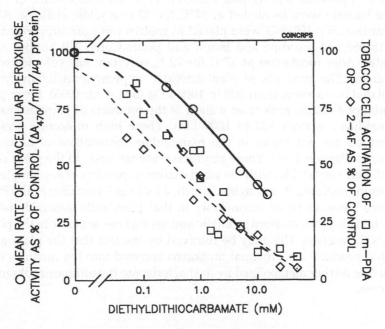

Figure 5 The relationship between the inhibition of TX1 cell peroxidase activity (○) and the repression of TX1 cell activation of 500 μM *m*-phenylenediamine (□) or 50 μM 2-aminofluorene (◇) vs. the log concentration of diethyldithio-carbamate.

Isolation of high molecular weight plant-activated aromatic amine products

Vacuoles and the cell wall are sites of deposition of many endogenous plant conjugates. Deposition is a detoxification mechanism. However, under suspension cell culture conditions, plant cells preferentially transport xenobiotic metabolites into the medium (Gareis et al. 1992). The tobacco cell-activated mutagenic products of 2-aminofluorene, *m*-phenylenediamine, benzidine, and 4-aminobiphenyl were isolated from the cell culture medium (Ju and Plewa 1993; Plewa et al. 1993; Seo et al. 1993). Early stationary-phase TX1 cells were exposed to the promutagen for 3 h at 28°C while shaking. The plant cells were removed by centrifugation at 4000 $\times g$ for 30 min at 4°C. The supernatant fluid was recovered and centrifuged at 100,000 $\times g$ for 1 h at 4°C. This ultracentrifuged supernatant was fractionated with sterile ultrafiltration membranes in an Amicon model 8400 stirred cell using N_2 at 20 to 25 psi at 4°C. The molecular weight limits were determined by membrane ultrafiltration; each retentate sample was concentrated approximately tenfold. The resulting filtrate and retentate fractions were tested for mutagenicity using a preincubation protocol. Each reaction tube contained 2×10^9 bacteria; varying amounts of the retentate or filtrate; and 100 mM potassium phosphate buffer, pH 7.4 in a total volume of 1 ml. The bacteria were incubated at 37°C for 30 min while shaking. After treatment, 5×10^8 cells were placed in molten top agar supplemented with 550 μM histidine and biotin and poured onto VB bottom agar plates. After incubation at 37°C for 72 h, revertant *his*⁺ colonies were counted. The products of plant-activated *m*-phenylenediamine were isolated in fractions from 300 to 1000 kDa (Seo et al. 1993). From preliminary data, the molecular weight of the products of plant-activated benzidine are from 100 to 1000 kDa. These high molecular weight products are not found in the mammalian metabolism of aromatic amines (Bartsch 1981). These products were not toxic to the plant cells and were very stable, with the plant-activated product of *m*-phenylenediamine retaining its mutagenic potency for over 1 year (Seo et al. 1993). There seems to be an incongruity in that plant cells generate stable, mutagenic products from aromatic amines that are not toxic to the plant cells themselves. This may be resolved by the fact that the plant-activated products are proximal mutagens secreted into the medium and must be further metabolized by the bacteria into the ultimate mutagenic agents.

Plant-activated aromatic amine products as substrates for acetyltransferases

The first step in the mammalian hepatic activation of aromatic amines is N-hydroxylation. Cellular acetyltransferases are essential in further metabolizing N-hydroxyarylamines to their ultimate mutagenic products (Saito et al. 1985). To study the effect of acetyltransferase activity on the plant-activated aromatic amine products, we used a series of *S. typhimurium* strains that differentially express acetyl-CoA:N-hydroxyarylamine O-acetyltransferase (OAT) (Wagner et al. 1994). The plant-activated products of 2-aminofluorene, benzidine, *m*-phenylenediamine, 4-aminobiphenyl, 2,4-diaminotoluene, and 2-naphthylamine served as substrates for bacterial OAT and induced mutation in *Salmonella*. Figure 6 illustrates a direct relationship between the mutagenic potency of the plant-activated products of 2-aminofluorene and benzidine and the amount of OAT expressed in the genetic indicator cells. The strain that has the highest expression of OAT (YG1024) is the most sensitive (Wantanabe et al. 1990).

NAT1 and *NAT2* are two human genes that encode functionally distinct but similar cytosolic CoASAc:arylamine N-acetyltransferase (Dupret and Grant 1992). Although *NAT1* and *NAT2* have diverse substrate specificities, they share over 80% amino acid sequence homology (Blum et al. 1990). Rat hepatic S9 activation of 2-aminofluorene, benzidine, and 2-amino-3,4-dimethylimidazo[4,5-*f*]quinoline was examined in *S. typhimurium* strains containing human *NAT1* and *NAT2*, DJ400 (TA1538/1,8-DNP:pNAT1), and DJ460 (TA1538/1,8-DNP:pNAT2), respectively (Grant et al. 1992). All three aromatic amines were mutagenic in DJ460. Only 2-aminofluorene and benzidine were mutagenic in DJ400. Our data demonstrated that the plant-activated benzidine products are exceedingly potent in the human *NAT2*-expressing *Salmonella* strains, while these metabolites are refractory in the strain expressing *NAT1* (Figure 7). These data indicate that the plant-activated benzidine products behave in a different manner than their mammalian-activated counterparts.

The plant activation of benzidine and 4-aminobiphenyl induced both frameshift and basepair substitution revertants in strains that overexpressed OAT, YG1024, and YG1029 (Figure 8). The plant-activated products were refractory in strains that overexpressed nitroreductase, YG1021, and YG1026. Thus, these plant-activated products are substrates for acetyltransferases but not for nitroreductases.

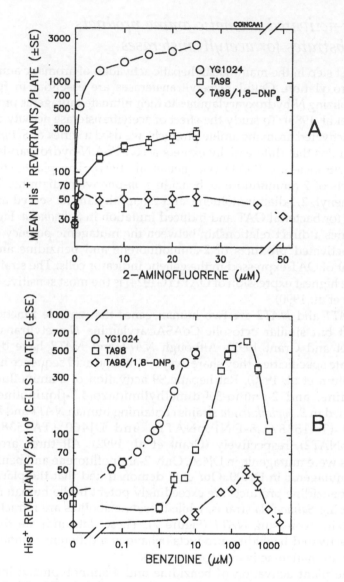

Figure 6 The effect of the expression of OAT in the genetic indicator cells exposed to the plant-activated products of 2-aminofluorene (A) or benzidine (B). *S. typhimurium* strain TA98/1,8-DNP$_6$ (◇) does not express OAT, while TA98 (□) and YG1024 (○) express the wild-type level and overexpress OAT, respectively.

Working model for the plant activation of aromatic amines

From the data presented here and in other studies, we have prepared a model of the TX1 cell activation of *m*-phenylenediamine that we

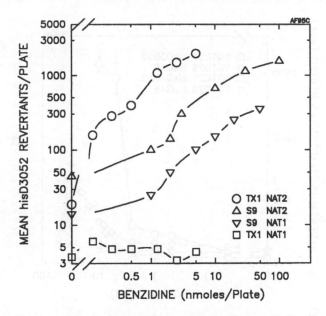

Figure 7 Log/log plot illustrating the mutagenic response of rat hepatic microsomal S9-activated benzidine assayed in *S. typhimurium* that expressed human *NAT1* (▽) or *NAT2* (△) (Grant et al. 1992). Mutagenic response of plant-activated benzidine in strains that expressed human *NAT1* (□) or *NAT2* (○).

extend to the plant activation of aromatic amine promutagens (Seo et al. 1993). The model — albeit simplistic and incomplete — integrates our data into a mechanistic framework and serves as a foundation for new experimental designs. The model has the following seven components: (1) the aromatic amine (R-NH$_2$) is transported into the plant (TX1) cell, (2) TX1 intracellular peroxidase oxidizes the molecule (R-NHOH) (refer to Figure 5), (3) the metabolite is conjugated to a macromolecule (R-NHOH-conjugate) (refer to previous section), (4) the amine conjugate is secreted into the extracellular medium (the spent medium also has the capacity to activate the aromatic amine promutagen), (5) the conjugate or a deconjugated plant-activated metabolite is absorbed by the *Salmonella* tester strains, (6) the plant-activated N-hydroxylated product is acetylated (R-NHO-COCH$_3$) and deacetylated by the bacterial OAT (refer to Figures 6, 7, and 8), and (7) the deacetylation results in a highly reactive nitrenium ion (R-NH$^+$).

Molecular effects of plant-activated promutagens

Mutant spectra analysis has been conducted on many of the *S. typhimurium* tester strains (Bell et al. 1991; Cebula and Koch 1990; Kupchella and Cebula 1991). These studies define the molecular basis of both spontaneous and induced revertants. To determine the molecular

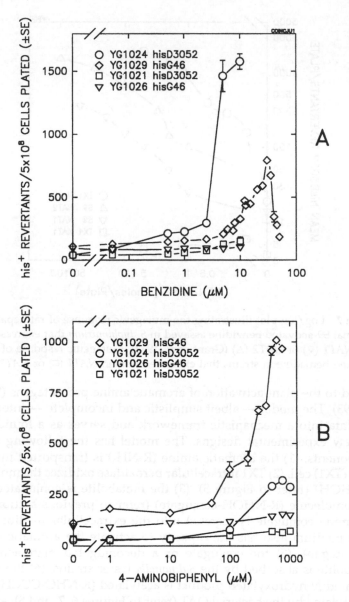

Figure 8 Tobacco cell activation of benzidine (A) and 4-aminobiphenyl (B) into mutagens that induce frameshift mutations at *hisD3052* and basepair substitution mutations at *hisG46* in *S. typhimurium* strains that overexpress *O*-acetyltransferase. Strains YG1021 and YG1026 overexpress nitroreductase.

effects of plant-activated promutagens, we analyzed revertants at the *hisD3052* allele which were collected after coincubation with TX1 cells and 2-aminofluorene. A rigorous three-round selection procedure was used to collect independently arising spontaneous *his+* revertants. Single colony isolates were collected from master plates and grown in 1 ml of LB (+ antibiotics) liquid medium overnight at 37°C. One hundred microliters of this suspension were added to VB minimal medium molten top agar supplemented with 550 μM histidine and biotin, poured onto a VB plate, and incubated at 37°C for 48 to 72 h. One revertant colony from each plate was isolated, streaked onto a VB + biotin quartet plate, and grown overnight. The quartet plates were stored at 4°C. For the collection of induced revertants, a concentration of 2-aminofluorene was chosen that yielded an approximately 15 times increase in mutation over the spontaneous frequency with no toxicity to the bacteria or plant cells. Following the procedures of the coincubation assay with a single concentration of 2-aminofluorene, 50 individual revertants were selected from each plate. These revertants were streaked onto VB + biotin quartet plates, grown overnight, and stored at 4°C. Revertant colonies from the quartet plates were grown in VB + biotin liquid medium overnight at 37°C. Genomic DNA was extracted using the DNA minipreparation procedure described in *Current Protocols in Molecular Biology* (Ausubel et al. 1994). The DNA was suspended in 100 μl Tris EDTA (TE) buffer and stored at –20°C.

A 635-basepair DNA fragment containing the *hisD3052* region was amplified for each genomic DNA sample using polymerase chain reaction (PCR). A tenfold dilution of the DNA served as a template. The primers used were AP1 (5'-CGTCTGAAGTACTGGTGATCGCA-3') and AP3 (5'-CGGGCTAAGTCAGCGACGCTGAG-3'). The concentration of primers was 0.2 μM. We optimized the thermal cycler for PCR amplification of this region; the program included an initial step at 94°C for 5 min followed by 30 cycles of 94°C for 1 min (denaturation), 55°C for 1 min (annealing), and 72°C for 30 s (extension). There was a final step of 72°C for 7 min. The PCR product was analyzed by agarose gel electrophoresis to verify the amplification of the correct region.

The Bio-Dot microfiltration apparatus (Bio-Rad®) was used to transfer the DNA to a nylon membrane. Each well of a 96-well microtiter plate was filled with 195 μl of buffer (TE pH 7.4, 400 mM NaOH, 10 mM EDTA) and 5 μl of PCR product. The plate was heated to 95°C for 10 min. The contents from each well were transferred to corresponding wells on the dot blot apparatus. A vacuum was used to draw each sample onto a charged nylon membrane. Five hundred microliters of 400 mM NaOH were added to each well and pulled through the membrane by a vacuum. The membrane was removed from the apparatus, rinsed in 2x sodium chloride/sodium citrate (SSC), and exposed to ultraviolet radiation to cross-link the DNA. The membrane was allowed to air dry and was stored at room temperature.

In the *hisD3052* allele, a high frequency of reversion occurs by a CG or GC deletion which is located in an alternating CG octamer (D878–885); a modified version of the Amersham nonradioactive ECL3'-oligolabeling and detection system was used to identify the –2 hotspot region (Cebula and Koch 1990). Each membrane was hybrid-ized for 30 min at 60°C with unlabeled TC-13 probe to linearize the target region of DNA. Fluorescein-dUTP-labeled TC5 (5′-CTGC-CGCGCGGACACCGC-3′) was added and incubated for 1 h at 60°C. The membranes were washed according to the procedure outlined by the Amersham kit RPN2131, with the final wash conditions at 0.75× SSC + 0.1% SDS at 62°C. The membranes were incubated for 30 min in block solution, rinsed, and incubated for 30 min with antifluorescein horseradish peroxidase conjugate. Following extensive washing, the reduction of the bound peroxidase was coupled to the oxidation of luminol, resulting in light emission (428 nm). This signal was detected on Kodak XAR-5 X-ray film (5 min exposure).

Mutants containing the specific –2 hotspot mutation appear as dark dots on the membrane. The dots that are absent indicate mutants with other mutational events. These remaining nonhotspot mutants were sequenced using dsDNA cycle sequencing with [33]P. Each membrane had a positive and negative control to insure that the hybridization procedure was working properly. The positive control and the negative control were both spontaneous mutants; their sequences were determined by DNA sequencing. The specificity of this hybridization procedure for the –2 hotspot is very high. The negative control is a –2 hotspot revertant that contains a basepair substitution within the probing region. Thus, the hybridization procedures are able to distinguish mutants that differ by only a single basepair.

Revertants at the *hisD3052* allele of strains YG1024 and TA98 were collected after coincubation with tobacco cells and 2-aminofluorene. Of the spontaneous TA98 revertants, 40% were due to –2 events at D878–885. In contrast, 94% of the plant-activated 2-aminofluorene-induced TA98 revertants probed as this –2 deletion. Cebula and Koch (1990) found that 92% of the revertants induced by mammalian S9-activated acetylaminofluorene were a –2 deletion in this hotspot region. YG1024 was more sensitive than TA98 to the plant-activated metabolites of aromatic amines due to elevated levels of *O*-acetyltransferase. Ninety-eight percent of the induced YG1024 revertants were the result of a –2 event. Despite the difference in mutagenic sensitivity, there was no difference in the frequency of –2 deletions at D878–885 of these plant-activated 2-aminofluorene-induced revertants.

Mutant spectra analysis was conducted on the *hisG46* allele (*S. typhimurium* strain YG1029) from spontaneous revertants, revertants induced by the mammalian (S9) activation of benzidine, and revertants induced by the XM300 retentate containing the plant-activated benzidine products. The collection of *his*+ revertants and DNA extraction were identical to that described previously.

For PCR amplification, a 187-basepair DNA fragment containing the *hisG46* region was amplified using the primers HISG46A (5′-GCCT-GATTGCGATGGCGG-3′) and HISG46B (5′-GTCAAGACG-GCGCTGGG-3′). The concentration of primers was 0.2 μ*M*. The thermal

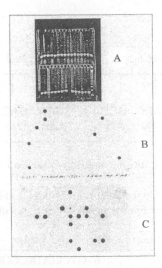

Figure 9   (A) PCR-amplified 187-basepair DNA from 30 independent *hisG46* revertants induced by plant-activated benzidine product visualized on a 1% agarose gel. The lower middle lane is the no DNA control and the lane in the lower right corner is the DNA size standard. (B and C) Dot blot analysis of the membranes containing the PCR-amplified DNA using two specific probes for specific basepair substitution mutations.

cycler was programmed for an initial step at 94°C for 5 min followed by 30 cycles of 94°C for 1 min, 60°C for 1 min, and 72°C for 30 s. The PCR product was evaluated by agarose gel electrophoresis (Figure 9A). Each PCR product was transferred to a charged nylon membrane, and the DNA was cross-linked.

The *hisG46* allele is a missense mutation (CTC→CCC) at *his*G codon 69 which renders the cell *his⁻*. In five of the six possible basepair substitutions at the first or second residue of the codon, the cell reverts to histidine prototrophy (Figure 10). Also, the occurrence of revertants expressing a CCC sequence at codon 69 indicates they are suppressor mutants. Thus, six possible events account for all of the *hisG46* revertants. A modified version of the Amersham nonradioactive ECL3′-oligolabeling and detection system was used to identify each of the six possible basepair substitution mutations (Table 2).

Each probe was end labeled with fluorescein-dUTP. Each membrane was prehybridized for 30 min at 34°C with no probe. Fluorescein-dUTP-labeled probe was added and incubated for 1 h at 34°C. The membranes were washed twice in 5× SSC, 0.1% SDS for 5 min at 34°C followed by two washes in 0.9× SSC, 0.1% SDS for 15 min at 39°C. The membranes were incubated for 30 min in block solution, rinsed, and incubated for 30 min with antifluorescein horseradish peroxidase conjugate. Following extensive washing, the membranes were exposed for 5 min on X-ray film.

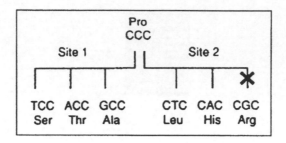

Figure 10   Basepair substitution mutations that lead to reversion at *hisG46*.

Table 2   Oligonucleotide Probes Used to Detect
Reversion Events at *hisG46* in *S. typhimurium*

Reversion event	Probe
CCC→ACC	5'-GTCGATACCGGTTAT
CCC→TCC	5'-GTCGATTCCGGTTAT
CCC→GCC	5'-GTCGATGCCGGTTAT
CCC→CTC	5'-GTCGATCTCGGTTAT
CCC→CAC	5'-GTCGATCACGGTTAT
CCC→CCC[a]	5'-GTCGATCCCGGTTAT

[a] Reversion due to an extragenic suppression mutation.

All mutants containing the basepair substitution identical to the
probe appear as dark dots (Figures 9B and C). The dots that are absent
indicate any of the other five mutational events. Each membrane had
six controls in the upper left corner representing each of the six possible
basepair substitutions. Only the control matching the probe should
appear as a dark dot. After each hybridization procedure, the membranes were stripped of all labeled probe by first washing in 5× SSC,
0.1% SDS for 5 min at room temperature followed by 0.1× SSC, 0.1%
SDS for 1 h at 65°C and a final wash in 5× SSC, 0.1% SDS for 5 min at
room temperature. The membranes were then wrapped in plastic and
stored at –20°C. The membrane was rehybridized with all six probes
without any loss in specificity.

Some of our preliminary data are presented in Figure 11 comparing
the mutant spectra of spontaneous (control), plant-activated, and S9-
activated benzidine (Ju et al. 1995). Of the spontaneous revertants,
transition mutations were 31.8% and transversion mutations were
68.2%. With plant-activated benzidine products, 22.5% of the recovered
revertants resulted from transitions and 77.6% from transversions. The
revertants induced by S9-activated benzidine were 14.4% transition and
85.8% transversion mutations. According to chi-square analysis, there
was a significant difference among these spectra at the level of transition mutations.

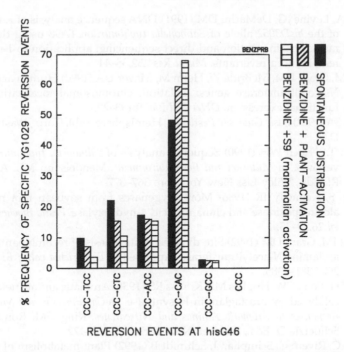

Figure 11 Mutant spectra analysis at reversion at *hisG46* in *S. typhimurium* strain YG1029 among the spontaneous control, plant-activated benzidine-induced, and mammalian-activated benzidine-induced revertants.

Conclusions

This chapter on the plant cell/microbe coincubation assay demonstrates its utility in investigating plant-activated promutagens. The assay is sensitive, and it can be used to determine the biochemical mechanisms of plant activation as well as the molecular genetic effects of the plant-activated products. The global impact of plant activation of environmental xenobiotics is unknown. However, the vast quantities of agents that find their way into the global "green sink" serve as a stimulus to heighten our curiosity and a challenge for future research.

References

Ausubel FM, Brent R, Kingston RE, Moore DD, Seidman JG, Smith JA, Struhl K (1994) *Current Protocols in Molecular Biology*, Green Pub. Assoc. and Wiley Inc., New York.

Bartsch H (1981) Metabolic activation of aromatic amines and azo dyes. In *Environmental Carcinogens Selected Methods of Analysis*, Egan H, Fishbein L, Castegnaro M, O'Neill IK, Bartsch H, Eds., International Agency for Research on Cancer, Lyon, France, 4:13–30.

Bell DA, Levine JG, DeMarini DM (1991) DNA sequence analysis of revertants of the *hisD3052* allele of *Salmonella typhimurium* TA98 using the poly-merase chain reaction and direct sequencing: application to 1-nitropy-rene-induced revertants. *Mutat Res* 252:35–44.

Blum M, Grant DM, McBride W, Heim M, Meyer UA (1990) Human arylamine N-acetyltransferase genes: isolation, chromosomal localization, and functional expression. *DNA Cell Biol* 9:193–203.

Briggs SA (1992) *Basic Guides to Pesticides,* Hemisphere Publ. Corp., Washington, D.C.

Cebula TA, Koch WH (1990) Sequence analysis of *Salmonella typhimurium* re-vertants. In *Mutation and the Environment,* Mendelsohn ML, Albertini RJ, Eds., Wiley-Liss New York, pp. 367–377.

Dennis S, Kennedy IR (1986) Monooxygenases from soybean root nodules: aldrin epoxidase and cinnamic acid 4-hydroxylase. *Pestic Biochem Phys-iol* 26:29–35.

Dupret JM, Grant DM (1992) Site-directed mutagenesis of recombinant human arylamine N-acetyltransferase expressed in *Escherichia coli. J Biol Chem* 267:7381–7385.

Eling TH, Petry TW, Hughes MF, Krauss RS (1988) Aromatic amine metabolism catalyzed by prostaglandin H synthase. In *Carcinogenic and Mutagenic Responses to Aromatic Amines and Nitroarenes,* King CM, Romano LJ, Schuetzle D, Eds., Elsevier, New York, pp. 161–172.

Gareis C, Rivero C, Schuphan I, Schmidt B (1992) Plant metabolism of xenobi-otics. Comparison of the metabolism of 3,4-dichloroaniline in soybean excised leaves and soybean cell suspension cultures. *Z Naturforsch* 47c:823–829.

Gichner T, Veleminsky J (1984) Inhibition of dimethylnitrosamine-induced mu-tagenesis in *Arabidopsis thaliana* by diethyldithiocarbamate and carbon monoxide. *Mutat Res* 139:29–33.

Gichner T, Veleminsky J, Rieger R (1988) Antimutagenic effects of diethyldithio-carbamate towards maleic hydrazide- and N-nitrosodiethylamine-in-duced mutagenicity in the Tradescantia mutagenicity assay. *Biol Plant* 30:14–19.

Gichner T, Lopez GC, Wagner ED, Plewa MJ (1994) Induction of somatic mu-tations in *Tradescantia* clone 4430 by three phenylenediamine isomers and the antimutagenic mechanisms of diethyldithiocarbamate and van-adate. *Mutat Res* 306:165–172.

Grant DM, Josephy PD, Lord HL, Morrison LD (1992) *Salmonella typhimurium* strains expressing human arylamine N-acetyltransferases: metabolism and mutagenic activation of aromatic amines. *Cancer Res* 52:3961–3964.

Guengerich FP, Butler MA, MacDonald TL, Kadlubar FF (1988) Oxidation of carcinogenic arylamines by cytochrome P-450. In *Carcinogenic and Mu-tagenic Responses to Aromatic Amines and Nitroarenes,* King CM, Romano LJ, Schuetzle D, Eds., Elsevier, New York, pp. 89–95.

Harvison PJ, Egan RW, Gale PH, Christian GD, Hill BS, Nelson SD (1988) Acetaminophen and analogs as cosubstrates and inhibitors of prosta-glandin H synthase. *Chem-Biol Interact* 64:251–266.

Ju YH, Plewa MJ (1993) Mutagenic characterization of the plant-activated prod-ucts of benzidine and 4-aminobiphenyl. *Environ Mol Mutagenesis* 21(suppl. 22):34.

Ju YH, Smith SR, Plewa MJ (1995) Mutant spectra analysis at *hisG46* in *Salmonella typhimurium* induced by mammalian S9- and plant-activated benzidine. *Environ Mol Mutagenesis* 25(suppl. 25):25.

Kawanishi T, Ohno Y, Takahashi A, Takanaka A, Kasuya Y, Omori Y (1985) Relation between hepatic microsomal metabolism of N-nitrosamines and cytochrome P-450 species. *Biochem Pharmacol* 34:919–924.

Kupchella E, Cebula TA (1991) Analysis of *Salmonella typhimurium hisD3052* revertants: the use of colony hybridization, PCR, and direct sequencing in mutational analysis. *Environ Mol Mutagenesis* 18:224–230.

Kuwahara T, Shigematsu H, Omura T (1988) Purification of an aromatic amine-dependent NAD(P)H oxidase from tomato fruit and its characterization as a peroxidase. *Agric Biol Chem* 52:2597–2603.

Lamoureux GL, Rusness DG (1986) Xenobiotic conjugation in higher plants. In *Xenobiotic Conjugation Chemistry*, Paulson GD, Caldwell J, Hutson DH, Menn JJ, Eds., American Chemical Society, Washington, D.C., 229:62–107.

Maron DM, Ames BN (1983) Revised methods for the *Salmonella* mutagenicity test. *Mutat Res* 113:173–215.

Murashige T, Skoog F (1962) A revised medium for rapid growth and bioassays with tobacco cultures. *Physiol Plant* 15:473–497.

Nelson SD, Dahlin DC, Rauckman EJ, Rosen GM (1981) Peroxidase-mediated formation of reactive metabolites of acetaminophen. *Mol Pharmacol* 20:195–199.

Nilan RA (1978) Potential of plant genetic systems for monitoring and screening mutagens. *Environ Health Perspect* 27:181–196.

Plewa MJ (1978) Activation of chemicals into mutagens by green plants: a preliminary discussion. *Environ Health Perspect* 27:45–50.

Plewa MJ, Gentile JM (1982) The activation of chemicals into mutagens by green plants. In *Chemical Mutagens: Principles and Methods for Their Detection*, de Serres FJ, Hollaender A, Eds., Plenum Press, New York, 7:401–420.

Plewa MJ, Wagner ED (1993) Activation of promutagens by green plants. *Annu Rev Genet* 27:93–113.

Plewa MJ, Weaver DL, Blair LC, Gentile JM (1983) Activation of 2-aminofluorene by cultured plant cells. *Science* 219:1427–1429.

Plewa MJ, Wagner ED, Gentile JM (1988) The plant cell/microbe coincubation assay for the analysis of plant-activated promutagens. *Mutat Res* 197:207–219.

Plewa MJ, Smith SR, Wagner ED (1991) Diethyldithiocarbamate suppresses the plant activation of aromatic amines into mutagens by inhibiting tobacco cell peroxidase. *Mutat Res* 247:57–64.

Plewa MJ, Gichner T, Xin H, Seo KY, Smith SR, Wagner ED (1993) Biochemical and mutagenic characterization of plant-activated aromatic amines. *Environ Toxicol Chem* 12:1353–1363.

Plewa MJ, Smith SR, Timme M, Cortez D, Wagner ED (1995) Molecular analysis of frameshift mutations induced by plant-activated 2-aminofluorene. Symposium on Molecular Biological Tools in Environmental Chemistry, Biology and Engineering, American Chemical Society, pp. 69–80.

Poulsen LL, Hyslop RM, Ziegler DM (1974) S-oxidation of thioureylenes cata-
    lyzed by a microsomal flavoprotein mixed-function oxidase. *Biochem
    Pharmacol* 23:3431–3440.
Saito K, Shinohara A, Kamataki T, Kato R (1985) Metabolic activation of mu-
    tagenic N-hydroxyarylamines by O-acetyltransferase in *Salmonella ty-
    phimurium* TA98. *Arch Biochem Biophys* 239:286–295.
Sandermann H Jr. (1992) Plant metabolism of xenobiotics. *Trends Biochem Sci*
    17:82–84.
Seo KY, Riley J, Cortez D, Wagner ED, Plewa MJ (1993) Characterization of
    stable high molecular weight mutagenic product(s) of plant-activated
    m-phenylenediamine. *Mutat Res* 299:111–120.
Sorsa M (1980) Somatic mutation theory. *J Toxicol Environ Health* 6:57–62.
Steele CM, Lalies M, Ioannides C (1985) Inhibition of the mutagenicity of
    aromatic amines by the plant flavonoid (+)-catechin. *Cancer Res*
    45:3573–3577.
Ullrich V, Formmer U, Weber P (1973) Differences in the O-dealkylation of
    7-ethoxycoumarin after pretreatment with phenobarbital and 3-meth-
    ylcholanthrene. *Hoppe-Seyler's Z Physiol Chem* 354:514–520.
Wagner ED, Gentile JM, Plewa MJ (1989) Effects of specific monooxygenase
    and oxidase inhibitors on the activation of 2-aminofluorene by plant
    cells. *Mutat Res* 216:163–178.
Wagner ED, Verdier MM, Plewa MJ (1990) The biochemical mechanisms of the
    plant activation of promutagenic aromatic amines. *Environ Mol Mu-
    tagenesis* 15:236–244.
Wagner ED, Smith SR, Xin H, Plewa MJ (1994) Comparative mutagenicity of
    plant-activated aromatic amines using *Salmonella* strains with different
    acetyltransferase activities. *Environ Mol Mutagenesis* 23:64–69.
Watanabe M, Ishidate M Jr., Nohmi T (1990) Sensitive method for the detection
    of mutagenic nitroarenes and aromatic amines: new derivatives of *Sal-
    monella typhimurium* tester strains possessing elevated O-acetyltrans-
    ferase levels. *Mutat Res* 234:337–348.
Weisburger JH (1988) Past, present, and future role of carcinogenic and mu-
    tagenic N-substituted aryl compounds in human cancer causation. In
    Carcinogenic and Mutagenic Responses to Aromatic Amines and Nitroarenes,
    King CM, Romano LJ, Schuetzle D, Eds., Elsevier, New York, pp. 3–19.
Woodwell GM, Whittaker RH, Reiners WA, Likens GE, Delwiche CC, Botkin
    DB (1978) The biota and the world carbon budget. *Science* 199:141–146.

chapter five

Statistical methods in plant environmental studies

Merrilee Ritter

Introduction

The goals when performing experiments on plants are to understand its responses to the experimental agent and to obtain this information in the most efficient way possible. Experimental design enables the investigator to collect information in a cost-effective manner to answer his/her questions. Statistical analyses summarize, display, quantify, and provide information for interpretation of the results for assessing the relationships of plant responses to the experimental agent. Statistical analysis also involves fitting a model to the data and making inferences with the model, determining differences between control and treatment groups (hypothesis testing), and describing the relationship between the level of treatment and the measured responses (concentration-response curves).

Experimental design

Statement of objective (Hicks 1973)

A statement of the question to be answered, the problem to be solved, or the objective to be met by the experiment is necessary to optimize the experimental design for the most efficient use of available resources. This statement will enable the investigator to determine the needed resources and some of the expected outcomes from the behavior of the experiment.

Choice of response variable (Box et al. 1978; Dixon and Massey 1983; Hahn and Meeker 1991; Hicks 1973; Rosner 1990)

The choice of response variables is critical to interpreting the effects of the experimental agent on plants. Response variables are those plant attributes that truly respond to the effects of the experimental agent and demonstrate change when the experimental agent is introduced into the system.

The most common types of variables are

1. Continuous — Measurements that are taken from a continuum, such as a measurement scale, and are divisible into infinite parts. Examples are shoot lengths, weights of plants, etc.
2. Count — Integer data that are counts of items, such as number of sunflower seeds or plant seedlings germinated. These data are whole numbers only, since a half count is not usually observed.
3. Discrete or categorical — Frequency counts of incidences that occur, which can be placed into categories of interest. Clinical observations such as mortality usually are discrete variables.
4. Calculated endpoints — Calculated endpoints consist of EC_{xx} (effective concentration at a particular percentage), NOEC (no-observed-effect concentration), threshold effects (LOEC or lowest-observed-effect concentration), percent changes from controls, slopes or rates, mean time to observed effect, and group

mean differences. These endpoints take the original responses from the experiment and relate them mathematically or statistically so that treatment effects are determined.

Any calculated endpoint should be presented with an estimate of its variability. This provides useful information in estimating the "true" value that the endpoint is representing. Confidence limits also allow judgment of the reliability of the measurements and suitability of endpoints.

Selection of factors (Hicks 1973)

Usually, there is only one factor in plant toxicity studies, and that is the agent being tested for its effects on the test species. However, as more complex toxicity phenomena are studied, the number of factors included in the designs will increase. For some experimental designs, both controlled and uncontrollable variables will be measured for their effects on plants. Two such variables might be ambient temperature and light cycles. Experiments that characterize a particular process will have several important factors to study, including their relationships to each other as well as to the response variable.

Levels of factors (Box et al. 1978; Hicks 1973; Rosner 1990)

The levels of each factor are chosen by the investigator to reflect the experimental space of interest. For a particular agent, the levels are usually chosen to represent a range of effects of the agent on plants. For other factors, the levels may represent different environmental conditions at which the agent is present. If levels chosen are the only ones available for that factor, out of the population of factors it represents, they are called fixed levels. If the levels are a subset of possibilities from a population of factors and have variability associated with their choice, they are called random levels.

How factor levels are to be combined (Box et al. 1978; Hicks 1973)

The way in which the factors are associated determines the treatment combinations for the experiment. These treatment combinations are called cells.

There are two major ways that experimental factors relate to each other: they can be crossed, where all levels of one factor are observed at all levels of the others, or they can be nested, where levels of one factor are restricted to being within the levels of the other. Within itself, a factor may possess an ordered structure, usually a monotonic one, either ascending or descending in order. One example of ordering is the selection of geometric exposure levels for an agent.

Decisions and power — sample size (Cohen 1988; Lipsey 1990)

When making decisions in testing for statistical significance, there are two errors that must be addressed. One is the alpha (α) error, or false positive, which is the error of calling an apparent treatment effect statistically significant when, in fact, no true treatment effect exists. The other error is the beta (β) error, or false negative, which is the error of calling an apparent treatment effect NOT statistically significant when, in fact, a true treatment effect exists.

In most experiments, the α-error is usually set at 0.05, by convention only, and the β-error is set between 0.10 and 0.20, depending on the resultant sample size needed to achieve these error levels. Along with the β-error, a difference from control is determined by the investigator, which is considered an inconsequential change between two or more groups. This difference is called delta (δ) and is the amount of change between two or more groups that is desirable to detect in the test system. It is also referred to as the sensitivity of the test.

Statistical power is the ability of an experiment to detect a chosen δ that is of interest to the investigator. The factors that control the statistical power of the study are the α-error, the sample size, the desired δ, the variability in the response variable, and the statistical tests used to analyze the data. Statistical power is calculated as 1-β.

Identification of the experimental units is important when determining the sample size necessary to achieve the desired power for detecting a particular δ. The agent is applied to plants, but the application may be done either independently or dependently. If each plant has its own pot and is exposed separately to the experimental agent, then these plants are independent of each other and represent individual responses to the agent. If several plants are sowed together in pots or chambers, these plants may influence each other's response to the agent because of their proximity. These plants are considered to be dependent and do not represent individual responses to the chemical. In this case, the pot or chamber becomes the individual response to the agent, and the plants within each pot or chamber are representative of the pot or chamber environment.

The effective sample size of each group is referred to as the number of experimental units (EUs) to which the treatment is applied independently of any other EUs. If the data collected are from independent individuals, then each individual is considered an EU and the numbers of these units per treatment become the sample sizes for analysis. If the data collected are from a container, then the container is the EU and the number of containers per treatment becomes the sample size for the analysis.

When the study is complete, the investigator should calculate the power for any nonsignificant comparisons to determine the actual power of detecting the desired maximum δ for which the experiment was designed.

Randomization — the order of experimentation (Hicks 1973; Milliken and Johnson 1984, 1989)

Randomization is the technique of ordering treatment combinations so that effects of extraneous variables are spread equally over the entire set of treatment combinations. Randomization assists in distributing the variability throughout the experiment so that the effects of the chemical agent or other independent factors can be more easily recognized and quantified. Randomization also assists in keeping errors of measurement independent, which is a requirement of many statistical analyses. There are several ways to randomize an experiment, and all of them depend on the availability of physical resources to actually conduct the experiment. Simple randomization, stratified randomization, incomplete block randomization, and others are techniques that will aid in distributing the study variability equitably across all treatment combinations.

Model for the hypothesis (Hicks 1973; Sage University Papers Series 1989; Snedecor and Cochran 1980; Winer 1971)

A mathematical model may be written to describe the appropriate experimental design. This model demonstrates the relationships among the independent factors as well as their relationship to the response variables. Any randomization restrictions are also included in the model. Thus, the model should be a complete mathematical representation of the experimental design. The model will allow us to obtain the results from the experiment through statistical analysis and also will help us interpret those results.

Statistical analysis

Exploratory data analysis (Tukey 1977)

Once the experiment has been conducted according to protocol, and all identified deviations have been examined and corrected, it is time to explore and characterize the data.

Plots (Cleveland 1994)

The FIRST step is to PLOT THE DATA. Histograms or stem-and-leaf plots are unidimensional plots that show the distributional shapes in the data and the frequencies of individual values. Box plots demonstrate the grouping of the data, and normality plots detect whether or not the data are normally distributed. These diagrams allow the investigator to check for outliers and also visually check the validity of some assumptions that are necessary for several statistical analyses which may be used to analyze the variables. Scatter plots of two or more variables demonstrate the relationships among the variables so that correlation can be observed and interactions can be identified and studied.

Outliers (ASTM 1975; Barnett and Lewis 1994)

On occasion, some data points in the histogram, scatter plot, or box plot appear to be too different from the majority of points. These data are called outliers and are tested to determine if they are truly different from the distribution of the experimental data. Z or t scores, with a confidence level of 99%, are usually used for testing. If these data are different and can be attributed to an error in the execution of the study (violation of protocol, data entry error, etc.), then they can be removed from the analyses. However, if there is no apparent reason to invalidate these data outliers, then they must be kept in the data set for further statistical analyses. If desired, the analyses are conducted on two data sets: the complete one and one with the outliers removed. In this way, the outliers' influence on the analyses can be studied.

Conditions that are associated with these outliers should be examined to determine any unusual events that may have occurred in the experiment. These conditions may also indicate situations in which the experimental agent may be unusually effective on the plants and could indicate future directions for additional experimentation.

Nondetected data values (Gilbert 1987)

Results that are below a chemical analysis threshold level of method detection limit of an analytical technique used to measure an analyte are identified as nondetected. Results that occur above the detection limit, but are below the limit of quantitation, are called nonestimable or nonquantifiable. Occasionally, these two terms are used interchangeably.

Several methods can be used to analyze data with nondetected values. If the amount of nondetected data is below 25% of the entire data set, then the nondetects are replaced by one half the method detection limit, and the analysis proceeds. If the amount of nondetected data is more than 25% of the entire data set, then the proportions of nondetects to total sample size for each treatment combination are calculated, and the analysis is done on the proportions.

Data summary (Dixon and Massey 1983; Hahn and Shapiro 1967; Rosner 1990; Snedecor and Cochran 1980)

The next step is to summarize the information contained in the data by means of descriptive statistics. The most common statistics used are measures of central tendency and measures of dispersion. Central tendency measures are mean, median, mode, trimmed mean, and 50th percentile. Dispersion measures are range, standard deviation, variance, and quantiles. Other descriptive statistical calculations are maximum and minimum values, sum, and coefficient of variation.

Tests of normality and homogeneity assumptions (Snedecor and Cochran 1980)

After examining the plots, histograms, and descriptive statistics, the statistical analysis assumptions of normality and homogeneity of variances among groups are tested. Normality is tested using Kolmogorov's test or Shapiro-Wilk's test, as well as visual evaluation of histograms, box, and normality plots. Homogeneity of variances across groups is tested using Levene's test, Cochran's test, or Bartlett's test. Levene's test and Cochran's test are preferred to Bartlett's test because they are more robust to non-normality and will perform more accurately in a wider range of analyses than Bartlett's test.

Performing statistical analyses

After the exploration of the data has been completed, facts are assembled and statistical analyses are performed. These analyses and α-levels were selected when the experimental design was chosen, prior to collection of data.

1. An α-level represents the willingness of the investigator to make a false–positive mistake (i.e., rejecting the null hypothesis when it is true).
2. The experimental design allows statistical model(s) to be determined, which will be fit to the data.
3. The type of data allows selection of the appropriate statistical methods to use (i.e., continuous, count, or categorical data).
4. If the assumptions for the selected analyses are violated, then transformations of data are needed. Violation of assumptions of the particular statistical analyses can lead to very erroneous results.
    a. Homogeneity of variance is an important assumption to meet with analysis of variance (ANOVA) and multiple compari-

sons of group means. If data plots or tests of homogeneity demonstrate that variance is not homogeneous across treatments, then variance-stabilizing transformations of the data might be necessary. The arc sine, square root, and logarithmic transformations are often used on dichotomous, count, and continuous data, respectively. Logarithmic transformations can be used with count data also, if the counts vary by orders of magnitude. An alternative approach is to use nonparametric procedures, which are actually rank transformations of the data and which make no assumptions about the original data distributions. Homogeneity of variances across groups is important for categorical data also. If nonhomogeneity occurs, then the data might be transformed into a normal distribution using the arc sine transformation or some other appropriate method and reexamined. If heterogeneity still persists, then nonparametric procedures on either the actual or transformed data will provide some assistance in analyzing the data (Agresti 1990; Bishop et al. 1975; Hollander and Wolfe 1973; Winer 1971).

b. Independence is another concern of a statistician analyzing the data. Many of the techniques used for analysis require the observations to be taken independently of one another to reduce the chance that some data could have influenced other data. For many designs, data are independent, and there are no difficulties. For other designs, such as repeated measures, the data are obtained from the same subjects repeatedly, making them dependent. It is also important to determine whether or not time is a factor in the analysis or if the data are autocorrelated. That is, the data are dependent upon one another for the end results that are measured. Designs with dependent measures require the appropriate statistical techniques to handle the dependence correctly so that erroneous interpretation of the results is avoided (Snedecor and Cochran 1980; Sokal and Rohlf 1981; Steel and Torrie 1980).

5. The validity of the experiment can be checked by a thorough analysis of the control group data. Data from the control group are compared with historical data, derived from previous experience with the organisms, or with data from other experiments using specified agent standards. If the control group values depart significantly from the expected range of historical or standard values, interpretation of the treatment group results are difficult, at best, and sometimes impossible. If the control values do not meet established criteria for an acceptable assay, normally specified in the design protocol, then the study should be repeated. If the control group is too variable, then the test method needs improvement in its procedure to make it more consistent

and reproducible. Otherwise, the experiment may not be very useful in evaluating effects of the experimental agent on plants. If both solvent and dilution-water control groups are included in the test, their results can be compared by using either a t-test for count or continuous data or a 2 × 2 contingency table test for categorical data. If there is a significant difference between the two control groups, then the solvent control group should be used for the control group comparisons with treatment groups (Gilbert 1987).

Selection of statistical analyses

The appropriate statistical analyses are normally selected prior to conducting the experiment, with the interests and goals of the investigator in mind. However, occasionally, ad hoc statistical analyses are necessary in order to extract whatever information possible from an experiment that did not optimally perform, but still has benefit. One example is a balanced design that lost some cells during the running of the experiment.

Differences among means analysis (Box et al. 1978; Dixon and Massey 1983; Hollander and Wolfe 1973; Miller 1981; Milliken and Johnson 1984, 1989; Toothaker 1991)

If the data are continuous, normally distributed, and have homogeneous variance, then ANOVA with suitable multiple mean comparison tests can be used to detect differences among group means. The particular ANOVA used is determined by the experimental design (nested, crossed, multiple factors, fractional factorial, repeated measures, multivariate ANOVA). The residuals from the model fitting are examined to determine how well the model describes the data and, if there are any surprises, such as latent variables exerting their influences, nonlinear effects that need to be modeled, etc. The particular multiple mean comparison test is determined by the investigator's main interests. If all groups are to be compared, then Tukey's Honestly Significant Difference test, Scheffe's test, or others suited for data snooping are used. If only the comparison of each treatment group to the control is of interest, then Dunnett's t-test is commonly used. If there are no more than three groups, including the control, in the experiment, then multiple t-tests can be used without seriously inflating the α-level.

If the data are continuous and homogeneous in variance, but not normally distributed, either nonparametric techniques analogous to ANOVA or a suitable data transformation and parametric techniques may be used to detect differences among groups.

If the data are continuous and normal, but nonhomogeneous in variance, then a variance-stabilizing transformation along with parametric analysis methods may be used. The natural logarithm transfor-

mation is often used here. Nonparametric methods (rank transformations) may be used if the variance proves intractable to other
transformations.

If the data are continuous, but both non-normal and nonhomogeneous, suitable transformations for both assumptions may be applied
if the conditions are not too severe. If the data prove to be too intractable, then methods such as smoothing, partitioning, or redefinition of
the response variable may be needed. For example, a continuous variable can be partitioned into categories and analyzed as an incidence
table using categorical analysis.

For categorical or frequency data, contingency table analysis is
used. ANOVA and regression can be used with proper transformation
of some data sets. However, clinical observations are usually analyzed
in incidence tables using the chi-square or likelihood ratio chi-square
statistics. Residuals obtained from comparing the model predicted
results to the actual results are examined to assist in evaluation of the
model, determination of fit, identification of outliers, etc. Multiple
means comparisons tests can be done on the group proportions in a
manner analogous to that done for continuous data means by assem-
bling the proportions into suitable tables and analyzing them using the
appropriate contingency table statistics.

Concentration-response curve analysis (Draper and Smith 1981; Hosmer and Lemeshow 1989)

To determine if a trend or a concentration-response relationship exists,
the response variable data are plotted against either the actual concentration levels or the log transform concentration levels. Statistical or
mathematical models are fit to the data, and the most suitable one is
identified. A statistically significant goodness of fit of the model at α
= 0.05 indicates that a real relationship exists between the response
variable and the treatment regimen. Examination of the residuals provides insight into the goodness of fit of the model and identifies any
misfits that need attention. If the model is acceptable, it is used to
describe the trend or concentration response in the data and to calculate
quantities of interest based on the model.

For determination of a concentration point of interest with categorical data (in particular, dichotomous data), contingency table analysis,
tests for trends in proportions, the logit model, or the probit model are
used, depending on the characteristics of each data set and if the criteria
for fitting the models are met. If a suitable model cannot be found,
moving average and nonlinear interpolation are mathematical distri-
bution-free methods which can be used to determine the estimates.
Regression analysis can be used on actual or transformed data that
meet the assumptions of the analysis. Again, examination of residuals
after model fitting will aid in obtaining the best model possible for the
data.

Reliability — time to event analyses (Lee 1992; Mann et al. 1974)
Many toxicity experiments are performed to determine the agent effects on time-related occurrences, such as survival time of the EU, the duration of a specific phenomenon, or the time necessary to reach a particular phase in the lifecycle of the EU. These objectives can be satisfied with life tests, and reliability techniques are used to analyze the life test data. The data in life tests are subject to censoring (premature exit of EUs from the experiment or ending the experiment before reaching the desired endpoint). Uncensored data arises when all the EUs in the experiment reach the endpoint prior to or at the termination of the experiment. Type I censored data occurs when the experiment is terminated prior to all EUs reaching the endpoint. Type II censored data occurs when the experiment is terminated after a specific number of EUs reach the endpoint. Progressive censored data occurs when plants are removed from the experiment at regular intervals, whether or not they have reached the endpoint.

When analyzing life data, distributions of the data are determined using graphical techniques. An appropriate model is fit to the data, and the mean time to the endpoint is estimated. Consideration of how the data are censored is important here so that the estimate is not severely biased. If there are several treatment groups, the mean times or the several slopes or both can be compared.

References

Agresti A (1990) *Categorical Data Analysis*, John Wiley & Sons, New York.
ASTM Designation: E178-75 (1975) Standard Recommended Practice for Dealing with Outlying Observations.
ASTM (1976) *ASTM Manual on Presentation of Data and Control Chart Analysis*, ASTM STP 15D, American society for Testing and Materials, Philadelphia, PA.
Barnett V, Lewis F (1994) *Outliers in Statistical Data*, 3rd ed., John Wiley & Sons, New York.
Beyer W, Ed. (1968) *CRC Handbook of Tables for Probability and Statistics*, CRC Press Inc., Boca Raton, FL.
Bishop Y, Fienberg S, Holland P (1975) *Discrete Multivariate Analysis*, MIT Press, Cambridge, MA.
BMDP Manual, 1990 Edition, BMDP, Los Angeles, CA.
Box GEP, Jenkins JM (1970) *Time Series Analysis*, Holden-Day, San Francisco, CA.
Box GEP, Hunter WG, Hunter JS (1978) *Statistics for Experimenters*, John Wiley & Sons, New York.
Cleveland WS (1994) *The Elements of Graphing Data*, Hobart Press, Summit, NJ.
Cohen J (1988) *Statistical Power Analysis for the Behavioral Sciences*, Lawrence Erlbaum Associates, Publishers, Hillsdale, NJ.
Dixon JW, Massey FJ Jr. (1983) *Introduction to Statistical Analysis*, 4th ed., McGraw-Hill, New York.

Draper W, Smith H (1981) *Applied Regression Analysis*, 2nd ed., John Wiley & Sons, New York.

Dunnett CW (1955) A multiple comparisons procedure for comparing several treatments with a control. *J Am Stat Assoc* 50:1–42.

Dunnett CW (1964) New tables for multiple comparisons with a control. *Biometrics* 20:482–491.

Finney DJ (1971) *Probit Analysis*, 3rd ed., Cambridge University Press, London.

Fleiss JL (1981) *Statistical Methods for Rates and Proportions*, 2nd ed., John Wiley & Sons, New York.

Fleiss JL (1986) *The Design and Analysis of Clinical Experiments*, John Wiley & Sons, New York.

Gilbert R (1987) *Statistical Methods for Environmental Pollution Monitoring*, Professional Books Series, Van Nostrand Reinhold Company, New York.

Grant EL, Leavenworth RS (1988) *Statistical Quality Control*, 6th ed., McGraw-Hill, New York.

Hahn G, Meeker WQ (1991) *Statistical Intervals*, John Wiley & Sons, New York.

Hahn G, Shapiro SS (1967) *Statistical Models in Engineering*, John Wiley & Sons, New York.

Hicks CR (1973) *Fundamental Concepts in the Design of Experiments*, 2nd ed., Holt, Rinehart and Winston, New York.

Hollander M, Wolfe DA (1973) *Nonparametric Statistical Methods*, John Wiley & Sons, New York.

Hosmer DW, Lemeshow S (1989) *Applied Logistic Regression*, John Wiley & Sons, New York.

Langley RA (1971) *Practical Statistics Simply Explained*, 2nd ed., Dover Publications, Inc., New York.

Lee ET (1992) *Statistical Methods for Survival Data Analysis*, 2nd ed., John Wiley & Sons, New York.

Lipsey MW (1990) *Design Sensitivity*, Sage Publications, Newbury Park, CA.

Mann NR, Schafer RE, Singpurwalla ND (1974) *Methods for Statistical Analysis of Reliability and Life Data*, John Wiley & Sons, New York.

Miller RG Jr. (1981) *Simultaneous Statistical Inference*, 2nd ed., Springer-Verlag, New York.

Milliken GA, Johnson, DE (1984) *Analysis of Messy Data, Vol I: Designed Experiments*, Van Nostrand Reinhold Company, New York.

Milliken GA, Johnson, DE (1989) *Analysis of Messy Data, Vol II: Nonreplicated Experiments*, Van Nostrand Reinhold Company, New York.

Ritter M (1991) An overview of experimental design. in *Plants for Toxicity Assessment*, Gorsuch S, Lower WR, Lewis MA, and Wang W, Eds., ASTM STP 1115, American Society for Testing and Materials, Philadelphia, PA, pp. 60–67.

Rosner B (1990) *Fundamentals of Biostatistics*, 3rd ed., PWS-Kent Publishing Company, Boston, MA.

Sage University Papers Series (1989) *Quantitative Applications in the Social Sciences*, Sage Publications, Newbury Park, CA.

SAS Institute (1989) *SAS/STAT User's Guide*, Vols. 1 and 2, Version 6, SAS Institute, Cary, NC.

Shapiro, SS (1990) *How to Test Normality and Other Distributional Assumptions*, American Society for Quality Control, Milwaukee, WI.

Snedecor GW, Cochran WG (1980) *Statistical Methods,* 7th ed., Iowa State University Press, Ames, IA.

Sokal RR, Rohlf FJ (1981) *Biometry,* 2nd ed., W.H. Freeman and Co., San Francisco, CA.

Steel RGD, Torrie JH (1980) *Principles and Procedures of Statistics, A Biometrical Approach,* 2nd ed., McGraw-Hill, New York.

Taylor JK (1990) *Statistical Techniques for Data Analysis,* Lewis Publishers, Boca Raton, FL.

Toothaker LE (1991) *Multiple Comparisons for Researchers,* Sage Publications, Newbury Park, CA.

Tukey JW (1977) *Exploratory Data Analysis,* Addison-Wesley Publishing Co., Reading, MA.

Williams DA (1971) A test for differences between treatment means when several dose levels are compared with a zero dose control. *Biometrics* 27:103–117.

Williams DA (1972) The comparison of several dose levels with a zero dose control. *Biometrics* 28:519–531.

Williams DA (1986) A note on Shirley's non-parametric test for comparing several dose levels with a zero dose control. *Biometrics* 42:183–186.

Winer BJ (1971) *Statistical Principles in Experimental Design,* 2nd ed., McGraw-Hill, New York.

chapter six

Water quality and aquatic plants

Michael A. Lewis and Wuncheng Wang

Introduction

Aquatic plants include algae; periphyton; phytoplankton; and various macrophytic emergent, submergent, and floating leaved macrophytic species. These plants are significant to the ecology of estuaries, lakes, and rivers, and their importance to the upper trophic biota such as fish has been documented thoroughly (Kilgore et al. 1993). Plants form the base of most aquatic food chains, produce oxygen, serve as refuge for

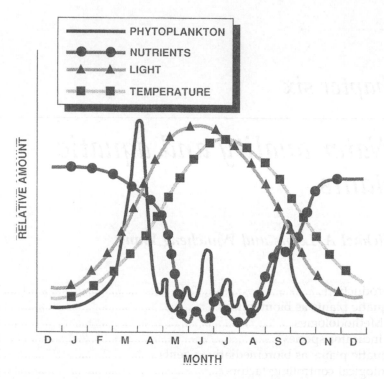

Figure 1 A generalized seasonal cycle for north temperate marine phytoplankton as related to nutrients, light, and temperature. (Adapted from Raymont JE [1963] *Plankton and Productivity in the Oceans,* Pergamon Press, Elmsford, NY.)

aquatic life, and are important in nutrient cycling. Plants interact with their environment through processes that include chemical bioconcentration and excretion, shading, and organic matter production and decomposition (Honnell et al. 1993; Carpenter and Lodge 1986). As a result of these interactions, aquatic plants may significantly affect water and sediment quality (Chambers and Prepas 1994).

Most aquatic plants have seasonal growth cycles which are controlled by a variety of ecological factors (Figure 1). The structure and function of aquatic plant communities are influenced directly or indirectly by changes in water quality resulting from the presence of nutrients and phytotoxicants. The input of nutrients and contaminants by anthropogenic activities can inhibit or stimulate plant growth and may cause ecosystem change. A good example of this is the decline in submersed aquatic vegetation in Chesapeake Bay due to urbanization in the watershed (Orth 1994). Excessive plant growths can have economic consequences because they often degrade recreational uses and the quality of drinking water (Lembi et al. 1990). For example, approximately 12,000 lakes in 38 states and 61% of drinking water supplies in the United States and Canada were adversely affected in 1983 by excessive growths of algae and macrophytes (American Water Works Association 1987; Welch and Lindell 1992). This impacted 38 million people.

Plants have been used frequently as biomonitors where the composition of the plant community has been related to water quality (Seddon 1972; Pip 1979; Hellquist 1980; Kadono 1982). Plants also have been used as bioremediative agents in water quality management, i.e., constructed wetlands. This chapter will discuss the interactions of aquatic plants and water quality in the context of their use as biomonitors and bioremediative agents.

Aquatic plants as biomonitors

Methodologies

Aquatic plants such as periphyton and phytoplankton respond rapidly to water quality changes. Therefore, considerable information concerning the environmental condition of an aquatic habitat can be obtained from their analysis. Patrick (1973), Collins and Weber (1978), Round (1981), Shubert (1984), Stevenson and Lowe (1986), and Sprecher and Netherland (1995) have reviewed the many field assessment techniques used to determine plant community structure and function as related to water quality.

A variety of physiological, morphological, and community parameters have been used to monitor the "health" of an ecosystem. Phytoplankton and periphyton can be collected in water samplers, nets, and on artificial substrates contained in a periphytometer and analyzed for pigment content, DNA content, ATP, and photosynthetic activity. Colonized algae after preservation can be analyzed also for species diversity using one of many available taxonomic keys. Several diversity indices can be calculated based on these taxonomic data, such as the Shannon diversity index (Shannon and Weaver 1949), Simpson's index (Simpson 1949), and the sequential comparison index (SCI) (Cairns et al. 1968). The SCI index is relatively simple; is based on differentiation by size, shape, and color; and requires only limited taxonomic expertise.

An excellent review of diversity indices and their use has been provided by Washington (1984), and the sensitivities of various indices have been compared (Brock 1977; Perkins 1983; Boyle et al. 1990). Numerical indices based on the presence and absence of algal were first presented by Knöpp (1954) and later, among others, by Dresscher and Mark (1976). Patrick (1950, 1957) was one of the first to recognize the usefulness of diversity as an indicator of ecological conditions.

Standing crop or biomass is a common measurement of the plant community, and a variety of indirect and direct methods have been used. These include ash-free dry weight, cell abundance, pigment content, and biovolume. Methods for these analyses are described in Weber (1973), Stephenson and Lowe (1986), the American Public Health Association et al. (1989), and the American Society for Testing and Materials (1992). A variety of numerical indices have been reported which are based on plant biomass such as the viability index (Clark et al. 1979). The autotrophic index (the ratio of biomass to chlorophyll *a*) has been used as an indicator of organic pollution (Weber 1973); low values indicate good water quality, and high values indicate polluted conditions.

Indicator species

Many attempts have been made to describe specific aquatic plants, primarily algae, that are indicative of specific water quality conditions. Kolkwitz and Marsson (1908) and Fjerdingstad (1950, 1963, 1964, 1965) were among the first to use presence and absence of algal species as an indicator of specific water quality zones in rivers. These zones (oligosaprobic, polysaprobic, and mesosaprobic) were based on the extent of organic pollution and the types of algae present. A generalized example of the water quality and floristic changes indicative of these river zones is shown in Figure 2. Diatoms in the unpolluted zone (oligosaprobic) are diverse, and filamentous green algae are present; however, as a result of the presence of organic pollution, the diversity and biomass of these algal groups decreases in the polysaprobic zone. In the mesosaprobic zone (polluted recovery zone), water quality improves; recovery begins; and diatoms such as *Gomphonema*, bluegreen species (*Oscillatoria* spp., *Phormidium* spp.), and filamentous green species (*Cladophora* spp., *Ulothrix* spp.) are present.

Many species indicator lists that describe the relationship and sensitivity of algal species to water pollution and nutrient enrichment have been published. Bratkovitch (1988) lists several species used for this purpose in marine environments, and Palmer (1962) published an illustrated guide that describes the marine and freshwater algae associated with taste and odor problems, filter clogging, and polluted and nonpolluted water. Several of the species representative of polluted and nonpolluted conditions appear in Table 1.

The "accuracy" of reported indicator species such as those in Table 1 is limited. It is very difficult to restrict a genus and species to a specific "pollution" zone. This has become very apparent in recent years due to the increase in scientific knowledge describing algal sensitivities to various types of toxicants and nutrients. This database is briefly discussed later in this chapter. Algal genera may include many species that are both tolerant and intolerant of the same type of pollution (Bringman and Kühn 1980). This also applies to different algal strains and varieties (Blanck et al. 1984). Therefore, it is difficult to conclude

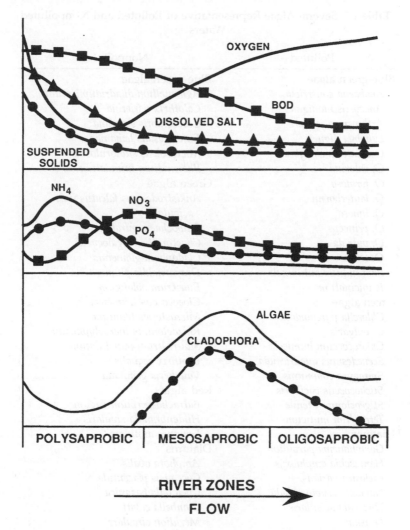

Figure 2   Water quality and change in the relative abundance of algae in a receiving water at various distances below an effluent discharge point. Effects are related to river zones as designated by Kolkwitz and Marsson (1908). (Adapted from Hynes HBN [1960] *The Ecology of Polluted Waters*, Liverpool University Press, Liverpool; Abel PD [1989] *Water Pollution Biology*, Ellis Horwood, Limited, Chichester, U.K.)

that the presence or absence of a single species in a specific habitat is the result of a single chemical factor. In contrast, the composition of plant communities has proven more useful in assessing water quality, and examples of the ability of community analysis to indicate impact have been presented by Whitton (1979), Haslam (1982), Levine (1984), and Sortkjaer (1984).

Table 1　Several Algae Representative of Polluted and Nonpolluted
Waters

Polluted	Nonpolluted
Blue-green algae	Blue-green algae
Anabaena constricta	*Agmenellum quadriduplicatum*
Anacystis montana	*Calothrix parietina*
Arthrospira jenneri	*Coccochloris stagnina*
Lyngbya digueti	*Entophysalis lemaniae*
Oscillatoria chalybea	*Microcoleus subtorulosus*
O. chlorina	*Phormidium inundatum*
O. formosa	Green algae
O. lauterbornii	*Ankistrodesmus falcatus* var.
O. limosa	*acicularis*
O. princeps	*Bulbochaete mirabilis*
O. putrida	*Chaetopeltis megalocystis*
O. tenuis	*Cladophora glomerata*
Phormidium autumnale	*Draparnaldia plumosa*
P. uncinatum	*Euastrum oblongum*
Green algae	*Gloeococcus schroeteri*
Chlorella pyrenoidosa	*Micrasterias truncata*
C. vulgaris	*Rhizoclonium hieroglyphicum*
Chlorococcum humicola	*Staurastrum punctulatum*
Scenedesmus quadricauda	*Ulothrix aequalis*
Spirogyra communis	*Vaucheria geminata*
Stichococcus bacillaris	Red algae
Stigeoclonium tenue	*Batrachospermum vagum*
Tetraedron muticum	*Hildenbrandia rivularis*
Diatoms	*Lemanea annulata*
Gomphonema parvulum	Diatoms
Hantzschia amphioxys	*Amphora ovalis*
Melosira varians	*Cocconeis placentula*
Navicula cryptocephala	*Cyclotella bodanica*
Nitzschia acicularis	*Cymbella cesati*
N. palae	*Meridion circulare*
Surirella ovata	*Navicula exigua* var. *capitata*
Flagellates	*N. gracilis*
Carteria multifilis	*Nitzschia linearis*
Chlamydomonas reinhardi	*Pinnularia nobilis*
Chlorogonium euchlorum	*P. subcapitata*
Cryptoglena pigra	*Surirella splendida*
Euglena agilis	*Synedra acus* var. *angustissima*
E. deses	Flagellates
E. gracilis	*Chromulina rosanoffi*
E. oxyuris	*Chroomonas nordstetii*
E. polymorpha	*C. setoniensis*
E. viridis	*Chrysococcus major*
Lepocinclis ovum	*C. ovalis*
L. texta	*C. rufescens*

Table 1 (continued)

Polluted	Nonpolluted
Pandorina morum	Dinobryon stipitatum
Phacus pyrum	Euglena ehrenbergii
Pyrobotrys gracilis	E. spirogrya
P. stellata	Mallomonas caudata
Spondylomorum quaternarium	Phacotus lenticularis
	Phacus longicauda
	Rhodomonas lacustris

Data from Palmer CM (1962) *Algae in Water Supplies*, U.S. Public Health Service, Public Health Service Publication No. 657, U.S. Government Printing Office, Washington, D.C.

Aquatic plants as bioremediative agents

Some algae and macrophytes accumulate high concentrations of potential contaminants such as metals in their roots, stems, and other tissues (Whitton 1984). This ability has led to their common use in traditional wastewater treatment systems such as trickling filters and sewage stabilization ponds to improve the quality of municipal and industrial effluents prior to discharge to receiving waters. Several plant species have been shown to reduce metals, phenols, biochemical oxygen demand (BOD), suspended solids, total alkalinity, nutrients, acidity, chemical oxygen demand (COD), hardness, and coliform bacteria in wastewater. Reductions up to 99% in several of these parameters have been reported, while dissolved oxygen may increase as much as 70% (Tripathi and Shukla 1991). The removal rates of contaminants by specific species have been reported. For example, Wolverton and McKnown (1976) reported that 1 ha of the water hyacinth *Eichhornia crassipes* was capable of removing 160 kg of phenol every 72 h.

In addition to traditional wastewater treatment processes, natural and constructed wetlands and their associated flora are being used increasingly to improve the water quality of wastewaters and nonpoint source runoff. Natural and constructed wetlands have been used to remove suspended solids, nutrients, heavy metals, toxic organics, and bacteria from municipal and industrial wastewater, landfill and agricultural runoff, and urban stormwater (for examples, see Hadden 1994, Lekven et al. 1994, and Olson and Preston 1990). There are approximately 300 constructed wetland treatment systems in North America and 500 in Great Britain that treat stormwater and municipal effluents (Knight 1994). Approximately 140 constructed wetlands are used to treat acid-mine drainage (Wieder 1989).

Wetlands are a very useful form of treatment, and considerable literature is available describing general principles, case histories, design criteria, and appropriate plant species needed for a successful system and their removal efficiencies (Knight 1994; Moshiri 1993; Adamus et al. 1991; Hammer 1989). The more common species used in North American wetland treatment systems are the cattails (*Typha* spp.) and bulrushes (*Scripus* spp.), although the duckweeds (*Lemna* spp.), reeds (*Sparangivm* spp.), and pickerel weed (*Pontederia* spp.) are also used frequently. In addition to these species, others have been tested for use in constructed wetlands (Table 2). Guntenspergen et al. (1989) and Johnston (1993) present excellent overviews of the effects of plants on sediments and their effect on contaminant bioavailability. Dortch (1992) has provided an analysis of the functional ability of wetlands to improve water quality and reported the removal efficiencies for several types of wastes (Table 3). Water quality standards and biocriteria for wetlands have been discussed by the U.S. Environmental Protection Agency (U.S. EPA 1990) and Nelson (1990).

Wetland plants improve water quality in a variety of ways, including binding soil and reducing the resuspension of muds. Furthermore, they increase the deposition of sediment by burial, decreasing the water

Table 2   Some Freshwater Plants that
Have Been Either Considered or Used in
Constructed Wetlands

> *Azolla caroliniana*
> *Ceratophyllum demersum*
> *Colocasia esculenta*
> *Canna flaccida*
> *Elodea nuttalli*
> *Eichhornia crassipes*
> *Eleocharis dulcis*
> *Egeria densa*
> *Hibiscus* spp.
> *Juncus effusus*
> *Leersia oryzoides*
> *Lemna* spp.
> *Phragmites australis*
> *Sagittaria rigida*
> *S. latifolia*
> *Scirpus acutus*
> *S. americanus*
> *Typha latifolia*
> *Vallisneria americana*
> *Myriophyllum aquaticum*
> *Potamogeton pectinatus*
> *Nuphar luteum*

Note: Only a few of the many species in Guntenspergen et al. (1989) and Mitsch and Gosselink (1993) are presented.

Table 3 The Removal Efficiencies for Selected Constituents in Wastewater by Constructed Wetlands and Their Associated Fauna and Flora

Source	Total suspended solids	BOD	Total nitrogen	Total phosphorus	Coliform bacteria	Metals	Toxic organics
Municipal effluent	55	80	25	12	80	0	ND
	72	90	70	40	95	50	
	94	95	93	57	99	99	
Stormwater runoff	43	40	36	43	ND	5	ND
	63	—	—	53		50—70	
	95	80	70	85		95	
Industrial effluent	42	29	69	30	ND	6	69
	—	—	—	—		70—85	95
	100	100	98	73		97	100
Agricultural runoff	54	75	ND	66	62	ND	ND
	—	—		—			
	82	97		81			
Mining waste	71	ND	ND	ND	ND	34	ND
	—					75—90	
	95					99	

Note: Values are reported as minimum, typical, and maximum. ND = No data.

Adapted from Dortch MS (1992) Literature Analyses of the Functional Ability of Wetlands to Improve Water Quality, Wetlands Research Program Bulletin Vol. 2, No. 4, U.S. Army Corps of Engineers, Vicksburg, MS.

velocity, and by facilitating biodegradation. The degradation of potential toxicants is enhanced by algae and microbes which are present in high densities on vegetation substrates.

Bioaccumulation by wetland plants can remove substantial quantities of potential toxicants and nutrients of water entering and passing through wetlands. The removal rates are variable and are affected by many physical and chemical factors, including shading, turbidity and sediment oxygen, and organic content. Accumulation and uptake are generally greater in roots than in leaves and stems. Typically, wetland plants remove nutrients from the water and sediment during the growing season and release them when light and temperature will not support algal growth. Free-floating plants remove nutrients from the water column, whereas rooted, submersed species remove nutrients primarily from the sediment. Nutrients in plant tissue usually are 10 to 100 times greater than those in surrounding water, but can be as much as 10,000 times greater (Peverly 1985; Nixon and Lee 1986). As much as 91% of the phosphorus and 94% of the nitrate nitrogen has been found to be removed in freshwater wetlands (Mitsch et al. 1977), some of which is due to bacterial denitrification. Despite these high removal rates, all wetlands should not be considered nutrient sinks (Mitsch and Gosselink 1993).

Ecological controlling factors

A variety of ecological factors influence and control the seasonal distribution and composition of natural floral communities. A variety of research has been conducted that has correlated the presence of plants with water quality (Pip 1979; Hellquist 1980). Aquatic plants are influenced by salinity, tidal fluctuations, sedimentation, light intensity, temperature, pH, hardness, and dissolved oxygen. Fewer studies have determined the effect of plants on the water environment. Plants interact with their environment by chemical uptake and excretion, physical processes such as shading and reduced circulation, organic matter production, and decomposition (Carpenter and Lodge 1986). In addition, anthropogenic activities, e.g., point and nonpoint source contaminants, can affect these controlling factors and aquatic plants by the addition of sediment, nutrients, and phytotoxicants. As a result of these contaminants, significant changes in freshwater and marine plant communities have occurred (Orth 1994). These impacts and the relationship of several controlling factors to plant community dynamics are discussed next.

Dissolved oxygen

Aquatic plants produce oxygen which is critical to the survival of most forms of aquatic life, particularly those inhabiting greater depths where mixing and diffusion of atmospheric oxygen is limited. Below the photic zone there is a net consumption of oxygen, and replenishment is accomplished by downward circulation. There are considerable diurnal fluctuations in the dissolved oxygen concentration which are due to plant photosynthesis and metabolism. During the day, light intensity, temperature, and photosynthesis attain maximum levels. Photosynthesis exceeds respiration and dissolved oxygen levels increase (light energy$+CO_2+H_2O\rightarrow(CH_2O)+O_2$). When the light intensity falls below the compensation level during the night and on cloudy days, respiration exceeds photosynthesis and dissolved oxygen concentrations decrease ($[CH_2O]+O_2\rightarrow CO_2+H_2O$). Other conditions that reduce oxygen concentrations are the decay of excessive growths of aquatic vegetation and the coverage of a waterbody with floating plants which reduces light penetration and photosynthesis. These conditions result from nutrient enrichment which is often a result of discharges of poorly treated municipal effluent or agricultural and stormwater (urban) run-off.

Temperature

The overall reported temperature range for survival of freshwater and marine algae is −40 to 75°C (Langford 1990). Aquatic plants, however, have species-specific minimum, optimum, and maximum temperature requirements for growth. For example, optimal temperature ranges for several major freshwater algal taxa are 15 to 25°C for diatoms, 25 to 35°C for green algae, and 30 to 40°C for blue-green algae (Hawkes 1969).

Water temperature is increased above seasonal ambient levels, primarily as the result of the addition of effluents from electric power-generating facilities. Langford (1990) provides an excellent historical perspective of the effects of these discharges on aquatic life. Typically, thermal effluents are discharged after a cooling process to lakes, rivers, and estuaries. The cooling process not only uses a significant amount of water, but also adds biocides to the receiving water. It has been estimated that in Britain 28% of the rainfall and 50% of the river flow have been used for cooling thermal effluent (Howells 1983). In many cases, no significant change in the plant community in rivers receiving thermal effluents has been seen. This is due to the usual rapid dilution in the receiving water and the rapid regeneration rate of algae. This minimal impact in receiving waters was observed in several cases reported in Langford (1990), where the thermal discharge represented 4 to 15% of the total flow of the receiving waters. In other studies, temperature increases up to 8°C were found to have no major effect on planktonic algae in European rivers (Appourchaux 1952; Swale 1964).

In contrast, some adverse effects of thermal effluents have been reported. Typically, an increase in productivity in the cooler months and a decrease in the warmer months is the most commonly observed effect, but decreases in species numbers and diversity also have been reported. Coutant (1962) showed that blue-green species increased while other types of algae decreased in the Delaware River below a power plant discharge. In addition, algal biomass and chlorophyll content were several times greater in the river areas affected by the thermal effluent.

The effects of thermal effluents in lakes, as in rivers, usually occur only in the vicinity of the discharge because of rapid dilution. Although it is difficult to generalize, Langford (1990) reported that algal photosynthesis was disrupted in lake water at temperatures of 37°C or greater and in river water at 38°C or greater — not much of a difference.

Blue-green algal species are generally more heat tolerant than other species. Patrick et al. (1969) reported a reduction in diatoms when the water temperature was between 35 and 40°C. Diatoms in this study were dominant at temperatures of 20 to 28°C, but at 30 to 35°C other algae were more numerous. This sequence is typical of that observed seasonally in temperate lakes where the algal community goes from one dominated by diatoms to one dominated by green and blue-green algae as the water temperature increases. Limited information is available concerning the thermal tolerance of macrophytes, although *Vallisneria spiralis* (wild celery) is found commonly in waters receiving thermal effluents (Hynes 1960).

pH

The effect of aquatic plants on pH is dependent upon the buffering capacity of the water and plant productivity. Algae remove carbon dioxide from the water column as a result of photosynthesis. This process shifts the equilibrium between the carbonic acid and less soluble bicarbonate and monocarbonates and consumes hydrogen ions. These changes affect the total hardness and pH of the water. Algal growths have been known to reduce hardness by up to one third of ambient levels. During maximum photosynthesis, when carbon dioxide is removed, the pH increases. When respiration exceeds photosynthesis, carbonic acid increases and the pH decreases.

The majority of aquatic plants are best suited for a pH of approximately 7.0, but some grow rapidly in waters having other pH values. Several examples of these species and their pH tolerances are in Table 4, and additional information can be found in Palmer (1962). Several acid-tolerant species can survive in waters with pH values between 2.0 and 3.0, which usually result from either the effects of acid rain or mining activities.

The effects of acid rain on aquatic plants have been studied extensively and reviewed by, among others, Adriano and Johnson (1989), Farmer (1990), and Howells (1992). Acid rain and acid-mine drainage

Table 4 Lowest pH Values at Which Several Freshwater Algae Have Been Collected

Green algae	pH	Diatoms	pH	Blue-green algae	pH
Chlamydomonas aplanata	<3	*Navicula nivalis*	3.0	*Anabaena* sp.	4.5
Ulothrix zonata	2.4–7.0	*Tabellaria* sp.	3.7–7.2	*Lyngbya* sp.	4.9
Mougeotia spp.	2.4–6.2	*Pinnularia* spp.	1.5–3.0	*Spirulina nordstedtii*	4.9
Stigeoclonium spp.	4.0–6.4	*Nitzschia palea*	2.7	*Oscillatoria* spp.	2.8
Desmidium spp.	1.8–3.2	*N. ovalis*	2.5	*Chroococcus rufescens*	4.5
Zygogonium spp.	2.1	*Eunotia exigua*	2.5–3.0		
Microthamnion spp.	3.0	*Synedra rumpens*	<3.0		
Penium jenneri	<3.0	*Fragilaria* sp.	2.8		
Scenedesmus abundans	3.3	*Melosira distans*	3.1		
Ankistrodesmus falcatus	3.4	*Cyclotella glomerata*	2.8		
Spirogyra sp.	3.3	*Gomphonema* spp.	3.1		
Stichococcus sp.	0.9–2.5	*Achnanthes* spp.	<4.1		

Data from Kelly M (1988) *Mining and Its Freshwater Environment*, Elsevier Applied Science, New York.

reduce the buffering capacity and the hardness of water, cause the alteration of phosphorus availability, change the form and quantity reference of inorganic carbon, and solubilize some heavy metals. The effects of these chemical changes can result in damage to plant morphology and physiology, such as leaching of mineral substances from plant tissues. More specifically, low pH can result in a loss of magnesium ions from the chlorophyll molecule, which results in its conversion to phaeophytin and other inactive photosynthetic components.

Changes in plant community composition have been reported on numerous occasions in streams recently acidified and having pH values less than 6. However, community biomass is not always reduced because of the increase in acid-tolerant plant species (Stokes et al. 1989). When the pH decreases below 5.0 in streams, an increase in algal biomass and primary productivity often has been observed. These changes have been attributable to reduced grazing pressure by macroinvertebrates, reduced microbial decomposition, increased micronutrient availability, and the low pH preference of some natural periphyton (Elwood and Mulholland 1989). Green algae often comprise a greater proportion of the algal community in acid streams.

The relative abundance of diatoms and blue-green species decreases as the pH decreases in lakes (Stokes et al. 1989). These algal species are more sensitive to pH changes than dinoflagellates. An increase in green filamentous forms is often observed for lake periphyton in low pH waters.

Farmer (1990) summarized the changes in macrophyte distribution associated with acidification and the liming of various acidified lakes. Some of the species changes are summarized in Table 5. As for phytoplankton and periphyton, total macrophyte biomass does not usually decrease, but species diversity does in low pH waters. Generally, calcicole species are replaced by more tolerant species such as *Eriocaulon* spp. (pipewort), Sphagnum moss, and *Juncus* spp. (rush).

Nutrients

Aquatic plants obtain their nutrients from the sediment and water. Phosphorus (P) and nitrogen (N) are the nutrients that usually limit plant growth in the environment. Phosphorus is important in the reproductive growth of plants and in the formation of ATP. Phosphorus usually occurs in the oxidized state as either inorganic orthophosphate ions or in organic compounds. Phosphorus is usually greater in the sediment than in the water column, and, as plants remove soluble phosphorus from the water column, sediment-bound phosphorus is released. The equilibrium of phosphorus concentrations between the sediment and water is controlled by several factors such as pH and dissolved oxygen. For example, low pH results in a release of phosphorus from the sediment to the water. Under anaerobic conditions,

Table 5 Changes in Plant Species Due to Acidification[a]

	Species loss (L) or decrease (D)	Species gain (G) or increase (I)
Canada (14 lakes)	*Lobelia dortmanna* (L)	*Leptodictium riparium* (I)
	Isoetes riparia (L)	*Eleocharis acicularis* (I)
	Myriophyllum tenellum (L)	
	Nuphar spp. (L)	
	Utricularia vulgaris (L)	
	Potamogeton epihydris L)	
Sweden (6 lakes)	*Lobelia dortmanna* (D)	*Sphagnum* (I)
	L. uniflora (D)	*Juncus bulbosus* (I)
	Isoetes lacustris (D)	
	I. echinospora (D)	
Sweden (Lake Gardsjon)		*Sphagnum* (I)
Sweden (2 lakes)	*L. dortmanna* (D)	*Sphagnum* (I)
	I. lacustris (D)	
Sweden (Lake Anketjarn)	*L. dortmanna* (D)	*Sphagnum* (I)
		J. bulbosus (I)
Norway (series of lakes)	*L. dortmanna* (D)	*Sphagnum* (I)
	Isoetes spp. (D)	
The Netherlands (41 lakes)	*L. uniflora* (L)	*Sphagnum* (I)
	Eleocharis acicularis (D)	*J. bulbosus* (I)
	Luronium natans (D)	
The Netherlands (147 lakes)	*Lobelia dortmanna* (L)	*S. denticulatum* (I)
	I. lacustris (L)	*S. cuspidatum* (I)
	I. echinospora (L)	
	L. uniflora (L)	
The Netherlands (2 lakes)	*L. uniflora*	*J. bulbosus*
	L. dortmanna	*Sphagnum* spp.
	P. natans	
	Sparganium spp.	
Scotland (9 lochs)	*I. lacustris* (L)	*Sphagnum* spp. (G)
	Juncus bulbosus (L)	
	P. pusillus (L)	
	P. lucens (L)	
United States (Lake Colden, NY)	*L. dortmanna* (D)	*Eriocaulon septangulare* (I)
	E. acicularis (D)	*Utricularia* spp. (I)
	N. luteum (D)	*S. pylaesii*

[a] Based on information reported by Farmer (1990) for a variety of lakes.

more phosphorus is available than under aerobic conditions. Important sources of phosphorus are nonpoint source runoff from agricultural lands, household detergents, and metabolic animal wastes. The increase in phosphorus levels in water from anthropogenic sources has been considered the single most important factor changing the balance of productivity in fresh water (Round 1981).

Nitrogen is important in vegetative growth and is an integral part of amino acids. Nitrogen becomes limiting in marine waters before phosphorus, but in fresh water the converse is true (Farnworth et al. 1979). There are several forms of nitrogen. These include nitrate, nitrite, and ammonium ions, as well as dissolved organic nitrogenous compounds such as urea. Nitrogen from the atmosphere is transformed by bacteria and blue-green algae into the forms that can be utilized by plants, NO_3 and NH_4. Nitrogen compounds are found typically in greater concentrations in the water column than in sediment. Nitrogen concentrations fluctuate more than phosphorus levels in the water column.

Plant growth is limited primarily by temperature, light, and the availability of nutrients. Nutrients from urban, agricultural, municipal, and industrial sources can cause excessive algal and macrophytic plant growth and reduced light penetration. The increased growth can be beneficial in oligotrophic waters, where primary productivity is nutrient limited, but not in eutrophic and mesotrophic waters, where increased growth can lead to increased respiration.

The most noticeable effect of excessive concentrations of nutrients is an acceleration in the natural eutrophication process in lakes and slow-moving streams. This effect has been documented and widely discussed (National Academy of Sciences 1969). Concern about eutrophication as a result of municipal effluents goes back to the 1940s, and the visibility of this problem increased in the 1960s due to evidence of the problem in the Great Lakes. One result of eutrophication is a reduction in algal diversity and a change in community composition, such as the dominance of green and blue-green species (Figure 3). In severe cases, there is a dramatic increase in growth of some species, such as blue-green algae, which often results in algal blooms. In lotic environments, excessive nutrients cause the periphyton to form slick coatings on the substrate. Rooted macrophytes are influenced more by nutrients in sediment than by those in the water; consequently, growth is greatest in nutrient-enriched sediments.

Accelerated eutrophication has become a major problem, and several methods have been developed to treat it. The restoration of affected water bodies has occurred since the 1970s. Techniques that have been used to reduce accelerated eutrophication include advanced wastewater treatment to remove nutrients, in-lake control measures such as hypolimnetic aeration, and dredging for sediment removal (Welch and Lindell 1992).

Algal blooms

Algal blooms can lead to water quality degradation, a buildup of sediment nutrient concentrations, loss of oxygen, and changes in food webs. In addition, there are several types of freshwater algal blooms

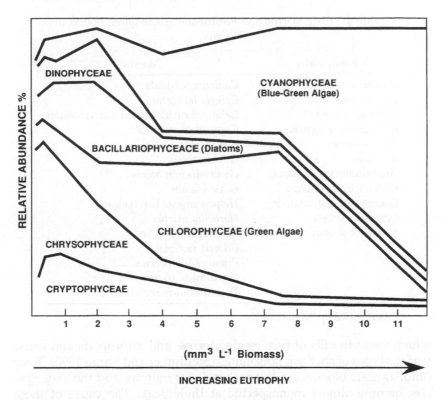

Figure 3 Relative abundance of phytoplankton in a lake as related to fertility. (Adapted from Skulberg [1980].)

which are comprised of algal species that sometimes are toxic to the flora and fauna (Table 6). Blue-green algal blooms are comprised typically of large filamentous species of *Anabaena* spp. and *Microcystis* spp., which are the most objectionable and undesirable since the toxins they produce adversely affect aquatic life and wildlife. Blue-green algal blooms occur worldwide, and their deleterious effects have been documented since 1878 (Francis 1878). The neurotoxins and/or hepatotoxins produced during bloom conditions have resulted in fatal poisonings of livestock, including horses, cattle, pigs, dogs, rabbits, and poultry. Neurotoxins are produced by strains of *Anabaena flos-aquae* and *Aphanizomenon flos-aquae*. Hepatotoxins are produced by, among others, *Microcystis aeuroginosa, Nodullaria spumigens,* and various *Oscillatoria* species. In some cases, humans have been adversely affected after contact with fresh water containing blue-green algal blooms. The hazards of freshwater blue-green algal blooms are discussed in greater detail by Gorham and Carmichael (1990).

Algal blooms also can be caused by diatoms and dinoflagellates, such as species of *Cyclotella, Synedra, Melosira, Fragilaria,* and *Stephanodiscus*. Marine dinoflagellates are best known for causing red tides

Table 6   Toxic Marine and Freshwater Algae Often Found in
Blooms[a]

Fresh water	Marine
Anabaena circinalis	*Caulerpa serrulata*
A. flos-aquae	*Egregia laevigata*
A. lemmermanni	*Gelidium cartilagineum* var. *robustum*
Microcystis aeruginosa	*Gonyaulax catenella*
M. flos-aquae	*G. polyedra*
M. toxica	*G. tamarensis*
Aphanizomenon flos-aquae	*Gymnodinium brevis*
Gloeotrichia echinulata	*G. veneficum*
Gomphosphaeria lacustris	*Hesperophycus harveyanus*
Lyngbya contorta	*Hornellia marina*
Nodularia spumigena	*Macrocystis pyrifera*
	Pelvetia fastigiata
	Prymnesium parvum
	Pyrodinium phoneus
	Trichodesmium erythraeum

[a] As reported by Palmer (1962).

which result in kills of fish, eichinoderms, and arthropods and cause
various types of shellfish poisonings (Steidinger and Vargo 1990). Toxic
dinoflagellate blooms can last for weeks or months, and the toxic species become almost monospecific at their peak. The cause of these
blooms have been investigated, but a single cause has not been identified.

Algal blooms die off eventually, decompose, and release nutrients.
Bloom subsidence has been attributed to grazing, cell death, parasitism,
advection, and shifts in the lifecycle. When a diatom bloom subsides,
the algal cells sink into the hypolimnion where the nutrients are
unavailable until the fall turnover. When blue-green algae die, the
release of nutrients is almost immediate, and the nutrients are rapidly
utilized by other algae and bacteria.

Algae also release organic compounds other than toxins into the
environment that can affect ecosystem function, community composition, and succession. These may be waste products and various internal
cellular components (Inderjit and Dakshini 1994; Duffy and Hay 1994).
Algae use the secretion of these compounds to gain dominance in lakes
by affecting other organisms through inhibition and stimulation of
growth, serving as a vitamin, food source, or chelator.

Management techniques

Algal blooms and excessive growths of macrophytes are undesirable
for many reasons and are discussed by Lembi et al. (1990) and the
Aquatic Plant Management Society (1993). Approximately 20 manage-

ment techniques have been utilized to eliminate nuisance plant growth (American Water Works Association 1987; Welch 1992). Mechanical equipment has been used to remove macrophytes, but this technique is inefficient, time consuming, and ineffective in controlling algae and duckweed such as *Lemna* spp. Habitat manipulation such as the use of shading to reduce plant growth has proven successful. Drawdown or periodic lowering of the water levels in reservoirs and rivers below dams can desiccate shoreline plants or expose them to freezing. Dredging also has been used to remove rooted plants and nutrient-enriched sediments. Several biological controls have been used: plant-eating fish such as grass carp (*Ctenopharyngodon idella*) and tilapia and some insects such as flea beetles feed on nuisance plants. Competitive plant species have been used for control purposes, but only with limited success. Finally, chemical control is the most successful and widely used management method. Many of the herbicides used, however, are toxic to some nontarget plant species as well as to invertebrates and fish (Nimmo 1985). This serious matter has received considerable regulatory and scientific attention which will be discussed later.

Phytotoxicants

A variety of chemicals enter either continually or intermittently into freshwater and estuarine waters as a result of atmospheric, point, and nonpoint sources. Many of these chemicals can be phytotoxic and/or phytostimulatory. Some are under regulatory control, such as agricultural pesticides (Federal Insecticide Fungicide and Rodenticide Act, FIFRA), nonpesticide chemicals (Toxic Substances Control Act, TSCA), and municipal and industrial effluents (National Pollutant Discharge Elimination System). These regulations require assessment of the environmental risk of potential toxicants to aquatic life prior to their use and disposal. Most environmental risk assessments formulated by the scientific and regulatory communities are based on toxicity data for animal test species, which in many cases have been assumed to be a suitable surrogate species for aquatic plants. This assumption, however, is not supported by the scientific literature (Lewis 1994). Nevertheless, phytotoxicity has not been the focal point of most environmental assessments for commercial chemicals. This trend is changing, however, and use of plants to assess the potential environmental effects of toxicants has increased since the 1970s when standardized algal toxicity tests were first published (U.S. EPA 1971, 1974).

The current standard procedure for determining the toxicities of pollutants on aquatic plants is the use of short-term (96 h) laboratory toxicity tests. A variety of test procedures have been described (Fletcher 1991; Lewis 1994) which use cultured freshwater green algae such as *Selenastrum capricornutum* (ASTM 1993a) and floating duckweed *Lemna minor* (ASTM 1993b). Tests have been conducted more frequently with

Table 7   Marine and Freshwater Algae and Macrophytes
Used in Single-Species Phytotoxicity Tests

Algae	Macrophytes
Fresh water	Fresh water
Selenastrum capricornutum	*Lemna minor*
Scenedesmus subspicatus	*L. gibba*
S. quadricauda	*L. perpusilla*
Chlorella vulgaris	*Myriophyllum spicatum*
Microcystis aeruginosa	*Ceratophyllum demersum*
Navicula seminulum	*Hydrilla* spp.
N. pelliculosa	*Potamogeton* spp.
Marine	Marine
Thalassiosira pseudonana	*Spartina alterniflora*
Skeletonema costatum	*Juncus* spp.
Dunaliella tertiolecta	
Champia parvula	

freshwater algae than with marine algae (Table 7). Although specific algal test methods are discussed in greater detail in other chapters of this book, the basic design of most standardized methods is similar. Algal populations are exposed during their exponential growth phase for 4 to 10 d to five concentrations of the test substance. The exposure is conducted in a nutrient-enriched medium and in a room or chamber controlled for light and temperature. Algal growth, as measured by changes in cell numbers, is monitored daily. The results are reported as the no-observed-effect concentration (NOEC), the 50% inhibitory concentration (IC$_{50}$ value), the stimulatory concentration SC$_{20}$, or the algistatic (no growth) and algicidal (lethal) concentrations.

Toxicity tests have been conducted with whole, rooted, macrophytic plants and their seeds, but less frequently than those using algae and duckweed. There are no standard test methods for rooted whole plants, and only a draft method exists for studies monitoring effects on seed germination and early seedling growth (APHA et al. 1994). Despite this difficulty, tests with rooted macrophytic plants are important and are best suited to determine the phytotoxic effects of contaminated sediments. However, for this to occur, considerable scientific attention is needed to identify sensitive test species and to develop better culture techniques and test methodologies. For more information, Sortkjaer (1984) provides a detailed summary of the use of macrophytes in toxicity testing.

An important issue related to the usefulness of the current toxicity test methodologies for plants is the environmental relevance of the results. The controlled experimental conditions and use of only a few, cultured, single test species limits the extrapolation of laboratory-derived phytotoxicity data to natural plant communities (Lewis 1990a).

To gain insight on this issue, more realistic multispecies toxicity tests have been conducted using "model ecosystems" such as mesocosms and microcosms. Examples of these types of studies using experimental ponds and streams can be found in Caquet et al. (1992), Boyle et al. (1985), and Mitchell et al. (1993). However, mesocosm methodologies have been used infrequently and only limited data exist. Therefore, the available phytotoxicity information for most commercial chemicals is based on laboratory-derived results for a few, cultured, freshwater algal species. A brief discussion of these data follows for several contaminants commonly found in freshwater, estuarine, and marine environments.

Pesticides

The use of pesticides has increased significantly since the 1940s when they were first used in substantial quantities. Approximately 430 million pounds of agricultural pesticides were applied in 1987 (NRC 1989). Herbicide use was about 20.2 million pounds, and about 69% of the pesticides were used in the estuarine drainage areas in the nation in 1987 (Pait et al. 1992). Pesticides are used to control selected target organisms, but their presence in the environment has had severe consequences for nontarget plants exposed during periods of high rainfall and subsequent runoff from urban and agricultural areas. Sprecher and Netherland (1995) summarized several of the methods used to monitor herbicide stress in submersed plants.

Because of FIFRA, a significant toxicity database for some pesticides and specific aquatic plants has been generated (for reviews of this database, see O'Kelley and Deason 1976, Stratton 1987, and Swanson et al. 1991). Several herbicides are phytotoxic at concentrations of less than 0.001 mg/l, although the majority are toxic at greater concentrations (Figure 4). For example, diuron is phytotoxic at 0.004 mg/l, and atrazine, a widely used herbicide, is toxic at 0.03 mg/l. Insecticides, as would be expected, are less phytotoxic than herbicides; most toxicity levels are between 0.1 and 100 mg/l. The insecticides dieldrin and DDT are phytotoxic at concentrations exceeding 100 mg/l, whereas chlorpyrifos is more phytotoxic at 1.2 mg/l. One interesting factor is that algae can biodegrade some pesticides such as malathion and methoxychlor and reduce their bioavailability for toxicity (O'Kelley and Deason 1976).

Based on the reported toxicity database, some trends in species sensitivity and toxicity are evident. Green and blue-green algae are particularly sensitive to herbicides, since they have common characteristics with higher terrestrial plants, the usual target organisms (Stratton 1987). In general, the phenoxyalkanoic acid-based herbicides are relatively nontoxic to algae, but the triazine and phenylurea herbicides are very toxic.

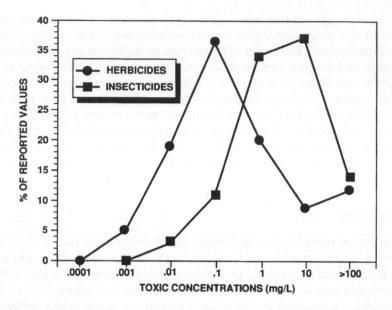

Figure 4 Distribution of phytotoxic concentrations reported for various herbicides and insecticides based on the results of 200 toxicity tests summarized in Stratton (1987).

Metals

Metals, unlike pesticides, are natural constituents in natural waters at low concentrations, and some are essential as trace elements for aquatic plant growth. However, increases in some heavy metals above ambient concentrations can be very toxic to aquatic plants. Copper is phytotoxic at relatively low concentrations and is commonly used as an algicide. When copper toxicity occurs, copper porphyrin is formed by the incorporation of copper rather than magnesium in the chlorophyll molecule. The copper form has little photosynthetic activity, and plant growth is reduced.

The toxicities of metals to aquatic life are affected by changes in ambient pH, water hardness, and nutrient availability. The effects of these factors on metal toxicity have been reported by Pagenkopf et al. (1974), Brown et al. (1974), and Shaw and Brown (1974). The presence of other contaminants (mixture toxicity) is also a factor. For example, zinc and copper in combination have a synergistic phytotoxic effect (Alabaster et al. 1988).

The primary source of metal pollution in fresh water is metal mining and the associated processing of ores. Mining not only increases concentrations of metals such as copper, lead, zinc, cadmium, nickel, and arsenic, but it also lowers the pH and adds oils, cyanides, acids, and alkalies to the receiving water. Acid-mine drainage, the result of oxidation of reduced iron and sulfur compounds during strip mining, can affect plants in other ways. In addition to decreasing the pH, the ferric precipitate characteristic of acid mine-affected waters can increase suspended solids which can reduce photosynthesis and also smother benthic periphyton.

The effects of metals on aquatic plants resulting from mining activities are discussed in Kelly (1988) and Stratton (1987). Many field studies have been conducted in rivers where, not surprisingly, the diversity of the plant community decreased as the heavy metal concentrations increased. This impact has been shown on numerous occasions, although some species of the genera *Plectonema, Mougeotia, Stigeoclonium,* and *Hormidium* have been shown to be tolerant of high heavy metal concentrations (Armitage 1980; Say et al. 1977). In addition to field observations, the toxicity of various metals have been determined in laboratory toxicity tests for various plant species. Figure 5 shows the relative distribution of some reported effect concentrations for mercury, copper, and zinc. Mercury is generally phytotoxic between 0.01 and 0.1 mg/l, and copper is more frequently toxic at 0.1 mg/l. The reported toxicities of zinc range from 0.01 to >100 mg/l. Additional details on heavy metal phytotoxicity can be obtained from Whitton (1970), Rai et al. (1981), and Guillizoni (1991).

Algae and macrophytic vegetation can bioconcentrate significant amounts of heavy metals in their tissues. This attribute has been used in a beneficial manner to remove toxicants from various types of waste-

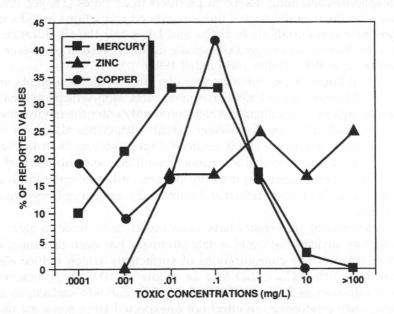

Figure 5 Distribution of phytotoxic concentrations reported for heavy metals based on the results of 101 (copper), 69 (zinc), and 114 (mercury) toxicity tests summarized by Stratton (1987) and Kelly (1988).

water and also allows plants to serve as biomonitors for metal-contaminated areas (Whitton 1984). In contrast, the tissue concentrations of metals, which can exceed those in the water column by several orders of magnitude, may be phytotoxic to the plant and also be a potential threat to grazer species.

Specific bioconcentration factors for several metals have been determined in laboratory tests for cultured algae and macrophytes. Some of these bioconcentrations factors are summarized in Chapter 13 and can exceed 1 million (Satake et al. 1989). Environmental factors found to affect the bioconcentration process include the presence of other pollutants and nutrients, water temperature, pH, salinity, photoperiod, and light intensity.

Detergents

The most recognized impact of detergents on aquatic plants has been their contribution of phosphate to the environment. Phosphorus (sodium tripolyphosphate) has been used in various laundry products (6 to 8% by weight) since the 1930s. Detergent phosphates are estimated to contribute 50% of the phosphorus in municipal wastewaters and 30 to 40% in aquatic environments. The recognition that the addition of phosphorus found in many detergent builders may increase the eutrophication process has resulted in the banning or restriction of phosphorus-containing detergent products in 23 states (Thayer 1992). However, the overall effect of these mandated restrictions in reducing phosphate concentrations in rivers and lakes and the significance of the contribution of detergent phosphate in the ecosystem is an issue of continuing debate (Welch and Lindell 1992).

In addition to phosphorus, granular and liquid detergents and household cleaning products also contain surfactants which are a major class of organic contaminants which commonly enter the environment in municipal effluents. To a lesser extent, surfactants also enter the environment as a result of their use in oil dispersants and as an adjuvant in some pesticide formulations. Anionic, nonionic, and cationic surfactants have been measured commonly in water and sediment from rivers and estuaries receiving treated and untreated wastewater from municipalities.

Commercial surfactants have been found to be toxic to algae in laboratory toxicity tests, and a data summary has been presented by Lewis (1990b). The concentrations of surfactants which reduce algal growth by 50% (EC_{50} value) may be as low as 0.001 mg/l; however, most values exceed 0.1 mg/l (Figure 6). The cationic surfactants are particularly phytotoxic, an effect not unexpected since some are used as algicides. As a result of their high toxicity, more realistic studies have been conducted with some surfactants to determine the ecological relevance of the laboratory-derived results. In these multispecies tests, the

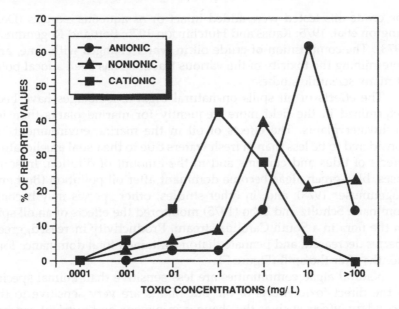

Figure 6 Distribution of surfactant concentrations found toxic to aquatic plants. The results are based on 100, 114, and 100 toxicity tests conducted with anionic, nonionic, and cationic surfactants, respectively, as reported in Lewis (1990b).

effects of cationic surfactants on natural algal assemblages were usually less than on single algal species (Lewis 1990b). With the exception of cationic surfactants, most surfactants are less phytotoxic than heavy metals and herbicides.

Oil

Reviews on the impact of oil and their dispersants on freshwater and marine algae have been presented by O'Brien and Dixon (1976), Lockwood (1976), and Shales et al. (1989). Overall, scientific understanding on the mechanism of entry and mode of action of oil on aquatic plants is limited. The effects may be physical, biophysical, or biochemical. Physical effects due to light exclusion can lead to reduced growth and photosynthesis. Penetration of the oil into plant tissues can cause damage, the extent of which depends upon the thickness of the cuticle and frequency of the stomata. Cell membranes can be damaged, resulting in the leakage of cell contents.

Most toxicological research on oil and refinery wastes has been done with single species of marine algae in laboratory tests (O'Brien and Dixon 1976). However, the majority of the more realistic microcosm studies have been conducted with freshwater species (Shales et al. 1989). The toxicity of oil has been found to vary because of differences in the sensitivity of the test species, the source of the oil, and whether the crude oils tested were added intact or as aqueous extracts (Dennington et al. 1975; Kauss and Hutchinson 1975; Bott and Rogenmuser 1978). The composition of crude oil in water changes with time, and determining the toxicity of the various fractions has been a focal point of many scientific studies.

The effects of oil spills on natural algal assemblages have been determined in the field more frequently for marine plants than for freshwater plants. The effects of oil in the marine environment are considered to be less than in fresh waters due to the usual ameliorating effects of tides and currents and to the amount of dilution. In some cases, blue-green algae become dominant after oil pollution (Bott and Rogenmuser 1978), and, in other studies, other species may become dominant. Schultz and Tebo (1975) monitored the effects of an oil spill on the flora in a South Carolina stream. Productivity increased, green species decreased, and pennate diatoms attained total dominance four months after the spill (Figure 7).

Natural algae communities are less sensitive than animal species to the direct toxic effects of oil, but plants are very sensitive to the secondary effects such as the changes in oxygen and nutrient concentrations caused by some types of oil (Werner et al. 1985). In some cases, particulate nitrogen and carbon increases resulting from the oil have caused phytoplankton blooms (Snow and Scott 1975). Schindler et al. (1975), in a dosed pond study, found that the largest dose of oil increased algal production.

Few reports describe the effects of oils spills on wetland macrophytes, and few data trends are evident. One of the few available is that reported by Burk (1977) for a freshwater marsh. There is evidence that some annual freshwater species such as *Sagettaria graminea* (coastal arrowhead), *Lemna minor* (duckweed), *Ceratophyllum demersum* (coontail), and *Elodea nuttalli* are more sensitive than perennial species to oil contamination (Baker 1970, 1971, 1973; Burk 1977). Lytle and Lytle (1987) reported that the marsh plants *Spartina alterniflora* (cordgrass) and, particularly, *Juncus roemerianus* (black rush) were resilient to oil spills. This latter species also had the capacity to bioaccumulate large quantities of petroleum hydrocarbons from sediment.

Conclusion

The seasonal community dynamics of aquatic plants are controlled by a variety of physical, chemical, and biological factors. The integrated effects of these factors on aquatic plants are not completely understood, and any of which, if altered, can change community structure and function. Maintenance of a balanced plant community is important, since they are the primary producers in most aquatic ecosystems. The diurnal and seasonal cycles in dissolved oxygen, nutrient concentra-

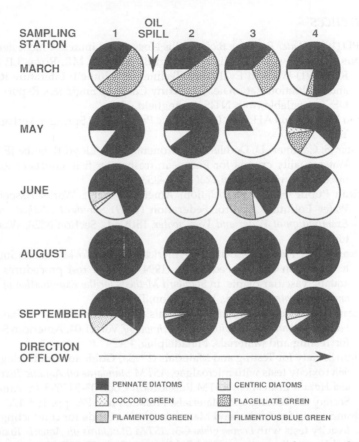

Figure 7 Monthly change in periphyton composition at four sampling stations above and below an oil spill in a South Carolina stream. (Adapted from Schultz D, Tebo LB [1975] *Conference on Prevention and Control of Oil Pollution*, San Francisco, CA.)

tions, and pH attributable to aquatic plants are important in maintaining balanced communities. Any change in community composition can alter trophic relationships, cause bloom conditions, alter water quality, and lead to ecosystem disruption, all of which can have economic consequences. This importance of plants to the functioning of the ecosystem has been recognized for many years by ecologists, and plants have been important components in biomonitoring and bioremediative programs.

The introduction of nutrients, toxics, and suspended solids into freshwater, marine, and estuarine environments as a result of anthropogenic activities has had significant ecological consequences to plants. This has led to the development of various field assessment techniques and laboratory toxicity test methods in the past 20 years to determine the potential phytotoxic effects of these contaminants.

References

Abel PD (1989) *Water Pollution Biology,* Ellis Horwood, Limited, Chichester, U.K.

Adamus PR, Stockwell LT, Ellis J, Clariain EJ, Morrow ME, Rozas LP, Smith RD (1991) Wetland Evaluation Technique Volume I: Literature Review and Evaluation Rationale, U.S. Army Corps of Engineers Report WRP-DE-Z. Available from NTIS, Springfield, VA.

Adriano DC, Johnson AH, Eds. (1989) *Acidic Precipitation,* Springer-Verlag, New York.

Alabaster JS, Calamari D, Dethlefsen V, Konemann H, Lloyd R, Solbe JF (1988) Water quality criteria for European freshwater fish: effects of toxicant mixtures in water, *Chem Ecol* 3:165–253.

American Public Health Association, American Water Works Association, Water Pollution Control Federation (1989) *Standard Methods for the Examination of Water and Wastewater,* 18th ed., Section 8220, Washington, D.C.

American Public Health Association, American Water Works Association, Water Pollution Control Federation (1994) Toxicity test procedures using aquatic vascular plants. In *Standard Methods for the Examination of Water and Wastewater,* 19th ed., Washington, D.C.

American Society for Testing and Materials (1992) *Annual Book of Standards, Section II, Water and Environment Technology,* Vol. 11.04, American Society for Testing and Materials, Philadelphia, PA.

American Society for Testing and Materials (1993a) Guide for conducting static 96h toxicity tests with microalgae, *ASTM Standards on Aquatic Toxicology and Hazard Evaluation,* ASTM Publication Code 03-547093-16, American Society for Testing and Materials, Philadelphia, PA, pp. 168–178.

American Society for Testing and Materials (1993b) Guide for conducting static toxicity tests with *Lemna gibba* G3, *ASTM Standards on Aquatic Toxicology and Hazard Evaluation,* ASTM Publication Code 03-547093-16, American Society for Testing and Materials, Philadelphia, PA, pp. 229–238.

American Water Works Association (1987) *Current Methodology for the Control of Algae in Surface Reservoirs,* American Water Works Association Research Foundation, Denver, CO.

Appourchaux M (1952) Effects de la temperature de l'eau sur la Faune et la Flore Aquatiques. *L'eau* 8:377.

Aquatic Plant Management Society (1993) Proceedings of International Symposium on the biology and management of aquatic plants. *J Aquat Plant Manag* 31.

Armitage PD (1980) The effects of mine drainage and organic enrichment on benthos in the River Nent system, Northern Pennies. *Hydrobiologia* 74:119–128.

Baker JM (1970) The effects of oils on plants. *Environ Pollut* 1:27–44.

Baker JM (1971) Seasonal effects of oil pollution on salt marsh vegetation. *Oikos* 22:106–110.

Baker JM (1973) Recovery of saltmarsh vegetation from successive oil spillages. *Environ Pollut* 4:223–230.

Blanck H, Wallin G, Wangberg S (1984) Species dependent variation in algal sensitivity to compounds. *Ecotox Environ Saf* 8:339–351.

Bott TL, Rogenmuser K (1978) Effects of No. 2 fuel oil, Nigerian crude oil and crankcase oil on attached algal communities: acute and chronic toxicity of water soluble constituents. *Appl Environ Microbiol* 36:643–682.

Boyle TP, Finger SE, Paulson RL, Rabeni CF (1985) Comparison of laboratory and field assessment of fluorene. Part II. Effects on the ecological structure and function of experimental pond ecosystems. In *Validation and Predictability of Laboratory Methods for Assessing the Fate and Effects of Contaminants in Aquatic Ecosystems*, Boyle TP, Ed., ASTM STP 865, American Society for Testing and Materials, Philadelphia, PA, pp. 134–151.

Boyle TP, Smillie GM, Anderson JC, Beeson DR (1990) A sensitivity analysis of nine diversity and seven similarity indices. *Res J Water Pollut Control Fed* 62:749–762.

Bratkovitch A (1988) The use of planktonic organisms distribution as an indicator of physical variability in marine environments. In *Marine Organisms as Indicators*, Soule DF, Kleppel GS, Eds., Springer-Verlag, New York, pp. 13–34.

Bringman G, Kühn R (1980) Comparison of toxicity thresholds of water pollutants to bacteria, algae and protozoa in the cell multiplication test. *Water Res* 14:231–241.

Brock D (1977) Comparison of community similarity indices. *J Water Pollut Control Fed* :2488–2495.

Brown VM, Shaw TL, Shurben DG (1974) Aspects of water quality and the toxicity of copper to rainbow trout. *Water Res* 8:797–803.

Burk CJ (1977) A four year analysis of vegetation following an oil spill in a freshwater marsh. *J Appl Ecol* 14:515–522.

Cairns J, Douglas WA, Busby F, Chaney MD (1968) The sequential comparison index — a simplified method for non-biologists to estimate relative differences in biological diversity in stream pollution studies. *J Water Pollut Control Fed* 40:1607–1613.

Caquet T, Thybaud E, Lebras S, Jonot O, Ramade F (1992) Fate and biological effects of lindane and deltamethrin in freshwater mesocosms. *Aquat Toxicol* 23:261–277.

Carpenter SR, Lodge DM (1986) Effects of submersed macrophytes on ecosystem processes. *Aquat Bot* 26:341–370.

Chambers PA, Prepas EE (1994) Nutrient dynamics in riverbeds: the impact of sewage effluent and aquatic macrophytes. *Water Res* 29:453–464.

Clark JR, Dickson KL, Cairns J Jr. (1979) Estimating aufwuchs biomass. In *Methods and Measurements of Periphyton Communities: A Review*, Weitzel RL, Ed., ASTM STP 690, American Society for Testing and Materials, Philadelphia, PA, pp. 116–141.

Collins GB, Weber DI (1978) Phycoperiphyton (algae) as indicators of water quality. *Trans Am Microsc Soc* 97:36–43.

Coutant CC (1962) The effect of a heated water effluent upon the macroinvertebrate riffle fauna of the Delaware River. *Proc Am. Acad Sci* 36:58–71.

Dennington VN, George JJ, Wyborn CHE (1975) The effects of oils on growth of freshwater phytoplankton. *Environ Pollut* 8:223–237.

Dortch MS (1992) Literature Analyses of the Functional Ability of Wetlands to Improve Water Quality, Wetlands Research Program Bulletin Vol. 2, No. 4, U.S. Army Corps of Engineers, Vicksburg, MS.

Dresscher GN, van der Mark H (1976) A simplified method for the biological assessment of the quality of fresh and slightly brackish water. *Hydrobiology* 48:199–201.

Duffy JE, Hay ME (1994) Herbivore resistance to seaweed chemical defense: the roles of mobility and predation risk. *Ecology* 75:1304–1319.

Elwood JW, Mulholland PJ (1989) Effects of acidic precipitation on stream ecosystems. In *Acidic Precipitation*, Vol. 2, Adriano DC, Johnsons AH, Eds., Springer-Verlag, New York.

Farmer AM (1990) The effects of lake acidification on aquatic macrophytes: a review. *Environ Pollut* 6:219–240.

Farnworth EG, Nichols MC, Vann CN, Wolfson LG, Bosserman RW, Hendrix PR, Gallery FB, Cooley JL (1979) *Impacts of Sediment and Nutrients on Biota in Surface Waters of the United States*, EPA-300/3-79-105, U.S. Environmental Protection Agency, Washington, D.C., 331 pp.

Fjerdingstad E (1950) The microflora of the River Mollea with special references to the relation of the benthal algae to pollution. *Folia Limnol Scand* 5:1–123.

Fjerdingstad E (1963) Limnological estimation of water pollution levels. *WHO/EBL* 10:1–29.

Fjerdingstad E (1964) Pollution of streams estimated by benthal phytomicroorganisms. I. A saprobic system based on communities of organisms and ecological factors. *Hydrobiologia* 49:63–131.

Fjerdingstad E (1965) Pollution of streams estimated by benthal phytomicroorganisms. II. Taxonomy and saprobic valency of benthic phytomicroorganisms. *Hydrobiologia* 50:475–604.

Fletcher RL (1991) Marine macroalgae as bioassay test organisms. In *Ecotoxicology and the Marine Environment*, Abel PD, Axiak V, Eds., Ellis Horwood, Limited, New York, pp. 111–131.

Francis G (1878) Poisonous Australian lakes. *Nature* 18:11–12.

Gorham PR, Carmichael WW (1990) Hazards of freshwater blue-green algae. In *Algae and Human Affairs*, Lembi CA, Waaland JR, Eds., Cambridge University Press, Cambridge, MA, pp. 403–431.

Guillizoni P (1991) The role of heavy metals and toxic materials in the physiological ecology of submersed macrophytes. *Aquat Bot* 41:87–109.

Guntenspergen GR, Stearns F, Kadlec JA (1989) Wetland vegetation. In *Constructed Wetlands for Wastewater Treatment*, Hammer DA, Ed., Lewis Publishers, Chelsea, MI, pp. 73–88.

Hadden DA (1994) A new slant on wetlands. *Water Environ Technol* February:46.

Hammer DA, Ed. (1989) *Constructed Wetlands for Wastewater Treatment*, Lewis Publishers, Chelsea MI, 831 pp.

Haslam SM (1982) A proposed method for monitoring river pollution using macrophytes. *Environ Technol Lett* 3:19–34.

Hawkes FB (1969) Thermal problems — old hat in Britain. Central Electricity Generating Board, London, CEGB Newsletter No. 83.

Hellquist CB (1980) Correlation of alkalinity and the distribution of *Potamogeton* in New England. *Rhodora* 82:331–344.

Honnell DR, Madsen JD, Smart RM (1993) Effects of Selected Exotic and Native Aquatic Plant Communities on Water Temperature and Dissolved Oxygen, U.S. Army Corps of Engineers Aquatic Plant Control Research Program Vol. A-93-2, U.S. Army Corps of Engineers, Vicksburg, MS.

Howells G (1992) *Acid Rain and Acid Water*, Ellis Horwood, Limited, Chichester, U.K., 215 pp.

Howells GD (1983) The effects of power station cooling water discharges on aquatic ecology. *Water Pollut Control* 82:10–17.

Hynes HBN (1960) *The Ecology of Polluted Waters*, Liverpool University Press, Liverpool.

Inderjit K, Dakshini MM (1994) Algal allelopathy. *Bot Rev* 60:182–192.

Johnston CA (1993) Mechanisms of wetland-water quality interaction. In *Constructed Wetlands for Water Quality Improvement*, Moshiri GA, Ed., Lewis Publishers, Ann Arbor, MI, pp. 293–300.

Kadono Y (1982) Occurrence of aquatic macrophytes in relation to pH, alkalinity, calcium, chloride, and conductivity. *Jpn J Ecol* 32:39–44.

Kauss PB, Hutchinson TC (1975) Studies on the susceptibility of *Ankistrodesmus* species to crude oil components. *Vehr Int Verein Limnol* 19:2155–2164.

Kelly M (1988) *Mining and Its Freshwater Environment*, Elsevier Applied Science, New York.

Kilgore J, Dibble ED, Hoover JJ (1993) Relationships between fish and aquatic plants: a plan of study, Miscellaneous paper A-93-1, U.S. Army Corps of Engineers WES, Vicksburg, MS.

Knight RL (1994) Treatment wetlands database now available. *Water Environ Technol* February:31.

Knöpp H (1954) A new method of displaying the results of biological examination of streams, illustrated by the quality of a longitudinal section of the Main. *Wasserwirtschaft* 45:9–15.

Kolkwitz R, Marsson M (1908) Oekologie der pflanzlichen Saprobien. *Ber Dtsch Bot Ges* 26A:505–519.

Langford TEL (1990) *Ecological Effects of Thermal Discharges*, Elsevier Applied Science, New York.

Lekven GC, Williams CR, Charney RD, Crites RW (1994) Wetlands put to the test. *Water Environ Technol* February:40.

Lembi CA, O'Neal SW, Spencer DF (1990) Algae as weeds: economic impact, ecology and management alternatives. In *Algae and Human Affairs*, Lembi CA, Waaland JR, Eds., Cambridge University Press, Cambridge, MA, pp. 455–481.

Levine HG (1984) The use of seaweeds for monitoring coastal waters. In *Algae as Ecological Indicators*, Schubert LE, Ed., Academic Press, New York, pp. 189–210.

Lewis MA (1990a) Are algal toxicity data worth the effort. *Environ Toxicol Chem* 9:1279–1284.

Lewis MA (1990b) Chronic toxicities of surfactants and detergent builders to algae: a review and risk assessment. *Ecotoxicol Environ Saf* 20:1279–128.

Lewis MA (1995) The use of freshwater plants in phytotoxicity testing. *Environ Pollut* 87:319–336.

Lockwood APM (1976) *Effects of Pollutants on Aquatic Organisms*, Cambridge University Press, Cambridge, MA.

Lytle JS, Lytle TF (1987) The role of *Juncus roemerianus* in cleanup of oil polluted sediments, *Proceedings of Oil Spill Conference American Petroleum Institute*, Baltimore, MD.

Mitchell GC, Bennett D, Pearson N (1993) Effects of lindane on macroinvertebrates and periphyton in outdoor artificial streams. *Ecotoxicol Environ Saf* 25:90–102.

Mitsch WJ, Gosselink JG (1993) *Wetlands,* Van Nostrand Reinhold, New York, 537 pp.

Mitsch WJ, Dorge CL, Wiemhoff JR (1977) Forested wetlands for water resources management in Southern Illinois, Water Resource Center, Urbana, Champaign, IL.

Moshiri GA, Ed. (1993) *Constructed Wetlands for Water Quality Improvement,* Lewis Publishers, Ann Arbor, MI, 632 pp.

National Academy of Sciences (1969) *Eutrophication: Causes, Consequences, Correctives,* SBN 309-01700-9, National Academy of Science Publishing Office, Washington, D.C.

National Research Council (1989) *Alternative Aquaculture: Committee on the Role of Alternative Farming Methods in Modern Production Agriculture,* National Academy Press, Washington, D.C., 448 pp.

Nelson RW (1990) National Guidance: Water Quality Standards for Wetlands Phase II-Long Term Development, Office of Wetlands Protection, U.S. Environmental Protection Agency, Washington, D.C.

Nimmo DR (1985) *Pesticides. Fundamentals of Aquatic Toxicology,* Hemisphere Publishing Corporation, New York.

Nixon SW, Lee V (1986) Wetland and water quality: a regional review of recent research in the United States on the role of freshwater and salt water wetlands as sources, sinks, and trans-formers of nitrogen, phosphorous, and various heavy metals, U.S. Army Engineer Waterways Experiment Station, Vicksburg, MS; prepared by the University of Rhode Island, Tech. Rep. Y-86-2.

O'Brien PY, Dixon P (1976) The effects of oil and oil components on algae: a review. *Br Phycol J* 11:115–142.

O'Kelley JC, Deason TR (1976) *Degradation of Pesticides by Algae,* EPA 60013-76-022, Office of Research and Development, U.S. Environmental Protection Agency, Athens, GA.

Olson R, Preston E (1990) *Wetland Treatment Systems,* U.S. Environmental Protection Agency, Corvallis, OR.

Orth RJ (1994) Chesapeake Bay submersed aquatic vegetation: water quality relationships. *Lake Reservoir Manage* 10:49–52.

Pait AS, DeSouza AE, Farrow DRG (1992) *Agricultural Pesticide Use in Coastal Areas: A National Summary,* National Oceanic and Atmospheric Administration, Rockville, MD.

Pagenkopf GK, Russo RC, Thurston RV (1974) Effect of complexation on toxicity of copper to fishes. *J Fish Res Board Can* 31:462–465.

Palmer CM (1962) *Algae in Water Supplies,* U.S. Public Health Service, Public Health Service Publication No. 657, U.S. Government Printing Office, Washington, D.C.

Patrick R (1950) A proposed biological measure of stream conditions. *Purdue Univ Eng Bull* 34:379–399.

Patrick R (1957) Diatoms as indicators of changes in environmental conditions. In *Biological Problems in Water Pollution,* U.S. Dept. Health, Education and Welfare, Public Health Service, Cincinnati, OH, pp. 71–83.

Patrick R (1973) Use of algae, especially diatoms, in the assessment of water quality. In *Biological Methods for the Assessment of Water Quality*, ASTM STP 528, American Society for Testing and Materials, Philadelphia, PA, pp. 76–95.

Patrick R, Crum B, Coles J (1969) Temperature and manganese as determining factors in the presence of diatoms or blue-green algal floras in streams. *Proc Natl Acad Sci USA* 64:472–478.

Perkins JL (1983) Bioassay evaluation of diversity and community comparison indices. *J Water Pollut Control Fed* 55:522–530.

Peverly JH (1985) Element accumulation and release by macrophytes in a wetland stream. *J Environ Qual* 14:137–143.

Pip E (1979) Survey of the ecology of submersed aquatic macrophytes in central Canada. *Aquat Bot* 7:339–357.

Rai LC, Gaur MJ, Po HD (1981) Phycology and heavy-metal pollution. *Biol Rev* 56:99–151.

Raymont JE (1963) *Plankton and Productivity in the Oceans*, Pergamon Press, Elmsford, NY.

Round FE (1981) *The Ecology of Algae*, Cambridge University Press, New York, 653 pp.

Satake K, Takamatsu T, Soma M, Shibat K, Nishikawa M, Say PJ, Whitton BA (1989) Lead accumulation and location in the shoots of the aquatic liverwort *Scapania undulata* (L.) in stream water at Greenside mine, England. *Aquat Bot* 33:111–122.

Say PJ, Diaz BM, Whitton BA (1977) Influence of zinc on lotic plants. I. Tolerance of *Hormidium* species to zinc. *Freshwater Biol* 7:357–376.

Schindler D, Scott BF, Carlisle DB (1975) Effects of crude oil on populations of bacteria and algae in artificial ponds subject to winter weather and ice formation. *Verh Int Verein Limnol* 19:2138–2144.

Schultz D, Tebo LB (1975) Boone Creek Oil Spill. In *Conference on Prevention and Control of Oil Pollution*, San Francisco, CA, pp. 583–588.

Seddon B (1972) Aquatic macrophytes as limnological indicators. *Freshwater Biol* 2:107–130.

Shales S, Thake BA, Frankland B, Khan DH, Hutchinson JD, Mason CF (1989) Biological and ecological effects of oils. In *The Fate and Effects of Oil in Freshwater*, Green J, Trett MW, Eds., Elsevier Applied Science, New York, pp. 81–173.

Shannon EE, Weaver E (1949) *The Mathematical Theory of Communication*, University of Illinois Press, Urbana, pp. 82–83, 104–107.

Shaw TL, Brown VM (1974) The toxicity of some forms of copper to the rainbow trout. *Water Res* 8:377–382.

Shubert E (1984) *Algae as Ecological Indicators*, Academic Press, London.

Simpson EH (1949) Measurement of diversity. *Nature* 163:688.

Skulberg OM (1980) Blue-green algae in Lake Mjosa and other Norwegian lakes. *Water Technol* 12:121–140.

Snow NB, Scott BF (1975) The effect and fate of crude oil spilt on two arctic lakes. In *Conference on Prevention and Control of Oil Pollution*, San Francisco, CA, pp. 527–534.

Sortkjaer O (1984) Macrophytes and macrophyte communities as test systems in ecotoxicological studies of aquatic systems. *Ecol Bull Stockholm* 36:75–80.

Sprecher SL, Netherland MD (1995) Methods for monitoring herbicide-induced stress in submersed aquatic plants: a review, Miscellaneous Paper 17-95-1, U.S. Army Corps of Engineers, Vicksburg, MS.

Steidinger KA, Vargo GA (1990) Marine dinoflagellate blooms, dynamics and impacts, in *Algae and Human Affairs*, Lembi CA, Waaland JR, Eds., Cambridge University Press, Cambridge, MA, pp. 373–401.

Stephenson RJ, Lowe RL (1986) Sampling and interpretation of algal patterns for water quality assessments. In *Rationale for Sampling and Interpretation of Ecological Data in the Assessment of Freshwater Ecosystems*, Isom BG, Ed., ASTM STP 894, American Society for Testing and Materials, Philadelphia, PA, pp. 118–149.

Stokes PM, Howell ET, Krantzberg G (1989) Effects of acidic precipitation on the biota of freshwater lakes. In *Acidic Precipitation, Vol. 2, Biological and Ecological Effects*, Springer-Verlag, New York, pp. 273–305.

Stratton GW (1987) The effects of pesticides and heavy metals towards phototrophic micro-organisms. In *Review in Environmental Toxicology 3*, Hodgson E, Ed., Elsevier, New York, pp. 71–147.

Swale EMF (1964) A study of the phytoplankton of a calcareous river. *J Ecol* 52:433–466.

Swanson SM, Rickard CP, Freemark KE, MacQuarrie P (1991) Testing for pesticide toxicity to aquatic plants: recommendations for test species. *Plants for Toxicity Assessment*, 2nd Vol., ASTM STP 115, Gorsuch JW, Lower WR, Wang W, and Lewis MA, Eds., American Society for Testing and Materials, Philadelphia, PA, pp. 77–97.

Thayer AM (1992) Soap and detergents. *Chem Eng News* 70:26–47.

Tripathi BD, Shukla SC (1991) Biological treatment of wastewater by selected aquatic plants. *Environ Pollut* 69:69–78.

U.S. Environmental Protection Agency (1971) *Algal assay: bottle test*. National Eutrophication Research Program, Pacific Northwest Environmental Research Laboratory, Corvallis, OR.

U.S. Environmental Protection Agency (1974) Marine algal assay procedure bottle test. Eutrophication and Lake Restoration Branch, National Environmental Research Center, Corvallis, OR.

U.S. Environmental Protection Agency (1990) *Water Quality Standards for Wetlands*, EPA 44015-90-011, U.S. Environmental Protection Agency, Washington, D.C.

Washington HG (1984) Diversity, biotic and similarity indices: a review with special reference to aquatic ecosystems. *Water Res* 18:653–694.

Weber CI (1973) *Biological Field and Laboratory Methods for Measuring the Quality of Surface Water and Effluents*, EPA 670/4-73-001, U.S. Environmental Protection Agency, Cincinnati, OH.

Welch EB, Lindell T (1992) *Ecological Effects of Wastewater*, Ellis Horwood, New York, p. 425.

Werner MD, Adams VD, Lamarra VA, Winters NL (1985) Responses of model freshwater ecosystems to crude oil. *Water Res* 19:285–292.

Whitton BA (1970) Toxicity of heavy metals to freshwater algae: a review. *Phykos* 9:116–125.

Whitton BA (1979) Algae and higher plants as indicators of river pollution. In *Biological Indicators of Water Quality*, James A, Evison L, Eds., John Wiley & Sons, Chichester, U.K.

Whitton BA (1984) Algae as monitors of heavy metals. In *Algae as Ecological Indicators*, Shubert LE, Ed., Academic Press, New York.

Wieder RK (1989) A survey of constructed wetlands for acid coal mine drainage treatment in the eastern United States. *Wetlands* 9:299–315.

Wolverton BC, McKnown MM (1976) Water hyacinths for removal of phenols from polluted waters. *Aquat Bot* 2:191–201.

chapter seven

Algal indicators of aquatic ecosystem condition and change

Paul V. McCormick and John Cairns, Jr.

Introduction

Three major factors will markedly alter the way in which environmental studies will be carried out in the next century. The first of these is the shift from an emphasis on environmental protection to an emphasis on ecosystem integrity or condition. This will place increased reliance on environmental monitoring programs capable of tracking changes in key biological characteristics, as opposed to merely comparing local water quality conditions to a set of standard criteria. The second is the rapid entry into the information age, which offers means whereby increasingly complex biological information obtained from these programs can be analyzed and integrated with that from other disciplines (e.g., economics) to facilitate holistic approaches to environmental management. Finally, the accelerating process of economic globalization has not only heightened awareness of the penalties of economic isolation, but also the need to arrive at a consensus on environmental quality that is shared by much of human society. All of these factors will influence the types of information that can and must be obtained from monitoring programs and, consequently, the types of ecological measures that will be used.

Failure of previous regulatory actions in developed countries such as the United States to achieve broad environmental goals can be traced to inadequacies in the existing capability to predict ecosystem risk accurately and to implement remedial action when actual damage or warnings of damage occur (Cairns and Niederlehner 1992; Knopman and Smith 1993). The traditional approach to risk assessment still used by many regulatory agencies and legislative bodies places inordinate reliance on information from simple laboratory bioassays low in environmental realism. This approach suffers from a number of shortcomings such as (1) the oversimplification and narrow definition of important ecosystem characteristics or attributes; (2) a sole focus on prevention of damage rather than a broader emphasis on ecosystem condition; (3) a failure to validate laboratory predictions in natural systems or reasonable surrogates thereof such that errors in prediction are not corrected; and (4) the fragmented approach of treating species and stressors separately rather than in the aggregate, which virtually ensures that complex interactions and cumulative effects will be either entirely missed or imprecisely predicted. Rectifying this process to allow for better prediction and estimation of ecological risk in the broader context of ecosystem condition will require greater emphasis on (1) field monitoring and assessment programs and (2) ecologically relevant indicators at multiple levels of biological organization.

Recognition of the spatial extent and magnitude of global environmental damage makes it abundantly clear that a series of generic attributes of ecosystem condition and ecological risk must be developed that will work reasonably well on the entire planet. This has implications for organism selection, since increased reliance will be placed on those organisms with global distributions that can be monitored in different types of aquatic ecosystems. Furthermore, if ecosystem integrity is the primary objective — that is, not only is the ecosystem protected, but in robust condition with optimal function — monitoring will inevitably focus more intently on the group of organisms that makes up the predominant biomass of aquatic ecosystems, the microorganisms. The information content of this group is extremely high because they mediate most fundamental ecosystem processes and, therefore, provide multiple signals relevant to system condition.

The significance of algae to aquatic ecosystems

The most compelling reason for including algal indicators in environmental monitoring programs is their key role in mediating the flux of energy into aquatic ecosystems. Algae contribute substantially to total ecosystem primary production in most freshwater and marine habitats. Even in aquatic ecosystems where energy inputs from terrestrial sources are considerable (e.g., headwater streams), algal biomass may support substantial secondary production (e.g., Mayer and Likens 1987). Algae mediate nutrient availability and spiraling in the aquatic environment. For example, empirical estimates and those from mass balance models indicate the importance of cyanobacterial (i.e., blue-green algal) nitrogen fixation as a significant source of bioavailable nitrogen in various ecosystems (Paerl et al. 1989; Craft and Richardson 1993). Macroalgae and microalgal mats provide both a breeding habitat and a refuge for various animals (Boaden et al. 1975; Power 1990). Thus, just as the condition of woody plants and other conspicuous vegetation is a focal point of terrestrial monitoring programs (Riitters et al. 1990), so too should the assessment of the composition and productivity of the algal assemblage be an integral part of aquatic monitoring programs.

Changes in both algal production and taxonomic composition can profoundly affect food web interactions and ecosystem dynamics. The ecological consequences of excessive algal growth (e.g., anoxia, fish kills, etc.) resulting from increased inputs of limiting nutrients such as phosphorus have been well documented (Edmondson 1961; Hynes 1969; Leach et al. 1977; Saether 1980). Algal taxa vary greatly in their edibility, and shifts in the nutritional value of this food resource can affect population growth and species composition at higher trophic levels in aquatic food webs (Porter 1976; Browder 1982; Lamberti and Moore 1984). Some cyanobacteria (e.g., *Anabaena*), which tend to dominate under eutrophic conditions, produce highly toxic secondary metabolites which prevent consumption (Gilbert 1990; DeMott and Moxter 1991). Thus, both structural and functional shifts in the algal assemblage can exert substantial impacts on other ecological compartments and ecosystem processes.

Fish and other commercial and recreational species (e.g., shellfish and game birds) are widely used as indicators of environmental condition, in part because of their economic value. Endangered species and those having some form of symbolism (e.g., the bald eagle in the United States) also engender a considerable degree of public and political concern and are frequently the focus of environmental impact assessments. Because of their key role in the aquatic ecosystem, shifts in the algal assemblage can also have substantial economic and esthetic impacts. Changes in the quantity and quality of algal production in response to increased nutrient loading can severely damage commercial and recreational fisheries. Water quality managers are frequently concerned with the effect that blooms of nuisance algae have on the taste and odor of water in municipal water supplies. Excessive growth of nuisance algae in response to impaired water quality can reduce both the esthetic appearance and use of lakes and swimming beaches (e.g., Auer et al. 1982). Conditions that favor the growth of toxin-producing algae can severely impact not only aquatic species such as fish, but livestock and other animals as well (Gorham 1964).

Specific attributes of algae as environmental indicators

Algal indicators are applicable to all aquatic habitats and a wide range of environmental stressors. Algae represent an abundant and diverse assemblage in most aquatic ecosystems. Furthermore, individual species generally exhibit wider distributions among ecosystems and geographical regions than most species of higher organisms, and many species are believed to be globally distributed (Cairns 1992). The widespread distribution of most species reduces problems associated with standardizing metrics among watersheds or other geographical units with dissimilar community types, such as those encountered when using fish (Karr et al. 1986; Leonard and Orth 1986; Oberdorff and Hughes 1992), and assure spatial continuity of indicators within regional and national monitoring programs.

Algae and other microbial taxa have long been used to detect impacts associated with organic enrichment and other forms of cultural eutrophication (e.g., Kolkwitz and Marsson 1908; Pantle and Buck 1955; Fjerdingstad 1964; Sládecek 1973; Whitton et al. 1991). Numerous laboratory and field studies have quantified the response of algal assemblages to a variety of other chemicals, including pesticides (Walsh 1972; Hawxby et al. 1977; Kosinski 1984; Krieger et al. 1988), heavy metals (Sigmon et al. 1977; Thomas and Seibert 1977; Pratt et al. 1987; Pratt and Bowers 1990; Scanferlato and Cairns 1990), polychlorinated biphenyls (PCBs) (Harding 1976), and complex effluents (Amblard et al. 1990). Changes in the algal assemblage have proven useful in assessing complex changes in aquatic ecosystems. For example, algal surveys have been useful in understanding the effects of multiple environmental changes associated with acidification in lakes in the northeastern United States (e.g., Stevenson et al. 1985; Cummings et al. 1992).

Algae respond rapidly to changes in ecosystem condition. While there is no compelling evidence that algae, as a group, are generally more sensitive to environmental change than other aquatic organisms, there is increasing evidence that changes in the algal-microbial assemblage provide accurate predictions of impacts on longer-lived organisms or ecosystem processes that respond over longer timescales. Algae have extremely short generation times, which allows these populations to exhibit rapid changes in response to environmental change. Algae are the principal biological receptors of certain impacts (e.g., increased nutrient loads), such that changes in the structure and activity of this assemblage largely determine secondary effects observed for higher organisms as a result of changes in the oxygen budget and the quality and quantity of the resource base for herbivores (see previous discussion). Numerous studies conducted in one of our laboratories (Cairns) at Virginia Tech have confirmed that the taxonomic structure and metabolism of naturally derived protist assemblages are rapidly altered by concentrations of toxic chemicals found to affect the growth and survival of longer lived organisms (for a summary, see Niederlehner and Cairns 1994). In reviewing several years of data collected in the Experimental Lakes Area in Canada, Schindler (1987) concluded that changes in phytoplankton species composition and the loss of sensitive species from this assemblage were among the earliest reliable indicators of ecosystem stress observed.

Collection and processing times for algal samples are comparable to those for other organisms. In order to provide useful information on changes in ecosystem condition in a timely manner (e.g., to support flexible ecosystem management practices), an indicator must not only respond rapidly to environmental deterioration, but must provide ecological information that can be rapidly analyzed and interpreted. Definitive comparisons of the collecting and processing times required to monitor algae as opposed to other aquatic organisms are lacking (see U.S. EPA 1993). However, there appears to be a common and, in our minds, unsubstantiated perception that it is more difficult to characterize the algal assemblage than other major aquatic groups. In fact, we would argue that the algal assemblage can be exhaustively collected in a quantitative manner in a shorter time period than required for comparable fish or macroinvertebrate sampling. For example, seining and electroshocking are both time consuming and labor intensive and, in large ecosystems (e.g., deep lakes), may provide neither a quantitative nor representative sample of the fish community. Sampling of stream macroinvertebrates has been simplified by the formalization of rapid, standardized protocols (e.g., Plafkin et al. 1989). Quantitative sampling of algae from the water column and the benthos can also be performed quickly and simply without specialized equipment or intensive labor in most cases. Unfortunately, methods for algal sampling and processing are scattered throughout the scientific literature, and few attempts have been made to standardize these methods and compile them in a single manual.

The time required for taxonomic analysis of algal samples is comparable to that for macroinvertebrates, although perhaps longer than for fish samples, which are generally processed in the field. As for any group of organisms, the rate at which samples can be sorted and taxa identified is ultimately limited by the availability of trained personnel. In this regard, the diatoms (Class Bacillariophyceae) are best suited for routine collection and processing. The taxonomy of this group is well understood, and the number of diatomists is large compared with the number of specialists for other algal groups. Unlike many other algae, diatoms can be identified routinely to the species level without the need for collecting special growth stages or laboratory culturing. Identification aids (e.g., iconographs) and other training and quality control techniques have been developed specifically for this group (e.g., Munro et al. 1990).

Algal assemblages can be used to establish benchmark conditions. Quantifying the current condition or "health" of an ecosystem requires a benchmark or standard for comparison. Historical information on ecological condition is lacking for most groups of aquatic organisms. Paleolimnological analysis of siliceous algal microfossils provides a means of establishing benchmark biological conditions and for inferring the rate and extent of past ecosystem changes (e.g., Kingston and Birks 1990). Paleoecological investigations of sediment microfossils in the Laurentian Great Lakes, for example, provided a record of changes in the open water diatom assemblage dating back to the preindustrial era, and changes in water quality since that time were reconstructed based on species autecologies (e.g., Wolin et al. 1988; Figure 1).

Although the paleolimnological approach is not suited for flowing waters and other habitats where sediments are periodically disturbed by flooding and/or dredging, valuable historical information on species assemblages in some watersheds may be obtained from permanent collections housed in museums, herbaria, and government depositories. Because of the ease with which entire diatom assemblages can be collected, preserved, and mounted, the completeness and permanence of these collections should be better than for larger organisms such as fish.

In the absence of reliable historical information, many investigators have adopted the use of regional reference sites for assessing the extent of ecosystem deterioration (Hughes et al. 1986). Ecological conditions in reference, or minimally impacted streams, are used as a benchmark against which to gauge the condition of other streams in the same physiographic province or "ecoregion" (sensu Omernick 1987). Algal assemblages are well suited to this approach, since diverse assemblages are present in most ecosystems within a region, regardless of size or condition. Assemblages of other organisms such as fish may be depauperate in smaller or ephemeral streams. The reference site approach may be combined with paleoecological and other historical methods to improve the characterization of baseline or nominal conditions using diatoms.

Algal assemblages provide a cost-effective monitoring tool. Karr (1991) estimated the costs of collecting, processing, and analyzing information from invertebrate and fish communities for monitoring purposes to be $824 and $740, respectively, per sample. We are unaware of any published comparison of costs that includes algal indicators, although, based on the cost estimates just cited for other organisms, we predict that algal assemblages will prove to be a cost-effective means of collecting ecological information. The cost-effectiveness of an indicator is ultimately determined by the amount of useful information gained per unit effort. The information content of algal assemblages is high because of the large number of species with various environmental tolerances encountered in most habitats. This feature is particularly advantageous when sampling locations containing depauperate assemblages of other organisms.

Algal indicators allow for nondestructive sampling. Sampling of algal assemblages rarely, if ever, results in perceptible environmental impact. Sampling of larger aquatic organisms is often more disruptive and may impact indigenous populations, particularly those of rare or threatened species (e.g., fish electroshocking and associated habitat disturbance in streams). Endangered species currently listed by the U.S. Fish and Wildlife Service include no algal or other microbial species. Animal rights organizations similarly exhibit little opposition to the removal, testing, or destruction of algae.

Limitations to the use of algal indicators

While algae possess several desirable qualities as ecological indicators, potential limitations should be recognized as well. Because almost all algal species rely principally on autotrophic nutrition, these assemblages are inevitably sensitive to changes in nutrient availability and limitation. While this sensitivity enhances the utility of algal indicators for predicting changes in trophic status, it may reduce algal sensitivity to other changes in water quality in some instances. Algal production in aquatic ecosystems is often dependent on the availability of one or a few macronutrients (Hansson 1992), although certain micronutrients may also control production in marine and some freshwater systems (e.g., Pringle et al. 1986). Basin-wide changes in the algal flora of the Laurentian Great Lakes, which have been subjected to numerous biological, chemical, and physical insults over the past century, are most closely associated with increased nutrient loading (Stoermer 1978; Wolin et al. 1988). Furthermore, increased availability of limiting nutrients can significantly reduce algal community sensitivity to toxic chemicals (e.g., Chen 1989; Wängberg and Blanck 1990). Given the close correspondence between algal condition and trophic status, further investigation of the interactive effects of nutrient enrichment and other stressors on algal indicators is certainly needed.

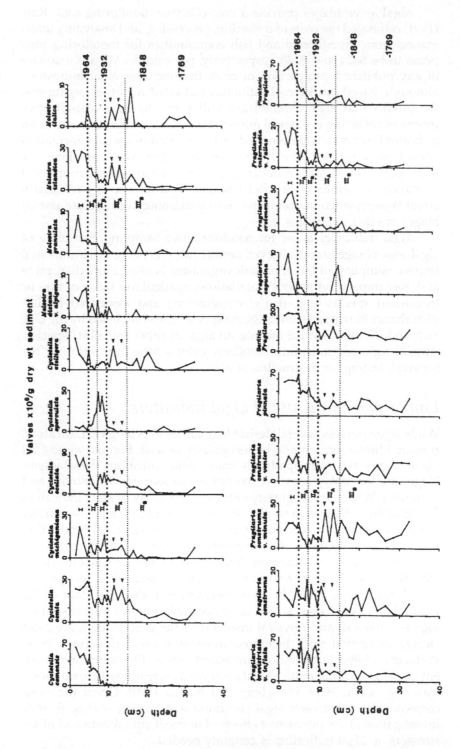

As with other groups of organisms, algal sensitivity to a given stressor may vary markedly in relation to other environmental factors. The response of algal growth and metabolism to heavy metals can vary considerably, depending on the presence of natural chelating agents (e.g., humic acids), nutrient concentration, as well as other abiotic and biotic parameters (e.g., Bentley-Mowat and Reid 1977; Wängberg and Blanck 1990). Prior exposure to environmental contaminants can alter the sensitivity of protist assemblages to a novel stressor (Niederlehner and Cairns 1993). Populations may exhibit decreased sensitivity or adaptation to a stressor as a result of chronic exposure (Foster 1982; Blanck and Wängberg 1988). Thus, as for other groups of aquatic organisms, the magnitude of shifts in the algal assemblage in response to changes in ecosystem condition is determined by several factors, including the nature and duration of exposure, as well as synergistic and antagonistic interactions.

While the early warning capability of algae and other microbes is significant, limitations must also be recognized. For example, chronic toxicity in higher organisms can occur in response to low environmental concentrations of chemical contaminants as a result of persistent bioaccumulation in vital organs; models used to explain such effects may not apply to algal cells. The trophic position of algae also limits their ability to predict accurately the impact of chemicals that biomagnify (e.g., pesticides). Thus, false negative signals, i.e., the indication that ecosystem conditions are acceptable when, in fact, significant environmental damage is occurring, may limit the utility of algal indicators in cases where persistent low-level toxicity and bioaccumulation or magnification processes are involved.

A second problem encountered with early warning indicators is the occurrence of false positive signals, i.e., indications that environmental damage is occurring when, in fact, ecosystem parameters are within an acceptable operating range. The same attributes that allow algal indicators to respond rapidly to anthropogenic stressors also makes them susceptible to fluctuations in environmental conditions unrelated to human impact. Background fluctuations in early warning

Figure 1 (opposite) Siliceous algal succession in Lake Huron as determined from depth profiles of diatom (Bacillariophyceae) cell (valve) abundance in a sediment core (Wolin et al. 1988). Dates provided on the right-hand side of each plate were determined from radioisotopic (cesium) dating of the core. Increases in the abundance of most dominant centric diatom (*Cyclotella* and *Melosira*) and pennate (*Fragilaria*) species indicate increases in algal productivity during recent history in association with increased nutrient loading. Further evidence of nutrient enrichment is indicated by an increase in abundance of species indicative of eutrophic conditions (e.g., *C. comensis*, *C. stelligera*, *F. capucina*, *M. distans* v. *alpigena*) in more recent sediments. (From Wolin JA et al. [1988] *Arch Hydrobiol* 114:175–198. With permission.)

indicators constitute extraneous variability, or "noise," that may be misinterpreted as a true signal of environmental damage. Measures chosen for use as early warning signals must exhibit sufficient insensitivity to normal environmental change to reduce the probability of false positive signals to an acceptable level. Problems associated with false positive signals are further reduced by selecting multiple biological measures for monitoring.

Types of algal indicators

Algal indicators can be broadly classified into those that provide structural or functional information concerning community and ecosystem condition. Structural indices include both assessments of algal biomass or standing crop and nontaxonomic measures (e.g., presence or abundance of a particular population or the species composition of the entire assemblage). Functional measures of algal condition include rates of productivity, nutrient flux, and other physiological parameters. Table 1 summarizes the major categories of indicators discussed in this chapter.

Various laboratory procedures have been standardized for assessing algal responses to chemical and other anthropogenic stressors (e.g., see Walsh 1988; Lewis 1992; Cairns et al. 1994). Some of these methods (e.g., algal growth potential tests and related bioassays) have been adapted for field use (e.g., Munawar and Munawar 1987). However, field assessments generally rely on characterizing the state of the indigenous algal assemblage in order to assess ecosystem condition. We limit our review and analysis to algal indicators most widely used in previous field ecological studies or monitoring programs.

Algal abundance and standing crop

The most fundamental assessment of the condition of any plant population or community is the number of individuals present in a given area or volume of habitat. A quantitative sample (i.e., collection of a known surface area or volume of water) is required to determine cell density or biomass per unit of habitat. Quantitative sampling of phytoplankton is generally more straightforward (Clesceri et al. 1989) than for benthic algae, which grow on a variety of different substrates (e.g., plants, rock, mud, and sand). Methods suitable for sampling different types of surfaces vary considerably and have been reviewed by Aloi (1990). Quantitative collection is often simplified through the use of artificial or introduced substrata of known size; the advantages and limitations of these collecting devices have been thoroughly discussed elsewhere (Weitzel et al. 1979; Tuchman and Stevenson 1980; Stevenson and Lowe 1986; Aloi 1990). Traditional methods for mounting and

Table 1 Community Attributes and Associated Algal Indicators Most
Commonly Used in Field Surveys and Monitoring Programs

Indicator category	Community attribute	Indicator
Nontaxonomic	Abundance	Cell density
structural indicators	Biomass	Ash-free dry mass
		Cell biovolume
		Chlorophyll *a*
Autecological indicators	Indicator species	Saprobien index
		pH index
		Trophic index
	Differentiating species	Pollution Tolerance Index
Community indicators	Biodiversity	Diversity indices
		Evenness indices
		Species richness
	Species composition	Similarity indices
	Ordination	Factor analysis
		Principal components analysis
		Detrended correspondence analysis
		Canonical correspondence analysis
Functional indicators	Net production	Change in biomass
	Productivity	Oxygen evolution
		Carbon (^{14}C) consumption
	Enzyme activity	Alkaline phosphatase

Note: Citations listed are intended to illustrate indicator use and performance. Methodological references for specific indicators are provided in text.

enumerating algae in field samples are discussed by Stevenson and
Lowe (1986). While improvements on many of these methods have
been proposed by various investigators, such refinements have not
been compiled in a single source and remain scattered throughout the
ecological literature.

Because cell size varies greatly among different species of algae,
estimates of algal cell density do not provide a reliable means of assessing changes in algal biomass or standing crop. Algal biomass is commonly estimated using one of three measures: (1) ash-free dry mass
(AFDM), (2) chlorophyll concentration, or (3) algal biovolume. The
utility of these alternative indices and methods for their determination
have been reviewed by Stevenson and Lowe (1986) and Steinman and
Lamberti (1996). Each of these measurements provides somewhat different information, and, because no single approach to estimating algal
biomass is most appropriate under all conditions, the use of multiple

measurements (e.g., measurement of chlorophyll *a* content and AFDM) is recommended to provide a more robust assessment of changes in biomass in space and time (Stevenson and Lowe 1986). Comparison of the ratio of AFDM to chlorophyll *a* in samples collected from different sites can be used to calculate the autotrophic index (Clesceri et al. 1989), which provides an indication of shifts in relative magnitude of heterotrophic and photoautotrophic processes within the larger microbial community. As for any integrative measurement, the limitations and range of applicability of individual biomass measurements apply to the autotrophic index as well.

The appeal of nontaxonomic algal measures such as biomass lies in their simplicity. Estimates of density and biomass can be obtained relatively quickly, and most require no taxonomic expertise. However, these measures also provide limited and potentially unreliable information on ecosystem condition unless used in conjunction with other algal measures (e.g., biomass-specific productivity or nutrient content). For example, reductions in biomass resulting from scouring or herbivory may be mistakenly interpreted as an impact from pollution. For such reasons, it is not recommended that biomass measurements be used alone to characterize the condition of the algal assemblage.

Taxonomic analyses of algal assemblages

The use of taxonomic analyses for evaluating ecosystem condition requires a definitive, commonly accepted classification scheme and the availability of qualified personnel for sample processing. Among the algae, diatoms (Class Bacillariophyceae) offer several advantages as environmental indicators. Compared with other algal groups (e.g., Cyanobacteria), the taxonomic classification of diatoms is well understood and widely accepted. Standardized methods for the collection and processing of diatom samples have been published in the peer-reviewed literature (e.g., see Stevenson and Lowe 1986). Diatom identification to the species level is based on the characteristics of the cell wall and is not dependent on observations of specialized structures (e.g., gametangia) which often are not present in field samples. Reliable identifications of algae to the species level are essential if species composition is to be used to infer environmental conditions because substantial autecological information may be lost if only coarser taxonomic identifications are obtained (Stoermer 1978). While Coste and co-workers (1991) found an overall highly significant correlation between a species- and genus-level diatom index of water quality in French watersheds, they also noted that the correspondence between these two indices decreased markedly with deteriorating water quality conditions.

Taxonomic assessments of the algal assemblage have historically been conducted for two reasons: (1) to classify the condition of a water-body based on the specific populations (i.e., indicator species) found to be present and (2) to compare algal diversity or taxonomic composition between reference (i.e., minimally impacted) sites and those suspected of being contaminated. Both population- and community-level approaches provide useful indications of ecological conditions, but each approach also has characteristic limitations (described later). Both types of information can usually be obtained from the same taxonomic data set and, when considered together, provide a more complete and robust reflection of the degree and causes of ecosystem change than when either is used alone.

Autecological indices

As for higher organisms, each algal species has a unique range of physiological tolerances to different environmental parameters. It follows that a given habitat will be dominated by those populations that find the environmental conditions most favorable for growth. Algal species have been found to vary in their sensitivity to certain anthropogenic perturbations, so that pollution-resistant species would be expected to dominate in degraded habitats, whereas pollution-sensitive species would be restricted to pristine locations. Thus, an understanding of the autecology (i.e., environmental tolerances) of individual species provides a potentially powerful tool for characterizing ecological conditions and for providing a preliminary diagnosis as to the causes of observed impacts.

The presence of certain algal species has long been used to classify aquatic ecosystems according to the degree of impact from organic enrichment (e.g., insufficiently or untreated wastewater). Various lists of taxa indicative of different degrees of impact have been compiled based on available information on species tolerances to this form of pollution, beginning with the Saprobien Index of Kolkwitz and Marsson (1908). Limited success in ranking ecosystem conditions using this approach was attributed to the flawed assumption that the mere presence of a particular species was sufficient for characterizing trophic state. In fact, many proposed "indicator" species were found to tolerate a broader range of water quality conditions than those characteristic of the zone to which they had been assigned (Patrick et al. 1954). Early classification schemes were improved by considering the proportional abundance of a species within the assemblage (Pantle and Buck 1955) and by weighting species based on the reliability with which their presence indicates a certain degree of pollution (Sládecek 1964). Lange-Bertalot (1979) proposed an alternative method for habitat assessment

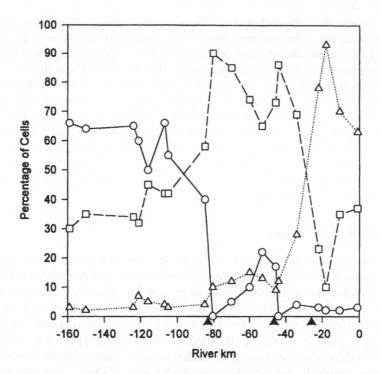

Figure 2 Changes in the relative abundance of pollution-sensitive (○), moderately pollution-tolerant (□), and extremely tolerant (△) diatom species along the course of the River Main, Germany from the upper reaches to the mouth at Wiesbaden as determined from the pollution tolerance index of Lange-Bertalot (1979). The location of major point source discharges into the river are shown by arrows along the bottom horizontal axis. Note the correspondence between such discharges and changes in the diatom flora downstream, specifically decreases in the relative abundance of pollution-sensitive taxa and increased dominance by moderately and extremely tolerant species. (From Lange-Bertalot H [1979] *Nova Hedwigia* 64:285–304. With permission.)

based upon the degree of tolerance of different diatom species to pollution rather than their presumed affinities to a particular level of impact (see Figure 2). Refinements of this approach continue to find use in many European countries (see Whitton et al. 1991).

Changes in populations of siliceous algae (diatoms and chrysophytes) have been used to evaluate ecosystem changes associated with specific anthropogenic impacts such as acid deposition (e.g., Battarbee et al. 1986; Dixit et al. 1989; Cummings et al. 1992) and increased nutrient loads (e.g., Agbeti 1992). Modern multivariate techniques used in these studies provide an objective means for identifying preferences of species across distinct regional environmental gradients (see Figure 3). Species preferences or optima derived from these analyses are used to develop autecological indices for classifying the condition of other waterbodies and for inferring historical changes in environmental conditions from paleoecological data (e.g., sediment cores). Autecological information combined with stratigraphic profiles of diatom and chrysophycean microfossils in sediment cores have provided a historical account of changes in major water quality parameters (e.g., trophic status, pH, and salinity) in North American lakes dating as far back as the time of European settlement and beyond (e.g., Wolin et al. 1988; Kingston et al. 1990; Stoermer et al. 1990). Such analyses have even found use in reconstructing historical relationships between environmental deterioration and changes in the condition of economically valued aquatic resources such as fisheries (Kingston et al. 1992). Application of these methods for assessing impacts associated with other stressors (e.g., toxic) will require information on species autecologies not currently available.

Autecological indices have been used principally to quantify changes in the ecological condition along a single anthropogenic gradient, and it is unlikely that indices developed to quantify impacts associated with one stressor are applicable for assessing other forms of impact. Because different classes of chemicals often affect cellular processes in different ways, it should not be expected that a species will exhibit similar tolerances to different types of pollution. Species of other organisms exhibit differential tolerances to different classes of chemicals (Mayer and Ellersieck 1986), and it is reasonable to expect that the same holds true for algae. Although some indices have been purported to reflect changes in pollution in general (e.g., Descy 1979), it is not apparent that any "generic" measure of impact has been validated under a range of pollutant regimes.

The application of autecological approaches has also been limited to circumstances where sampling locations are arranged along a specific and rather steep gradient of impact. However, many ecosystems are increasingly affected by the cumulative effects of many anthropogenic stressors, although each may be present at lower levels than in the past due to increasingly stringent environmental regulations. Exposure to one stressor can alter algal and other microbial responses to additional impacts (Niederlehner and Cairns 1993). For example, increased availability of a limiting nutrient can affect algal responses to toxic inputs at the ecosystem as well as the physiological level (Stoermer et al. 1980; Chen 1989). Findings such as these suggest that changes in specific populations may not consistently correspond to changes in specific stressors. Clearly, evaluations of the performance of both stressor-specific and general autecological indices in ecosystems exposed to different cumulative impacts are needed to understand more thoroughly the applicability and limitations of these approaches.

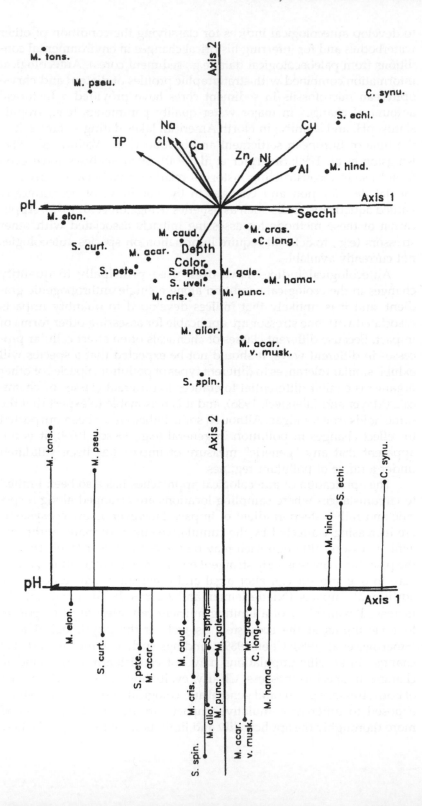

Community structure

Measures described here also involve the identification and enumeration of individual species, but, in contrast to population-based approaches just described, require little or no understanding of the autecology of individual species in order to be implemented. Therefore, the use of community analyses is not limited by the availability and reliability of information concerning species tolerances to different stressors. However, a reference condition is typically required in order to interpret shifts in community structure among sites within a watershed or other geographic unit. This benchmark can be obtained by sampling minimally impacted sites within the study area and/or by reconstructing historical conditions through paleoecology or historical collections as discussed earlier.

Species number or richness is the most fundamental measure of community structure and has been used routinely in environmental monitoring programs. Changes in species richness reflect the differential tolerances of individual populations to pollution without requiring that these tolerances be quantified. Species richness may increase or decrease in response to modest changes in water quality depending upon whether these changes represent a subsidy or a stress to the organisms, but highly impacted sites typically have very few species compared to reference sites (see Figure 4).

Accurate estimation of algal species richness is complicated by at least three factors. First, characterization of an entire algal assemblage demands a high level of taxonomic expertise because hundreds of species usually must be identified accurately and many may be represented by only a few specimens. Rare taxa and those distinguished on the basis of reproductive structures or other ephemeral features may not be identifiable to the species level in all samples. Second, considerable time is required to estimate species richness accurately. Third,

Figure 3 (opposite) Relationship between chrysophycean species distributions and environmental parameters in 72 lakes affected by smelters in the Sudbury region of Ontario, Canada as determined by Dixit et al. (1989). Canonical correspondence analysis, a multivariate statistical technique, was used to plot the optimal occurrence of different taxa along axes that represent linear combinations of measured environmental variables (top). Changes in specific environmental variables are presented on the same plot as arrows, the length of which indicate the importance of that variable in explaining changes in species composition among the lakes. Species optima along major environmental gradients (e.g., pH in this example) are located by drawing a line from the species point perpendicular to the arrow and noting the point of intersection. Analyses such as these provide information for characterizing the autecology and indicator potential of different algal species with respect to specific anthropogenic impacts. (From Dixit SS et al. [1989] *Can J Fish Aquat Sci* 46:1667–1676. With permission.)

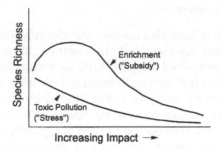

Figure 4 Contrasting response of algal assemblages to increasing impacts from nutrient enrichment (subsidy effects) and toxic pollution (stress effects). Note that species richness (and diversity) are often enhanced by low levels of enrichment, whereas severe impacts by any chemical stressor typically depress these community measures.

species richness is a function of sampling effort. Based on detailed and laborious taxonomic counts of stream diatom assemblages, Patrick and co-workers (1954) concluded that several thousand cells must be counted and correctly identified to measure accurately the species richness of a sample because most species are present at very low abundances. Much shorter counts, e.g., 500 cell counts recommended by Stevenson and Lowe (1986), are widely used in ecological studies to estimate species richness, although these procedures only consider species that are at least moderately abundant. If rare species are of interest, more exhaustive counting (e.g., 1000 cells per sample) is required. Standardized counts of the type just described probably yield better estimates of species evenness, i.e., the relative dominance of different species in a community, than of species richness. For example, increasing levels of impact typically result in algal assemblages that are increasingly dominated by a few species such that the number of species identified in a 500-cell count will be less than for an assemblage where species are present in more equitable abundances. Regardless of which aspect of community structure is to be estimated, any reasonable counting procedure will be biased in some manner. For the purposes of a standardized monitoring program, it is most important that a consistent counting scheme (e.g., same number of cells counted in all samples) be adopted for an entire study to ensure comparability among samples.

Community diversity indices (e.g., Shannon-Weaver) have been widely used as an indicator of ecosystem condition. Diversity indices are attractive because they allow for information concerning both species richness and species evenness to be integrated into a single expression of ecological condition. The usefulness of diversity indices in assessing ecosystem condition was originally predicated upon a long held assumption that communities exhibiting greater species diversity also possessed greater ecological stability and, by extension, were more "healthy" than those with lower species richness. In fact, no conclusive relationship between diversity and stability has been established, and some pristine habitats (e.g., boreal forests) exhibit extremely low diversity. Algal assemblages exhibit both increases and decreases in species diversity in response to anthropogenic impacts (e.g., Archibald 1972; Stevenson 1984). Thus, there is little empirical support for characterizing habitat quality on the basis of the magnitude of the diversity index. Rather, the *magnitude of change* in diversity from the reference condition is most indicative of deterioration in water quality (e.g., Stevenson 1984). The mathematical formulations and properties of different estimators of species diversity and evenness are discussed in depth by Pielou (1975). As recommended by Stevenson and Lowe (1986), a minimum of 1000 cells should be enumerated per sample for the estimation of species diversity.

Measures of similarity in species composition between sampling sites have been used increasingly to detect ecological impacts from pollution. Similarity indices make greater use of community information than do diversity indices because the identities of individual species, as well as their abundance, are considered in the mathematical calculations. Also, similarity indices are often more sensitive to changes in water quality than diversity indices (Boyle et al. 1984; Pontasch and Brusven 1988). Various similarity (and dissimilarity) indices are available, and the choice of an index can influence the conclusions derived from the analysis (e.g., Czarnecki 1979; Hruby 1987). There are few guidelines for selecting the algorithm most suitable for a particular purpose, although the following general recommendations have been made (Hruby 1987; Pontasch et al. 1989): (1) use of indices based on species abundance rather than mere presence or absence, (2) use of absolute abundance rather than proportional abundance when possible, (3) use of indices that *do not* consider the absence of a species from two samples as a point of similarity, and (4) use of data transformations (e.g., logarithmic) to avoid excessive influence of dominant taxa on the index. While noting that no index is necessarily most suitable in all cases, both of the studies just cited indicate that the Bray-Curtis index (Bray and Curtis 1957) provided the most reliable summation of macroinvertebrate species-sample matrices obtained from environmental impact studies. Confidence intervals and other statistics for detecting differences among sites can be calculated for similarity values if replicate samples are available from each site being compared (Smith et al. 1986, 1990; Pontasch et al. 1989). Statistical inference of this sort is preferable to the use of clustering procedures, which require subjective decisions to be made concerning differences in community composition among sites.

Multivariate statistical analyses can provide a useful tool for associating changes in community composition and water quality from some data sets. These procedures allow for multidimensional species-sample matrices to be collapsed to a few derived variables that retain much of the information in the original matrix. These derived variables can then be correlated with water quality information to identify significant species–environmental relationships (e.g., Varis et al. 1989; Guzkowska and Gasse 1990). Multivariate analyses are also useful for distinguishing taxonomic structure associated with natural vs. anthropogenic gradients. For example, Swift and Nicholas (1987) used factor analysis to distinguish changes in the taxonomic composition of the Florida Everglades periphyton community across natural environmental gradients of water hardness and pH from those associated with increased phosphorus loading from agricultural runoff. The influence of differences in water quality among tributaries of the Ohio River on phytoplankton species composition were separated out from seasonal influences using various multivariate procedures (Peterson and Stevenson 1989). The utility of multivariate analyses is reduced either when many anthropogenic gradients coincide or when natural and anthropogenic gradients are strongly correlated, for example, as in some longitudinal surveys of streams (Chessman 1987). In such instances, controlled experimental studies are required to establish important relationships between community structure and water quality. Useful overviews of the application of multivariate analyses in ecology are provided by Gauch (1982) and Jongman et al. (1987). The use of various multivariate techniques in environmental studies is illustrated by Green (1979).

Functional indices

Functional measures of algal condition provide valuable information concerning rates of important ecosystem processes, but are not routinely used in water quality assessments. Although a conceptual debate over the relative merits of structural and functional measures for evaluating ecosystem stress persists (e.g., Cairns and Pratt 1986; Crossey and LaPoint 1988), structural indicators are generally easier to standardize and collect under field conditions as part of a routine monitoring program. Substantial field time is required to prepare, incubate, and process samples in order to obtain functional measures *in situ*, thus limiting the number of sampling stations that can be evaluated in a single day. Logistical problems associated with field incubations can be reduced through the use of shipboard or laboratory incubations, although simulating ambient light and water quality conditions in the laboratory is difficult and often costly.

Intuitively, it can be reasoned that metabolic responses to environmental change, which reflect changes in the physiology of individual cells, should precede shifts in species composition, which reflect differential changes in population growth of different taxa resulting from these physiological effects. However, physiological measurements such as algal productivity are susceptible to rapid and erratic fluctuations in environmental conditions unrelated to anthropogenic impacts (e.g., daily changes in light intensity or the timing and frequency of natural disturbances). Even slight differences in current velocity among experimental channels, for example, can contribute to substantial error in productivity estimates for stream periphyton (e.g., Bott et al. 1985; Crossey and LaPoint 1988). In summarizing findings from a series of laboratory studies, Pratt and Bowers (1992) concluded that, while variability of structural and functional measures was similar, functional parameters were generally more susceptible to changes in resource supply rates, which may confound responses to toxic chemicals and other stressors. For such reasons, changes in taxonomic composition often provide a more reliable early warning indication of environmental damage (e.g., Stoermer 1984; Eaton et al. 1986; Schindler 1987).

Primary productivity is by far the most widely used functional indicator of algal condition. Algal productivity is typically measured by quantifying either the rate of oxygen evolution or carbon dioxide consumption per unit time within a closed system. Oxygen production is quantified using either titrimetric methods or a polarographic oxygen probe (Clesceri et al. 1989). Probes provide the most expedient means of quantifying changes in oxygen concentrations within experimental chambers, but are generally less accurate than titrations, particularly in instances where oxygen bubbles become trapped in benthic algal mats. Carbon dioxide consumption is usually quantified by measuring the uptake of radiolabeled carbon (^{14}C) as bicarbonate by the algae assemblage (Clesceri et al. 1989). A comparison of the utility and limitations of oxygen and carbon dioxide methods for quantifying algal productivity is provided by Aloi (1990) and Wetzel and Likens (1991). Whereas changes in ^{14}C concentration can be measured much more accurately than changes in dissolved oxygen, carbon uptake measurements provide an estimate of net primary productivity only. Measurements of oxygen fluxes in light and dark enclosures provide simultaneous estimates of net algal productivity, community respiration, and gross productivity. Rate estimates should be standardized for biomass and cumulative radiation received during incubation to allow for meaningful comparisons among sites.

Short-term measurements of algal productivity are considerably more difficult in flowing water systems than in lakes and ponds. Because relatively small changes in current velocity can substantially influence algal metabolism (Whitford and Schumacher 1961), it is essential that a consistent and environmentally realistic current regime be maintained within experimental chambers used to quantify and compare algal productivity at different sites. For this reason, chambers used to estimate lotic productivity must be more elaborate than those used in lakes to allow for water recirculation (e.g., Rodgers et al. 1978; Keithan and Lowe 1985; Crossey and LaPoint 1988). Open stream methods provide an alternative technique for estimating system productivity in streams where algal growth is patchy, but require estimates of gas exchange rates between water and air, which may be difficult if flow is turbulent (see Marzolf et al. 1994).

A number of indices (e.g., changes in storage composition, ratios of structural components, etc.) have been used in ecological investigations to assess the physiological status of algal assemblages, particularly with regard to nutrient limitation (e.g., Rosen and Lowe 1984; Pick 1987; Flynn et al. 1989). However, these methods are generally not suited for routine monitoring since changes in these indicators are not always easily interpreted and many of the techniques have yet to be standardized. For example, the activity of specific enzymes (e.g., alkaline phosphatase) has been proposed as a measure of the nutrient status of algae (Fitzgerald and Nelson 1966). While increases in alkaline phosphatase activity are affected by phosphorus availability, production of this enzyme is also affected by changes in temperature, light, growth rate, senescence, and the presence of other pollutants such as heavy metals (see Wynne et al. 1991), thus making it difficult to infer phosphorus limitation from phosphatase activity alone. Because the algal assemblage coexists with bacteria and other heterotrophic microbes, most measures of physiological activity indicate impacts on the broader microbial community and not just the algae.

Summary and conclusions

Regulatory agencies and other organizations involved in environmental monitoring have often embraced monitoring techniques with which they are most familiar and indicator organisms which elicit the greatest public recognition and concern. It is not surprising, therefore, that large, conspicuous organisms such as birds and fish are widely used for biological assessment and monitoring programs designed to gauge ecosystem condition. The disproportionate use of a few groups of organisms for a wide range of monitoring activities is inconsistent with available scientific evidence, which indicates that no single group of organisms provides the most reliable assessment of ecosystem condition under all circumstances. Compared with the many heterotrophic assemblages used for monitoring purposes, algae provide a relatively unique perspective on the nature of environmental impact, one which provides direct evidence of changes in the resource base, i.e., aquatic primary production. Because of their short response times, algal assemblages often provide one of the first signals of ecosystem impacts, thereby allowing for corrective regulatory and management actions to be taken before other undesirable impacts occur.

The increased use of algae in monitoring efforts is hindered by a lack of commonly accepted indicators and procedures for their measurement. For example, there is no consensus in the form of a published document concerning the types of field indicators that are most appropriate and reliable for different uses. Development of standardized procedures for sample collection and processing and, most importantly, for analyzing information generated from algal samples lags behind that for other groups of aquatic organisms. Although the objectives of individual monitoring programs and, therefore, the types of indicators required are somewhat unique, an array of commonly accepted measures should be available from which to select. Methodologies for conducting these measurements must be validated under different conditions to ensure adequate quality control and to enhance the comparability of information among monitoring programs and regions.

For reasons stated in this chapter, we advocate the use of structural rather than functional measures of algal condition for routine monitoring purposes. Although taxonomic analyses are more costly to perform, they provide a more reliable estimate of ecosystem condition than simple measures such as biomass. It is not always necessary to characterize the entire algal assemblage; focusing on a single diverse assemblage such as the diatoms should provide sufficient information for assessing the nature and extent of ecosystem change. While measures of physiological condition hold promise as sensitive indicators of ecosystem change, further research is required to develop functional measures that are both interpretable and reliable under a wide range of circumstances.

Regardless of the group of organisms selected for monitoring, it is essential that reliance be placed on more than one type of indicator. For example, while analysis of changes in specific algal populations can be useful for classifying ecosystem condition with respect to well-defined anthropogenic gradients, available autecological information limits the application of this approach to a few classes of pollution. Furthermore, autecological indices may yield inconclusive or even erroneous results in cases where habitats are subject to chronic, low-level impacts from multiple stressors. Community-level analyses are capable of quantifying ecosystem change in response to a broad range of impact scenarios, including those involving cumulative effects, but require that a suitable benchmark or reference condition be established. Analysis and interpretation that considers both levels of biological organization derive the greatest amount of information from taxonomic data sets and should allow for a more complete and reliable assessment of environmental change than if either approach is used alone.

Acknowledgments

This work was sponsored in part by a grant from the U.S. Air Force Office of Scientific Research (AFOSR-91-0379) to the University Center for Environmental and Hazardous Materials Studies at Virginia Tech, Blacksburg, VA. However, the views expressed herein are solely those of the authors and are not necessarily endorsed by the U.S. Air Force or the South Florida Water Management District. We are indebted to Susan Gray, B. R. Niederlehner, Barry Rosen, and Alan Steinman for comments on earlier drafts of the manuscript. The authors also thank Darla Donald for editorial assistance in the preparation of this manuscript for publication and Teresa Moody for transcribing dictation.

References

Agbeti MD (1992) Relationship between diatom assemblages and trophic variables: a comparison of old and new approaches. *Can J Fish Aquat Sci* 49:1171–1175.

Aloi JE (1990) A critical review of recent freshwater periphyton field methods. *Can J Fish Aquat Sci* 47:656-670.

Amblard C, Couture P, Bourdier G (1990) Effects of a pulp and paper mill effluent on the structure and metabolism of periphytic algae in experimental streams. *Aquat Toxicol* 18:137–162.

Archibald REM (1972) Diversity in some South African diatom associations and its relation to water quality. *Water Res* 6:1229-1238.

Auer MT, Canale RP, Grundler HC, Matsuoka Y (1982) Ecological studies and mathematical modeling of *Cladophora* in Lake Huron: 1. Program description and field monitoring of growth dynamics. *J Great Lakes Res* 8:73–83.

Battarbee RW, Smol JP, Meriläinen J (1986) Diatoms as indicators of pH: an historical review. In *Diatoms and Lake Acidity*, Smol JP, Battarbee RW, Davis RB, Meriläinen J, Eds., W. Junk Publishers, Dordrecht, pp. 5–14.

Bentley-Mowat JA, Reid SM (1977) Survival of marine phytoplankton in high concentrations of heavy metals, and uptake of copper. *J Exp Mar Biol Ecol* 26:249–264.

Blanck H, Wängberg SÅ (1988) Induced community tolerance in marine periphyton established under arsenate stress. *Can J Fish Aquat Sci* 45:1816–1819.

Boaden PJS, O'Connor RJ, Seed R (1975) The composition and zonation of a *Fucus serratus* community in Strangford Lough, Co. Down. *J Exp Mar Biol Ecol* 17:111–136.

Bott TL, Brock JT, Dunn CS, Naiman RJ, Ovink RW, Petersen RC (1985) Benthic community metabolism in four temperate stream systems: an interbiome comparison and evaluation of the river continuum concept. *Hydrobiologia* 123:3–45.

Boyle TP, Sebaugh J, Robinson-Wilson E (1984) A hierarchical approach to the measurement of changes in community induced by environmental stress. *J Test Eval* 12:241–245.

Bray JR, Curtis JT (1957) An ordination of the upland forest communities of southern Wisconsin. *Ecol Monogr* 27:325–349.

Browder JA (1982) Biomass and primary production of microphytes and macrophytes in periphyton habitats of the southern Everglades. South Florida Research Center Report T-662, National Park Service, Everglades National Park, Homestead, FL.

Cairns J Jr. (1992) Probable consequences of a cosmopolitan distribution. *Spec Sci Technol* 14:41–50.

Cairns J Jr., Pratt JR (1986) On the relation between structural and functional analyses of ecosystems. *Environ Toxicol Chem* 5:785–786.

Cairns J Jr., Niederlehner BR (1992) Adaptation and resistance of ecosystems to stress: a major knowledge gap in understanding anthropogenic perturbations. *Spec Sci Technol* 12:23–30.

Cairns J Jr., McCormick PV, Niederlehner BR (1994) Bioassay and field assessment of pollutant effects. In *Algae and Water Pollution*, Rai LC, Gaur JP, Soeder CJ, Eds., E. Schweizerbart'sche Verlagsbuchhandlug, Stuttgart, pp. 267–282.

Chen CY (1989) The effects of limiting nutrient to algal toxicity assessment: a theoretical approach. *Toxic Assess* 4:35–42.

Chessman BC (1987) Diatom flora of an Australian river system: spatial patterns and environmental relationships. *Freshwater Biol* 16:805–819.

Clesceri LS, Greenberg AE, Trussell RR (1989) *Standard Methods for the Examination of Water and Wastewater*, 17th ed., American Public Health Association, Washington, D.C.

Coste M, Bosca C, Dauta A (1991) Use of algae for monitoring rivers in France. In *Use of Algae for Monitoring Rivers*, Whitton BA, Rott E, Friedrich G, Eds., Institut für Botanik, Innsbruck, Austria, pp. 75–88.

Craft CB, Richardson CJ (1993) Peat accretion and N, P, and organic C accumulation in nutrient-enriched and unenriched Everglades peatlands. *Ecol Appl* 3:446–458.

Crossey MJ, La Point TW (1988) A comparison of periphyton community structural and functional responses to heavy metals. *Hydrobiologia* 162:109–121.

Cummings BF, Smol JP, Birks HJB (1992) Scaled chrysophytes (Chrysophyceae and Synurophyceae) from Adirondack drainage lakes and their relationship to environmental variables. *J Phycol* 28:162–178.

Czarnecki DB (1979) Epipelic and epilithic diatom assemblages in Montezuma Well National Monument, Arizona. *J Phycol* 15:346–352.

DeMott WR, Moxter F (1991) Foraging on cyanobacteria by copepods: responses to chemical defenses and resource abundance. *Ecology* 72:1820–1834.

Descy JP (1979) A new approach to water quality estimation using diatoms. *Nova Hedwigia* 64:305–323.

Dixit SS, Dixit AS, Smol JP (1989) Relationship between chrysophyte assemblages and environmental variables in seventy-two Sudbury lakes as examined by canonical correspondence analysis (CCA). *Can J Fish Aquat Sci* 46:1667–1676.

Eaton J, Arthur J, Hermanutz R, Kiefer R, Mueller L, Anderson R, Erickson R, Nordling B, Rodgers J, Pritchard H (1986) Biological effects of continuous and intermittent dosing of outdoor experimental streams with chlorpyrifos. In *Aquatic Toxicology and Hazard Assessment*, Eighth Symposium, Bahner RC, Hansen DJ, Eds., American Society for Testing and Materials, Philadelphia, PA, pp. 85–118.

Edmondson WT (1961) Changes in Lake Washington following an increase in the nutrient income. *Verh Int Verein Limnol* 14:167–175.

Fitzgerald GP, Nelson TC (1966) Extractable and enzymatic analyses for limiting or surplus phosphorus in algae. *J Phycol* 2:32–37.

Fjerdingstad E (1964) Pollution of streams estimated by benthal physomicroorganisms. I. A saprobic system based on communities of organisms and ecological factors. *Int Rev Hydrobiol Hydrogr* 49:63–131.

Flynn KJ, Dickson DMJ, Al-Amoudi OA (1989) The ratio of glutamine: glutamate in microalgae: a biomarker for N-status suitable for use at natural cell densities. *J Plankton Res* 11:165–170.

Foster PL (1982) Metal resistances of Chlorophyta from rivers polluted by heavy metals. *Freshwater Biol* 12:41–61.

Gauch HG Jr. (1982) *Multivariate Analysis in Community Ecology*, Cambridge University Press, Cambridge, U.K.

Gilbert JJ (1990) Differential effects of *Anabaena affinis* on cladocerans and rotifers: mechanisms and implications. *Ecology* 71:1727–1740.

Gorham PR (1964) Toxic algae. In *Algae and Man*, Jackson DF, Ed., Plenum Press, New York, pp. 307–336.

Green RH (1979) *Sampling Design and Statistical Methods for Environmental Biologists*, John Wiley & Sons, New York.

Guzkowska MAJ, Gasse F (1990) The seasonal response of diatom communities to variable water quality in some English urban lakes. *Freshwater Biol* 23:251–264.

Hansson L-A (1992) Factors regulating periphytic algal biomass. *Limnol Oceanogr* 37:322–328.

Harding LW (1976) Polychlorinated biphenyl inhibition of marine phytoplankton photosynthesis in the north Adriatic Sea. *Bull Environ Contam Toxicol* 16:559–566.

Hawxby K, Tubea B, Ownby J, Bosler E (1977) Effects of various classes of herbicides on four species of algae. *Pestic Biochem Physiol* 7:203–209.

Hruby T (1987) Using similarity measures in benthic impact assessments. *Environ Monit Assess* 8:163–180.

Hughes RM, Larsen DP, Omernik JM (1986) Regional reference sites: a method for assessing stream potentials. *Environ Manage* 10:629–635.

Hynes HBN (1969) The enrichment of streams. In *Eutrophication: Causes, Consequences, Correctives*, National Academy of Sciences, Washington, D.C., pp. 188–196.

Jongman RHG, ter Braak CJF, van Tongeren OFR (1987) *Data Analysis in Community and Landscape Ecology*, Pudoc, Wageningen, The Netherlands.

Karr JR (1991) Biological integrity: a long neglected aspect of water resource management. *Ecol Appl* 1:66–84.

Karr JR, Fausch KD, Angermeier PL, Yant PR, Schlosser IJ (1986) Assessing biological integrity in running waters: a method and its rationale. Ill Nat Hist Surv Spec Publ 5, Urbana, IL, 28 pp.

Keithan ED, Lowe RL (1985) Primary productivity and spatial structure of phytolithic growth in streams in the Great Smoky Mountains National Park, Tennessee. *Hydrobiologia* 123:59–67.

Kingston JC, Birks HJB (1990) Dissolved organic carbon reconstructions from diatom assemblages in PIRLA project lakes, North America. *Phil Trans R Soc London B* 327:279–288.

Kingston JC, Cook RB, Kreis RG Jr., Camburn KE, Norton SA, Sweets PR, Binford MW, Mitchell MJ, Schindler SC, Shane LCK, King GA (1990) Paleoecological investigation of recent lake acidification in the northern Great Lakes states. *J Paleolimnol* 4:153–201.

Kingston JC, Birks HJB, Uutala AJ, Cumming BF, Smol JP (1992) Assessing trends in fishery resources and lake water aluminum from paleolimnological analyses of siliceous algae. *Can J Fish Aquat Sci* 49:116–127.

Knopman DS, Smith RA (1993) Twenty years of the Clean Water Act. *Environment* 35:17–41.

Kolkwitz R, Marsson M (1908) Ökologie der pflanzlichen Saprobien. *Ber Dtsch Bot Ges* 26:505–519.

Kosinski RJ (1984) The effect of terrestrial herbicides on the community structure of stream periphyton. *Environ Pollut Ser A* 36:165–189.

Krieger KA, Baker DB, Kramer JW (1988) Effects of herbicides on stream *Aufwuchs* productivity and nutrient uptake. *Arch Environ Contam Toxicol* 17:299–306.

Lamberti GA, Moore JW (1986) Aquatic insects as primary consumers. In *The Ecology of Aquatic Insects*, Resh VH, Rosenberg DM, Eds., Praeger Scientific Publishers, New York, pp. 164–195.

Lange-Bertalot H (1979) Pollution tolerance of diatoms as a criterion for water quality estimation. *Nova Hedwigia* 64:285–304.

Leach JH, Johnson MG, Kelso JRM, Hartmann J, Numann W, Entz B (1977) Response of persid fishes and their habitats to eutrophication. *J Fish Res Board Can* 33:1964–1971.

Leonard PM, Orth DJ (1986) Application and testing of an index of biotic integrity in small, coolwater streams. *Trans Am Fish Soc* 115:401–415.

Lewis MA (1992) Periphyton photosynthesis as an indicator of effluent toxicity: relationship to effects on animal tests species. *Aquat Toxicol* 23:270–288.

Marzolf ER, Mulholland PJ, Steinman AD (1994) Improvements to the diurnal upstream-downstream dissolved oxygen change technique for determining whole-stream metabolism in small streams. *Can J Fish Aquat Sci* 51:1591–1599.

Mayer FL Jr., Ellersieck MR (1986) Manual of aquatic toxicity: interpretation and data base for 410 chemicals and 66 species of freshwater animals. Resource Publication 160, U.S. Department of Interior, Washington, D.C.

Mayer MS, Likens, GE (1987) The importance of algae in a shaded headwater stream as food for an abundant caddisfly (Trichoptera). *J N Am Benthol Soc* 6:262–269.

Munawar M, Munawar IF (1987) Phytoplankton bioassays for evaluating toxicity of *in situ* sediment contaminants. *Hydrobiologia* 149:87–105.

Munro MAR, Kreiser AM, Battarbee RW, Juggins S, Stevenson AC, Anderson DS, Anderson NJ, Berge F, Birks HJB, Davis RB, Flower RJ, Fritz SC, Haworth EY, Jones VJ, Kingston JC, Renberg I (1990) Diatom quality control and data handling. *Phil Trans R Soc London B* 327:257–261.

Niederlehner BR, Cairns J Jr. (1993) Effects of previous zinc exposure on pH tolerance of periphyton communities. *Environ Toxicol Chem* 12:743–753.

Niederlehner BR, Cairns J Jr. (1995) Naturally derived microbial communities as receptors in toxicity tests. In *Ecological Toxicity Testing: Scale, Complexity, and Relevance*, Cairns J Jr., Niederlehner BR, Eds., Lewis Publishers, Chelsea, MI, pp. 123–147.

Oberdorff T, Hughes RM (1992) Modification of an index of biotic integrity based on fish assemblages to characterize rivers of the Seine Basin, France. *Hydrobiologia* 228:117–130.

Omernick JM (1987) Ecoregions of the conterminous United States. *Ann Assoc Am Geograph* 77:118–125.

Paerl HW, Bebout BM, Prufert LE (1989) Naturally occurring patterns of oxygenic photosynthesis and N_2 fixation in a marine microbial mat: physiological and ecological ramifications. In *Microbial Mats: Physiological Ecology of Benthic Microbial Communities*, Cohen Y, Rosenberg E, Eds., American Society for Microbiology, Washington, D.C., pp. 326–341.

Pantle R, Buck H (1955) Die biologische Überwachung der Gewässer und die Darstellung der Ergebnisse. *Gas Wassfach* 96–604.

Patrick R, Hohn MH, Wallace JH (1954) A new method for determining the pattern of the diatom flora. *Not Nat* 259:1–12.

Peterson CG, Stevenson RG (1989) Seasonality in river phytoplankton: multivariate analyses of data from the Ohio River and six Kentucky tributaries. *Hydrobiologia* 182:99–114.

Pick FR (1987) Carbohydrate and protein content of lake seston in relation to plankton nutrient deficiency. *Can J Fish Aquat Sci* 44:2095–2101.

Pielou EC (1975) *Ecological Diversity*, John Wiley & Sons, New York.

Plafkin JL, Barbour MT, Porter KD, Gross SK, Hughes RM (1989) *Rapid Bioassessment Protocols for Use in Streams and Rivers: Benthic Macroinvertebrates and Fish,* EPA/444/4-89-001, U.S. Environmental Protection Agency, Washington, D.C.

Pontasch KW, Brusven MA (1988) Diversity and community comparison indices assessing macroinvertebrate recovery following a gasoline spill. *Water Res* 22:619–626.

Pontasch KW, Smith EP, Cairns, J Jr. (1989) Diversity indices, community comparison indices and canonical discriminant analysis: interpreting the results of multispecies toxicity tests. *Water Res* 23:1229–1238.

Porter KG (1976) Enhancement of algal growth and productivity by grazing zooplankton. *Science* 192:1332–1334.

Power ME (1990) Effects of fish in river food webs. *Science* 250:811–814.

Pratt JR, Bowers NJ (1990) A microcosm procedure for estimating ecological effects of chemicals and mixtures. *Toxic Assess* 5:189–205.

Pratt JR, Bowers NJ (1992) Variability in community metrics: detecting changes in structure and function. *Environ Toxicol Chem* 11:451–457.

Pratt JR, Niederlehner BR, Bowers N, Cairns J Jr. (1987) Prediction of permissible concentrations of copper from microcosm toxicity tests. *Toxic Assess* 2:417–436.

Pringle CM, Paaby-Hansen P, Vaux PD, Goldman CR (1986) In situ nutrient assays of periphyton growth in a lowland Costa Rican stream. *Hydrobiologia* 134:207–213.

Riitters KH, Law B, Kucera R, Gallant A, DeVelice R, Palmer C (1990) Indicator strategy for forests. In *Environmental Monitoring and Assessment Program Ecological Indicators,* Hunsaker CT, Carpenter DE, Eds., EPA/600/3-90/060, U.S. Environmental Protection Agency, Research Triangle Park, NC, pp. 6-1–6-13.

Rodgers JH, Dickson KL, Cairns J Jr. (1978) A chamber for in situ evaluations of periphyton productivity in lotic systems. *Arch Hydrobiol* 84:389–398.

Rosen BH, Lowe RL (1984) Physiological and ultrastructural responses of *Cyclotella meneghiniana* (Bacillariophyta) to light intensity and nutrient limitation. *J Phycol* 20:173–183.

Saether OA (1980) The influence of eutrophication on deep lake benthic invertebrate communities. *Prog Water Technol* 12:161.

Scanferlato VS, Cairns J Jr. (1990) Effect of sediment-associated copper on ecological structure and function of aquatic microcosms. *Aquat Toxicol* 18:23–34.

Schindler DW (1987) Detecting ecosystem responses to anthropogenic stress. *Can J Fish Aquat Sci* 44(suppl 1):6–25.

Sigmon CF, Kania HJ, Beyers RJ (1977) Reductions in biomass and diversity resulting from exposure to mercury in artificial streams. *J Fish Res Board Can* 34:493–500.

Sládecek V (1964) Zur Ermittlung des Indikations-Gewichtes in der Gewasseruntersuchlung. *Arch Hydrobiol* 60:242–243.

Sládecek V (1973) System of water quality from the biological point of view. *Arch Hydrobiol Beih* 7:1–218.

Smith EP, Genter RB, Cairns J Jr. (1986) Confidence intervals for similarity between algal communities. *Hydrobiologia* 139:237–245.

Smith EP, Pontasch KW, Cairns J Jr. (1990) Community similarity and the analysis of multispecies environmental data: a unified statistical approach. *Water Res* 24:507–514.

Steinman AD, Lamberti GA (1996) Biomass and pigments of benthic algae. In *Methods in Stream Ecology,* Hauer FR, Lamberti GA, Eds., Academic Press, San Diego, CA, pp. 295–313.

Stevenson RJ (1984) Epilithic and epipelic diatoms in the Sandusky River, with emphasis on species diversity and water pollution. *Hydrobiologia* 114:161–175.

Stevenson RJ, Lowe RL (1986) Sampling and interpretation of algal patterns for water quality assessments. In *Rationale for Sampling and Interpretation of Ecological Data in the Assessment of Freshwater Ecosystems,* Isom BG, Ed., American Society for Testing and Materials, Philadelphia, PA, pp. 118–149.

Stevenson RJ, Singer R, Roberts DA, Boylen CW (1985) Patterns of epipelic algal abundance with depth, trophic status, and acidity in poorly buffered New Hampshire lakes. *Can J Fish Aquat Sci* 42:1501–1512.

Stoermer EF (1978) Phytoplankton assemblages as indicators of water quality in the Laurentian Great Lakes. *Trans Am Microsc Soc* 97:2–16.

Stoermer EF (1984) Qualitative characteristics of phytoplankton assemblages. In *Algae as Ecological Indicators,* Shubert LE, Ed., Academic Press, London, pp. 49–70.

Stoermer EF, Sicko-Goad L, Lazinsky D (1980) Synergistic effects of phosphorus and heavy metals loading on Great Lakes phytoplankton. In *Proceedings of the Third USA-USSR Symposium on the Effects of Pollutants upon Aquatic Ecosystems,* Swain WR, Shannon VR, Eds., EPA-600/9-80-034, pp. 171–186.

Stoermer EF, Schelske CL, Wolin JA (1990) Siliceous microfossil succession in the sediments of McLeod Bay, Great Slave Lake, Northwest Territories. *Can J Fish Aquat Sci* 47:1865–1874.

Swift DR, Nicholas RB (1987) Periphyton and Water Quality Relationships in the Everglades Water Conservation Areas, Tech Pub 87-2, South Florida Water Management District, West Palm Beach, FL.

Thomas WH, Seibert LR (1977) Effects of copper on the dominance and the diversity of algae: controlled ecosystem pollution experiment. *Bull Mar Sci* 27:23–33.

Tuchman ML, Stevenson RJ (1980) A comparison of clay tile, sterilized rock, and natural substrate diatom communities in a small stream in southeastern Michigan. *Hydrobiologia* 75:73–79.

U.S. Environmental Protection Agency (1993) Stream Indicator and Design Workshop, EPA/600/R-93/138, Environmental Research Laboratory, U.S. Environmental Protection Agency, Corvallis, OR.

Varis O, Sirvió H, Kettunen J (1989) Multivariate analysis of lake phytoplankton and environmental factors. *Arch Hydrobiol* 117:163–175.

Walsh GE (1972) Effects of herbicides on photosynthesis and growth of marine unicellular algae. *Hyacinth Control J* 10:45–48.

Walsh GE (1988) Principles of toxicity testing with marine unicellular algae. *Environ Toxicol Chem* 7:979–987.

Wängberg SÅ, Blanck H (1990) Arsenate sensitivity in marine periphyton communities established under various nutrient regimes. *J Exp Mar Biol Ecol* 139:119–134.

Weitzel RL, Sanocki SL, Holecek H (1979) Sample replication of periphyton collected from artificial substrates. In *Methods and Measurements of Periphyton Communities: A Review,* Weitzel RL, Ed., ASTM STP 690, American Society for Testing and Materials, Philadelphia, PA, pp. 90-115.

Wetzel RG, Likens GE (1991) *Limnological Analyses,* 2nd ed., Springer-Verlag, New York.

Whitford LA, Schumacher GF (1961) Effect of current on mineral uptake and respiration by freshwater algae. *Limnol Oceanogr* 6:423–425.

Whitton BA, Rott E, Friedrich G (1991) Use of Algae for Monitoring Rivers. Institut für Botanik, Innsbruck, Austria.

Wolin JA, Stoermer EF, Schelske CL, Conley DJ (1988) Siliceous microfossil succession in recent Lake Huron sediments. *Arch Hydrobiol* 114:175–198.

Wynne D, Kaplan B, Berman T (1991) Phosphatase activities in Lake Kinneret phytoplankton. In *Microbial Enzymes in Aquatic Environments,* Chróst R, Ed., Springer-Verlag, New York.

chapter eight

Photosynthetic electron transport as a bioassay

Joanna Gemel, Beth Waters-Earhart, Mark R. Ellersieck, Amha Asfaw, Gary F. Krause, Vivek Puri, and William R. Lower

Introduction

During the last 15 years, interest has developed on the practical application of the electron transport of chlorophyll fluorescence as a rapid, sensitive, and nondestructive bioassay for the determination of photosynthetic response of plants to different physical or chemical factors (Papageorgiou 1975; Lavorel and Etienne 1977). The advantage of this technique is that it permits the rapid noninvasive determination of changes in the photosynthetic apparatus much earlier than the appearance of visible injury (Baker et al. 1983). Recently, portable instruments with microcomputers capable of measuring changes in fluorescence yield and features for data processing, analysis, and storage have become available (Schreiber et al. 1975; Ogren and Baker 1985; Schreiber 1986). These features make this technique readily applicable both as a laboratory test and an environmental bioassay in the field under real world conditions (Öquist and Wass 1988; Greaves et al. 1992).

Previous studies from our laboratory have demonstrated the potential of chlorophyll fluorescence measurements as a bioassay (Lower et al. 1984; Schaeffer et al. 1987; Judy et al. 1990a,b, 1991). The current study is a continuation of the development of a useful laboratory and field bioassay for the effects of environmental pollution. We are investigating the response of seven fluorescence parameters of the electron transport of photosynthesis measured by the Morgan CF-1000 fluormeter for different species of plants exposed to different chemicals in order to determine which of the seven parameters are the most useful. Moreover, the responses of particular species to all of the chemicals used is of interest.

The laboratory experiments reported assessed the effects of four chemicals on the chlorophyll fluorescence of five species of plants. The field study is from an area adjacent to Interstate 70 (I-70) near Columbia, MO and measures the photosynthetic response of three species of plants with decreasing distance to the heavily traveled roadway.

A simple description of chlorophyll fluorescence follows: electron transport of photosynthesis begins in the photosystem II (PSII, P680) reaction center where photons produce an oxidizing potential to oxidize water into oxygen, protons, and electrons ($2 H_2O + 4$ photons $\rightarrow O_2 + 4H^+ + 4e^-$). Subsequent energy conversion occurs in photosystem I (PSI, P700) with the reduction of $NADP^+$ and carbohydrate synthesis by CO_2 fixation. *In vivo*, about 3% (Vredenberg and Slooten 1967; Miles 1990) of the energy transferred to the two photosystems is lost as fluorescence emission. After a few minutes of dark acclimation of a leaf and then sudden illumination by light, there is a characteristic onset of kinetic fluorescence emission, known as the Kautsky effect. Figure 1 presents a characteristic chlorophyll fluorescence induction curve and includes nomenclature of Papageorgiou (1975) to describe fluorescence induction phases (namely, the fluorescence levels O, I, D, P, S, M, T). The fast induction kinetics indicated by OIPD are mainly related to the primary photochemistry of PSII. The fluorescence quenching indicated by SMT, respectively, is attributed primarily to the transfer of electrons to PSI, the formation of trans-thylakoid pH gradient (Krause and Weis 1984), phosphorylation of the light harvesting chlorophyll a/b complex (LHCP) (Horton and Black 1983), and, finally, dissipation of excess energy by carotenoids (Demmig-Adams et al. 1989).

After dark adaptation of a leaf, followed by illumination, seven fluorescence parameters can be measured. They are defined in the following (Van Kooten and Snel 1990):

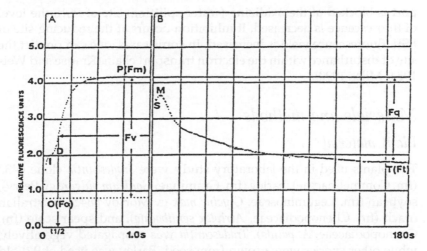

Figure 1 Characteristic chlorophyll fluorescence induction kinetics of a dark adapted maize leaf: (A) fast kinetics, (B) slow kinetics. Light intensity was 450 μmol m^{-2} s^{-1}, and the temperature was 25°C. (From Greaves JA et al. [1992] CF-1000 Instruction Manual, Morgan Instruments, Inc., Andover. With permission.)

1. Fo is the minimal, nonvariable, or initial fluorescence level that is reached immediately after illumination. This value characterizes a completely dark-adapted leaf when most of PSII reaction centers are open and the primary acceptor of PSII, Q$_A$, is fully oxidized.
2. Fm is the maximal fluorescence measured at the P-peak of the induction curve when the Q pool becomes fully reduced.
3. Fv, variable fluorescence, is the difference between Fo and Fm.
4. Fv to Fm ratio is related to the efficiency with which electrons are processed in photosynthesis and commonly has a value around 0.8.
5. t$_{1/2}$ is half the length of time of the rise from the value Fo to Fm and is a measure of the electron acceptor pool and the photochemical reaction of PSII.
6. Ft is the terminal fluorescence value.
7. Fq, the fluorescence quenching (i.e., reduction in the emission of light) capacity of the system, is the difference between Fm and Ft.

Commonly, Ft and Fq are measured at 180 s (3 min). However, Ft and Fq, as used in this study, are approximations because they are measured at 20 s.

Any disturbance in electron transport is reflected by changes in the fluorescence pattern and provides insight into the functioning of the photosynthetic apparatus. Some inhibitors or stress conditions (e.g., cold, heat, drought) change fluorescence parameters. If electron transport is blocked at the oxidizing (water-splitting) site of PSII, the level of fluorescence is decreased. If inhibition occurs at the reducing site of PSII, fluorescence yield is increased. In some cases, one can predict the site of disturbance within the electron transport chain (Krause and Weis 1984; Miles 1990).

Materials and methods

Plant material

The plants used in the laboratory study were *Tradescantia* clone 4430 (fm. Commelinaceae), barley (fm. Graminae, *Hordeum vulgare* cv Perry), soybean (fm. Leguminoseae, *Glycine max* cv Morsoy 8842), Australian orach (fm. Chenopodiaceae, *Atriplex semibacata*), and spear scale (fm. Chenopodiaceae, *A. patula*). *Tradescantia* was propagated vegetatively, while other species were grown from seed. Barley was used at 9 d old (day 1 equals the day planted), soybean at 14 d old, and *A. semibacata* and *A. patula* at 5 to 6 weeks old. The plants were grown in Promix in a plant growth chamber 20.6 m^3 under a cycle of 16 h light/8 h dark. General Electric High Output Cool White fluorescent lights provided a light intensity of 250 μmol m^{-2} s^{-1}. The temperature was approximately 19°C at night and 26°C during the day.

In field studies, three species of plants were examined next to Exit 137 on I-70 at distances of 2.5, 20, 30, and 55 m from the highway: cattail (fm. Typhaceae, *Typha latifolia*), milkweed (fm. Asclepiadaceae, *Asclepias purpurascens* L.), and nightshade (fm. Solanaceae, *Solanum carolinense* L.).

Chemicals

The chemicals were chosen to represent some diversity and include inorganic and organic compounds, priority pollutants (lead chloride, sodium pentachlorophenate, and β-naphthalamine), and a commonly used pesticide (Ortho®*). For laboratory experiments, the chemicals lead chloride (Baker Reagette, 2308-04); sodium pentachlorophenate (Fluka AG, 76480); β-naphthylamine (Sigma, N-8381); and the pesticide Ortho Outdoor Ant, Flea and Cricket spray were used. Ortho is a commercial complex mixture of chlorpyrifos [O,O-diethyl O-(3,5,6-trichloro-2-pyridyl) phosphothioate], 5.3% by weight, and undefined inert ingredients containing petroleum distillate, 94.7% by weight. Concentration of all chemicals, except Ortho, is expressed in micromoles per liter. The concentration of Ortho is expressed in parts per million of commercially available solution and is converted to micromoles of

*Registered trademark of Chevron Chemical Company, San Ramon, CA.

active ingredient per liter. A control group using double-distilled water was included for each experiment conducted.

Exposure of plants to chemicals

Plant cuttings were exposed to chemicals for 24 h in a growth chamber. Exposure was conducted in 4-oz Ball Quilted Crystal Jelly Jars wrapped in aluminum foil on all four sides to control photodegradation of chemicals and placed in six-cup Ekco Bakers Secret Texas Size Muffin Pans with each cup 8.8×4.4 cm ($3^1/_2 \times 1^3/_4$ in.). Holes were punched in the top foil of each jar through which the plant stem or blade was inserted into the solution. The leaves or stems of plants were cut above the roots using a razor blade which was first rinsed with alcohol and then with Milli Q water. Leaves or stems were immediately placed cut side down in the treatment solution. A six-cup muffin pan was used to hold jars and contain spills. Either two or three jars were used for each concentration to provide space for 20 plant cuttings. Thirty milliliters of solution were used in each jar, except for experiments with *A. patula* and *A. semibacata* when 90 ml of the solution was applied. These plants tended to tip slightly after being placed in the jars; therefore, a larger volume was required to prevent them from emerging from the solution. The chemicals were administered to the plants at three to ten concentrations ranging from 0.005 to 9000 mM. Twenty-four hours after cutting, the plants looked healthy with well-preserved turgor except at the highest concentrations of chemical. In this case, partial wilting was due to an effect of the tested substance and not from the cutting of the plant.

On the basis of statistical calculations (Snedecor and Cochran 1989) from preliminary data, a sample size of 20 was chosen for experiments (except when plants died at high concentrations, resulting in a lower sample size). With 20 assays it is possible to detect differences of several percentage points between various concentrations of tested chemicals for all measured parameters. For example, with Fv/Fm, the detectable difference was always less than 7% in all experiments performed to date.

Fluorescence measurements

A Morgan CF-1000 chlorophyll fluorescence measurement system with dark acclimation cuvettes was used to record and measure fluorescence parameters. This instrument is a portable microprocessor and computer-operated instrument for the measurement and detailed analysis of chlorophyll fluorescence induction kinetics. As mentioned earlier, the microprocessor calculates seven fluorescence parameters. The CF-1000 has a fiber-optic probe which inserts into a dark acclimation cuvette. A leaf is placed in the acclimation cuvette, and the fiber-optic probe is then inserted into the cuvette. The dark acclimation cuvette has a shutter gate, allowing the fiber-optic probe connected to the CF-1000 to be inserted through the shutter gate without exposing the leaf to external light. The light source is a halogen lamp. The CF-1000 provides actinic light levels of 30 to 1000 μmol m^{-2} s^{-1} and sample times of 2 to 300 s to determine fluorescence induction curves. The leaves were illuminated for 20 s with 200 or 400 μmol m^{-2} s^{-1} of light, depending on the species, after 2 min of dark acclimation of each leaf. Fluorescence was measured at room temperature from the middle abaxial (upper) surface of plant leaves. For field experiments, illumination of leaves of all species was conducted at 400 μmol m^{-2} s^{-1} between 9 a.m. and 1 p.m. At each experimental site, depending on availability, two to twelve measurements were taken.

Data from the CF-1000 were printed using a DPH-411 thermal printer and were transferred from the CF-1000 to an IBM PCAT 486 for storage and statistical analysis.

For practical reasons, all fluorescence values are presented as readings from the CF-1000. Values for Figure 2 were divided by 100 to convert them into relative units usually used in publications on chlorophyll fluorescence (Greaves et al. 1992), except $t_{1/2}$ which is expressed in milliseconds.

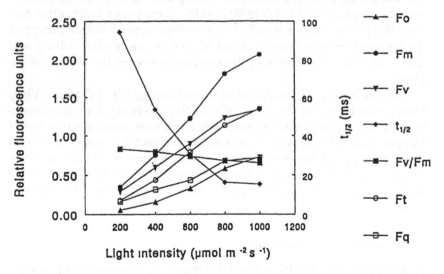

Figure 2 Fluorescence parameters of barley leaves as functions of the light intensity.

Statistical analysis

The statistical package SAS release 6.04 (SAS Institute Inc. 1987) was used for all analyses to provide means, error terms, percent vs. control, analyses of variance (ANOVA), and regression.

A one-way ANOVA using the general linear model procedure was used to evaluate the effect of concentration on each of the seven fluorescence parameters for each experiment conducted.

Regression analysis was completed, and linear, quadratic, cubic, and logarithmic models were evaluated (Zar 1984). In the logarithmic models, the dependent variables consisting of the seven fluorescence parameters were transformed. Statistical significance of models was determined based upon type I sum of squares p-values concerning the F-value.

Results and discussion

There are three aspects to our studies:

1. Determination of response of the seven parameters of control plants not exposed to chemicals, but illuminated at four to five light intensities: 200, 400, 600, 800, or 1000 µmol m^{-2} s^{-1}.
2. Preliminary field study evaluating the response of three species to pollution generated by traffic on I-70.
3. The response of the seven parameters to a variety of concentrations of four chemicals in five species of plants.

Fluorescence parameters under different light intensities

The conditions of measurements for application of photosynthetic electron transport capacity as a bioassay need to be standardized. First, it requires a decision as to the best light intensity needed for each species. Actinic light should be bright enough to saturate all electron acceptors of PSII at the P-peak (see Figure 1). When photon flux densities are too low, not all electron acceptors of PSII remain in a reduced state of the P-peak, resulting in underestimation of the true Fv/Fm value. If the quantum flux density of the actinic light is too high, the resulting rate of fluorescence rise is too high for a satisfactory resolution of Fo; therefore, Fo is overestimated (Öquist and Wass 1988). A Fv to Fm ratio of approximately 0.8 is typical for a well-functioning photosynthetic apparatus measured at an appropriate light intensity in most, but not all, species (Björkman and Demmig 1987).

For barley, soybean, *Tradescantia*, and *A. patula*, measurements were taken at light intensity of 200 to 1000 µmol m^{-2} s^{-1} and for *A. semibacata* at 400 to 1000 µmol m^{-2} s^{-1}. Two hundred micromoles m^{-2} s^{-1} were insufficient for a normal response of *A. semibacata*. To illustrate the results of these experiments, a typical curve for fluorescence parameters at different light intensities for barley is shown in Figure 2. Fv/Fm is relatively constant over the whole range of light intensities, in comparison with the rest of the parameters tested, but it tends to decrease with increasing photon flux densities. As the actinic light intensity is increased, the rate of photochemical reaction also increases, resulting in a decrease in $t_{1/2}$; proportional rises in Fo, Fv, Fm, and Ft; and a minor increase in Fq. If the light intensity is lower than mentioned previously, then the CF-1000 chlorophyll fluorescence measuring system is incapable of taking a measurement. The lower limit of light intensity for *A. semibacata* was higher than for the rest of the tested species. The reason is that *A. semibacata* is a C_4 plant, and C_4 plants require higher actinic light intensity for sufficient saturation of PSII electron acceptors. However, depending on the age of the plants, most of the experiments could be run at 200 µmol m^{-2} s^{-1}. Whenever Fo was too low, it was necessary to take additional readings in order to maintain a sample size of 20.

Even though *A. patula* is a C_3 plant, occasionally, depending on the age of the plant tested, Fm is too low to allow measurements at 200 µmol m^{-2} s^{-1}. This is caused, presumably, by low chlorophyll content in PSII reaction center (see Introduction). Therefore, 400 µmol m^{-2} s^{-1} is the light level chosen for measurements in both *A. patula* and *A. semibacata*; 200 µmol m^{-2} s^{-1} is appropriate for barley, soybean, and *Tradescantia*.

For field experiments, 400 µmol m^{-2} s^{-1} was chosen for all measurements. Plants in the field grow at light intensities that vary depending on the weather conditions, but are usually in excess of 1000 µmol m^{-2} s^{-1}. For comparison, our growth chamber conditions are approximately 250 µmol m^{-2} s^{-1}. Clearly, 200 µmol m^{-2} s^{-1} is not sufficient for the field experiments.

Field data

Three species were studied at 2.5, 20, 30, and 55 m from I-70 — milkweed, nightshade, and cattail — to evaluate changes in the seven parameters. In general, milkweed exhibited the greatest differences between distances from I-70, followed by nightshade and cattail. The results of milkweed from a typical field experiment are reported (Table 1). The values for all parameters at distances of 30, 20, and 2.5 m were compared with the values measured at 55 m. Each mean was divided by the mean at 55 m to provide a percent referenced to this distance.

The percents and mean values decreased significantly in four of the seven parameters with proximity to I-70. Fm, Fv, Fv/Fm, and Ft all had slopes (b) signigicant at $p < 0.01$. The parameters Fo, $t_{1/2}$, and Fq did not show significant changes.

Fv/Fm values for all sites were low, even at 55 m (0.604). Usually, healthy plants have an Fv/Fm value of approximately 0.8. Low Fv/Fm readings were also observed in the other species throughout the sampling period. Perhaps the values were higher earlier in the season. The low Fv/Fm measurements observed may have been due to the additive effects of pollution or to the approaching end of the growing season.

Table 1 Number of Samples (N), Percents, Means, Coefficients of Regression, Slopes, and p-Values
for Milkweed on September 15, 1993

Distance	N	Fo	Fm	Fv	$t_{1/2}$	Fv/Fm	Ft	Fq
55	10	**100** (174)	**100** (464)	**100** (289)	**100** (31)	**100** (0.604)	**100** (229)	**100** (235)
30	12	**89** (154)	**77** (355)	**69** (201)	**123** (38)	**90** (0.542)	**72** (165)	**81** (190)
20	12	**103** (179)	**75** (348)	**58** (169)	**6** (1.8)	**78** (0.473)	**80** (184)	**70** (164)
2.5	7	**107** (186)	**70** (326)	**48** (140)	**126** (40)	**68** (0.409)	**67** (152)	**74** (174)
r^2		0.018	0.171	0.282	0.003	0.426	0.407	0.073
b		0.140	-0.593	-1.03	-0.410	-0.625	-0.599	-0.588
$p_b > t$		0.406	0.007	0.0003	0.725	0.0001	0.0001	0.087

Note: Percents are in bold and means are in parenthesis.

At 20 m, $t_{1/2}$ decreased dramatically. All readings were low at this location. Perhaps this was caused by some environmental stress factors at this specific site. These plants were located in a low area along a service road. Evidence of flooding in this area was observed twice during the sampling period.

Exposure of plants to chemicals

For the purpose of developing a useful bioassay, it is necessary to select sensitive plant species which can be used as standards. Our choice of the five species tested so far represent some of the diversity in the angiosperms.

Fluorescence techniques have already been shown to be useful as a bioassay in determining changes in photosynthetic efficiency in barley and soybean exposed to herbicides (Judy et al. 1991). The seven different fluorescence parameters may be useful to varying degrees for determining disturbance in photosynthetic electron transport caused by different chemicals. From the five species of plants used in laboratory experiments, barley was determined to be the most responsive to the chemicals tested; *Tradescantia*, *A. patula*, soybean, and *A. semibacata* were more resistant.

Tables 2 to 5, including data from a number of experiments, show mean percent changes for all fluorescence parameters in barley exposed to four different chemicals. A similar series of experiments with the same four chemicals were performed with other species mentioned earlier: *Tradescantia*, *A. patula*, *A. semibacata*, and soybean. However, the results were not included in this chapter because these species were less responsive than barley.

Table 2 Mean Percent Changes in Fluorescence Parameters in Barley Exposed to Lead Chloride

Conc. (μM)	Fo	Fm	Fv	$t_{1/2}$	Fv/Fm	Ft	Fq	N
0.0	100	100	100	100	100	100	100	2
0.005	107	110	110	93	101	120	102	1
0.01	111	110	110	87	100	122	102	1
0.05	98	94	93	104	99	99	90	1
0.1	95	95	95	93	98	110	83	2
0.5	89	93	94	93	101	105	84	1
1.0	92	99	100	104	102	116	85	2
5.0	92	96	96	99	101	104	90	1
10.0	75	90	93	104	103	120	67	1
100.0	99	98	98	80	100	112	87	2
1000.0	117	90	85	66	94	114	73	2
2000.0	128	71	59	60	84	115	34	1

Table 3 Mean Percent Changes in Fluorescence Parameters in Barley
Exposed to Ortho

ppm	μM	Fo	Fm	Fv	$t_{1/2}$	Fv/Fm	Ft	Fq	N
0.0	0.0	100	100	100	100	100	100	100	5
12.5	2.0	106	105	104	95	100	106	104	5
62.5	10.0	108	107	107	89	100	111	104	3
125.0	20.0	108	102	101	93	97	106	104	5
625.0	100.0	120	110	109	80	98	116	105	3
1250.0	200.0	123	106	102	75	97	127	89	4
2500.0	400.0	144	94	85	69	89	106	83	3
6250.0	1000.0	145	86	75	73	87	102	69	3
12500.0	2000.0	124	81	73	75	89	109	53	2

Sometimes changes in fluorescence parameters may seem to be inconsistent. However, it is important to realize that the changes in the fluorescence parameters in one direction at low concentrations of tested chemical and in the opposite at higher concentrations has been reported and explained before for lead chloride in tomato chloroplasts (Miles et al. 1972). At the lowest concentrations of lead, Fo was not changed and Fm decreased. At intermediate levels of lead, Fo slightly increased and Fm was reduced. At high concentrations of lead, Fo increased. This phenomena may be explained as follows. Low concentrations of lead may result in a disturbance of the photosynthetic electron transport at the oxidizing site of PSII, and then, with increasing concentration, the effect may be extended to the reducing site also.

According to Table 2, 10 μM lead chloride was the concentration when parameters started to respond to its harmful effect. Fo decreased at low concentrations (up to 10 μM), then it increased up to 128%. At 2000 μM, Fv/Fm, Fv, Fm, $t_{1/2}$, and Fq decreased by 16 to 66%; Ft increased by 15% at 2000 mM.

Table 4 Mean Percent Changes in Fluorescence Parameters in
Barley Exposed to β-Naphthylamine

Conc. (μM)	Fo	Fm	Fv	$t_{1/2}$	Fv/Fm	Ft	Fq	N
0.0	100	100	100	100	100	100	100	5
1.0	99	102	102	82	101	98	105	2
10.0	100	105	101	95	100	100	109	5
50.0	102	109	102	101	101	110	107	3
100.0	104	107	110	86	101	110	105	5
200.0	117	114	107	92	100	122	106	2
500.0	120	105	97	84	97	107	104	3
1000.0	114	97	98	79	97	95	99	4
1500.0	113	91	84	85	96	85	99	3
2000.0	104	91	88	84	96	80	103	4

Table 5 Mean Percent Changes in Fluorescence Parameters in
Barley Exposed to Sodium Pentachlorophenate

Conc. (μM)	Fo	Fm	Fv	$t_{1/2}$	Fv/Fm	Ft	Fq	N
0.0	100	100	100	100	100	100	100	9
0.1	108	108	108	98	100	94	108	7
0.5	106	105	105	99	100	88	106	5
1.0	111	106	106	102	100	105	111	7
5.0	109	104	104	100	99	87	109	4
10.0	114	103	101	100	98	87	116	8
25.0	151	96	87	99	90	78	115	4
50.0	156	82	66	94	82	66	96	5
75.0	184	88	70	95	80	79	105	4
100.0	180	83	64	90	77	72	97	8
200.0	165	71	52	78	68	63	84	2

At the concentration 100 μM of Ortho (Table 3), barley started to react to this chemical. Fo increased by 20%, and $t_{1/2}$ decreased to 80%. All other parameters changed only by 2 to 16%. With increasing concentrations of this chemical, Fo increased 124 to 145%, and Ft rose only by several percentage points. Fv/Fm, Fm, $t_{1/2}$, Fv, and Fq declined by 11 to 47%.

β-Naphthylamine (Table 4) caused very small percentage changes in fluorescence parameters (20% was the highest change). As with the other chemicals, a typical pattern was repeated since Fo increased and the rest of parameters decreased or showed little change, except Ft which increased at low concentrations and then decreased at high concentrations.

In the presence of sodium pentachlorophenate (Table 5), at low concentrations (0 to 10 μM), fluorescence values were not changed. At 200 μM, Fo was increased by over 50%; all others, Fq, $t_{1/2}$, Fm, Fv/Fm, Ft, and Fv, declined by 16 to 48%.

Some interpretations of changes in fluorescence parameters values are available from previous studies. An Fo increase reflects an inhibition of the reaction center, and a decrease of Fv results from damage of the water-splitting system (Weis and Berry 1988).

The intercepts (I), slopes (b_L, b_Q), standard errors of the slopes, and coefficients of determination (r^2) for the linear and quadratic regression models of the percent change in fluorescence parameters in barley leaves exposed to sodium pentachlorophenate are presented in Table 6. Only the linear and quadratic models are presented since they represent the best fit of the data. The most reliable parameter in terms of predicting the effect of sodium pentachlorophenate on barley was Fv/Fm, which had r^2 values of 0.830 to 0.848. These regression curves are presented in Figure 3. The parameter with the second highest r^2 values was Fo. These ranged from 0.627 to 0.711. All other fluorescence parameters had lower values and were less consistent in their behavior.

Table 6 The Intercepts, Slopes, Standard Errors of Slope, and Coefficients of Determination for Linear and Quadratic Regression Models of the Percent Change in Fluorescence Parameters in Barley Leaves Exposed to Sodium Pentachlorophenate

		Linear				Quadratic					
Date	Variable	I	b_L	SE	r^2	I	b_L	SE_L	b_Q	SE_Q	r^2
2/18	Fo	96.9	1.12*	0.055	0.703	94.2	1.566*	0.205	−0.005*	0.002	**0.711**
3/4	Fo	135.3	1.280*	0.073	0.622	132.9	1.692*	0.270	−0.005	0.003	**0.627**
3/16	Fo	101.6	1.16*	0.060	0.668	96.5	2.08*	0.225	−0.010*	0.002	**0.698**
2/18	Fm	97.0	−0.113*	0.034	0.060	99.3	−0.486*	0.124	0.004*	0.001	0.110
3/4	Fm	117.1	−0.183*	0.044	0.084	118.2	−0.376*	0.164	0.002	0.002	0.091
3/16	Fm	99.2	−0.170*	0.039	0.094	99.7	−0.257*	0.152	0.001	0.002	0.096
2/18	Fv	97.0	−0.336*	0.034	0.351	100.2	−0.855*	0.124	0.006*	0.001	0.415
3/4	Fv	114.0	−0.440*	0.044	0.349	115.6	−0.739*	0.162	0.003*	0.002	0.362
3/16	Fv	98.8	−0.431*	0.040	0.381	100.4	−0.716*	0.157	0.003	0.002	0.392
2/18	$t_{1/2}$	102.5	−0.227*	0.033	0.212	103.7	−0.421*	0.124	0.002	0.001	0.224
3/4	$t_{1/2}$	97.8	−0.114*	0.024	0.112	97.3	−0.038*	0.087	−0.001	9.3×10^{-4}	0.116
3/16	$t_{1/2}$	99.2	−0.166*	0.030	0.141	99.4	−0.207*	0.118	4.0×10^{-4}	0.001	0.142
2/18	Fv/Fm	100.1	−0.253*	0.009	0.813	101.3	−0.445*	0.031	0.002*	3.3×10^{-4}	**0.848**
3/4	Fv/Fm	97.4	−0.259*	0.009	0.823	98.0	−0.367*	0.032	0.001*	3.4×10^{-4}	**0.834**
3/16	Fv/Fm	99.6	−0.312*	0.011	0.814	100.5	−0.478*	0.041	0.002*	4.4×10^{-4}	**0.830**
2/18	Ft	102.0	−0.355*	0.047	0.248	107.4	−1.231*	0.162	0.010*	0.002	0.362
3/4	Ft	110.4	−0.433*	0.047	0.315	115.1	−1.260*	0.162	0.009*	0.002	0.406
3/16	Ft	92.3	−0.246*	0.040	0.167	95.1	−0.742*	0.154	0.005*	0.002	0.215
2/18	Fq	92.5	0.103*	0.036	0.046	92.0	0.180*	0.134	−0.001	0.001	0.048
3/4	Fq	125.6	0.130*	0.065	0.021	122.2	0.725	0.238	−0.007*	0.003	0.055
3/16	Fq	106.7	−0.087	0.054	0.014	104.7	0.263	0.213	0.004	0.002	0.029

Note: Concentrations are 0, 0.1, 1, 5, 10, 25, 50, 75, and 100 μM. *$p < 0.05$.

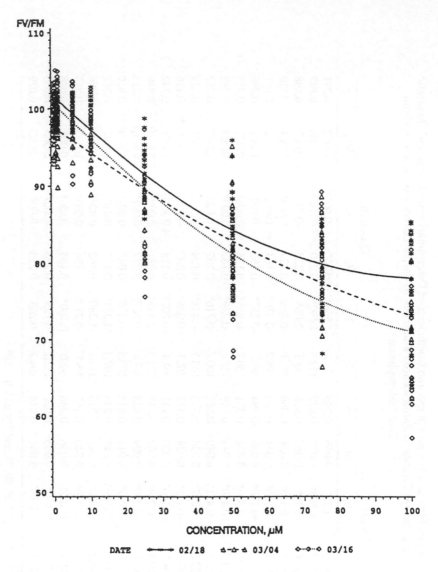

Figure 3 Regression of percent changes for Fv/Fm in barley exposed to sodium pentachlorophenate (quadratic model).

Conclusions

Chlorophyll fluorescence measurements in leaves were shown to be useful as a laboratory test and an environmental bioassay for detection of pollution. From five plant species (barley, soybean, *Tradescantia*, *Atriplex semibacata*, and *A. patula*) tested in the laboratory phase of the study, barley was the most responsive one. Sodium pentachlorophenate was the most toxic chemical tested, lead chloride and Ortho (containing chlorpyrifos) were less powerful, and β-naphthylamine had minimal influence on chlorophyll fluorescence. In barley exposed to sodium pentachlorophenate, fluorescence parameters measurements showed variability, except Fv/Fm and Fo which showed initial fluorescence. In most cases, these two parameters were proven to respond with the best reproducibility to all tested chemicals in barley. High coefficients of determination allow us to recommend their use in bioassay studies utilizing this species.

Acknowledgments

We gratefully acknowledge the technical assistance of Jack Lonigro. We also thank Dr. Robin Kennedy for assistance in the identification of wild plant species used in field experiments.

This research was supported by U.S. Environmental Protection Agency Grant No. R81963-01-1 awarded to Dr. W.R. Lower.

References

Baker NR, East TM, Long SP (1983) Chilling damage to photosynthesis in young *Zea mays*. II. Photochemical function of thylakoids *in vivo. J Exp Bot* 34:189–197.

Björkman O, Demmig B (1987) Photon yield of O_2 evolution and chlorophyll fluorescence characteristics at 77K among vascular plants of diverse origins. *Planta* 170:489–504.

Demmig-Adams B, Winter K, Kruger A, Czygan F-C (1989) Zeaxanthin and the induction and relaxation kinetics of the dissipation of excess excitation energy in leaves in 2% O_2 0% CO_2. *Plant Physiol* 90:887–893.

Greaves JA, Blair BG, Russotti RM, Law EA, Cloud NP (1992) Measurement of chlorophyll fluorescence kinetics in photosynthesis research with a new portable microprocessor and computer operated instrument. CF-1000 Instruction Manual, Morgan Instruments, Inc., Andover, pp. 43–62.

Horton P, Black MT (1983) A comparison between cation and protein photo-synthesis effects on the fluorescence induction curve in chloroplasts treated with 3-[3,4,-dichlorophenyl]-1-dimethylurea. *Biochim Biophys Acta* 722:214–218.

Judy BM, Lower WR, Thomas MW (1990a) The chlorophyll fluorescence assay as a rapid indicator of herbicide toxicity. *Proc South Weed Sci Soc* 43:358–365.

Judy BM, Lower WR, Miles CD, Thomas MW, Krause GF (1990b) *Chlorophyll Fluorescence of a Higher Plant as an Assay for Toxicity Assessment of Soil and Water*, ASTM STP 1091, Wang W, Gorsuch JW, Lower WR, Eds., American Society for Testing and Materials, Philadelphia, PA, pp. 308–318.

Judy BM, Lower WR, Ireland FA, Krause GF, (1991) *A Seedling Chlorophyll Fluorescence Toxicity Assay*, ASTM STP 1115, Gorsuch JW, Lower WR, Wang W, Eds., American Society for Testing and Materials, Philadelphia, PA, pp. 146–158.

Krause GH, Weis E (1984) Chlorophyll fluorescence as a tool in plant physiology. II. Interpretation of fluorescence signals. *Photosynth Res* 5:139–157.

Lavorel J, Etienne A-L (1977) *In vivo* chlorophyll fluorescence. In *Primary Processes of Photosynthesis*, Elsevier, Amsterdam, pp. 203–268.

Lower WR, Underbrink AG, Yanders AF, Roberts K, Ranney TK, Lombard G, Hemphill DD, Clevenger T (1984) New methodology for assessing mutagenicity of water and water related sediments. Proceedings of the Second International Conference on Ground-Water Quality Research, Tulsa, OK, pp. 194–196.

Miles CD, Brandle JR, Daniel DJ, Chu-Der O, Schnare PD, Uhlik DJ (1972) Inhibition of photosystem II in isolated chloroplasts by lead. *Plant Physiol* 49:820–825.

Miles D (1990) *The Role of Chlorophyll Fluorescence as a Bioassay for Assessment of Toxicity in Plants*, ASTM STP 1091, Wang W, Gorsuch JW, Lower WR, Eds., American Society for Testing and Materials, Philadelphia, PA, pp. 297–307.

Öquist G, Wass R (1988) A portable microprocessor operated instrument for measuring chlorophyll kinetics in stress physiology. *Physiol Plant* 73:211–217.

Ogren E, Baker NR (1985) Evaluation of a technique for the measurement of chlorophyll fluorescence from leaves exposed to continuous light. *Plant Cell Environ* 8:539–547.

Papageorgiou G (1975) Chlorophyll fluorescence: an intrinsic probe of photosynthesis. In*Bioenergetics of Photosynthesis*, Academic Press, New York, pp. 319–371.

SAS Institute Inc. (1987) SAS proprietary software release 6.04, SAS Institute Inc., Cary, NC.

Schaeffer DJ, Novak EW, Lower WR, Yanders AF, Kapila S, Wang R (1987) Effects of chemical smokes on the flora and fauna under field and laboratory exposures. *Ecotoxicol Environ Saf* 13:301–315.

Schreiber U (1986) Detection of rapid induction kinetics with a new type of high frequency modulated chlorophyll fluorometer. *Photosynth Res* 9:262–272.

Schreiber U, Grobermann L, Vidaver W (1975) A portable, solid state fluorometer for the measurement of chlorophyll fluorescence induction in plants. *Rev Sci Instrum* 46:538–542.

Snedecor GW, Cochran WG (1989) *Statistical Methods*, 8th ed., Iowa State University Press, Ames.

Van Kooten O, Snel JFH (1990) The use of chlorophyll fluorescence nomenclature in plant stress physiology. *Photosynth Res* 25:147–150.

Vredenberg WJ, Slooten L (1967) Chlorophyll a fluorescence and photochemical activities of chloroplast fragments. *Biochim Biophys Acta* 143:583–594.

Weis E, Berry JA (1988) Plants and high temperature stress. In *Plants and Temperature*, The Company of Biologists Ltd., Cambridge, U.K., pp. 329–346.

Zar JH (1984) *Biostatistical Analysis*, Prentice-Hall, Englewood Cliffs, NJ.

chapter nine

Laboratory bioassays with microalgae

Niels Nyholm and Hans G. Peterson

Introduction

Ecological importance

Planktonic microalgae are primary producers and constitute a major part of the lowest level of the food chain in aquatic systems. Many species are used directly as a food source for zooplankton, which are subsequently consumed by other invertebrates, fish, or birds. Phytoplankton microalgae are therefore of utmost importance to aquatic life, and the entire aquatic ecosystem may be influenced by changes in algal populations. If the biomass of algae becomes too high or if certain species become abundant, water quality may be negatively impacted; decreased water transparency and consumption of oxygen in bottom waters after settling are two principal consequences of algal overproductivity. Decreases in water transparency may affect growth and survival of vascular aquatic plants and cause changes in fish populations. Blue-green algae (cyanobacteria) are poor food sources for zooplankton, and, if green algae and diatoms are outcompeted by cyanobacteria, zooplankton populations may decrease. These few examples of ecological relationships indicate the close interdependence of algal population structure and both water quality and aquatic ecosystem functioning.

Human and domestic animal use of surface waters, primarily direct consumption, may be substantially influenced by algae. Algae produce dissolved organic matter, and both particulate algal cells and the dissolved organic products they produce may cause problems in drinking water treatment processes. Dissolved organics react with disinfectants such as chlorine, forming carcinogenic byproducts and making water difficult to disinfect. In addition, disinfection residuals may be lost during distribution. Some algae produce specific organics that are significant. Certain cyanobacteria release organics with a strong odor (including geosmin and methylisoborneol), while others produce highly toxic compounds (some as toxic as cobra snake venom). Taste, odor, and toxin-producing algae are most common among the cyanobacteria. Algae can also produce slimes and foams which make water treatment especially difficult.

Response to pollution

Phytoplankton communities respond quickly to anthropogenic inputs of nutrients and toxic substances, making them good indicators of changes in environmental water quality. Increased fertility of water-bodies usually results in both increased primary production and increased algal standing crops. As the balance between phytoplankton species is driven by nutrients, light, and preferential grazing by zooplankton, increases in nutrient inputs will upset this balance. Changes in phytoplankton communities, in turn, may give rise to secondary structural effects at all levels of the ecosystem. Ecosystem changes resulting from nutrient overfertilization of a waterbody are referred to as cultural eutrophication. Excess nutrients may enter the aquatic ecosystem through point sources, such as sewage effluent, or through nonpoint sources, primarily agricultural runoff. The control of point sources may be accomplished on an individual basis and may be monitored, as inputs are generally recognized. More challenging is controlling nonpoint source nutrient inputs. This may be considered the greatest remaining large-scale pollution problem in many industrial countries.

Although most phytoplankton species are generally sensitive to toxicants, they constitute taxonomically diverse groups and exhibit large interspecific differences in sensitivity. This intertaxa variation in response is most pronounced with chemicals that have specific modes of action, including many pesticides. The initial response of a natural phytoplankton community to toxic pollution is typically an alteration of its community structure, and only at higher toxic doses will functional parameters such as overall productivity be affected. The initial loss of a few, particularly sensitive species can potentially occur at extremely low doses and remain undetected in field observations. It is possible that, depending on the toxicant, such early changes in community structure could be the most sensitive response of an aquatic ecosystem to toxic pollution. At higher toxic doses, initial (transient) effects on functional characteristics (decreased productivity and standing crop) may be observed, followed rapidly by the establishment of a new and more tolerant community and the return of functional characteristics to preexposure levels. This phenomenon has been termed "pollution induced community tolerance" (PICT) (Blanck et al. 1988).

The initial response to pollution by both nutrients and toxicants can be qualitatively similar since changes in algal community structure may be brought about before changes in productivity can be ascertained. It is probable that consequences for individual species would differ between the two situations, but, with the present insufficient knowledge of these phenomena, it is difficult to differentiate toxic and nutrient effects relying on field observations alone.

Use of bioassays with algae

Bioassays with both natural populations and laboratory cultures of microalgae have been crucial in the understanding of the eutrophication phenomenon and the impacts of toxicants on aquatic ecosystems. Algal bioassays have also become practical tools which can be used for a variety of purposes in environmental management and control relating to both toxicant pollution and eutrophication. At a time when bioassays are being increasingly used in environmental work, the need for the development of such assays with algae is obvious in view of the important role these organisms play in aquatic environments.

Algal bioassays have long been used in eutrophication studies to assess the amount of available nutrients in surface waters or in effluents; they have also been used to identify factors that may be limiting growth (for reviews, see Maestriniet al. 1984; Schelske 1984; Trainor 1984). Recent applications include assessments of nutritional status of indigenous phytoplankton through measurements of various physiological indicator parameters such as differential nutrient uptake rates, alkaline phosphatase activity, or alternatively through short-term growth rate bioassays with nutrient spikings (Nyholm and Lyngby 1988).

The field of algal toxicity testing has been developed relatively recently compared to that of nutrient testing (for reviews, see Wong and Couture 1986; Walsh 1988,b; Nyholm and Källqvist 1989; Peterson and Nyholm 1993). As with other toxicity tests, algal bioassays may serve a range of purposes. Results from algal tests are included in the base set of information required for environmental hazard evaluation of chemicals as recommended internationally by the Organization for Economic Cooperation and Development (OECD) and demanded in various national legislative framework, e.g., in European Union chemical directives and in U.S. Toxic Substance Control Act (TSCA) and other legislation. Similarly, algae are often included among the species used in biotest batteries for hazard assessment of chemically contaminated wastes, such as industrial effluents or leachates from chemical dumpsites, or combined with site-specific exposure assessments for environmental impact evaluations of waste discharges.

Algal bioassays may be used for monitoring purposes to detect changes in the toxicity of wastewaters or to test compliance with discharge permits. A final important application is for simple detection of toxicity in environmental samples or in effluent fractions or substreams, e.g., in toxicity reduction evaluations (TREs) (U.S. EPA 1988, 1989). The potential uses of algal tests extend even to monitoring toxicity of chemically contaminated soil (e.g., Greene et al. 1989) and groundwater, although these latter applications have no immediate ecological relevance. Nevertheless, algal tests can be useful monitors of toxic chemical contaminants because of sensitivity and cost-effectiveness.

The development of innovative testing and measurement techniques, including fluorometric *in vivo* measurements of chlorophyll as an estimate of algal biomass, automated dilutions, and computerized instrument readings, has allowed the use of miniscale algal toxicity testing, offering highly cost-effective methods of examining large numbers of samples. Such methods may allow for the cost-effective use of multispecies batteries in routine hazard assessment, monitoring, and commercial chemical registration testing.

History of algal bioassays

Bioassays with cultured algae

Studies with cultured algae date back to the last century and have addressed both algal physiology and ecology (for review, see Rodhe 1978). Fogg (1965) suggested that a synthesis of results from physiological and biochemical laboratory investigations with cultured algae and from field studies was needed to fully understand algal growth phenomena in nature. Algal bioassays were introduced in water pollution studies in the 1960s (Skulberg 1964, 1966) and were used to assess the nutrient status of surface waters and the fertilizing potential of effluents or streams entering waterbodies. Conducted with spikings of nutrients in different amounts and proportions, the assays were also useful in identifying potential growth-limiting factors. Cultured algal species were inoculated into field water samples after removal of any indigenous plankton, and the final algal biomass yield after a 1- to 2-week incubation period was recorded as the response parameter. This type of algal assay is referred to as a growth potential test (Algal Growth Potential, AGP) because it is used to determine the maximum amount of algae that can be produced in a field water sample. The "algal assay bottle test" AGP method conducted with the freshwater green alga *Selenastrum capricornutum*, introduced by Skulberg (1964, 1966), was further developed into a comprehensive protocol by the U.S. Environmental Protection Agency (EPA) (U.S. EPA 1971; Miller et al. 1978), which has been adopted worldwide. An example of results from an AGP test is shown in Figure 1.

Algal toxicity test methods were later developed, based on AGP protocols (e.g., Chiaudani and Vighi 1978; Greene et al. 1978; Källqvist 1978; Payne and Hall 1978; Christensen et al. 1979; Damgaard and Nyholm 1980). Other similar tests were developed in which the endpoint was some measure of growth inhibition, quantified either as reduced yield (or biomass) or as reduced specific growth rate (e.g., Bringmann and Kühn 1977; Walsh and Alexander 1980; Trevors 1982; Klaine and Ward 1983; Blanck et al. 1984; Jensen 1984; Yamane et al. 1984; Geyer et al. 1985; Blaise et al. 1986; Walsh 1988a,b; ISO 1989; Kühn and Pattard 1990). Other researchers used tests with alternative endpoints, mostly photosynthesis (essentially equivalent to growth) measured as ^{14}C assimilation (e.g., Steemann-Nielsen et al. 1969; Wium-Andersen 1974; Soto et al. 1975; Hutchinson et al. 1980; Giddings and Washington 1981; Nyholm et al. 1981; Eloranta and Haltunen-

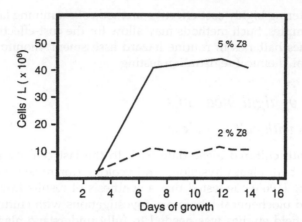

Figure 1 Example of results from an AGP test with the green alga *Selenastrum capricornutum*. AGP tests are used to assess the nutrient status of water samples (in this example distilled water with two different concentrations of a synthetic nutrient solution, Z8 [Kotai 1972]). AGP is recorded as the maximum biomass (standing crop) that can be produced from the amounts of bioavailable nutrients present in the sample. (From Gargas E, Pedersen K [1974] Contributions from the Water Quality Institute, No. 1, Water Quality Institute, Hoersholm, Denmark. With permission.)

Keyriläinen 1984; Jensen et al. 1984; Lewis and Hamm 1986; Nyholm and Damgård 1990; Kusk and Nyholm 1991; Peterson et al. 1994) or as oxygen evolution (e.g., Kusk 1978, 1981a; Turbak et al. 1986; Versteeg 1990) or indicated by short-term changes in chlorophyll flourescence (Rehnberg et al. 1982; Wong and Couture 1986; Samson et al. 1988). Other endpoints used in toxicity studies include inhibition of nutrient uptake (Peterson and Healey 1985; Nyholm 1991) and changes in morphology (Soto et al. 1979), pigmentation, and other cellular components including carbohydrate, lipid, and protein content (Thompson and Couture 1991). Even the ability of volume regulation in naked marine flagellates exposed to osmotic shocks has been suggested as a toxic stress indicator (Riisgård 1979).

The utility of results from early algal toxicity studies was the focus of considerable controversy. Some concluded that algal toxicity tests were insensitive (Kenaga and Moolenaar 1979). In round-robin tests among different laboratories, tests were initially revealed to be irreproducible (see Figure 2) (ISO 1980; Källqvist and Ormerod 1981; Laake 1982). At least some of these early discrepancies among results were due to insufficient control of experimental factors (Nyholm and Källqvist 1989; Lewis 1990; Peterson and Nyholm 1993), as well as a lack of adequate standardized test protocols. Standardization of test variables including air/water carbon dioxide mass transfer rate, pH, nutrient source (especially carbon or nitrogen source), and trace metal complexation is still a concern (Peterson and Nyholm 1993). Neither

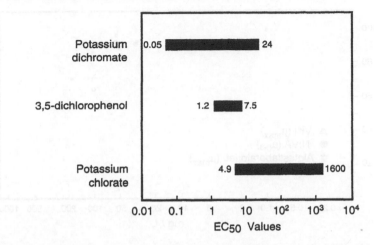

Figure 2 Toxicity of potassium dichromate and 3,5-dichlorophenol toward the green algae *Scenedesmus subspicatus* and *Selenastrum capricornutum*. Results were obtained in the first interlaboratory comparison program with algal toxicity growth inhibition tests organized by the International Organization for Standardization (ISO 1980). EC_{50} (50% inhibition) is reported by the participating laboratories using their own test method and way of calculating the EC_{50} value.

the International Standards Organization (ISO 1989), the current EPA (Greene et al. 1989), or the American Standards for Testing Materials (ASTM Standard D 3978-80, reapproved 1993) test protocols are completely satisfactory in this respect, as some test conditions or procedures have not been strictly specified or remain in question. It has, however, been shown that with the same type of algal toxicity test or bioassay, appropriately conducted tests are both highly reproducible and generally sensitive. More recent interlaboratory test reproducibility has been demonstrated in round-robin exercises organized by the ISO (Hanstveit 1982) using the protocol for toxicity tests with freshwater algae that has since been accepted as an international standard (ISO 1989). In addition, Källqvist and Ormerod (1981) compared results of tests on three chemicals using *Selenastrum* conducted in three experienced Scandinavian laboratories. Although incubation conditions (light intensity and temperature) and experimental setups differed, similar tests conducted with the same medium in the three laboratories produced equivalent results (Figure 3).

It has also been shown that, among short-term toxicity bioassays, algal tests with commonly used, standard species are frequently the most sensitive indicators of the toxicity of complex wastes and industrial chemicals (e.g., Walsh et al. 1980, 1982; Adema et al. 1983; Sloff et al. 1983; Walsh and Merrill 1984; Miller et al. 1985; Benenati 1990; Nyholm et al. 1991).

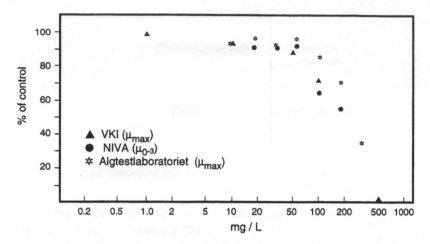

Figure 3 Demonstration of algal toxicity test reproducibility. Results are from
growth inhibition tests with anisole and the freshwater green alga *Selenastrum
capricornutum* obtained in three Scandinavian laboratories using the same test
medium, but incubating at different light intensities and temperatures, which
resulted in different absolute growth rates. Responses are expressed as relative
specific growth rates (specific growth rates normalized by growth rates recorded in control cultures). Laboratories are Water Quality Institute, Denmark
(VKI); Norwegian Water Research Institute (NIVA); and Algal Assay Laboratory, Inst. of Physiological Botany, Uppsala, Sweden. (From Källqvist T, Ormerod K [1981] Ekotoxikologiska metoder för akvatisk miljö, Report No. 25,
Nordic Cooperative Organization for Applied Research, Helsinki, Finland.
With permission.)

Bioassays with natural phytoplankton

Field or *in situ* enclosure studies with natural phytoplankton and laboratory microcosm studies are two algal test approaches that have been
advocated by many ecologists since the early 1980s. Supporters of
model ecosystem studies have argued that single-species tests cannot
be used for predicting ecological effects because they cannot detect
interactions between organisms or impacts between higher levels of
biological organizations (e.g., National Research Council 1981; Cairns
1980, 1983; Giesy 1985; Taub 1989; Swartzman et al. 1990). It has been
emphasized that ecosystems have emergent properties, meaning that
the whole system is greater than the sum of its parts. Therefore, only
toxic effects criteria based on studies at the ecosystem level are believed
to provide a sufficiently safe protection of the real ecosystem. Microcosms and mesocosms are suggested as empirical models of nature that
may bridge the gap between laboratory experiments and full-scale field
observations.

There has been a lot of debate on these views, particularly relating
to regulatory testing. The objective of most of this testing is to provide

estimates of safe exposure levels: PNECs (predicted no effect concen-trations) which are derived from experimental LOEC (lowest concen-tration with an observable effect) or NOEC (highest applied concen-tration with no significant effect observed) estimates. For this purpose, ecosystem-level type of responses need not be evaluated unless they are more sensitive than responses obtained with single species. Of course, using just one single algal test species such as *Selenastrum capricornutum* to represent the highly diverse taxonomic group of algae in real ecosystems may not always be sufficient, but why should a battery of many representative and important single species not be? This question has been addressed, e.g., by Mount (1985), who also expressed the view that the only justification for using complex mul-tispecies tests rather than tests with several single species was the possibility of identifying interaction or compensation related to toxic effects which would be missed in single-species tests. However, expo-sure levels potentially affecting interspecies interactions may also be assessed in single-species tests by selecting appropriate endpoints. In algal tests, such ecologically relevant endpoints are the commonly mea-sured reductions in growth rate and reductions in potential nutrient uptake rates under nutrient-deficient conditions. Both parameters are crucial for the competitive ability of a species.

For routine hazard assessments a question of interest is which techniques are technically more appropriate and cost-effective — mul-tispecies tests or multiple tests with single species? Microcosm and mesocosm studies are generally laborious and expensive. The results often depend on a number of factors which may be complicated to control and interpret, and replicability is a major problem. It must also be realized that it is technically impossible to measure all relevant properties and responses of complicated multispecies systems. Because observations on species composition are laborious, only summary parameters (such as rate of photosynthesis or chlorophyll concentra-tion) are normally observed in routine studies, and, therefore, many effects will not be detected (Kooijman 1985; Van leeuwen 1989). In many microcosm studies, pH and carbon supply have not been controlled. Experiments conducted following the well-known synthetic aquatic microcosm (SAM) test protocol developed by Taub (1989) suffer from uncontrolled carbon limitation resulting in large pH increases. This protocol has been demonstrated to provide similar sequences of events (although different timing of events) in an interlaboratory comparison exercise with copper (Taub et al. 1989), but such a demonstration of reproducibility does not in itself ensure environmental relevance or predictive capability.

In addition, the exposure and history of exposure may also be difficult to describe quantitatively in model ecosystems, due to complex physicochemical behavior of organic chemical substances and heavy metals. The partitioning of toxicants among biotic and abiotic elements may be highly variable and depends not only on abiotic factors such as water hardness, particulate, colloidal and dissolved organic matter, pH, etc., but also on such complex biotic factors as the algal species composition, the algal density (total biomass as well as the biomass of individual species), and the physiological states of the organisms. Complex niche partitioning (Kooijman and Metz 1984) may be an additional complicating factor. Therefore, with absorbable toxicants, it may be difficult to generalize the results of model ecosystem studies for predictive uses. Such studies may be conducted essentially for the purpose of qualitatively elucidating ecosystem level types of responses or to verify predictions based on mathematical fate models combined with the use of toxicity data for individual key species.

Having put forward these severe criticisms of model ecosystem studies as environmental management tools, it must not be forgotten that such studies have contributed much valuable general information of the more phenomenologically oriented nature on how ecosystems and algal communities respond to inputs of toxicants or nutrients. Also, their role in identifying sensitive species and in validations of model-based predictions can hardly be overemphasized. A further use is that of identifying the potential of ecosystem recovery after exposure to toxic stress. For a review of different opinions on the uses of multispecies tests, see, e.g., the monograph edited by Cairns (1985) and published studies reviewed by Gearing (1988) and Crane (1990).

A few toxicity studies have included the aim of comparing the responses of the complex systems with those obtained in single-species tests. Several such studies that focused on microalgae have revealed toxic effects at comparable dose levels in different systems, ranging from single-species algal test batteries to microcosm studies and field studies, e.g., in extensive studies with the herbicide atrazine (deNoyelles et al. 1982; Larsen et al. 1986). Crane (1990) concluded from a limited, general survey of multispecies tests vs. single-species tests that neither test system was consistently more sensitive than the other: a single-species test was more sensitive in 14 examples, and a multispecies test was more sensitive in 13 examples. With algae, such comparisons based on past literature data can be difficult, because results generated with nonstandardized methods can be extremely variable (Nyholm and Källqvist 1989; Peterson and Nyholm 1993).

Recent comparative studies with pesticides showed that for some substances structural changes in the phytoplankton communities of pelagic, oligothrophic lake mesocosms could occur at lower pesticide application levels (1 µg/l) than expected from the results of laboratory toxicity tests with six different algal species (Källqvist and Romstad 1994; Källqvist et al. 1994). However, these findings do not rule out that the low effect levels seen in the mesocosms might also have been seen in laboratory tests if other species had been selected or if tests had been conducted with as low an algal biomass as in the oligotrophic lake. It could be concluded that applying a safety factor of 10 to 100 on the EC_{50} values obtained with the most sensitive species, as normally used in ecotoxicological hazard assessment, did in fact result in safe exposure levels.

Both model ecosystem studies and single-species tests have their place in ecotoxicology, but serve different purposes; likewise, both type of studies must be conducted following sound principles with adequate control of significant experimental factors.

Laboratory bioassays with natural water samples from which zooplankton have been removed lie between those two approaches and are reviewed herein as a special subset of laboratory algal bioassays. Early advocates of the use of natural phytoplankton bioassays in eutrophication studies, including Goldman (1978) and Shapiro and Ribeiro (1965), argued that these tests more closely resembled natural systems than did bioassays using standard species in pure culture. Bioassays with natural phytoplankton were conducted in a similar fashion to the growth potential tests with cultured species; algal biomass was recorded after a long-term incubation. An estimate of the amount of available nutrients in the sample was obtained by means of various spikings aimed at identifying which nutrient was in stoichiometric minimum. With such long-term assays, the information derived related to the algal population that developed in the course of the experiment and not to the starting population of indigenous phytoplankton. Therefore, the information obtained could be described as a kind of eutrophication potential, reflecting both a potential maximum standing crop and changes in community structure following nutrient enrichment. Experimental techniques and interpretation of results have been discussed by Maestrini et al. (1984) and Schelske (1984).

Another quite different objective of bioassays with natural phytoplankton samples is that of assessing the actual physiological status of the field population from which samples are taken. Such information relates to actual kinetic nutrient limitation, i.e., one attempts to assess whether deficiency of one or more nutrients may suppress the *in situ* growth rate of the phytoplankton. In order to assess such nutrient deficiency, different physiological indicator parameters can be measured directly on the samples, including alkaline phosphatase activity or instantaneous nutrient uptake rates after spiking with phosphate, ammonium, or nitrate, preferably using radioisotope (^{32}P or ^{15}N) techniques. However, such measurements have not been extensively applied yet, and the results can be difficult to interpret in quantitative terms, although they may be useful in detecting nutrient deficiency and in identifying limiting nutrient(s) (e.g., Chiaudani and Vighi 1976). These types of measurements are performed with a minimum of disturbance to the original phytoplankton population and assayed quickly after collection.

Most researchers who have conducted laboratory toxicity tests with phytoplankton field samples have used ^{14}C assimilation as the endpoint (e.g., Erickson and Hawkins 1980; Giddings et al. 1983; Lumsden and Florence 1983; Södergren and Gelin 1983; Eloranta and Halttunen-Keyriläinen 1984; Kuivasniemi et al. 1985; Lewis and Hamm 1986; Davis et al. 1988; Kusk and Nyholm 1991, 1992). With natural phytoplankton, the ^{14}C technique appears to be the only feasible technique because of its sensitivity, at least with low algal concentrations. A proposed standard method using a 6-h incubation time was evaluated by Kusk and Nyholm (1991) for marine phytoplankton. They found that the method was highly reproducible when used with cultured species, but the sensitivity of natural phytoplankton to potassium dichromate, pentachlorophenol, and an industrial effluent varied considerably between both the sampling locations and the time of year. Similar variations in the sensitivity of lake phytoplankton to surfactants were reported by Lewis (1986).

Although toxicity tests with natural phytoplankton have immediate, site-specific ecological relevance, their variable sensitivity may rule out a number of regulatory applications of such tests. To make meaningful impact or hazard evaluations based on tests with natural phytoplankton, it would be necessary to test a large number of samples representative of the yearly cycle.

Growth and physiology of algae

Basic growth kinetics and limiting nutrients

Microorganisms, including microalgae, in free suspension follow the basic growth pattern of exponential growth:

$$dX/dt = \mu \cdot X$$

or the integrated equivalent:

$$X = X_o \cdot e^{\mu \cdot t}$$

where X is the biomass (X_o at time zero), and μ is the specific growth rate (d^{-1}). Usually, X is expressed using dry weight as a reference and should not be confused with cell numbers because the ratio of cell numbers to biomass may vary with changes in cell size. Under continuous light, unlimited carbon dioxide, and balanced growth conditions, the specific growth rate can be expressed as a function of light, temperature, and nutritional status according to the following equation:

$$\mu = f(I, T, C_{i-n})$$

where I is the effective light intensity to which the algal cells are exposed (see below); T is temperature; and C_i is the intracellular concentration (per unit of biomass dry weight) of limiting nutrients such as phosphorus, nitrogen, silicon (with diatoms), or essential microelements (trace metals, vitamins, or other organic growth factors, if required). It is important to note that the growth rate is not proportional to the extracellular concentrations of substrate components as implied by the Monod kinetic description of bacterial growth under limitation by carbonaceous substrates (Monod 1942), the kinetic equation typically used for growth of microorganisms. In contrast to carbonaceous substrates that are immediately metabolized, limiting nutrients are rapidly taken up and stored (Nyholm 1976).

Growth can indeed continue long after external nutrients have been exhausted from the aqueous phase as the cells now use internal stores. The internal nutrient concentrations that determine the growth rate comprise not only concentrations of nutrient stored in excess of immediate requirements, but also nutrients incorporated into functional compounds, such as phosphorus in RNA, which can later be utilized for synthesis of "structural" cell components necessary for maintaining the structural integrity and basic metabolic functioning of the cell. Letting this minimum nutrient concentrations be termed C_a (a for absolute requirement), a so-called "excess" intracellular nutrient concentration, C_e, which is available for growth, can then be defined as $C_e = C_i - C_a$.

The relationship between specific growth rate and the intracellular excess concentration of a nutrient can be of simple first order, as demonstrated for nitrogen by Nyholm (1977a) and shown or indicated for other nutrients such as silicon (Paasche 1973a,b), vitamin B_{12} (Droop 1968), and iron (Davis 1970). Phosphorus, on the other hand, can be stored internally in algal cells in large amounts, and for phosphorus the relationship between specific growth rate and internal concentration follows saturation kinetics (Rhee 1973; Nyholm 1977b). With green algae, a range of internal phosphorus concentrations of more than a factor of 20 has been reported (Nyholm 1977b) (from about 3.5% phosphorus on a dry-weight basis during balanced nutrient sufficient growth to 0.15% in outgrown batch cultures with zero growth rate). The biomass of an algal batch culture can thus increase by a factor of more than 20 after extracellular phosphate has been exhausted from the medium, through the use of intracellular reserves. The corresponding factor for nitrogen is only about 2.5 (the internal nitrogen concentration may typically vary from about 4 to 10% of the dry weight [Nyholm 1977a]).

For simultaneous limitation by two or more nutrients, Rhee (1978) and Droop (1974) reported threshold-type kinetics, in which a single nutrient is limiting at a given time, rather than multiplicative-type kinetics. Others have used a "resistance in series" construct to describe multiplicative nutrient limitation (Bloomfield et al. 1974; Nyholm 1978):

$$\mu = n/\left[1/f(C_1) + 1/f(C_2) + \ldots\ldots + 1/f(C_n)\right]$$

With this construct, a gradual transition between thresholds for each nutrient is assumed.

Another implication of classical Monod kinetics is that substrate utilization or uptake is proportional to growth:

$$dX = -Y \cdot dS$$

where S is the extracellular substrate concentration, and Y is the yield coefficient. From this equation it should be clear that only under conditions of nutrient-sufficient, balanced growth is the increase in algal biomass proportional to the uptake (or consumption) of extracellular nutrients (substrates). The yield coefficient here is equal to the reciprocal of the intracellular concentration, $C_{i,max}$, characteristic of nutrient sufficient growth under the particular conditions (temperature, light, etc.; see, e.g., Rhee and Gotham 1981a,b; Rhee 1982), that is, $Y = 1/C_{i,max}$.

Under nutrient-deficient conditions, on the other hand, there is no such direct coupling between growth and nutrient uptake. Under deficient conditions, nutrient uptake is a rapid process with time constants much smaller than those for growth, and virtually all available nutrient will be taken up. Thus, in batch culture, the extracellular nutrient concentration will approach zero rapidly after C_i has dropped below $C_{i,max}$. In a continuous culture (with a continuous external input of nutrient) or in a natural system with additional nutrient supply by mineralization processes, the nutrient will be taken up almost instantaneously, and a very low residual (quasistationary) nutrient concentration results from a dynamic equilibrium between input flux and uptake. The specific growth rate, however, is controlled solely by the intracellular nutrient concentrations, which result from the fluxes of nutrients into the biomass and the dilution by biomass growth. In laboratory cultures with synthetic medium, the intracellular concentration, C_i, is approximately equal to the total concentration divided by the biomass, $C_i = S_t/X$, because the small residual dissolved concentration, S, is much smaller than the total (or nominal) nutrient concentration, S_t. With first order kinetics with respect to the intracellular excess nutrient concentration, the overall growth rate becomes

$$dX/dt = \mu \cdot X = k \cdot C_e \cdot X = k \cdot (S_t - C_a \cdot X)$$

which is basically a linear growth pattern (logistic when $C_a \bullet X$ is no longer small) rather than the exponential growth pattern seen under nutrient sufficient conditions. Also, the saturation kinetics of phosphate-limited growth become linear (and eventually logistic) under severe phosphate limitation; at small (intracellular) concentrations, saturation kinetics transform into first order kinetics.

Nutrient uptake by nutrient-deficient algae can be described in terms of Michaelis–Menten (saturation) kinetics:

$$v = v_{max} \cdot S/(K_S + S)$$

where v is the specific rate of uptake per unit of biomass, S is an extracellular substrate concentration, and K_S is the half saturation constant for growth. Both kinetic parameters v_{max} and K_S, however, seem very dependent on the physiological conditions of the algae (see, e.g., Nyholm 1977b, 1978, 1991; Healey 1978, 1979; Zevenboom 1980).

Light

Normally, the growth rate of algae increases with light intensity along a saturation curve, finally declining at an upper threshold if the light reaches inhibitory high levels. Light intensity interacts with other chemical and physical factors. Greater light intensities are required to reach saturation as temperature increases toward the optimum. A similar trend is apparent in the relationship with nutrient concentrations. Different algal species have different light optima: cyanobacteria, for example, usually thrive at lower light intensities than green algae and diatoms (e.g., van Liere 1979). Light intensity is properly expressed in terms of the quantal flux in the photosynthetically active wavelength range of approximately 400 to 700 nm (lux units are not directly proportional to quanta, but for "universal white light" and unidirectional light an approximate conversion factor from lux units to quantal units is 1 klux [2π collector] = about 15 $\mu E/m^2/s$ or $9.0 \cdot 10^{18}$ photons per square meter per second) (Steemann-Nielsen and Willemoes 1971; ISO 1989). Another consideration is "self-shading" in cultures with high algal densities and/or with long light paths. Self-shading may reduce the effective light intensity to which algae are exposed. With saturating light at the surface, good mixing reduces that effect and improves light use because algae are able to benefit from light absorbed in short flashes. It is also worth noting that with all incoming light absorbed (dense cultures and/or long light paths) the growth pattern becomes linear (Tamiya et al. 1953; Pipes and Koutsoyannis 1962; Hanson et al. 1977):

$$dX/dt = k$$

where k is a constant proportional to the number of light quanta received (which may increase with mixing), and X is the biomass.

With a natural (day/night) light cycle rather than continuous light, growth oscillates and becomes unbalanced; cells become partially synchronized, leading to fluctuations in the ratio of cell numbers to biomass. During the first part of the dark period, some biomass growth will take place through the use of stored energy reserves. Also, cell division occurs in the dark, resulting in larger cell numbers but equal biomass. Therefore, laboratory bioassays with light/dark cycles may not be practical. Little is known, however, about differences in the expression of toxic effects in the presence of light/dark cycles as compared to continuous light, and more research is needed. Gavies et al. (1976), for example, found that different light regimes (continuous or day/night cycle) could either increase or decrease the toxicity of copper to marine phytoplankton, depending on the clone studied.

Temperature

The specific growth rate of algae increases exponentially with temperature, with the rate of change per 10°C being in the order of 2 to 3 (Canale and Vogel 1974) until an optimum is reached, after which the growth rate rapidly declines. As outlined earlier, temperature, light, and nutritional status are interacting factors.

Carbon dioxide and pH changes

Microalgae primarily use physically dissolved carbon dioxide, $(CO_2)_{aq}$, as their carbon source for photosynthesis. While the gas is quite soluble in water (about 1450 mg/l in fresh water at 25°C and 1 atm partial pressure), the concentration of dissolved CO_2 in equilibrium with atmospheric air is rather low (about $1.072 \cdot 10^{-5}$ M for low ionic strength fresh water) due to the low CO_2 content in air (0.033%) (figures for the carbonate system are quoted from Stumm and Morgan 1981). Therefore, at neutral or basic pH, the HCO_3^- and CO_3^- species of the carbonate system constitute a much larger pool of inorganic carbon than dissolved CO_2. In natural waters, the carbonate system is generally the dominating buffer, the speciation of which determines pH (in saltwater borate also contributes to the buffering capacity). This is similar in most algal test media, unless very high concentrations (one to several millimolars) of artificial buffers such as phosphate, HEPES ((N-[2-hydroxyethyl]piperazine-N'[2-ethanesulfonic acid), or TES (N-tris[hydroxymethyl]methyl-2-aminoethane-sulfonic acid) are added instead of bicarbonate. The acidity constants for carbonic acid and bicarbonate at 25°C in low ionic strength water (fresh waters and most freshwater algal test media) are $pK_1' = 6.33$ and $pK_2' = 10.25$, respectively, and, consequently, the HCO_3^- species dominates in a pH range from neutral to slightly basic. Provided that no other significant buffers are present, and taking only the CO_2/HCO_3^- buffer system into account as a first approximation, the pH of a 25°C low ionic strength freshwater medium in equilibrium with the atmosphere becomes

$$(pH)_{eq} = 11.30 + log[HCO_3^-]$$

For seawater, the situation is more complex, and the borate concentration has to be taken into account. In equilibrium with the atmosphere, pH of oceanic seawater is about 8.2 at 15°C.

In such an equilibrium situation, the dissolved CO_2 concentration is determined by the CO_2 concentration in air, according to Henry's law. While the process of photosynthesis, converting CO_2 into organic carbon, is alkalinity neutral, the utilization of bicarbonate as a CO_2 (or carbon) source for photosynthesis results in increased pH:

$$HCO_3^- \rightarrow OH^- + CO_2$$

It is sometimes erroneously stated that the occurrence of algal growth or photosynthesis as such causes an increase in pH; however, this only occurs if physically dissolved CO_2 is depleted and CO_2 (or carbon) is subsequently derived from bicarbonate.

A secondary factor which affects both alkalinity and pH is the nitrogen source, ammonium, or nitrate. The respective reaction schemes for algal uptake of ammonium and nitrate ions are

$$NH_4^+ \rightarrow H^+ + algal - N$$
$$NO_3^+ + H_2O \rightarrow OH^- + algal - N$$

In natural waters, algal growth is normally not carbon limited, and physically dissolved CO_2 is available for photosynthesis because the supply of CO_2 from mineralization processes and gas exchange with the atmosphere can usually accommodate the algal CO_2 demand. Shortages of CO_2 can occur, however, both in acid waters of low alkalinity (where carbon can be limiting) and during heavy algal blooms in eutrophic waters, where extremely high pH values (pH 10 to 11) may be observed due to bicarbonate utilization. Laboratory algal cultures, with their unnaturally high cell densities, also experience CO_2 limitation quite readily, along with the concomitant increase in pH. Adequate carbon supply and pH control are a major practical problem with algal bioassays, particularly toxicity tests.

When the pool of dissolved CO_2 in an algal culture has been depleted through algal utilization, and algal demand is greater than the CO_2 mass transfer rate from the gas phase (zero in closed, filled bottles) plus the rate of CO_2 formation from heterotrophic microbial activity (insignificant in synthetic algal media), bicarbonate will be utilized, resulting in a pH increase. However, algae can usually tolerate high pH, and exponential growth may prevail until the bicarbonate is exhausted. When this occurs, growth becomes linear and proportional to the mass transfer rate of CO_2 from the gas phase into the culture (Pipes 1962):

$$dX/dt = Y \cdot r(CO_2)$$

where $r(CO_2)$ is the volumetric carbon dioxide supply rate and Y the yield coefficient (about 2 mg dry weight per milligram of carbon, as carbon usually makes up approximately 50% of the algal dry weight).

Declines in algal growth rate due to nutrient or light limitation result in a reduced algal demand eventually less than the mass transfer rate. At this point, pH decreases as excess CO_2 reacts with OH^- (and CO_3^-) ions, restoring the bicarbonate pool. Had no other pH changing reactions occurred, pH would drop back to the starting value when equilibrium with the atmosphere is reached. The new equilibrium pH may either be lower or higher than the initial pH, however, depending on which nitrogen source is used in the medium (see earlier).

Moderate pH changes can be tolerated in eutrophication-related algal assays used to assess limiting nutrients. However, algal toxicity tests must be conducted at a well-defined pH, as this parameter may affect both speciation and inherent toxicity of test materials (e.g., Saarikoski and Viluksela 1981; Peterson et al. 1984). A pH drift of less than 0.5 should be a requirement or validity criterion of any toxicity test conducted with test materials of pH-sensitive toxicity (see later). Details on pH control in algal cultures for toxicity testing are covered by Nyholm and Källqvist (1989).

Cultivation methods

Batch cultures

Simple batch culturing in shake flasks is undoubtedly the most commonly used algal test system. Batch-type, short-term tests are the only feasible method of larger-scale testing due to the great number of flasks that may be needed (e.g., for a toxicity test with a battery of species aimed at establishing a dose-response relationship with each). There are many drawbacks to the changing conditions that are inherent in batch culture methods. Changes include initial lag phase, increasing cell density, decline of growth after nutrient exhaustion, changes in medium pH, and release of dissolved organics from the algae. Growth potential tests for nutrient availability using cultured algae do not suffer much from such problems (algae tolerate pH changes, for example), and the short duration of similar tests with natural phytoplankton to assess physiological nutrient limitation minimizes algae-induced problems.

The performance of toxicity tests, on the other hand, is more critical, but there are a number of ways to minimize unavoidable physicochemical changes in the medium. Many problems can be avoided by conducting tests of short duration (1 to 3 d) using nutrient-sufficient, exponentially growing algal stock inoculated at low initial concentrations (10^3 to 10^4 cells per milliliter with the standard freshwater green alga *Selenastrum capricornutum*). The lag phase can be avoided with the use of exponentially growing cells that have been propagated under test conditions. With a low inoculum concentration and a short test time, growth can proceed exponentially at a constant specific rate throughout the test period with the level of biomass sufficiently restricted such that changes in the medium and absorption of test material are sufficiently small to be ignored. To limit pH increase, it is essential that the mass transfer rate of CO_2 into the culture is sufficient to meet the demand for photosynthesis so that CO_2 equilibrium exists between water and gas phase. In most situations, this implies continuous shaking or aeration (with high cell densities, CO_2-enriched air and increased bicarbonate concentrations must be used and should also be considered with lower cell densities for testing volatiles in closed flasks headspace systems). Further control of pH drift may be gained through the use of ammonium as the nitrogen source. It is possible, nevertheless, to conduct static tests of 1 to 2 d without aeration if the inoculated cell density is low and the air–liquid interphase area of the culture is large (1 cm liquid height may be feasible [Nyholm, unpublished data]) and in addition a moderate specific growth rate is maintained (low light intensity and low temperature may be chosen to reduce the growth rate).

Although low initial inoculum density and moderate growth rates are desirable in the success of static algal bioassays, their use presents the problem of accurate and precise measurement of low biomass concentrations. Although it is not necessary to be able to measure the initial concentration in bioassay cultures because the nominal inoculated concentration can be used, it is important to consider that, even if the final biomass in control cultures may be accurately measured with available techniques, the final biomass in severely affected cultures may even be lower than the starting concentration. At a minimum, the final biomass should be measurable in cultures suffering a 50% reduction in growth rate. At this level of inhibition, the biomass, termed X_{50}, is equivalent to the square root of the corresponding control biomass times the initial biomass:

$$X_{50} = \sqrt{(X \cdot X_0)}$$

The accurate measurement of this figure may be used as the design criterion determining the lower feasible limit of inoculum density and/or test time.

Continuous cultures

With continuous algal culture systems, chemostats, or turbidostats, it is possible to maintain algae in a constant environment, at a well-defined physiological state, for extended periods of time. A chemostat is operated with a constant rate of inflow of culture medium and an equal rate of outflow of culture liquid. Under steady-state operation, the specific growth rate becomes equal to the inflow rate divided by the culture volume (the reciprocal retention time) and is termed the dilution rate, D (d^{-1}). Chemostat cultures are typically nutrient and/or light limited. In a turbidostat, the medium pump is controlled by a turbidity-measuring device and maintains cell density at a constant, preset level. The turbidostat is suitable for cultivation of nutrient-sufficient algae growing near their maximal specific growth rate at the prevailing light and temperature. Unless cell density is kept low and the light path short, however, growth in a turbidostat readily becomes light limited. In both types of continuous culture, limitation of growth by insufficient CO_2 supply can be a problem but can be remedied using CO_2 enriched air. Operation with controlled carbon limitation is possible, but is neither practical nor environmentally relevant. Light limitation, on the other hand, is ecologically relevant and readily achieved in algal culture. As biomass growth is linear and proportional to the number of light quanta absorbed under light-limited conditions, steady-state biomass is inversely proportional to the dilution rate (Pipes and Koutsoyannis 1962). In a nutrient-limited chemostat, the steady-state biomass yield decreases with increasing dilution rate, and the intracellular concentration of the limiting nutrient is equal to the reciprocal yield, because the residual dissolved concentration is negligibly small (Nyholm 1977a,b). With phosphorus-limited green algae, this yield variation between zero and maximum growth rate may be a factor of 20.

Continuous cultures have been used extensively in eutrophication studies for the examination of growth kinetics (e.g., Droop 1968; Caperon and Meyer 1972; Rhee 1973; Goldman et al. 1974; Nyholm 1977b; Healey 1978; van Liere and Mur 1978; Zevenboom and Mur 1978; Gons and Mur 1979; Kunikane et al. 1984). Their use for bioassays on water samples have also been investigated (Torien et al. 1970). Peterson and Nyholm (1993) provide details of the techniques and use of chemostat and turbidostat systems for algal toxicity studies, and their use in practical toxicity testing is reported by Conway (1978), Lederman and Rhee (1982), Bennett and Brooks (1989), and Hall et al. (1989).

Results of growth inhibition studies using continuous cultures must be interpreted in light of relevant growth kinetics. If, for instance, light is limiting, no reduction of the steady-state biomass will be observed until treatment-induced inhibition is very large, leading to cessation of light limitation. This is because the production of biomass is not determined by biological process rates, but by the influx of light quanta. In a nutrient-limited chemostat (where the specific growth rate is preset), the steady-state biomass yield can be expected to decrease with increasing toxic inhibition (higher intracellular concentration required to sustain the given specific growth rate). With turbidostatic cultures that are not limited by light or CO_2, the interpretation of results is obvious as the specific growth rate (recorded as the pumping rate) is the dependent variable.

It should be emphasized that toxicity studies with continuous cultures are laborious, and the number of units available is usually small. Therefore, the major potential use of continuous cultures in testing is providing algal culture material of a well-defined physiological state for use in smaller-scale, short-term assays, as described in Peterson et al. (1984), Peterson and Healey (1985), and Peterson (1991).

Biomass measurements

The most widely used methods for the determination of algal biomass in pure cultures are typically indirect measurements. These include cell volume and cell number determinations by means of electronic particle counters, optical density (light absorption in a spectrophotometer, e.g., at 600 nm where the absorption by chlorophyll is small), and fluorescence (*in vivo* chlorophyll fluorescence). These measurements are referenced to direct algal quantification and achieved either through cell enumeration using microscopy, biomass dry weight determination, or spectrophotometric or fluorometric determinations of extracted chlorophyll. Cell volume usually bears the most direct relationship to biomass dry weight, the most common reference for algal biomass. Measurements of cell size and chlorophyll content, in particular, may vary considerably with the physiological condition of the culture (as influenced by light conditions, nutritional state, temperature, etc.). Fluorescence per unit biomass is influenced by chlorophyll content and fluorescence yield per unit chlorophyll, and cell number is determined by the average cell size. Optical density is influenced both by cell size and cell chlorophyll content. In toxicity bioassays, test materials may cause changes in chlorophyll content, cell size, and fluorescence yield (Walsh 1988b). Such variability is of minor importance in bioassays, however, because measurements are taken relative to controls and typically only the initial part of the response curve, up to 50% relative response, is of interest. Nevertheless, sometimes corrections are necessary.

The various indirect methods of algal measurement have differing levels of sensitivity. The sensitivity limit of electronic particle counters is approximately 10^3 cells per milliliter with the standard freshwater algal species *Selenastrum capricornutum* (40 to 60 μm^3 cell volume); thus, accurate measurements require about ten times that level or in the order of 10^4 cells per milliliter (equivalent to about 0.2 mg dry weight per liter for *Selenastrum*). Fluorescence measurements can be of equal or even greater sensitivity than particle counters, but optical density is much less sensitive. With dilute cultures, a larger volume (5 cm cuvettes) may be required to achieve sufficiently accurate optical density measurements.

The use of electronic particle counters may be problematic if non-algal particulates are present and cannot be separated out or if algae are irregularly shaped or not freely suspended. With modern counters, however, particles can often be screened out by adjusting the size interval counted. Corrections may also be performed by counting test material blanks. Finally, difficult algae can sometimes be counted after gentle treatment with ultrasound. In some situations, however, particle counts are not feasible, and then either *in vivo* chlorophyll fluorescence or extracted chlorophyll are the preferred methods of biomass estimation. Of course, an alternative to measuring biomass is the determination of ^{14}C assimilation, in which case the biomass needs only be estimated for characterizing the biological reagent.

Current test methods used in eutrophication studies
Algal growth potential tests

Growth potential tests measure the growth response of algae in environmental samples of water to assess eutrophication of waterbodies. Water samples are membrane filtered (0.45 μm) to remove indigenous algae, zooplankton, and detritus particles that may interfere with the enumeration of algal cells using electronic particle counters. Filtered samples are then inoculated with a nutrient-deficient cultured alga (nutrient deficient to minimize nutrient carryover) and incubated for 14 d to achieve the maximum algal yield, recorded as the growth response. The freshwater green alga *Selenastrum capricornutum* is the most commonly used test species and has gained almost universal acceptance as the principal freshwater test alga. In addition, *Phaeodactylum tricornutum*, *Skeletonema costatum*, and *Thallassiosira pseudonana* (formerly *Cyclotella nana*) (diatoms) or *Dunaliella tertiolecta* (green alga) are the most frequently used marine test species.

Miller et al. (1978) presented a comprehensive test protocol for bioassay techniques with *Selenastrum capricornutum* which has served as a basis in most subsequent freshwater studies (Maloney et al. 1972; Greene et al. 1975) and has been adopted as an ASTM standard (ASTM Standard D 3978-80, reapproved 1993). With respect to marine algal assays, reference can be made to the U.S. EPA (1974), Maestrini et al. (1984), and Bonin et al. (1986).

Although growth potential tests were used extensively throughout the 1970s (Källqvist 1975; Kotai et al. 1978) when interest in the problem of freshwater eutrophication was at its peak, their use has declined, in part due to the reduction in eutrophication research, but also because it has been demonstrated that chemical analyses alone may be sufficient to assess the fertility of waters (Cain et al. 1979; Grobler and Davies 1979). These bioassays, however, contribute important complementary information on bioavailability of nutrients, while also serving as indicators for presence of toxic substances or lack of essential growth factors, but they do not provide information on kinetic (physiological) nutrient limitation and thus do not elucidate which nutrient(s) are controlling instantaneous *in situ* algal growth rate. A mathematical analysis of the algal assay emphasizing both kinetic limitation and potential (Liebig's law of limitation) of the maximum yield is discussed by Nyholm and Lyngby (1988).

Physiological indicator assays

Several physiological indicator variables have been proposed to assess the nutrient deficiency of phytoplankton. Riegman et al. (1990) has described the use of short-term phosphorus and nitrogen uptake rates to assess nutrient limitation in the North Sea. Chiaudani and Vighi (1976) have provided a thorough discussion on the use of alkaline phosphatase activity and the amounts of extractable "surplus" phosphorus. Such indicator parameters reliably reveal the nutrient status of phytoplankton. However, they are insufficient to actually quantify the degree of nutrient deficiency. The work of Healey (1978) has shown that there is no unique pattern of variation relating directly to deficiency in terms of intracellular concentration.

Nyholm and Lyngby (1988) suggested an alternative short-term (half day) "growth rate" bioassay method with natural phytoplankton which allows a semiquantitative assessment of physiological nutrient deficiency. An indigenous phytoplankton sample is taken to the laboratory and incubated for 12 to 15 h in shake flasks under light with excess nutrients spiked in various combinations. At the end of the incubation, an approximate estimate of the instantaneous growth rate (evaluated relative to controls) is obtained by measuring the ^{14}C assimilation rate during a short incubation time (e.g., 2 h). Healey (1979) found that short-term photosynthetic responses by nutrient-deficient algae to nutrient enrichment were unreliable means of identifying nutrient deficiencies. He also emphasized the inadequacies of long-term experiments for this purpose and suggested further that a test duration of 24 h seemed to be a reasonable division between short-term and long-term experiments. The "medium" incubation time of 12 to 15 h for growth rate assays used by Nyholm and Lyngby (1988) is, however, long enough to overcome lag phases following nutrient enrichment yet short enough to avoid large changes in the physiological state of the controls as well as in the algal population structure. Examples of results from such growth rate assays obtained during a season with samples from a coastal location near Copenhagen, Denmark are shown in Figure 4 (Lyngby and Mortensen 1995). Typical of coastal waters, nitrogen is seen to be growth limiting during the summer period, while during the rest of the season nutrients are excess.

Current algal toxicity test methods

Growth inhibition tests

The most widely used algal toxicity test method is a growth inhibition test in which a laboratory culture of algae is exposed to the test material in a batch culture while growing exponentially for 2 to 4 d under continuous light. Test protocols currently in use include (1) the ISO standards for toxicity testing with the freshwater green algae *Scenedesmus subspicatus* or *Selenastrum capricornutum* (ISO 1989) and the marine diatoms *Skeletonema costatum* and *P. tricornutum* (ISO 1993), (2) the EPA test protocol (Greene et al. 1989) with *Selenastrum* developed from the eutrophication growth potential test, and (3) an ASTM method for both freshwater and marine testing with a freshwater part similar to the EPA test (ASTM 1990). The ISO freshwater test has also been adopted by the OECD (1984) for chemicals testing. A thorough discussion of the differences among the standards for freshwater testing is provided by Nyholm and Källqvist (1989) and Peterson and Nyholm (1993) and is summarized as follows. ISO tests are designed to ensure exponential growth with excess nutrients throughout a test period of 3 d. The nitrogen (N) to phosphorus (P) ratio is maintained at about 11 on a weight basis, and ammonium is used as the nitrogen source to counteract pH increase during growth. The EPA/ASTM test lasts 4 d, and during the last day the growth rate may decline somewhat due to phosphorus limitation (the P concentration is 0.186 mg/l and the N to P ratio is 23). The pH of the ISO medium is 8.1 (50 mg of HCO_3^- per liter) and that of the EPA medium is 7.5 (15 mg of HCO_3^- per liter). In both media, EDTA is used as a complexing agent to keep iron and trace metals in solution (78 μg/l in the ISO medium and 234 μg/l in the EPA medium). The composition of the ISO test medium is given in Table 1. Figure 5 provides an example of the results of a typical ISO growth inhibition toxicity test. A severe practical problem of both methods is that of pH increase, a problem which has to be mitigated by modifying

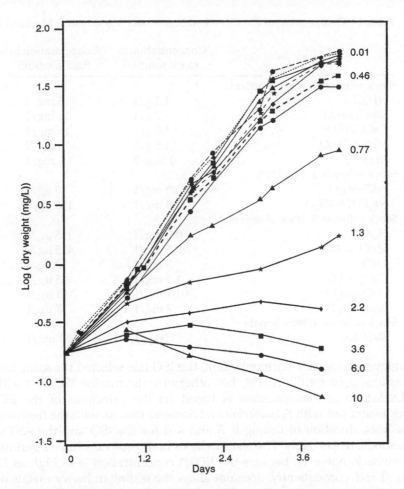

Figure 4 Example of results from "growth rate" bioassays to assess physiological nutrient deficiency of natural phytoplankton sampled from the coastal location of Stege Bay, Denmark (outer Baltic Sea). The response to nutrient enrichment is recorded as the rate of ^{14}C assimilation (assayed during 2 h) and measured after 15 h preincubation under continuous illumination in shake flasks with various nutrient spikings. The results demonstrate severe physiological nitrogen limitation during the summer season and otherwise nutrient sufficiency or borderline deficiency. (From Lyngby JE, Mortensen E [1995] Biomonitoring of eutrophication levels in shallow ecosystems, in press. With permission.)

tests in the future. Unless the culture system allows a very efficient CO_2 mass transfer and/or the algal growth rate is small, pH will increase one to two units by day 3 of the tests.

Marine tests can be performed using either a basal synthetic seawater medium or a natural seawater-based medium, both of which are amended with macro- and micronutrients and trace elements. To

Table 1   Composition of the ISO Algal Freshwater Toxicity Test Medium

Nutrient	Concentration in stock solution	Concentration in final medium
Stock solution 1: macronutrients		
NH_4Cl	1.5 g/l	15 mg/l
$MgCl_2 \cdot 6H_2O$	1.2 g/l	12 mg/l
$CaCl_2 \cdot 2H_2O$	1.8 g/l	18 mg/l
$MgSO_4 \cdot 7H_2O$	1.5 g/l	15 mg/l
KH_2PO_4	0.16 g/l	1.6 mg/l
Stock solution 2: Fe-EDTA		
$FeCl_3 \cdot 6H_2O$	80 mg/l	80 µg/l
$Na_2EDTA \cdot 2H_2O$	100 mg/l	100 µg/l
Stock solution 3: trace elements		
H_3BO_3	185 mg/l	185 µg/l
$MnCl_2 \cdot 4H_2O$	415 mg/l	415 µg/l
$ZnCl_2$	3 mg/l	3 µg/l
$CoCl_2 \cdot 6H_2O$	1.5 mg/l	1.5 µg/l
$CuCl_2 \cdot 2H_2O$	0.01 mg/l	0.01 µg/l
$Na_2MoO_4 \cdot 2H_2O$	7 mg/l	7 µg/l
Stock solution 4: bicarbonate		
$NaHCO_3$	50 g/l	50 mg/l

achieve interagency harmonization, the ISO has selected the same test medium used by the ASTM, but otherwise the marine ISO test with *Skeletonema* or *Phaeodactylum* is based on the principles of the ISO freshwater test with *Selenastrum* and *Scenedesmus*. As with the freshwater tests, duration of testing is 3 and 4 d for the ISO and the ASTM methods, respectively. The test media of both protocols have a serious drawback, however, because the EDTA concentration is as high as 13 mg/l and, consequently, does not allow the testing of heavy metals or complex wastes, where the presence of heavy metals is inevitable. This high EDTA concentration has been found necessary by several investigators to achieve sufficient growth. Recent findings by Källqvist (1993), however, suggest that this may be due to a high and possibly toxic zinc concentration of 150 µg of zinc per liter. Further research at the Norwegian Institute of Water Research (Källqvist 1993) and at the Water Quality Institute, Denmark (Kusk and Nyholm 1992) suggests that marine testing can be conveniently performed using a medium based on natural seawater enriched with the nutrient mixture of the ISO freshwater medium (resulting in an added EDTA concentration of only 78 µg/l). Others, however, have reported variable results using natural seawater-based media (Walsh 1988a; Thomson 1993). Media prepared from commercial sea salt mixtures may be an alternative to using natural seawater, but Walsh et al. (1987) has reported large differences in responses to the same toxicant using different formulations.

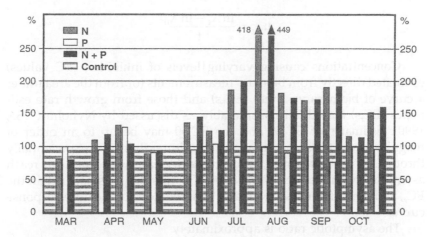

Figure 5 Growth curves obtained in a toxicity growth inhibition test with potassium dichromate and the freshwater green alga *Scenedesmus subspicatus* carried out according to ISO standards (ISO 1989), but using a draft version with less nitrogen in the medium (2 mg of N per liter) than used in the current standard (4 mg of N per liter). Growth is exponential in control cultures throughout the test period of 3 d, but declines rapidly thereafter due to exhaustion of the nitrogen source.

Commercial formulations of seawater apparently contain too high levels of heavy metals, and their use for culturing and testing marine algae is not feasible without the addition of EDTA in the milligram range (Steele 1994). It may be feasible to use the EDTA-based medium as a culturing medium only and make up the test medium with no EDTA, iron, or trace metals, but it must be investigated first to see how many generations the algal test species can sustain growth at maximal rate growing on internal stores of iron or trace metals. The duration of tests with no EDTA, iron, and trace metals added may have to be shortened considerably (refer also to Peterson and Nyholm 1993).

The problems of batch culture toxicity testing have been addressed earlier in Cultivation Methods under the Batch Cultures section. There remains the question of selection of an appropriate test endpoint for establishing a concentration-response curve for quantitative description of growth inhibition relative to controls in algal toxicity testing. The choice of response variable has been controversial in test standardization, and a consensus has not been reached yet. Basically, two different approaches are currently used. One is to use the biomass (or cell number), X_t, recorded by the end of the test or related to the area under the growth curve in a linear plot, A_t. The alternative is to use some estimate of the specific growth rate, μ (d^{-1}). This growth rate estimate can conveniently be the "average specific growth rate" μ_{av} (Nyholm and Källqvist 1989).

$$\mu_{av} = \frac{\ln X_t - \ln X_0}{t}$$

Concentrations causing varying levels of inhibition (EC values) estimated directly from biomass measurements (or from the area under a curve of biomass vs. time) (EC_bs) and those from growth rate estimates (EC_rs) may differ considerably, as discussed by Nyholm (1985, 1990). Estimates of EC_b50 and the EC_r50 may be up to an order of magnitude apart in some situations. In cultures that grow exponentially throughout the test period, EC_b figures decrease with time, but reach asymptotic values after several days. The final EC_b is smaller than the EC_r, and the differences are greater than the slope of the dose-response curve.

The asymptotic ratio is approximately

$$\frac{EC_b50}{EC_r50} = 10^{\frac{-0.5}{\alpha}}$$

where α is the slope of the concentration-response curve for growth rates linearized around EC_r50. As EC_b values are calculated before the asymptotic ratio is reached and depend on test duration and absolute magnitude of the specific growth rate, it is recommended that EC_r values be used. This recommendation should be seen as a simple mathematical consequence of exponential growth and involves no arguments in terms of "ecological relevance." The growth pattern in inhibited cultures may deviate from strict exponential growth ("distorted" growth curves may be observed), but the basic growth pattern in virtually all cultures having positive growth is closer to exponential than linear. The use of biomass as an endpoint logically requires linear growth in the test system.

For calculating μ_{av}, an accurate estimate of the starting biomass must be known. While accurate direct measurements of the low initial biomass may be difficult, it may be readily calculated from the biomass in the preculture and its dilution during inoculation.

^{14}C assimilation tests

^{14}C assimilation is commonly used as an indicator of the photosynthetic rate in assessing algal toxicity. This method is based on techniques used for the measurement of primary productivity and is particularly well suited for tests with low algal biomass and of short duration. Short tests with marginal or restricted growth may be needed for various purposes, such as for testing heavy metals in chelator-free medium, for testing volatiles in closed flasks, for testing highly absorbable substances where the dose per biomass must be kept constant, or for examination of unstable test materials. ^{14}C-fixation measurements may also be preferred with samples that are difficult to count by means of electronic particle counters, including natural phytoplankton samples containing detrital particles, filamentous or colony-forming algal species, benthic algae attached to a solid substrate, or with test materials containing particulates. Kusk and Nyholm (1991) described and discussed a protocol for a 6-h ^{14}C assimilation toxicity test conducted in closed and filled bottles. Another test protocol was suggested by Giddings et al. (1983). Peterson et al. (1994b) developed a 24-h miniature test using nutrient-deficient chemostat algal culture incubated in vials with a large air headspace. The CO_2 supply from the air in the headspace in combination with the restricted growth rate caused by nutrient deficiency prevented the pH increase that is inevitable when algae are incubated in light for 24 h in closed, filled bottles.

Not surprisingly, short-term ^{14}C assimilation tests (2 to 6 h) are generally less sensitive than longer-term growth inhibition tests (Nyholm and Damgaard 1990). Nevertheless, for some substances, the sensitivity of short-term ^{14}C assimilation tests has been found equal to that of growth tests (Kusk and Nyholm 1992). It can be expected that with longer-term ^{14}C assimilation tests (24 h) the sensitivity of the two types of tests is equivalent, since photosynthesis is roughly proportional to growth. Longer-term tests can be carried out in flasks with a sufficiently large headspace (as described by Peterson et al. 1994a) if the algae are nutrient deficient or the biomass is low. As mentioned earlier, it may also be possible to enrich the headspace air with CO_2 and supplement the medium with bicarbonate to achieve the desired pH in equilibrium with the gas phase.

Alternative endpoints

Inhibition of growth or photosynthesis are the most common test endpoints and are of obvious ecological relevance. The nutrient uptake rate of nutrient-deficient algae is also of ecological importance; nevertheless, studies on toxic inhibition of nutrient uptake are few. It has been shown by Nyholm (1991) that phosphate uptake rates by phosphorus-deficient algae after differential spiking with ^{32}P-labeled phosphate were more affected by potassium dichromate and 3,5-dichlorophenol than were growth rates, but the differences were relatively small. In investigations of metal speciation and toxicity, Peterson and Healey (1985) and Peterson (1991) studied the inhibition of the nitrate, ammonium, and phosphate uptake systems by cadmium and copper as a function of pH using chemostat-grown algae. Obviously, more research is needed to explore the general utility of nutrient uptake rates as endpoints in toxicity studies.

While the previous endpoints reflect sublethal effects, the measurement of cell death may also be of interest, particularly for testing algicides. Some toxicants may kill cells, while others merely slow down cell processes. Payne and Hall (1978) classified algal toxicity responses into three categories: (1) reduction in growth rate, (2) cessation of cell division (algistatic), and (3) cell death (algicidal). They also described a graphical method for calculating the algistatic concentration of a test material. Walsh (1983) described a method for determining dead cells in algal toxicity tests using Evans blue stain (see also Crippen and Perrier 1974).

Toxicants may also affect cell morphology; therefore, microscopic examinations of the algal cells are advisable. Morphologic effects are normally reported only as secondary observations, however, and are difficult to quantify.

Continuous culture studies

For routine testing, continuous cultures can be used to supply algal material in a well-defined physiological state for use in short-term tests. Nutrient-limited chemostat cultures can thus provide nutrient-limited algae for use in ^{14}C assimilation tests or for assessing toxic effects on nutrient uptake rates. The use of continuous cultures to determine the toxicity of different materials is of limited use because of cost and time requirements, and it may not always be easy to interpret the data generated (see Continuous Cultures under the Cultivation Methods section). However, there are some advantages with direct tests on continuous cultures. For example, with highly absorbable test materials, the dose (concentration per unit of biomass) can be kept constant for long time periods using continuous cultures. A feasible way of testing such materials in longer-term tests is also to use semicontinuous cultures with partial replacement of the algal suspension with fresh medium (plus toxicant) each day or to use "pseudodynamic" tests (Jouany et al. 1983) where fresh medium is added periodically to large test vessels without overflow.

Statistical evaluation of concentration-response curves

Traditional statistical methods for bioassays, such as probit analysis (Finney 1971), deal with quantal data (e.g., death) from assays with animals and, therefore, cannot be used for algal toxicity tests (Nyholm et al. 1992). Response in algal tests is continuous and must be evaluated relative to control response. Its variance is related to handling differences between test flasks and to the errors in measurement of the response (e.g., growth rate, biomass, or ^{14}C assimilation). Unlike animal bioassays, there is no underlying assumption of a binomial tolerance distribution between individual test subjects in algal tests because a population of algae is essentially infinitely large, with properties that can be regarded as those of a continuum having no relation to individual cells. While dose-response relationships in both types of bioassays may be sigmoid-shaped curves, the underlying variance structure differs.

The EC_{50} (the concentration causing 50% inhibition in the response variable) is generally considered to be the single most important number derived from a test. Also, the LOEC and/or the NOEC are of particular interest. The EC_{10} (10% inhibition) can usually be determined with acceptable precision in algal tests and may therefore be used as an LOEC estimate. The proper statistical method for algal tests is traditional regression analysis or curve fitting using an appropriate mathematical representation of a sigmoid curve, such as the logistic equation, the probit model (the log-normal probability function), or the Weibull model (see Figure 6 for an example), and with consideration of the nature of the error on the response (constant and/or proportional). Often, the assumption of a constant variance is a sufficient approximation with algal tests. Nonlinear regression can be used directly on the data (without weighting if the variance is constant). A number of excellent computer programs are available for nonlinear regression; a dose-response curve may readily be fitted, and the EC_{50} and other descriptors such as the EC_{10} or the EC_{20} can subsequently be estimated from the inverse function. While confidence limits are easily calculated for the response and are provided as an option in several commercial computer programs, the inverse estimation of confidence intervals for the corresponding concentrations (the ECs) is complicated and not provided in any commercial programs known to the authors.

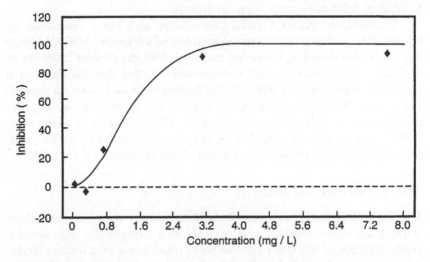

Figure 6 Concentration-response curve (linear scale) from a growth inhibition toxicity test with the green alga *Selenastrum capricornutum* and potassium dichromate. Direct nonlinear curve fit using the Weibull equation (Christensen and Nyholm 1984).

The problem of inverse estimation has been described by Vølund (1978). Recently, Andersen (1994) developed a method using the log-normal probability or the Weibull function to fit the data. An approximate method is to reparameterize the problem, fitting the model with the desired endpoint (EC_{50} or EC_{10}) as one of its parameters (Bruce and Versteeg 1992). Parameter confidence intervals are automatically calculated in many commercial, nonlinear regression programs.

Alternatively, weighted linear regression can be carried out on linear transformations of the data (Christensen and Nyholm 1984; Nyholm et al. 1992), applying weighting factors that compensate for the alteration of the variance distribution caused by the transformation ("transformation weighting"). Confidence intervals on the concentration axis must be calculated by inverse estimation, and their calculation is discussed in detail by Nyholm et al. (1992).

Nonlinear regression on untransformed data, however, is more robust in handling data irregularities inherent in the initial portion of the curve where some responses at low toxicant concentrations may, at random, exceed control response. Also, concentration-response curves from algal tests frequently exhibit real stimulation at low concentrations, which may be caused by some physiological imbalance. This phenomenon is usually referred to as hormesis (Stebbing 1982). The resulting nonmonotone concentration-response curves can be described by nonlinear functions with four or more parameters, such as the four-parameter logistic equation (e.g., Vølund 1978; Bahner and Oglesby 1982; Van Ewijk and Hoekstra 1993). However, this response may not always be of particular interest, and it may be more practical to analyze inhibitory responses separately.

Traditionally, NOECs and LOECs have also been estimated by statistical hypothesis testing using analysis of variance (ANOVA) techniques. Such estimates, however, merely reflect the design (number of replicates and choice of test concentrations) and the accuracy of a particular experiment. Therefore, the figures obtained have no general value and should not be used as a basis for extrapolation, e.g., by multiplication with application factors. The problem of estimating LOECs and NOECs by ANOVA techniques vs. calculating an EC_x estimate by regression analysis has recently been reviewed by Pack (1993).

Test species and test batteries

The freshwater test species used most commonly are the green algae *Selenastrum capricornutum* and *Scenedesmus subspicatus*. For several years, cultures of the latter species were used in toxicity testing under the name *S. quadricauda*, but the mistake was corrected during ISO ring-test work. Several other green algae are commonly used, and, in addition, some toxicity testing work has been carried out using diatoms, cyanobacteria, and some other taxonomic groups (see, e.g., Blanck et al. 1984; Wong and Couture 1986; Källqvist and Romstad 1994; Peterson et al. 1994a). A list of freshwater phytoplankton that have been used in toxicity testing work is given in Table 2.

Marine species that have been used for toxicity testing other than *Skeletonema* and *Phaeodactylum* include the diatoms *T. pseudonana* (Bonin et al. 1986) and *Minutocellus polymorphus* (Walsh et al. 1988) and the flagellates *D. tertiolecta* (commonly used), *Rhodomonas baltica*, and *Pavlova lutheri* (Nyholm et al. 1991).

Species may differ greatly in their sensitivity to toxicants having specific modes of action (many pesticides, for example). In the extreme, differences as large as three or four orders of magnitude between effect concentrations have been observed among species (Blanck 1984; Blanck et al. 1984; Wängberg and Blanck 1988; Swanson et al. 1991; Källqvist and Romstad 1994; Peterson et al. 1994a). Thus, for hazard evaluation and testing for the registration of pesticides, the use of test batteries with a minimum of five to ten algal species from different phyla seems necessary. Even the use of a test battery provides information on a relatively small data set, given the overall diversity of the aquatic ecosystem. It is possible to extrapolate from an available data set, using distribution models for the observed interspecies sensitivity variation, to calculate concentration levels that protect a given percent of the species in the aquatic ecosystem (Stephan et al. 1985; Kooijman 1987; Van Straalen and Denneman 1989; Wagner and Løkke 1991). Experience with test batteries is still limited, however, and for regulatory purposes, standardization of battery size and species composition is necessary to create a comprehensive and uniform reference database. However, it should be recognized that toxicity tests may serve a wide range of purposes, and, for some of these (e.g., for simple detection of toxicity or for monitoring toxicity changes), tests with only one species may still be adequate.

Pure cultures vs. mixed communities

The problem of interspecific sensitivity differences and interest in ecological realism has prompted the use of microcosm and mesocosm testing as an alternative to single-species tests. The utility of microcosm and mesocosm testing was reviewed earlier. In this section, only test technical factors are addressed. Two major drawbacks to the microcosm and mesocosm methods are emphasized.

1. It can be extremely difficult to control the experimental conditions (exposure, nutrient status, pH, CO_2 supply, and species composition) with mesocosm and, in particular, with microcosm studies, and it is often impossible to reproduce these conditions. Therefore, the results may be site and time specific and are seldom generalized.

Table 2 Freshwater Phytoplankton Used in Toxicity Testing Work

Green algae	Diatoms	Cyanobacteria	Other
Chlorella vulgaris	Nitzschia palea	Microcystis aeroginosa	Klebsormidium marinum
C. emersonii	Nitzschia sp.	Anabaena flos-aquae	Raphidonema longiseta
Scenedesmus quadricauda	Cyclotella meneghiana	Anabaena inaequalis	Bumilleriopsis filiformis
Selenastrum capricornutum	Navicula pelliculosa	A. variabilis	Monodus subterraneus
S. subspicatus		Pseudoanabaena sp.	Tribonema aequale
S. obtusiusculus		Oscillatoria sp.	
S. acutus		Aphanizomenon flos-aquae	
Chlamydomonas reinhardii		Synechocous leopoliensis	
C. cf. obesa			
C. dysomos			
Kirchneriella contorta			
Monoraphidium pusillum			

2. Available measuring techniques for assessing impacts on mixed field communities are crude and essentially restricted to certain functional parameters and population extensive parameters, such as primary productivity and chlorophyll. Species composition can only be assessed by labor-intensive counting and identification work by taxonomic specialists, yet subtle changes in species composition are still unlikely to be detected. The properties of a mesocosm or microcosm that may be affected by toxicants are numerous, yet it is only technically possible to assess the most crude effects.

It appears to be more practical to use the system in its components and to study the effects on critical and, at the same time, sensitive components and processes. The ecological aspect should not be forgotten, and mesocosm and microcosm studies may help us identify which components and processes to select for further study (e.g., with algae those factors that are decisive for interspecific competition, including growth rate and nutrient uptake ability).

Finally, it should be mentioned that perhaps the PICT phenomenon (see earlier) may be utilized for easy monitoring of structural changes in mesocosms. If more tolerant species have replaced sensitive species as a result of exposure to a given compound, this might be ascertained by means of a simple toxicity test with this compound, e.g., a simple ^{14}C assimilation toxicity test performed on a water sample from the mesocosm. However, studying effects of pesticides, Källqvist et al. (1994) could only detect decreased sensitivity in mesocosms that were exposed to a pesticide chemical toward which great interspecies sensitivity differences had been shown from single-species laboratory tests.

Test media

Discussions of media for the cultivation of algae are included in Rodhe (1978) for freshwater species and in Morel et al. (1979) and Walsh (1988b) for saltwater species. Some algae require the addition of certain organic growth factors (vitamins, etc.) (can also be accomplished by using natural water derived media), while others, including the standard freshwater test algae *Selenastrum* and *Scenedesmus*, grow well in simple inorganic salt media made with highly purified water. Trace elements and iron must be kept available in longer-term tests; to achieve this, chelators (such as EDTA) must be added. Even so, the trace metal stock solution must be inspected regularly for precipitates and discarded if any precipitates are present. Tests using chelator-free media must be of short duration (e.g., 1 d), where the algae can grow using stored reserves of trace elements. Having prepared a test medium, it is advised to equilibrate it with air before use in order to achieve the desired (equilibrium) pH. This can be accomplished by bubbling with air for a couple of hours or with shake flasks by leaving flasks on the shaker overnight.

Characteristics of the medium may greatly affect toxicity, particularly of heavy metals. Heavy metal toxicity is affected by a number of chemical factors, including pH (e.g., Peterson et al. 1984), hardness (or salinity), general ionic composition, and concentration of chelators (Peterson and Nyholm 1993). In contrast, the toxicity of organic chemicals is not affected by chelators (Walsh and Alexander 1980), and other physicochemical factors are less important. The bioavailability and hence the toxicity of highly absorbable compounds though may be reduced by dissolved, colloidal, and particulate organic matter, including exudates released from algae during the test.

Difficult test materials

Generally, a number of test materials cause problems in bioassays, but with algal tests some additional problems are encountered concerning growth, cell enumeration, and pH control. Difficult test materials include heavy metals; volatile chemicals; slightly soluble and highly adsorbable chemicals; unstable test materials; and complex effluents and environmental samples, such as contaminated sediment and soil.

To assess the toxicity of heavy metals quantitatively, all of the following significant factors must be controlled: pH, chelator concentrations (added chelators; algal exudates; and with natural water-based media, other natural dissolved and colloidal organic compounds), concentration of particulates including algae to which metals may sorb, water hardness and major ionic composition (for review, see Peterson and Nyholm 1993). In the ISO standard freshwater tests, EDTA is added in a relatively low concentration of 78 µg/l to simulate the chelation capacity of some standard surface water and, at the same time, to maintain a relatively constant chelation capacity during the test, because the added EDTA is believed to be the dominant chelator. If maximum sensitivity toward heavy metals is desired, algal cells can be grown in a normal medium which includes chelator, then washed in chelator-free medium, and subsequently assayed in that medium in a short-term test (Peterson et al. 1984).

Toxicity of volatile compounds can be examined in short-term tests (1 to 2 d) using closed, shaken flasks with a large headspace (Galassi and Vighi 1981) from which CO_2 is available. The equilibrium toxicant concentration in the aqueous phase is calculated from Henry's law or analyzed. Completely filled, closed bottles can only be used for short duration (hours), low inoculum tests to avoid increases in pH. Such tests can be conveniently conducted by measuring the ^{14}C assimilation rate. A newly developed test (Halling-Sørensen et al. 1996) is a head-space method with 1–2% CO_2 in the headspace air and bicarbonate enrichment of the medium.

Slightly soluble chemicals may be brought into solution using solvent carriers such as acetone or dimethylformamide, which in low concentrations (e.g., less than 1 ml/l, but preferably less than 0.1 ml/l [ISO 1982]) do not exert toxicity to algae. However, solvents may alter the phase distribution and bioavailability of the toxicant, thus affecting toxicity. Interactions between solvents and pesticide toxicity have been studied by Stratton and co-workers (Stratton 1986). Another problem with the testing of slightly soluble compounds is their propensity for sorption onto algal biomass. Sorption may increase with time because of biomass growth, resulting in a reduction in available toxicant dose. Sorption may be minimized using a sufficiently low inoculated biomass and a short test duration. Alternatively, continuous or semicontinuous cultivation techniques may be used. The phenomena of sorption and accumulation of organic chemicals in algae has received little research attention. However, it appears that bioconcentration factors (partition coefficients) increase with decreasing algal biomass concentration (Cox 1970; Rice and Sikka 1973; Lederman and Rhee 1982; Yu-yun et al. 1993; Nyholm, unpublished results, 1993), and it is assumed that surface adsorption contributes significantly in the overall sorption or bioaccumulation of chemicals in microscopic algae. Bioconcentration factors (BCFs) in algae appear to be considerably higher than those in aquatic animals. Casserly et al. (1983) estimated BCFs for *Selenastrum capricornutum* (assayed at 9 to 24 mg dry weight per liter [dwt/l]) that were 10 to 100 times greater than those for fish and proposed the following relationship between BCFs (on a dry weight basis) and *n*-octanol–water partition coefficient (K_{ow}):

$$\log BCF = 0.46 \cdot \log K_{ow} + 2.36$$

Geyer et al. (1985) found the following equation for another green algae, *Chlorella fusca*:

$$\log BCF = 0.681 \cdot \log K_{ow} + 0.863$$

In planning experimental protocol, it may be decided that the sorbed fraction of toxicant should be less than 10% by the end of the test. The sorbed fraction can be roughly estimated from the K_{ow} of the compound and one of the above equations (BCF = K_d = partition coefficient), assuming the equations are valid for the algal concentrations typically found by the end of an algal test (e.g., 10 to 100 mg dwt/l).

The sorped fraction, α, of the total amount of toxicant equals

$$\alpha = \frac{C_{sorped}(mg/l)}{C_{total}(mg/l)} = \frac{X \cdot 10^{-6} \cdot BCF}{1 + X \cdot 10^{-6} \cdot BCF}$$

where X is the dry weight of the biomass (mg/l), and C is the concentration of chemical studied.

As an example, consider a batch test control culture with *S. capricornutum* inoculated at 10^4 cells per millimeter (about 0.2 mg dwt/l) and growing at a rate of 1.8 d^{-1}. After the ISO standard test duration of 3 d, the biomass would reach about 44 mg dwt/l. For two test compounds with log K_{ow} of 3 and 5, respectively, the BCF estimates would be $5.5 \cdot 10^3$ and $4.6 \cdot 10^4$ (mg/kg dwt)/(mg/l) using the equation of Casserly et al. (1983), with sorped fractions of 19 and 67%, respectively. By reducing the test duration to 2 d, the final biomass would be 7 mg dwt/l, with sorped fractions of 4 and 24% for the two compounds, respectively. Thus, the compound with the lower log K_{ow} of 3 could be tested in a standard batch test by reducing test duration to 2 d. The other compound with log K_{ow} = 5 would be difficult to investigate in a batch test. By reducing the inoculum density to $3 \cdot 10^3$ cells per millimeter, the final biomass after 2 d would become about 2 mg dwt/l (10^5 cells per millimeter), resulting in sorption of approximately 8% with an unchanged BCF. Nevertheless, at the lower algal density, the BCF may in fact be substantially higher than that calculated from the equation of Casserly et al. (1983), and the criterion of less than 10% may therefore not be met.

Testing of complex effluents and environmental samples may cause problems with algal enumeration by means of electronic particle counters (refer to the Biomass Measurements section) and may cause alkalinity changes affecting ^{14}C measurements. However, such problems can usually be overcome by various corrective means or through the use of other methods and by correcting for the alkalinity changes in ^{14}C assimilation tests (Kusk and Nyholm 1991). It is worth noting, for instance, that with flourometric measurements even suspensions of soil, sediment, or solid waste material may be tested and bioavailable sorped toxicants thus assayed.

Future uses of algal tests

Recent advances in instrumentation and techniques have made algal toxicity testing an ideal screening method in the assessment of environmental and wastewater samples, as well as in the registration process of toxic chemicals. Bioassay methods are cost-effective, algae are sensitive test organisms, and test conditions are generally less demanding than with aquatic animals because algae can tolerate ammonia and particulate matter and do not require oxygen.

Reduced assay volumes are important in increasing the cost-effectiveness of algal toxicity tests. They may be scaled down to 1 to 2 ml size or less and used with automated fluorescence readers and serial diluters for preparing test concentrations (e.g., Källqvist and Romstad 1994). Batteries with several algal species can be used, and large number of samples can be conveniently processed. Even microtests in microplates with culture volumes of 250 μl^3 have been proposed (Blanck 1984; Blaise et al. 1986). The validity of such tests in terms of pH control and adequate control of growth has not been demonstrated, however. In the tests described, microplates have been incubated under static conditions, and, even though the mass transfer rate of CO_2 per culture volume may be relatively high, pH drift probably cannot be avoided if the biomass increases to levels high enough to measure with a spectrophotometer as has been suggested. Blanck (1984) proposed it would be possible, however, to use such tests for simple screening by taking the EC100 (no growth), recorded by simple visual inspection of the plates, as the endpoint. Automated or semiautomated fluorescence measurements or, alternatively, ^{14}C assimilation measurements further increase the efficiency of sample processing.

The future of pesticide registration and hazard assessment of other toxic compounds requires the selection of appropriate species for algal test batteries upon which an international consensus may be reached. Species must represent a wide taxonomic range and be suitable for laboratory cultivation. Ideally, some species should also be sensitive indicator species, although experience suggests that no species appear to be universally sensitive to toxicants (Blanck 1984; Blanck et al. 1984; Wängberg and Blanck 1988; Swanson et al. 1991; Källqvist and Romstad 1994; Peterson et al. 1994a).

In future eutrophication monitoring programs, assays with natural phytoplankton samples, reflecting physiological nutrient deficiency, are believed to hold promise. Changes in patterns of algal response to nutrient limitation may reflect changes in nutrient loadings and the eutrophic state of natural waterbodies better than changes in nutrient concentrations measured chemically.

Summary

Bioassays with microalgae, both in pure cultures and in natural communities, may be used for a number of different purposes in environmental studies. Assays can be divided into two major categories: (1) algal assays for use in eutrophication studies to assess limiting nutrients and (2) algal toxicity tests.

Nutrient bioavailability can be determined by means of the growth potential test with cultured algal species. Monitoring nutrient limitation of surface waters using samples of indigenous phytoplankton to assess physiological nutrient deficiency is being developed. Although this method has been used relatively little, it is believed to hold promise as a future monitoring tool that sensitively reflects trends in trophic status.

Toxicity tests are used in hazard evaluation schemes for chemicals, chemical products, or wastewaters or for screening, monitoring, or toxicity identification of complex test materials, including effluents as well as various environmental samples such as soil, wastes, sediments, surface water, or groundwater.

Testing using algal growth is complicated because growth depends on many interacting factors, such as light, nutrients, CO_2 and the carbonate system, and various trace elements. In toxicity studies, many of the factors may interact with the toxic response. A thorough understanding of these factors is necessary for conducting proper algal assays beyond the routine application of standard test protocols.

It is concluded that algal toxicity tests have a great potential in environmental studies because they are sensitive, cost-effective, and can be used easily with various complex test materials. Because of the degree of intertaxa variability in sensitivity, particularly to pesticides, it is concluded that there is a need for research leading to the selection and subsequent international agreement upon a range of suitable standard test species to be used in test batteries.

Acknowledgments

We would like to thank the Danish Strategic Environmental Research Progam and several Canadian government departments (Environment Canada, the Department of Supply and Services Environmental Innovation Progam of Canada's Green Plan, and Natural Resources Canada) for financial support which has made collaboration between the Technical University of Denmark and the Saskatchewan Research Council possible. Rick Scroggins of Environment Canada was instrumental in facilitating this collaboration. Pamela Martin edited the manuscript, and Yvonne Wilkinson provided technical assistance.

References

Adema DMM, Kuiper J, Hanstveit AO, Canton HH (1983) Consecutive system of tests for assessment of the effects of chemical agents in the aquatic environment. In *IUPAC Pesticide Chemistry: Human Welfare and the Environment*, Pergamon Press, Elmsford, NY, pp. 537–544.

American Society for Testing and Materials (ASTM) (1990) *Standard Guide for Conducting Static 96-h Toxicity Tests with Microalgae,* Designation: E-1218-90, American Society for Testing and Materials, Philadelphia, PA, 11 pp.

Andersen H (1994) Statistiske metoder til vurdering af spildevands toxicitet. M.Sc. Thesis. The Technical University of Denmark, Copenhagen (in Danish).

ASTM Standard D 3978-80, reapproved 1993 (1996) Practice for algal growth potential testing with *Selenastrum capricornutum*. In *ASTM Annual Book of Standards*, Vol 11.05, American Society for Testing and Materials, Philadelphia, PA, pp. 29–33.

Bahner LH, Oglesby JL (1982) Models for predicting bioaccumulation and ecosystem effects of Kepone and other materials. In *Environmental Risk Analysis for Chemicals*, Conway RA, Ed., Van Nostrand Reinhold, New York, pp. 461–473.

Benenati F (1990) Keynote address: plants — keystone to risk assessment. In *Plants for Toxicity Assessment*, Wang W, Gorsuch JW, Lower WR, Eds., ASTM STP 1091, American Society for Testing and Materials, Philadelphia, PA, pp. 445–459.

Bennett WN, Brooks AS (1989) Measurement of zinc amelioration of cadmium toxicity in *Chlorella pyrenoidosa* using turbidostat culture. *Environ Toxicol Chem* 8:877–882.

Blaise C, Legault R, Bermingham N, Van Coillie R, Vasseur P (1986) A simple microplate algal assay for technique for aquatic toxicity assessment. *Toxic Assess* 1:261–281.

Blanck H (1984) Species-dependent variation among aquatic organisms in their sensitivity to chemicals. *Ecol Bull* 36:107–119.

Blanck H, Wallin G, Wängberg S (1984) Species dependent variation in algal sensitivity to chemical compounds. *Ecotoxicol Environ Saf* 8:339–351.

Blanck H, Wängberg SÅ, Molander S (1988) *Pollution Induced Community Tolerance — A New Ecotoxicological Tool*, ASTM STP 988, American Society for Testing and Materials, Philadelphia, PA, pp. 219–230.

Bloomfield JA, Park RA, Scavia D, Zahorcak CS (1974) Aquatic modelling in the eastern deciduous forest biome U.S. *Modelling the Eutrophication Process*, Ann Arbor Science, Ann Arbor, MI.

Bonin DJ, Droop MR, Maestrini SY (1986) Physiological features of six microalgae to be used as indicators of seawater quality. *Cryptogamie, Algologie* 7:23–83.

Bringmann G, Kühn R (1977) Grenzwerte der Schadwirkung wassergefärdender Stoffe gegen Bakterien (*Pseudomonas putida*) und Grünalgen (*Scenedesmus quadricauda*) im Zellvermehrungshemmtest. *Z Wasser Abwasser Forsch* 10:87–98.

Bruce RD, Versteeg DJ (1992) A statistical procedure for modelling continuous toxicity data. *Environ Toxicol Chem* 11:1485–1494.

Cain JR, Klotz RL, Trainor FR, Costello R (1979) Algal assay and chemical analysis: a comparative study of water quality assessment techniques in a polluted river. *Environ Pollut* 13:215–224.

Cairns J Jr. (1980) Beyond single species toxicity testing. *Mar Environ Res* 3:157–159.

Cairns J Jr. (1983) Are single species toxicity tests alone adequate for estimating environmental hazard? *Hydrobiologica* 100:47–57.

Cairns J Jr., Ed. (1985) *Multispecies Toxicity Testing*, Pergamon Press, New York.

Canale RP, Vogel AH (1974) Effects of temperature on phytoplankton growth. *ASCE Env Engng Div* 100(EE1):231–241.

Caperon J, Meyer J (1972) Nitrogen limited growth of marine phytoplankton I. Changes in population characteristics with steady-state growth rate. *Deep-Sea Res* 19:601–618.

Casserly DM, Davis EM, Downs TD, Guthrie RK (1983) Sorption of organics by *Selenastrum capricornutum*. *Water Res* 17:1591–1594.

Chiaudani G, Vighi M (1976) Comparison of different techniques for detecting limiting or surplus nitrogen in batch cultures of *Selenastrum capricornutum*. *Water Res* 10:725–729.

Chiaudani G, Vighi M (1978) The use of *Selenastrum capricornutum* batch cultures in toxicity studies. *Mitt Int Verein Limnol* 21:316–329.

Christensen ER, Nyholm N (1984) Ecotoxicological assays with algae: Weibull dose-response curves. *Environ Sci Technol* 18:713–718.

Christensen ER, Scherfig J, Dixon PS (1979) Effects of manganese, copper and lead on *Selenastrum capricornutum*, and *Chlorella stigmatophora*. *Water Res* 13:79–92.

Conway HL (1978) Sorption of arsenic and cadmium and their effects on growth, micronutrient utilization and photosynthetic pigment composition of *Asterionella formosa*. *J Fish Res Board Can* 35:286–294.

Cox JL (1970) Low ambient level uptake of ^{14}C-DDT by three species of marine phytoplankton. *Bull Environ Contam Toxicol* 5:218–221.

Crane M (1990) A review of the use of aquatic multispecies systems for testing the effects of contaminants (ES 9416). WRc plc. Henley Road, Medmenham, PO Box 16, Marlow, Buckinghamshire SL7 2HD, U.K.

Crippen RW, Perrier JL (1974) The use of neutral red and Evans blue for live-dead determinations of marine plankton. *Stain Technol* 49:97–104.

Damgaard BM, Nyholm N (1980) Inter laboratory programme on toxicity tests on the freshwater planktonic algae *Selenastrum capricornutum* and *Scenedesmus quadricauda*. Water Quality Institute, Hoersholm, Denmark, 15 pp.

Davis AG (1970) Iron, chelation and the growth of marine phytoplankton. I. Growth kinetics and chlorophyll production in cultures of the Eurohaline flagellate *Dunaliella tertiolecta* under iron-limiting conditions. *J Mar Biol Assoc UK* 50:65–86.

Davis TM, Vance BD, Rodgers JH (1988) Productivity response of periphyton and phytoplankton to bleach-kraft mill effluent. *Aquat Toxicol* 12:83–106.

deNoyelles F, Kettle WD, Sinn DE (1982) The responses of plankton communities in experimental ponds to atrazine, the most heavily used herbicide in the United States. *Ecology* 63:1285–1293.

Droop MR (1968) Vitamin B_{12} and marine ecology IV. The kinetics of uptake, growth and inhibition in *Monochrysis lutheri*. *J Mar Biol Assoc UK* 48:689–733.

Droop MR (1974) The nutrient status of algal cells in continuous culture. *J Mar Biol Assoc UK* 54:825–855.

Eloranta VA, Halttunen-Keyriläinen L (1984) A comparison of the *Selenastrum capricornutum* bottle test and the natural phytoplankton assay in algal toxicity tests. *Arch Hydrobiol Suppl* 67:447–459.

Erickson SJ, Hawkins CE (1980) Effects of halogenated organic compounds on photosynthesis in estuarine phytoplankton. *Bull Environ Contam Toxicol* 24:910–915.

Finney DJ (1971) *Probit Analysis*, 3rd ed., Cambridge University Press, London, U.K.

Fogg GE (1965) *Algal Cultures and Phytoplankton Ecology,* University of Wisconsin Press, Madison, 126 pp.

Galassi S, Vighi M (1981) Testing toxicity of volatile substances with algae. *Chemosphere* 10:1123–1126.

Gargas E, Pedersen K (1974) Algal assay procedure. Batch technique. Contributions from the Water Quality Institute, No. 1, Water Quality Institute, Hoersholm, Denmark.

Gavies J, Guillard RRI, Woodward BL (1976) Cupric ion activity and the growth of phytoplankton isolated from different marine environments. *J Mar Res* 39:315–333.

Gearing JN (1988) The role of aquatic microcosms in ecotoxicological research as illustrated by large marine systems. In *Ecotoxicology: Problems and Approaches,* Levin S.A, Harwell MA, Kelly JF, Kimball KD, Eds., Springer-Verlag, New York, pp. 411–470.

Geyer H, Scheunert I, Korte F (1985) The effects of organic environmental chemicals on the growth of the alga *Scenedesmus subspicatus*: a contribution to environmental biology. *Chemosphere* 14:1355–1369.

Giddings JM, Washington JN (1981) Coal-liquefaction products, shale oil and petroleum. Acute toxicity to freshwater algae. *Environ Sci Technol* 15:106–108.

Giddings JM, Stewart AJ, O'Neill RV, Gardner RH (1983) An efficient algal bioassay based on short-term photosynthetic response. In *Aquatic Toxicology and Hazard Assessment: Sixth Symposium,* Bishop WE, Cardwell RD, Heidolph BB, Eds., ASTM STP 802, American Society for Testing and Materials, Philadelphia, PA, pp. 445–459.

Giesy JP (1985) Multispecies tests: research needs to assess the effects of chemicals on aquatic life. In *Aquatic Toxicology and Hazard Assessment: Eighth Symposium,* ASTM STP 891, American Society of Testing Materials, Philadelphia, PA, pp. 67–77.

Goldman CR (1978) The use of natural phytoplankton populations in bioassay. *Mitt Int Verein Limnol* 21:364–371.

Goldman JC, Oswald WF, Jenkins D (1974) The kinetics of inorganic carbon limited growth. *J Water Pollut Control Fed* 46:554–574.

Gons HJ, Mur LR (1979) Growth of *Scenedesmus protuberans* Fritsch in light-limited continuous cultures with a light-dark cycle. *Arch Hydrobiol* 85:41–56.

Greene JC, Miller WE, Shiroyama T, Maloney TE (1975) Utilization of algal assays to assess the effects of municipal, industrial, and agricultural waste water effluents upon phytoplankton production in the Snake River system. *Water Air Soil Pollut* 4:415–434.

Greene JC, Miller WE, Shiroyama TS, Soltero RA, Putnam K (1978) Use of laboratory cultures of *Selenastrum, Anabaena* and the indigenous isolate *Sphaerocystis* to predict effects of nutrients and zinc and interactions upon phytoplankton growth in Long Lake, Washington. *Mitt Int Verein Limnol* 21:372–384.

Greene JC, Bartels CL, Warren-Hicks WJ, Parkhurst BR, Linder GL, Peterson GA, Miller WE (1989) *Protocols for Short-Term Toxicity Screening of Hazardous Waste Sites.* EPA 600/3-88/029, U.S. Environmental Protection Agency, Washington, D.C.

Grobler DC, Davies E (1979) The availability of sediment phosphate to algae. *Water SA* 5:114–122.

Hall J, Healey FP, Robinson GGC (1989) The interaction of chronic copper toxicity with nutrient limitation in chemostat cultures of *Chlorella. Aquat Toxicol* 14:15–26.

Halling-Sørensen B, Nyholm N, Baun A (1996) Algal toxicity tests with volatile and hazardous substances in air-tight test flasks with CO_2 enriched headspace. *Chemosphere* 32:1513–1526.

Hanson DT, Fredrickson AG, Tsuchia HM (1977) Continuous propagation of microalgae: Part III. Material balance relation. *Chem Eng Prog Symp Ser* 67:114–151.

Hanstveit AO (1982) Evaluation of the results of the third ISO-interlaboratory study with an algal toxicity test. ISO/TC 147/SC5/WG 5. Nederlands Normalisatie Instituut, Delft, the Netherlands.

Healey FP (1978) Physiological indicators of nutrient deficiency in algae. *Mitt Int Verein Limnol* 21:34–41.

Healey FP (1979) Short-term responses of nutrient deficient algae to nutrient addition. *J Phycol* 15:289–299.

Hutchinson TC, Hellebust JA, Tam D, Mackay D, Mascharenhas RA, Shiu WY (1980) The correlation of the toxicity to algae of hydrocarbons and halogenated hydrocarbons with their physico-chemical properties. In *International Symposium on the Analysis of Hydrocarbons,* Plenum Press, New York.

International Organisation for Standardization (ISO) (1980) Evaluation of the results of the European ISO-test programme with algae toxicity tests (by A. O. Hanstveit). ISO/TC 147/SC5/Wg 5. Nederlands Normalisatie Instituut, Delft, the Netherlands.

International Organisation for Standardization (ISO) (1982) Water quality — determination of the inhibition of the mobility of *Daphnia magna* Straus (*Cladocera Crustacea*). ISO 6341-1982(E). Geneva, Switzerland, 6 pp.

International Organisation for Standardization (ISO) (1989) Water quality — freshwater algal growth inhibition test with *Scenedesmus subspicatus* and *Selenastrum capricormutum*. ISO 8692:1989(E). Geneva, Switzerland, 5 pp.

International Organisation for Standardization (ISO) (1993) Water quality — marine algal growth inhibition test with *Skeletonema costatum* and *Phaeodactylum tricornutum*. ISO/DIS 10253. Nederlands Normalisatie Instituut, Delft, the Netherlands.

Jensen A (1984) Marine ecotoxicological tests with phytoplankton. In *Ecotoxicological Testing for the Marine Environment*, Vol. 1, Persoone G, Jaspers E, Claus C, Eds., State Univ. Gent and Inst. Mar. Scient. Res., Bredene, Belgium, pp. 195–213.

Jensen K, Pedersen SM, Nielsen GÆ (1984) The effect of aromatic hydrocarbons on the productivity of various marine planktonic algae. *Limnologica* (Berlin) 15:581–584.

Jouany JM, Ferard JF, Vasseur P, Gea J, Truhaut R, Rast C (1983) Interest of dynamic tests in acute ecotoxicity assessment in algae. *Ecotoxicol Environ Saf* 7:216–228.

Källqvist T (1975) Algal growth potential of six Norwegian waters receiving primary, secondary and tertiary sewage effluents. *Verh Int Verein Limnol* 19:2070–2081.

Källqvist T (1978) Noen erfaringer af algetoksisitetstester ved NIVA. Publication 1978:2. Nordic Cooperative Organization for Applied Research, Secretariat of Environmental Sciences, Helsinki, Finland, pp. 147–166.

Källqvist T (1993) Comments from the NIVA on ISO/DIS 10253. ISO/TC 147/SC 5/WG 5 N142. Netherlands Normalisatie Instituut (NNI), Delft, the Netherlands.

Källqvist T, Ormerod K (1981) Ringtest med 48 mikrobielle toksikologiske metoder over for: phenol, anisol, Na-laurylsulfat og kobbersulfat. Ekotoxikologiska metoder för akvatisk miljö, Report No. 25, Nordic Cooperative Organization for Applied Research, Helsinki, Finland.

Källqvist T, Romstad R (1994) Effects of agricultural pesticides on planktonic algae and cyanobacteria — examples of interspecies sensitivity variations. *Norw J Agric Sci* Suppl. 13:117–131.

Källqvist T, Abdel-Hamid MI, Berge D (1994) Effects of agricultural pesticieds on freshwater plankton communities in enclosures. *Norw J Agric Sci* Suppl. 13:133–152.

Kenaga EE, Moolenaar RJ (1979) Fish and *Daphnia* toxicity as surrogates for aquatic vascular plants and algae. *Environ Sci Technol* 13:1479–1480.

Klaine SJ, Ward CH (1983) Growth-optimized algal bioassays for toxicity evaluation. *Environ Toxicol Chem* 2:245–249.

Kooijman SALM (1985) Toxicity at the population level. In *Multispecies Toxicity Testing*, Cairns J Jr., Ed., Pergamon Press, New York, pp. 143–164.

Kooijman SALM (1987) A safety factor for LC_{50} values allowing for differences in sensitivity among species. *Water Res* 21:269–276.

Kooijman SALM, Metz JAJ (1984) On the dynamics of chemically stressed populations: the deduction of population consequences from effects of individuals. *Ecotoxicol Environ Saf* 8:254–274.

Kotai J (1972) Instructions for preparation of modified nutrient solution Z8 for algae. NIVA publication B-11/69, April 1972. Norwegian Institute for Water Research, Oslo, Norway.

Kotai J, Krogh T, Skulberg OM (1978) The fertility of some Norwegian inland waters assayed by algal cultures. *Mitt Int Verein Limnol* 21:413–436.

Kühn R, Pattard M (1990) Results of the harmful effects of water pollutants to green algae (*Scenedesmus subspicatus*) in the cell multiplication inhibition test. *Water Res* 24:31–38.

Kuivasniemi K, Eloranta V, Knuuitinen J (1985) Acute toxicity of some chlorinated phenolic compounds to *Selenastrum capricornutum* and phytoplankton. *Arch Environ Contam Toxicol* 14:43–49.

Kunikane S, Kaneko M, Maehara R (1984) Growth and nutrient uptake of green alga, *Scenedesmus dimorphus*, under a wide range of nitrogen/phosphorus ratio. I. Experimental study. *Water Res* 18:1299–1311.

Kusk KO (1978) Effects of crude oil and aromatic hydrocarbons on the photosynthesis of the diatom *Nitzschia palea*. *Physiol Plant* 43:1–6.

Kusk KO (1981) Effects of hydrocarbons on respiration, photosynthesis and growth of the diatom *Phaeodactylum tricornutum*. *Bot Mar* 24:413–418.

Kusk KO, Nyholm N (1991) Evaluation of a phytoplankton toxicity test for water pollution assessment and control. *Arch Environ Contam Toxicol* 20:375–379.

Kusk KO, Nyholm N (1992) Toxic effects of chlorinated organic compounds and potassium dichromate on growth rate and photosynthesis of marine phytoplankton. *Chemosphere* 25:875–886.

Laake M, Ed. (1982) Ekotoxikologiska metoder för akvatisk miljö. In *Miljövårds-serien*, Publication 1982:2, Nordic Cooperative Organization for Applied Research, Helsinki, Finland.

Larsen DP, deNoyelles F Jr., Stay F, Shiroyama T (1986) Comparisons of single-species, microcosm and experimental pond responses to atrazine exposure. *Environ Toxicol Chem* 3:47–60.

Lederman TC, Rhee G-Y (1982) Influence of hexachlorobiphenyl in great lakes phytoplankton in continuous culture. *Can J Fish Aquat Sci* 39:388–394.

Lewis MA (1986) Comparison of the effects of surfactants on freshwater phytoplankton communities in experimental enclosures and on algal population growth in the laboratory. *Environ Toxicol Chem* 5:319–332.

Lewis MA (1990) Are laboratory-derived toxicity data for freshwater algae worth the effort? *Environ Toxicol Chem* 9:1279 –1284.

Lewis MA, Hamm BG (1986) Environmental modification of the photosynthetic response of lake plankton to surfactants and significance to a laboratory-field comparison. *Water Res* 20:1575–1582.

Lumsden BR, Florence TM (1983) A new algal assay procedure for the determination of the toxicity of copper species in seawater. *Environ Technol Lett* 4:271–276.

Lyngby JE, Mortensen E (1995) Biomonitoring of eutrophication levels in shallow coastal ecosystems. In *Biology and Ecology of Shallow Coastal Waters*, 28th EMBS Symposium. Eleftheriou A, Ansell AD, Smith CJ, Eds., Olsen Olsen, Fredensburg, Denmark.

Maestrini SY, Bonin DJ, Droop MR (1984) Phytoplankton as indicators of seawater quality: bioassay approaches and protocols. In *Algae as Ecological Indicators*, Schubert LE, Ed., Academic Press, New York, pp. 3–14.

Maloney TE, Miller WE, Shiroyama T (1972) Algal responses to nutrient additions in natural waters I. Laboratory assays. In *Nutrients and Eutrophication*, Vol. 1, The American Society of Limnology and Oceanography, Allan Press, Inc., Lawrence, KS, pp. 134–140.

Miller WE, Greene JC, Shiroyama T (1978) The *Selenastrum capricornutum* Printz algal assay bottle test. Experimental design, application and data interpretation protocol. EPA-600/9-78-018, U.S. Environmental Protection Agency, Corvallis, OR.

Miller WE, Peterson SA, Greene JC, Callahan CA (1985) Comparative toxicology of laboratory organisms for assessing hazardous waste sites. *J Environ Qual* 14:569–574.

Monod J (1942) *La Cruissance des Cultures Bacteriennes*, Herman, Paris.

Morel FMM, Rueter JD, Anderson DM, Guillard RRI (1979) Aquil: a chemically defined phytoplankton culture medium for trace metal studies. *J Phycol* 15:135–141.

Mount DI (1985) Scientific problems in using multispecies toxicity tests for regulatory purposes. In *Multispecies Toxicity Testing*, Cairns J Jr., Ed., Pergamon Press, New York, pp. 13–18.

National Research Council (1981) *Testing for the Effects of Chemicals on Ecosystems*, National Academy Press, Washington, D.C.

Nyholm N (1976) A mathematical model for microbial growth under limitation by conservative substrates. *Biotechnol Bioeng* 18:1043–1056.

Nyholm N (1977a) Kinetics of nitrogen-limited algal growth. *Prog Water Technol* 8:346–358.

Nyholm N (1977b) Kinetics of phosphate-limited algal growth. *Biotechnol Bioeng* 19:467–492.

Nyholm N (1978) A mathematical model for growth of phytoplankton. *Mitt Int Verein Theor Angew Limnol* 21:193–206.

Nyholm N (1985) Response variable in algal growth inhibition tests — biomass or growth rate? *Water Res* 19:273–279.

Nyholm N (1990) Expression of results from growth inhibition toxicity tests with algae. *Arch Environ Contam Toxicol* 19:518–522.

Nyholm N (1991) Toxic effects on algal phosphate uptake. *Environ Toxicol Chem* 10:581–584.

Nyholm N (unpublished, 1993) Niels Nyholm, Ph.D., Associate Professor, Institute of Environmental Science and Engineering, Technical University of Denmark.

Nyholm N, Damgaard BM (1990) A comparison of the algal growth inhibition toxicity method with the short-term ^{14}C assimilation test. *Chemosphere* 21:671–679.

Nyholm N, Källqvist T (1989) Methods for growth inhibition toxicity tests with freshwater algae. *Environ Toxicol Chem* 8:689–703.

Nyholm N, Lyngby JE (1988) Algal bioassays in eutrophication research — a discussion in the framework of a mathematical analysis. *Water Res* 10:1293–1300.

Nyholm N, Kusk KO, Damgaard BM (1981) Water quality — toxicity test based on measurements of the carbon assimilation of phytoplankton. Water Quality Institute, Hørsholm, Denmark.

Nyholm N, Bach H, Birklund J, Jensen TL, Kusk KO, Schleicher O, Schrøder H (1991) Environmental studies of a marine wastewater discharge from a sulphite pulp mill — exemplification of a general study approach for marine industrial discharges. *Water Sci Technol* 23:151–161.

Nyholm N, Settergren PS, Kusk KO, Christensen ER (1992) Statistical treatment of data from microbial toxicity tests. *Environ Toxicol Chem* 11:157–167.

Organization for Economic Cooperation and Development (OECD) (1984) Guideline for testing of chemicals. No. 201. Alga growth inhibition test. OECD, Paris.

Paasche E (1973a) Silicon and the ecology of marine plankton diatoms. I. *Thalassiosira pseudonana (Cyclotella nana)* grown in a chemostat with silicate as limiting nutrient. *Mar Biol* 19:117–126.

Paasche E (1973b) Silicon and the ecology of marine plankton diatoms. II. Silicate-uptake kinetics in five diatom species. *Mar Biol* 19:262–269.

Pack S (1993) A review of statistical data analysis and experimental design in OECD aquatic toxicology test guidelines. Shell Research Ltd., Sittingbourne, Kent, U.K.

Payne AG, Hall RH (1978) Application of algal assays in the environmental evaluation of new detergent materials. *Mitt Int Verein Limnol* 21:507–520.

Peterson HG (1991) Toxicity testing using a chemostat grown green alga, *Selenastrum capricornutum*. In *Plants for Toxicity Assessment: Second Volume,* Gorsuch JW, Lower WR, Wang W, Lewis MA, Eds., ASTM STP 1115, American Society for Testing and Materials, Philadelphia, PA, pp. 107–117.

Peterson HG, Healey FP (1985) Comparative pH dependent metal inhibition of nutrient uptake by *Scenedesmus quadricauda* (Chlorophyceae). *J Phycol* 21:217–222.

Peterson HG, Nyholm N (1993) Algal bioassays for metal toxicity identification. *Water Pollut Res J Can* 28:129–153.

Peterson HG, Healey FP, Wageman R (1984) Metal toxicity to algae: a highly pH dependent phenomenon. *Can J Fish Aquat Sci* 41:974–979.

Peterson HG, Boutin C, Martin PA, Freemark KE, Ruecker NJ, Moody M (1994a) Aquatic phyto-toxicity of 23 pesticides applied at expected environmental concentrations. *Aquat Toxicol* 28:275–292.

Peterson HG, Boutin C, Freemark KE, Martin PA (1994b) Toxicity of hexazinone and diquat to green algae, diatoms, cyanobacteria and duckweed. *Environ Toxicol Chem* (submitted).

Pipes WO (1962) Carbon dioxide limited growth of *Chlorella* in a continuous culture. *Appl Microbiol* 10:281–288.

Pipes WO, Koutsoyannis SP (1962) Light limited growth of *Chlorella* in continuous cultures. *Appl Microbiol* 10:1–5.

Rehnberg BG, Schulz DA, Raschke RL (1982) Limitations of electronic particle counting in reference to algal assays. *J Water Pollut Control Fed* 54:181–186.

Rhee G-Y (1973) A continuous culture study of phosphate uptake, growth rate and polyphosphate in *Scenedesmus* sp. *J Phycol* 9:495–506.

Rhee G-Y (1978) Effects of N:P atomic ratios and nitrate limitation on algal growth, cell composition, and nitrate uptake. *Limnol Oceanogr* 23:10–25.

Rhee G-Y (1982) Effects of environmental factors and their interactions on phytoplankton growth. *Adv Microbiol Ecol* 6:33–74.

Rhee G-Y, Gotham IJ (1981a) The effect of environmental factors on phytoplankton growth: temperature and interactions of temperature with nutrient limitation. *Limnol Oceanogr* 26:635–648.

Rhee G-Y, Gotham IJ (1981b) The effect of environmental factors on phytoplankton growth: light and interactions of light with nitrate limitation. *Limnol Oceanogr* 26:649–659.

Rice CP, Sikka HC (1973) Uptake and metabolism of DDT by six species of marine algae. *J Agric Food Chem* 21:148–152.

Riegman R, Colijn F, Malscharert JFP, Kloosterhuis HT, Cadée GC (1990) Assessment of growth rate limiting nutrients in the North Sea by the use of nutrient-uptake kinetics. *Neth J Sea Res* 26:53–63.

Riisgård HU (1979) Effect of copper on volume regulation in the marine flagellate *Dunaniella marina*. *Mar Biol* 50:198–193.

Rodhe W (1978) Algae in culture and nature. *Mitt Int Verein Limnol* 21:7–20.

Saarikosky J, Viluksela M (1981) Influence on pH on the toxcity of substituted phenols in fish. *Arch Environ Contam Toxicol* 10:747–753.

Samson G, Morisette J, Popovic R (1988) Copper quenching of the variable fluorescence in *Dunaliella tertiolecta*. Evidence for a copper inhibition effect on the oxidizing side of PSII. *Photochem Photobiol* 48:329–332.

Schelske CL (1984) *In situ* and natural phytoplankton assemblage bioassays. In *Algae as Ecological Indicators*, Schubert LE, Ed., Academic Press, New York, pp. 3–14.

Shapiro J, Ribeiro R (1965) Algal growth and sewage effluent in the Potomac estuary. *J Water Pollut Control Fed* 37:1034–1043.

Skulberg OM (1964) Algal problems related to the eutrophication of European water supplies and a bioassay method to assess fertilizing influences of pollution on inland waters. In *Algae and Man*, Jackson E, Ed., Plenum Press, New York, pp. 262–299.

Skulberg OM (1966) Algal cultures as a means to assess the fertilizing influence of pollution. Third International Conference on Water Pollution Research, Section 1:6, Washington, D.C., pp. 1–15.

Sloff W, Canton JH, Hermens JLM (1983) Comparison of the susceptibility of 22 freshwater species to 15 chemical compounds: I. (Sub)acute toxicity tests. *Aquat Toxicol* 4:113–128.

Södergren A, Gelin C (1983) Effect of PCbs on the rate of carbon-14 uptake in phytoplankton isolates from oligotrophic and eutrophic lakes. *Bull Environ Contam Toxicol* 30:191–198.

Soto C, Hellebust JA, Hutchinson TC (1975) Effect of naphthalene and aqueous crude oil extracts on the green flagellate *Chlamydomonas angulosa*. II. Photosynthesis and the uptake and release of naphthalene. *Can J Bot* 53:118–126.

Soto C, Hutchinson TC, Hellebust JA, Sheath RG (1979) The effect of crude oil on the morphology of the green flagellate *Chlamydomonas angulosa*. *Can J Bot* 57:2717–2728.

Stebbing ARD (1982) Hormesis — the stimulation of growth by low levels of inhibitors. *Sci Total Environ* 22:213–234.

Steele R (1994) Personal communication. U.S. Environmental Protection Agency (USEPA), Newport, OR.

Steemann-Nielsen E, Willemoes M (1971) How to measure the illumination rate when investigating the rate of photosynthesis of unicellular algae under various light conditions. *Int Rev Gesamten Hydrobiol* 56:541–556.

Steemann-Nielsen E, Kamp-Nielsen L, Wium-Andersen S (1969) The effect of deleterious concentrations of copper on the photosynthesis of *Chlorella pyrenoidosa*. *Physiol Plant* 22:1121–1133.

Stephan CE, Mount DI, Hansen DJ, Gentile JH, Chapman GA, Brungs WA (1985) Guidelines for deriving numerical national water quality criteria for the protection of aquatic organisms and their uses. PB-85-227049, U.S. Environmental Protection Agency, Washington, D.C., 98 pp.

Stratton GW (1986) Medium composition and its influence on solvent pesticide interactions in laboratory bioassays. *Bull Environ Contam Toxicol* 36:807–814.

Stumm W, Morgan JJ (1981) *Aquatic Chemistry*, Wiley, New York.

Swanson SM, Rickard CP, Freemark KE, MacQuarrie P (1991) Testing for pesticide toxicity to aquatic plants: recommendations for test species. In *Plants for Toxicity Assessment: Second Volume*, Gorsuch JW, Lower WR, Wang W, Lewis MA, Eds., ASTM STP 1115, American Society for Testing and Materials, Philadelphia, PA, pp. 77–97.

Swartzman GL, Taub FB, Meador J, Huang C, Kindig A (1990) Modeling the effect of algal biomass on multispecies aquatic microcosms response to copper toxicity. *Aquat Toxicol* 17(2):93–117.

Tamiya H, Hase E, Shibata K, Mituya A, Iwamura T, Nihei T, Sasa T (1953) Kinetics of growth of *Chlorella* with special references to its dependence on quantity of available light and on temperature. In *Algal Culture from Laboratory to Pilot Plant*, Burlew JS, Ed., Publication 600, Carnegie Institution of Washington, Washington, D.C., pp. 204–323.

Taub FB (1989) Standardized aquatic microcosm — development and testing. In *Aquatic Ecotoxicology: Fundamental Concepts and Methodologies*, Vol. 2, CRC Press, Boca Raton, FL, pp. 47–92.

Taub FB, Kindig AC, Conquest LL, Meador JP (1989) Results of interlaboratory testing of the standardized aquatic microcosm protocol. In *Aquatic Toxicology and Environmemntal Fate*, Vol. 11, Suter GW II, Lewis MA, Eds., ASTM STP 1007, American Society for Testing and Materials, Philadelphia, PA, pp. 368–394.

Thompson PA, Couture P (1991) Short and long-term changes in growth and biochemical composition of *Selenastrum capricornutum* populations exposed to cadmium. *Aquat Toxicol* 21:135–144.

Thomson RS (1993) The growth of *Skeletonema costatum* in ISO/DIS 10253 medium with modified levels of iron, zinc and EDTA. ISO/TC 147/SC 5/WG 5 N 144. Netherlands Normalisatie Instituut (NNI), Delft, the Netherlands.

Torien DF, Huang CH, Radimsky J, Pearson EA, Scherfig J (1970) Final report. Provisional algal assay procedure. SERL report No. 71-6, University of California, Berkeley, CA.

Trainor FR (1984) Indicator algal assays: laboratory and field approaches. In *Algae as Ecological Indicators*, Schubert LE, Ed., Academic Press, New York, pp. 3–14.

Trevors JT (1982) A comparison of methods for assessing toxicant effects on algal growth. *Biotechnol Lett* 4:243–246.

Turbak SC, Olson SB, McFeters GA (1986) Comparison of algal assay systems for detecting waterborne herbicides and metals. *Water Res* 20:91–96.

U.S. Environmental Protection Agency (1971) Algal assay procedure: bottle test. National Eutrophication Research Programme. Pacific Northwest Water Laboratory, Corvallis, OR.

U.S. Environmental Protection Agency (1974) Marine algal assay procedure: bottle test. Eutrophication and Lake Restoration Branch. Pacific Northwest Environmental Research Laboratory, National Environmental Research Center, Corvallis, OR.

U.S. Environmental Protection Agency (1988) *Methods for Aquatic Toxicity Identification Evaluations. Phase I Toxicity Characterization Procedures*, EPA/600/3-88/034, Environmental Research Laboratory, Duluth, MN.

U.S. Environmental Protection Agency (1989) *Methods for Aquatic Toxicity Identification Evaluations. Phase II Toxicity Characterization Procedures*, EPA/600/3-88/035, Environmental Research Laboratory, Duluth, MN.

Van Ewijk PH, Hoekstra JA (1993) Calculation of the EC50 and its confidence interval when subtoxic stimulus is present. *Ecotoxicol Environ Saf* 25:25–32.

van Leeuwen K (1989) Ecotoxicological effects assessment in the Netherlands. Recent developments. Directorate for Environmental Protection, Chemical Substances Division, Leidschendam, the Netherlands, 34 pp.

van Liere L (1979) On *Oscillatoria agardhii* gomont, experimental ecology and physiology of a nuisance bloom-forming cyanobacterium. Thesis. Universiteit van Amsterdam, the Netherlands.

van Liere L, Mur LR (1978) Light-limited cultures of the blue-green alga *Oscillatoria agardhii*. *Mitt Int Verein Limnol* 21:158–167.

Van Straalen NM, Denneman GAJ (1989) Ecotoxicological evaluation of soil quality criteria. *Ecotoxicol Environ Saf* 18:241–251.

Versteeg DJ (1990) Comparison of short- and long-term toxicity test results for the green alga, *Selenastrum capricornutum*. In *Plants for Toxicity Assessment*, Wang W, Gorsuch JW, Lower WR, Eds., ASTM STP 1091, American Society for Testing and Materials, Philadelphia, PA, pp. 40–48.

Vølund A (1978) Application of the four parameter logistic models to bioassay: comparison with slope ratio and parallel line models. *Biometrics* 34:357–365.

Wagner C, Løkke H (1991) Estimation of ecotoxicological protection levels from NOEC toxicity data. *Water Res* 25:1237–1248.

Walsh GE (1983) Cell death and inhibition of population growth of marine unicellular algae by pesticides. *Aquat Toxicol* 3:209–214.

Walsh GE (1988a) *Methods for Toxicity Tests of Single Substances and Liquid Complex Wastes with Marine Unicellular Algae*, EPA/600/8-87/043, U.S. Environmental Protection Agency, Washington, D.C.

Walsh GE (1988b) Principles of toxicity testing with marine unicellular algae. *Environ Toxicol Chem* 7:979–987.

Walsh GE, Alexander SV (1980) A marine algal bioassay method: results with pesticides and industrial wastes. *Water Air Soil Pollut* 13:45–55.

Walsh GE, Merill RG (1984) Algal bioassays of industrial and energy process effluents. In *Algae as Ecological Indicators*, Schubert LE, Ed., Academic Press, London, pp. 329–360.

Walsh GE, Bahner LH, Horning WB (1980) Toxicity of textile mill effluents to freshwater and estuarine algae, crustaceans and fishes. *Environ Pollut* 21:169–179.

Walsh GE, Duke KM, Forster RB (1982) Algae and crustaceans as indicators of bioactivity of industrial wastes. *Water Res* 16:879–883.

Walsh GE, Yoder MJ, McLauglin LL, Lores EM (1987) Responses of marine unicellular algae to brominated organic compounds in six growth media. *Ecotoxicol Environ Saf* 14:215–222.

Walsh GE, McLauglin LL, Yoder MJ, Moody PH, Lores EM, Forester J, Wessin-Duval PB (1988) *Minutocellus polymorphus*: a new marine diatom for use in algal toxicity tests. *Environ Toxicol Chem* 7:925–929.

Wängberg S, Blanck H (1988) Multivariate patterns of algal sensitivity to chemicals in relation to phylogeny. *Ecotoxicol Environ Saf* 16:72–82.

Wium-Andersen S (1974) The effect of chromium on the photosynthesis and growth of diatoms and green algae. *Physiol Plant* 32:308–310.

Wong PTS, Couture P (1986) Toxicity screening using phytoplankton. In *Toxicity Testing Using Microorganisms*, Vol. 2, Dutka BJ, Bitton G, Eds., CRC Press, Boca Raton, FL, chap. 4.

Yamane AN, Okada M, Sudo R (1984) The growth inhibition of planktonic algae
    due to surfactants used in washing agents. *Water Res* 18:1101–1105.
Yu-yun T, Thumm W, Jobelius-Korte M, Attar A, Freitag D, Kettrup A (1993)
    Fate of two phenylbenzoylurea insecticides in an algae culture system
    (*Scenedesmus subspicatus*). *Chemosphere* 26:955–962.
Zevenboom W (1980) Growth and nutrient uptake kinetics of *Oscillatoria agar-
    dhii*. Ph.D. Thesis. Universiteit van Amsterdam, the Netherlands.
Zevenboom W, Mur LR (1978) N-uptake and pigmentation of N-limited chemo-
    stat cultures and natural populations of *Oscillatoria Agardhii*. *Mitt Int
    Verein Limnol* 21:261–274.

chapter ten

Aquatic plant communities for impact monitoring and assessment

Brian H. Hill

Introduction

During the past two decades, biological monitoring has risen to the forefront of environmental impact assessments and routine monitoring programs. The scope of biological monitoring has evolved from the collection of biological data in support of determinations of the toxicity of waters to the collection of biological information for the prediction of the quality of these waters. Cairns (1990) describes three uses of biological information related to water quality assessments: (1) biological surveys, which he defines as the collection of a set of standardized observations from a site(s) over a short time period for the purpose of establishing descriptive data; (2) biological surveillance, defined as the continued, systematic collection of biological data from a site or sites through time; and (3) biological monitoring, defined as surveillance undertaken to ensure that regulatory standards are being met.

Recently, the U.S. Environmental Protection Agency (EPA) mandated that states establish water quality criteria based on biological information (U.S. EPA 1990). The impetus for this action stems from the Clean Water Act which directs the EPA to develop programs that will evaluate, restore, and maintain the chemical, physical, and biological integrity of the nation's waters. Biological integrity, as related to the Clean Water Act, has been defined as "the ability of an aquatic ecosystem to support and maintain a balanced, integrated, adaptive community of organisms having a species composition, diversity, *and functional organization* comparable to that of the natural habitats within a region" (Karr and Dudley 1981).

Cairns (1990) lists three goals of biomonitoring: (1) to maintain balanced biological communities, (2) to protect both structural and functional integrity of ecosystems, and (3) to protect biodiversity. While structure-based criteria (e.g., community composition, species diversity, and derived indices of environmental quality) are well established for macroinvertebrate and fish assemblages, those based on plant communities (algae, mosses, and aquatic vascular plants) are rarely used, in spite of evidence supporting their utility (e.g., Newman and McIntosh 1989 and Guzkowska and Gasse 1990 for algae; Burton 1990 for mosses; Franzen and McFarlane 1980 and Dennison et al. 1993 for aquatic vascular plants).

Plant communities, especially algal communities, are well suited to environmental monitoring. Aquatic plants are relatively short-lived, have rapid response times to many perturbations, are often cosmopolitan in their distributions, and are relatively easily identified.

Recently, researchers have begun to criticize the reliance on structural measures of biotic conditions to assess aquatic ecosystem integrity. From an ecosystem management perspective, structural metrics such as the Index of Biotic Integrity (IBI) for fish (Karr 1981) and the Invertebrate Community Index (ICI) are proving to be less reliable than previously thought. Grossman et al. (1990) argue that the high spatial and temporal variability exhibited by many stream fish assemblages preclude the use of population data alone as indicators of anthropogenic disturbances. They urge resource managers to exercise caution in the use of assemblage data. Pratt and Cairns (1985) presented evidence that apparently chaotic data from invertebrate and protozoan assemblages reveal significant patterns when the species are assigned to functional groups.

Ecosystem research over the past several years has increasingly focused on functional parameters rather than the more traditional structural metrics. Emphasis on system-level functional roles may not answer population-level questions, but it permits clustering of genetically and taxonomically diverse groups into functional guilds. In spite of the mounting evidence in support of the need for criteria aimed at assessing ecosystem functions (i.e., energy and material cycling), these approaches to biological monitoring are poorly accepted within the regulatory community (Cummins 1991).

Functional indicators are less likely to be constrained by regionally restricted biota. Thus, functional approaches lead to a more global view of stream ecosystems, a view that is much less variable than one based only on taxa inhabiting stream communities (Cummins 1988). Hunsaker et al. (1990) have stated that for regional ecological risk assessments to be effective, the system must be functionally defined, with the spatio-temporal boundaries of the system set by functional attributes of the communities inhabiting the system. Assessments that are functionally based are likely to have greater applicability across regions (Hunsaker et al. 1990).

Biological monitoring, the use of the biota to determine the biological integrity of a community, and biological criteria, the establishment of regulatory standards based on the biological community, are presently being used throughout the United States and the world for assessing impacts and setting goals for maintenance and remediation of aquatic ecosystems. Most of these efforts are based only on macroinvertebrate and fish assemblages. The purpose of this chapter is to present evidence for the inclusion of plant community data, including functional measures, in the monitoring and assessment of impacts in aquatic ecosystems.

Physiological ecology: an organism approach to biomonitoring

Physiological measures used in biomonitoring can be categorized as those involving the cellular integrity (structure) of test organisms, organism survival, organism or population growth, and metabolic functions of an organism.

Cellular integrity

Measures of cellular integrity fall into two broad categories: those related to morphological changes in cell structure and those related to changes in cell membrane permeability. Few researchers have used changes in cellular structure to monitor physiological condition of the cells or to predict water quality. Lanza and Cairns (1972) used general cell structure as one of their indicators of stress in indigenous and culture populations of diatoms exposed to elevated temperatures. At the highest levels of temperature increases (20°C above ambient), they reported coagulation of cell components. Abnormalities in the frustules of certain diatoms have been used to infer environmental conditions. Analysis of five species of fossil diatoms collected from Mono Lake, CA revealed a large percentage of deformed individuals, possibly related to the transition of this lake from fresh water to alkaline, brackish waters (Solladay personal communication). Recent studies in our lab of three diatom genera (*Synedra, Hannaea,* and *Cocconeis*) collected from rivers receiving heavy metal mine wastes have shown a weak correlation between dissolved zinc concentrations and percentage of deformed individuals (McFarland et al. 1997). A similar teratological response was reported for *Eunotia* growing in copper-mine wastes (Carter 1990). Changes in chlorophyll structure (possibly through allomerization) have been assessed as changes in chlorophyll absorbance spectra, and changes in fluorescence in response to elevated temperatures, heavy metals, and herbicides. Lanza and Cairns (1972) reported decreased autofluorescence in diatoms exposed to elevated temperatures. Likewise, Judy et al. (1990) reported decreased chlorophyll fluorescence in *Tradescantia* exposed to heavy metals or herbicides. Puckett and Burton (1981) used shifts in the chlorophyll absorption spectra of several species of mosses exposed to heavy metals. These authors reported marked deviations from control spectra, including shifts of both the red and blue peaks toward shorter wavelengths and a loss of absorbance at these two peaks.

Measures of changes in cell membrane permeability have been used more commonly to assess the integrity of stressed cells. These measures include plasmolysis (usually as loss of pigment from chloroplasts), potassium efflux, the ratio of sodium to potassium in leaf tissue, and induced lipid fluorescence. Diatoms exposed to thermal stress exhibited loss of pigment from chloroplasts, diffusion of pigment throughout the cell, and a shift from green to brown coloration of the pigment within the cell (Lanza and Cairns 1972). These authors also reported increases in cellular lipid fluorescence in heat-stressed diatoms. Induction of plasmolysis is generally used as the criterion of injury in studies of metal toxicity on bryophytes (Puckett and Burton 1981). Since plant cell membranes are the first metabolic sites that encounter environmental contaminants, they are often impaired. Acute toxicity may be assessed by measuring potassium efflux from intact cells. Metal-induced losses of potassium from lichens has been demonstrated for most metals, with results ranging from little or no potassium loss caused by nickel or cobalt relatively independent of metal concentration to metal concentration dependent losses of potassium caused by copper, mercury, and silver (Puckett 1976; Puckett and Burton 1981). Assessment of past toxic exposures may be assessed by measuring cellular ratios of sodium (Na) to potassium (K). Gilfillan et al. (1989) measured elevated Na to K ratios in aquatic vascular plant leaf tissue in response to crude oil spills. They hypothesized that the crude oil extracted lipids from cell membranes, resulting in potassium losses in excess of losses of sodium. Since no mortality of these plants had occurred as a result of these exposures to oil, the authors suggested that an elevated Na to K ratio is a sublethal stress response in plants.

Growth

Growth and survival studies have been the mainstay of ecotoxicological assessments of waters and wastewaters for nearly 30 years. Most of the work in this area has employed the algal bottle assay procedure (U.S. EPA 1978, 1989a) using *Selenastrum capricornutum* in static culture. The literature on ecotoxicological studies using algae is extensive and can be summarized as such: perturbations to algal populations can be monitored by measuring changes in impacted population densities relative to changes in control populations. In the case of nutrient additions, impacted populations increase relative to control populations. The reverse is true for toxic impacts (e.g., Cairns 1968; Payne and Hall 1979; Hughes and Vilkas 1983; Klaine and Ward 1983; Ram and Plotkin 1983; Christensen and Nyholm 1984; U.S. EPA 1989a; Thompson and Couture 1990; Wong and Chang 1991).

Studies of natural periphyton community responses to perturbations have been carried out in artificial streams and *in situ*. The use of natural populations allows for a more realistic assessment of community-level effects, though the variance associated with dynamic populations may mask the effects of perturbations. Most community-level assessments focus on changes in biomass rather than cell density. Periphyton biomass accrual has been inhibited by chlorine, chromium, copper, zinc, and mixed effluents (Clark et al. 1979; Rodgers et al. 1979; Boston et al. 1991). I have measured reduced periphyton accrual rates in zinc-contaminated streams. Sigmon et al. (1977) found reduced periphyton density in artificial stream communities in response to elevated mercury.

Growth responses of aquatic macrophytes have been studied using seed germination, tuber sprouting, and root and shoot growth. Results from these studies generally follow the same pattern as the algal assays: stimulation by nutrient additions and inhibition by toxicants. While the use of aquatic plant germination and growth are relatively new, the results are consistent with those reported from terrestrial studies (e.g., Walsh et al. 1991). Duckweed growth studies have received more attention because of their ease of application (Wang 1990). The inhibition of frond development has been studied for industrial effluents, metals, oil, and pesticides (e.g., Wang 1986, 1987, 1988).

Photosynthesis

One of the most direct measures of plant physiology is photosynthesis. This nontaxonomic integrator of physiological condition is responsive to changes in environmental condition and can be accurately measured either by carbon incorporation or oxygen evolution. The actual mechanism of photosynthetic inhibition varies by chemical, but most inhibitors fall into three categories: those that interrupt electron transport activity in the Hill reaction, those that alter the structure of chloroplasts, and those that reduce chlorophyll concentrations within the chloroplast. Most herbicides and organochlorine and organophosphate pesticides inhibit photosynthesis by blocking electron transport in the Hill reaction. Photosynthesis by four aquatic vascular plants was reduced by as much as 90% by atrazine concentrations ranging up to 600 µg/l (Jones and Winchel 1984; Jones et al. 1986). Organochlorine pesticides at concentrations as low as 50 µg/l inhibited photosynthesis in marine macrophytes (Ramachandran et al. 1984). Heavy metals have been shown to inhibit photosynthesis by causing structural damage to the chloroplasts. Photosynthetic inhibition in the presence of elevated metal concentration has been demonstrated for periphyton, mosses and liverworts, and aquatic vascular plants (e.g., Puckett and Burton 1981; Boyle 1984; Guilizzoni 1991; Hill et al. 1997). Halogens have been correlated with reductions in cellular chlorophyll *a* and resulting decreases in photosynthesis in periphyton communities (Boyle 1984; Boston et al. 1991). Boston et al. (1991) reported photosynthetic rates significantly below control values in the presence of as little as 250 µg of total residual chlorine per liter, while measuring no differences in biomass between impacted and control streams.

One of the main concerns with the use of photosynthesis and its resulting primary productivity is the sense that natural variability in the data will mask any responses to perturbations (e.g., Crossey and LaPoint 1988). Adjustment of photosynthesis for chlorophyll *a* per unit mass allows for comparisons of communities with differing levels of biomass, resulting in lower variance components to these measures (Boston et al. 1991). They found that chlorophyll-adjusted photosynthesis was as sensitive to chemical perturbations as were fish and macroinvertebrates.

Respiration

Respiration in aquatic plants has often been overlooked in the assessment of physiological condition and responses to perturbations. Respiration integrates most cellular functions and indirectly measures impacts to all cell systems. The mechanisms of inhibition from chemical substances are poorly understood, but electron transfer within glycolysis and the Krebs cycle seem to be likely points of interference. As with the photosynthetic studies reported previously, respiration in aquatic plants is inhibited by herbicides, pesticides, heavy metals, and halogens (Puckett and Burton 1981; Boyle 1984; Guilizzoni 1991; Hill et al. 1997).

Plant community structure for biomonitoring

Measurement and analysis of community structure is the mainstay of biomonitoring programs. Community structure can be measured as lists of the total number of species present within a community or of indicator species. Alternatively, an aggregate metric, one derived from other measures of community structure, may be used to integrate pollution tolerance, species density and diversity, and relative abundances (Patrick 1973; Descy 1979; Lange-Bertalot 1979; Bahls 1993; Mills et al. 1993).

The need to develop ecological indicators as tools for assessment of biological integrity of ecosystems has been established as a priority by the U.S. Environmental Protection Agency (EPA) so that states must adopt narrative biological criteria into water quality standards (U.S. EPA 1990). Currently, structural indicators based on attributes of communities of aquatic organisms are used to address conditions in aquatic systems (U.S. EPA 1989b). Numerical metrics include data on the presence and absence of species at specific sites or within drainages; measures of species richness, diversity, or stability (Shannon's H, Kendall's W); biomass estimates per unit area; and Karr's (1981) IBI. The Modified Index of Well-Being (Ohio EPA 1988; after Gammon 1976; Gammon et al. 1981) reflects community productivity (numbers, biomass) and diversity, excluding exotic species, hybrids, and tolerants. The ICI (Ohio EPA 1988) and Hilsenhoff Biotic Index (HBI) (U.S. EPA 1989b) are multimetric evaluations of macroinvertebrate community structure patterned after the IBI. Bahls (1993) has proposed several indices of water quality based on periphyton community structure. Likewise, Mills et al. (1993) have developed a diatom bioassessment index based on pollution tolerance, abundance of sensitive species, and similarity to reference communities. While most examples of structure-based community metrics are from fish and macroinvertebrate studies, they do indicate the utility and limitations of the use of metric-based assessments.

Structurally based metrics have the advantages of constituting a standardized, consistent application of methods and procedures. Data is rapidly and easily collected and readily compared with known ranges of species (in the case of presence-absence data) and with other sites for which data has already been obtained. Generalizations about trophic structure are possible, and proportions of sensitive or tolerant species can be expressed. The disadvantages of structure-based metrics are that they are oriented to measures of abundance and stability. Thus, these metrics may fail to address the inherent instability of natural populations. These metrics are also affected by the continued presence of rare species (Grossman et al. 1990).

Periodic changes of community structure in response to disturbance may go undetected. In the use of structure-based metrics, questions arise as to the nature of the factors which contribute to the composition of stream assemblages (Grossman et al. 1982; Herbold 1984; Yant et al. 1984). Natural perturbations can result in deviations of assemblage composition from expected values. Where biogeographic and evolutionary processes obscure assemblage structure in areas of high endemism or paucity of species, community metrics require judgements about indicator "species equivalence" (Schlosser 1990). Thus, taxon-dependent indicators are less applicable for use in upstream areas because of greater temporal variability.

The impacts of alterations in physical habitat and species introductions may be reflected in the reduced diversity of species because of reduced habitat, increased siltation, or biotic interactions with non-native species (Berkman and Rabeni 1987; Schlosser 1991). The resulting abiotic variability causes disruption of structural heterogeneity and elevated primary productivity (Schlosser 1982).

In the assessment of the response of a community to disturbance, it is necessary to ask if recovery is based on the reappearance of species or a return to their previous abundances. Accurate assessment of the impact of disturbance is dependent on the characterization of the variability of the community assemblage on a spatio-temporal scale (Connell and Sousa 1983). It is essential to determine that the spatial scale encompasses the minimum range of the dominant taxa so that community composition is not dependent on migration. The study must also be of sufficient duration to ensure that the stability observed is based on a stable age structure. The high variability of biotic assemblages may make the distinction between disturbance-induced population changes and natural environmental variation difficult.

Species richness and relative abundance

Species richness, the number of species in a community, is often a useful indicator of pollution. While this data is limited to presence or absence of a species and not its relative abundance, most studies have reported decreases in species richness with increasing pollution. Patrick (1973) demonstrated decreases in diatom species richness with increasing organic pollution. Several researchers have documented decreases in diatom species richness as a result of metal contamination (e.g., Crossey and LaPoint 1988; Sudhakar et al. 1991). Declines in aquatic vascular plant richness have been documented for lakes and streams experiencing eutrophication, siltation, and metal contamination (e.g., Besch and Roberts-Pichette 1970; Dale and Miller 1978; Dennison et al. 1993).

The relative abundance of a species is the number of organisms of that species relative to all other species in an assemblage and indicates how well that species is suited to, or tolerates, present environmental conditions. Decreases in the relative abundance of a species compared to previous measurements or reference conditions suggests changes in environmental conditions. For example, Weber and McFarland (1981) documented shifts in algal dominance in response to additions of copper to a stream. Dennison et al. (1993) reported several incidences of dominance shifts in aquatic macrophyte communities related to siltation and suspended solids in tributaries to the Chesapeake Bay.

Indicator species

The autecology of a species, including relative pollution tolerance, is the basis of the indicator species concept. Originating with the Saprobien concept (Kolkwitz and Marsson 1908), indicator species are used to assess current or recent environmental conditions. While there are examples of the use of indicator species in mosses (e.g., copper mosses, Puckett and Burton 1981) and higher plants (e.g., Seddon 1972; Hutchinson 1975; Haslam 1978), the concept is most advanced for diatoms (Descy 1979; Lange-Bertalot 1979). Diatom autecology is fairly well documented and has been used extensively in predicting current and past environmental conditions (Patrick 1973; Lowe 1974; Charles 1985; Baron et al. 1986; Dixit et al. 1992; Bahls 1993).

An extension of the indicator species concept involves using the relative proportion of sensitive or tolerant species for such environmental conditions as nutrient concentration (Hall and Smol 1992), pH (Charles 1985; ter Braak and van Dam 1989), organic enrichment (Patrick 1973; Lange-Bertalot 1979), salts (Fritz 1990), siltation (Bahls 1993), and toxic substances (Lange-Bertalot 1979; Kingston et al. 1992).

Data analysis techniques associated with indicator species have evolved from correlation and regression models to weighted-average models. Weighted-average models have the advantages of being conceptually sound, objective, and lacking species biases, problems which have plagued earlier autecological models (Agbeti 1992). Weighted averaging is based on species distributions along environmental gradients, assuming unimodal distributions along those gradients. Environmental optima for each species are estimated as the average value from all sites in which that species occurs weighted for its relative abundance (ter Braak and Barendregt 1986; Agbeti 1992).

Species diversity

Species diversity, the relative number of different species in a community, has two components. The first of these is species richness, which was described earlier. The second component of diversity is the evenness of distribution of individual organisms among different species. In theory, natural, unstressed communities tend toward both an increase in the kinds of species present and the equitable distribution of individuals among species. These two aspects of species diversity present two separate measures of stress in disturbed communities. Stress may cause a reduction of species or loss of species richness, or it may change the distribution of organisms among species, a decrease in evenness. Boyle et al. (1990) reviewed nine commonly used indices of diversity and found that some indices were more sensitive to changes in taxa number while others responded more to changes in the number of organisms. The Shannon index (Shannon and Weaver 1963) increases with both increasing numbers of species and increasing evenness of distribution of organisms among those species.

Species diversity has proven to be a valuable indicator of water quality because it responds to changes in community composition, including the number of species and the density and distribution of organisms among those species. While it is generally assumed that algal diversity will decrease as stress increases (e.g., Patrick 1973; Sigmon et al. 1977; Weber and McFarland 1981; Stevenson and Lowe 1986; Crossey and LaPoint 1988; Carrick et al. 1988; Sudhakar et al. 1991; Bahls 1993), we must not assume that these changes will be similar for all environmental stresses.

Contrary to the extensive data supporting the idea that diversity decreases as a result of stress, several studies of community responses to acid-mine drainage, metals, nutrients, oil spills, and pesticides have failed to detect changes in species diversity (Archibald 1972; Eisle and Hartung 1976; Letterman and Mitsch 1978; Thomas 1978; Osborne et al. 1979).

The use of diversity indices to describe changes in aquatic vascular plants is limited by the depauperate nature of macrophyte communities, which are dominated by a few species and are generally low in taxa richness (Sculthorpe 1967; Haslam 1978). Diversity in depauperate systems is dominated by the evenness component, resulting in erroneous interpretations of the community based on that data (Letterman and Mitsch 1978; Marshall and Mellinger 1980; Stevenson and Lowe 1986).

Community similarity

Community similarity, the degree of compositional agreement among the species in two or more communities along an environmental gra-

dient, is particularly suited for identifying changes in community structure relative to the distance from the source of perturbation (Hellawell
1977). Sheehan (1984a) suggested that similarity indices are more sensitive to low-level stress than are diversity indices. Marshall and Mellinger (1980) were able to detect shifts in similarity among zooplankton
communities exposed to cadmium, even though diversity did not
change. I have found similar changes in periphyton communities collected along a metal impacted stream (Hill et al. 1992c, 1993). Bahls
(1993) found good agreement between species diversity and community similarity when control sites were adequately characterized.

Community similarity and related ordination techniques have been
used extensively to classify lakes based on aquatic vascular plant composition (Seddon 1972; Jensen and van der Maarel 1980), though there
has apparently been no application of this concept to pollution assessment.

Biomass

One of the simplest measures of aquatic plant community structure is
standing crop of biomass. The relationship between standing crop and
water quality, however, is not easily interpreted. There are a number
of methods available for biomass determinations. Biomass may be
measured directly as dry mass or ash-free dry mass. Dry mass estimates
include all organic and inorganic material collected in the sample without differentiation into living and nonliving and autotrophic or heterotrophic biomass. As plant size increases, these limitations decrease.
For periphyton communities, the proportion of dry mass that is inorganic matter and the proportion of the living mass that is heterotrophic
can be considerable. Ash-free dry mass corrects for the inorganic component of the sample, but is unable to differentiate autotrophic and
heterotrophic biomass (e.g., Clark et al. 1979; Stevenson and Lowe
1986). Most analyses of mass in aquatic macrophyte communities
employ either dry mass or ash-free dry mass (e.g., Westlake 1965; Hill
1981, 1987; Hill and Webster 1984).

Algal biomass may be estimated from measurement of chlorophyll
a (APHA 1992). While this method has been widely used to estimate
algal standing crops and assess the impact of water pollution (Clark et
al. 1979), it has been demonstrated that the chlorophyll to mass ratio
varies with habitat and cell morphology (Stevenson and Lowe 1986).

ATP analysis provides an estimate of total living mass within periphyton samples (APHA 1992). The utility of this measure lies in the
fairly constant ATP to mass ratios for periphytic organisms (Clark et
al. 1979).

Clark et al. (1979) compared these methods of estimating periphyton biomass in response to chemical perturbations in stream mesocosms. They found that no one method was consistently better than

any of the others in detecting the impact of copper, chromium, and chloride contaminations. In general, they found that biomass coloniz-ing clean substrates was depressed in response to these contaminants. Sigmon et al. (1977) found significant reductions in algal biomass with increasing mercury concentrations in stream mesocosms. No consistent correlation was found between periphyton biomass on natural and artificial substrates and and metal concentrations in streams. However, mass accrual rates on tiles were significantly lower at those sites with zinc concentrations in excess of 100 μg/l (Hill et al. 1992c, 1993). Boston et al. (1991) reported reduced biomass accumulation on tiles in response to chlorine and mixed industrial effluent in streams.

Aquatic vascular plant biomass has rarely been used to indicate environmental contamination, with the exception of nutrient enrich-ment (e.g., Seidel 1976; Haslam 1978). Hill and Webster (1984) were able to detect the shift from gneiss and schist to limestone and dolomite bedrock underlying the New River, VA, where biomass of *Podostemum ceratophyllum* was significantly higher in the hardwater reach.

Chlorophyll *a* concentration has been used both to predict biomass and as a covariant of biomass. As mentioned earlier, the prediction of biomass from chlorophyll concentrations is subject to much error. How-ever, chlorophyll can be used in conjunction with biomass to assess water quality or toxic impacts. Weber (1973) proposed using the autotrophic index (biomass/chlorophyll) to assess water quality. He found that both organic enrichment and copper contamination increased the autotrophic index. Specific chlorophyll content (chloro-phyll/biomass) on tiles have been used to detect the impact of elevated zinc concentrations on the periphyton community (Hill et al. 1992c, 1993).

The ratio of chlorophyll *a* to phaeophyton *a*, a degradation product of chlorophyll, has been suggested as a measure of stress in periphyton communities (Weber 1973).

Plant community function for biomonitoring

Ecosystems are complex, self-regulating, functional units. Unlike com-munities and populations which are structurally defined, ecosystems are defined by rates and processes, such as energy flow or material cycling, which are mediated by the trophic structure of the ecosystem. Functional indicators are those metrics which directly or indirectly measure energy flow and material cycling within ecosystems.

The argument over the use of structural or functional indicators is linked to two basic approaches to studying ecosystems. On one hand, the population-community approach views ecosystems as networks of interacting populations. On the other hand, the process-functional approach views ecosystems as the sum of the physical, chemical, and biological process active within a space-time unit (O'Neill et al. 1986).

Separation of structural and functional parameters of ecosystems misses the point that the two are interdependent (O'Neill et al. 1986).

Cairns and Pratt (1986) listed three possible relationships between ecosystem structure and function: (1) structure and function are so tightly linked that one cannot change without changing the other; (2) structure can change without a detectable change in function; and (3) function may change without any change in structure. Stream ecosystem monitoring has usually operated under the assumption that the first two relationships are true, while at the same time acknowledging that ecosystem function may, in fact, be diminished by the collective reduction of individual or population function without changing ecosystem structure.

Ecosystem function has been a major focus of stream research over the past decade. In spite of these advances in stream ecology, regulatory assessment and monitoring of streams are still limited to collection of structural data (Cairns and Pratt 1986; Cummins 1991). While "critter counts" are relatively easy and employ standardized methods, ecologists generally agree that within a community variability hampers their ability to assess stream ecosystem integrity (Cairns and Pratt 1986). While functional metrics such as community metabolism or nutrient spiralling are also subject to variability, their ability to integrate diverse populations into a single attribute allows easier comparisons within a system through time (temporal heterogeneity) and among diverse systems (spatial heterogeneity) (O'Neill et al. 1986; Cummins 1988).

Community metabolism

The most commonly measured functional attributes of ecosystems are gross and net primary productivity (GPP, NPP) and community respiration (CR_{24} or R). These two metrics are termed community metabolism when considered together. Two additional metrics are calculated from GPP and CR_{24}: the ratio of GPP and CR_{24}, usually referred to as P/R, and daily GPP-CR_{24} which yields an estimate of net daily metabolism (NDM). These four metrics have been shown to be sensitive indicators of ecosystem stress (Matthews et al. 1982; Bott et al. 1985; Hill and Gardner 1987; Hill et al. 1997). Niemi et al. (1993) analyzed the ability of several measures of chemical and biological structure and function to detect impact and recovery in stream ecosystems. They found that GPP and respiration were able to detect impacts at lower levels than most structural measures.

The addition of biomass estimates (B), indirectly a measure of net primary productivity (NPP = GPP – R) of the autotrophic community, allows the calculation of two additional metrics of ecosystem energetics. The ratios of GPP to B and CR_{24} to B have been suggested as indicators of ecosystem response to stress. Odum (1985) predicted that community respiration would be the first early warning sign of stress in ecosystems. As more energy is diverted toward maintaining existing biomass, community CR_{24}/B and community GPP/B should increase (Odum 1985; Rapport et al. 1985).

Adjustment of primary productivity to chlorophyll concentration, termed the assimilative ratio (AR), has been useful in reducing variance of these measures. While Crossey and LaPoint (1988) were unable to detect differences in GPP between metal-impacted and control sites, they were able to measure significant differences in AR, with higher values at the control site.

While functional measures have not been extensively used in environmental monitoring, several researchers have demonstrated the utility of functional assessments of perturbations. Stressors such as metals, chlorine, pesticides, oil, and channel desiccation have been shown to depress primary productivity and/or respiration (Maki and Johnson 1976; Bott and Rogenmuser 1978; Rodgers et al. 1979; Matthews et al. 1982; Hill and Gardner 1987; Crossey and LaPoint 1988; Hill et al. 1997).

Nutrient uptake and spiralling

Many of the transformations which nutrients undergo while being transported into, through, and out of a stream segment are determined by difference between input and output. The actual pathway an atom of nutrient takes through this system, usually inferred from movements of radioactive tracers from one ecosystem component to another, is described as spiralling (Webster and Patten 1979).

Nutrient spiralling, defined as spatially dependent nutrient cycling in stream ecosystems (e.g., Elwood et al. 1983), links the concept of nutrient cycling with unidirectional flow. Nutrient cycles of ecosystems are viewed as either closed (i.e., an atom of nutrient is continuously recycled within the ecosystem) or open (i.e., an atom of a nutrient is cycled within the system, but is eventually exported from the system). Most ecosystems are considered to be open, though the degree of openness may depend on the relative timescale used to analyze nutrient cycles. Streams are considered extremely open ecosystems. Because of the unidirectional flow, nutrient cycling was never considered as an attribute of streams. Leopold (1941), addressing the problem of nutrient losses from ecosystems, described the rolling, downslope movement of nutrients from the hills to the seas. It is this concept which Webster (1975) intended to describe with the term spiralling. The spiralling concept acknowledges downstream displacement of a nutrient while it passes through each cycle. With the aid of this concept, we are able to simultaneously view nutrient cycles as closed and open (Elwood et al. 1983). A basic difference between spiralling and cycling in an open system is that spiralling moves the nutrient downstream within the same system rather than losing it from the system (Elwood et al. 1983).

That is, transport occurs as a part of the nutrient cycle rather than as an alternative to it.

Elwood et al. (1983) list three aspects of the spiralling concept which make it useful for the analysis of stream ecosystems. First, it is a simple, useful means of measuring nutrient (including carbon) flow through streams. Second, this concept emphasizes the interaction of hydrologic transport and ecological processes governing nutrient cycling. Third, the interaction of both spatial and temporal dynamics of streams is emphasized. Additionally, spiral length may be indicative of increases in nutrient turnover, transport, and loss from streams as predicted for stressed ecosystems (Odum 1985).

Newbold et al. (1981) developed an index of nutrient spiralling known as spiral length, defined as the average downstream distance associated with one complete cycle of a nutrient atom. In developing this index they have assumed that a nutrient atom passes through water, particulate, and consumer compartments while being transported. These components of the spiral can be quantified, and the sum of their average distances is the spiral length. Under steady-state conditions, spiral length is expressed as the ratio of total downstream transport of a nutrient to nutrient utilization. Generally speaking, the shorter the spiral length the more efficiently nutrients are used within a stream. These authors emphasized that abiotic, rather than biotic, factors predominate in regulation of spiral length.

Recent investigation of the factors controlling spiral length have shown a greater role by the biotic component (e.g., Elwood et al. 1981; Newbold et al. 1982, 1983; Mulholland et al. 1985a,b). These authors found that spiral length is decreased by those organisms that act to retain organic matter and nutrients within a stream (e.g., periphyton, filter-feeding insects, and grazers) and increased by organisms which promote particle transport (e.g., shredders). Especially important is the role of periphyton in nutrient uptake. Elwood et al. (1981) found that 91% of the exchangeable phosphorus uptake by fine particulate organic matter (FPOM) was due to periphyton. Similarly, about 80% of the exchangeable phosphorus uptake by stream sediments was due to microorganisms. These findings are in agreement with those reported by Gregory (1978), who reported that less than 20% of the phosphorus uptake by stream sediments and organic matter was abiotic.

The aforementioned studies indicate that spiral length is dominated by uptake length. Newbold et al. (1983) reported that uptake length accounted for nearly 90% of the total spiral length. Mulholland et al. (1985a, 1990) found uptake length to be as much as 98% of spiral length. If spiral length can be estimated accurately ($\pm 10\%$) from uptake length, then pulse additions of nonradioactive phosphorus and ensuing downstream depletion of this phosphorus should serve as a measure of phosphorus spiralling in those streams where the use of radioactive tracers would be undesirable (Stream Solute Workshop 1990; Webster et al. 1991).

Enzyme activity

Material cycles within ecosystems are dominated by the microbial community. Through its role in detritus processing, microbial communities integrate carbon and nutrient cycles (e.g., Rich and Wetzel 1978) within the scheme of energy flow through ecosystems. As such, the microbial community may be the best indicator of overall ecosystem integrity. Because of their critical role in ecosystem functions, any reduction in microbial metabolic rates may be construed as an impact (Burton and Lanza 1985; Burton et al. 1987). This has been demonstrated for numerous stream, lake, and sediment systems (e.g., Burton and Lanza 1985; Burton et al. 1987; Newman et al. 1987; Perry et al. 1987; Dougherty 1989).

The use of microbial communities to assess ecosystem response to disturbances is a relatively new idea. The lack of a substantial microbial history in assessment of ecosystem perturbation stems largely from the lack of understanding of the microbial community within the ecosystem. Traditionally, microorganisms have been studied as individual colonies in laboratories and not viewed as components of ecosystems (Margulis et al. 1986). This view is beginning to change. Several recent studies have focused on such traditional ecological topics as structure, function, diversity, competition, succession, and stability among microbial communities (e.g., Atlas 1984; Weibe 1984; Federle et al. 1986; Margulis et al. 1986). Generally, it has been shown that microbial communities function much like communities of larger organisms.

Since most microbial metabolic pathways are dependent on respiration, electron transport activity would be a critical process to measure. Dehydrogenases play an important role in organic matter processing and indirectly reflect electron transport function. Dehydrogenase activity has been used to measure the effects of many different toxicants in soil and sediment microbial communities (e.g., Lenhard 1968; Bitton 1983; Burton and Lanza 1987; Burton et al. 1987).

Acid and alkaline phosphatase activities have also been demonstrated to be sensitive indicators of microbial community disturbance from toxicants (Sayler et al. 1979, 1982). Levels of phosphatase activity have been correlated to microbial numbers and biomass in response to perturbation and recovery. Depressed alkaline phosphatase activity has been reported from habitats which have been impacted by pollution (e.g., Tyler 1976; Tu 1981; Baker and Morita 1982; Sayler et al. 1982; Burton and Lanza 1987).

Hydrolytic enzymes such as amylase, cellulase, and protease are essential for the breakdown of organic matter and the cycling of carbon and nitrogen in sediment systems (Skujins 1976). These hydrolytic enzymes have been shown to be inhibited by metals, oil, pesticides, and industrial effluents in studies of pollution impact on sediment systems (e.g., Little et al. 1979; Griffiths and Morita 1981; Burton and Lanza 1987; Burton et al. 1987; Burton and Stemmer 1988).

Decomposition

Leaf litter decomposition is the result of processing by physical forces and chemical interactions and use by microbial and larger invertebrate communities. Numerous studies have described the effects of substrate type, current velocity, nutrient availability, and consumption by microbial and macroinvertebrate communities (see Brinson et al. 1981; Webster and Benfield 1986). Less-often studied are the effects of perturbations on the decomposition process, though these studies have proven to be quite reliable in the assessment of impacts to aquatic ecosystems (Niemi et al. 1993). Several natural and anthropogenic perturbations have been shown to inhibit decomposition of terrestrial and aquatic leaf litter. Both channel desiccation and reservoir drawdown inhibit microbial activity, resulting in decreased breakdown rates (Hill 1985; Hill et al. 1988, 1992a). There are several reports of the reduction of decomposition rates due to acidification of streams and lakes (e.g., Carpenter et al. 1983; Burton et al. 1985). Heavy metals, either alone or in conjunction with pH and chemical flocs, inhibit leaf decay through both microbial and macroinvertebrate inhibition (Giesy 1978; Guthrie et al. 1978; Forbes and Magnuson 1980; Gray and Ward 1983; Leland and Carter 1985). Newman et al. (1987) and Perry et al. (1987) both reported inhibition of aquatic macrophyte decay by elevated chlorine. Sheehan (1984b) summarized research on the effects of acid precipitation, heavy metals, pesticides, and oil, all of which depressed decomposition, usually through a combination of inhibition of microbial and macroinvertebrate processing.

Ecosystem stability

The concept of ecological stability has been considered for several decades, with most researchers linking stability to system complexity (e.g., MacArthur 1955; Margalef 1963; Waide and Webster 1975; Van Voris et al. 1980). Two main ideas have emerged from these and other studies on ecological stability. First, ecosystems which exhibit little temporal variability are stable. Second, ecosystems which can recover rapidly from disturbances are stable. An ecosystem is considered stable if its response to disturbance is small or if its return to its original state is rapid. An ecosystem is unstable if it is markedly changed by a disturbance or if it returns slowly or not at all to its original state (Webster et al. 1983).

Webster et al. (1975) stated that the usefulness of this stability concept is as a starting point for more germane discussions of the "relative" stability of different ecosystems. Three approaches to studying stream ecosystem stability have been identified: (1) relationships between diversity, maturity, and stability (traditional stability concepts); (2) the ability of streams to ameliorate the effects of pollution (assimilative capacity); and (3) indices of relative stability.

Traditional indices of stability, such as diversity and maturity, have mainly addressed how stability changes along stream continua (e.g., Margalef 1960; Harrel et al. 1967; Harrel and Dorris 1968; Sheldon 1968; Vannote et al. 1980). Generally, these authors reported downstream increases in faunal diversity through the mid-order streams, then a decrease in diversity farther downstream. Several researchers have studied energy relationships in streams and suggested that low order streams are immature and unstable because they export organic matter to downstream ecosystems (e.g., Margalef 1960, 1963; Odum 1969; Hall 1972). Other researchers have argued that because of their detrital energy base, high benthic organic matter storage, and low P to R ratios, low-order streams are mature and relatively stable (Motten and Hall 1972; Fisher and Likens 1973; Naiman and Sedell 1979).

Webster et al. (1983) indicated that the best index of ecosystem maturity, and therefore stability, is ecosystem energy efficiency, the ratio of carbon respired to carbon inputs (Fisher and Likens 1973). This point is echoed by Cummins et al. (1983), who maintained that ecosystem efficiency and an outgrowth of ecosystem efficiency, stream metabolism index (Fisher 1977), are the only valid measure for comparing organic matter dynamics between stream ecosystems.

The use of stability measurements to assess impacts to ecosystems has received little attention. Webster et al. (1983) used periphyton productivity, along with organic matter transport, leaf litter decomposition rates, and chemistry, to assess the relative ecological stability of streams draining old growth and clearcut watersheds. Hill et al. (1992b) measured organic matter transport in intermittent and perennial prairie streams to assess the impact of channel desiccation on ecological stability.

Summary and conclusions

Biological monitoring and biological criteria for water quality are used throughout the world for assessing impacts and setting goals for the maintenance and remediation of aquatic ecosystems. While most of these efforts rely primarily on aquatic insect and fish assemblages, aquatic plants are proving to be useful for these purposes as well. This is especially true for the use of diatom assemblages, which have been extensively used as indicators of past and present water quality.

Several approaches to biological monitoring using aquatic plants have be employed, ranging from ecotoxicological determinations of water quality to functional assessments of aquatic ecosystems. Physiological approaches include measurement of (1) cellular integrity, such as changes in membrane permeability or chlorophyll absorbance spectra; (2) growth; (3) photosynthesis and respiration; and (4) enzyme activity.

As has been true with faunistic surveys, most environmental assessments using aquatic plants have been focused on comparisons of plant community structure at impacted sites with community structure at reference sites. Such measures as species richness, relative abundance, species diversity, indicator species, and community similarity are commonly used in floristic assessments of environmental impacts and water quality.

Less commonly used are the functional approaches to ecosystem assessment. While such measurements as primary productivity, community respiration, and organic matter degradation rates are often dismissed as being too variable to detect impacts, recent studies are proving that this is not true and that functional assessments are often more sensitive than structural assessments of aquatic ecosystem integrity.

The studies reviewed here suggest that both structural and functional assessments of aquatic plant communities are valuable tools in the determination of environmental impacts and water quality. I am not suggesting that aquatic plants be used in lieu of macroinvertebrates or fish for biological assessments of aquatic ecosystems. Rather, I am proposing that all three approaches be employed for monitoring ecological integrity and environmental impacts.

References

Agbeti MD (1992) Relationship between diatom assemblages and trophic variables: a comparison of old and new approaches. *Can J Fish Aquat Sci* 49:1171–1175.

APHA (1992) *Standard Methods for the Analysis of Water and Wastewater,* 18th ed., American Public Health Association, Washington, D.C.

Archibald REM (1972) Diversity in some South African diatom associations and its relation to water quality. *Water Res* 6:1229–1238.

Atlas RM (1984) Use of microbial diversity measurements to assess environmental stress. In *Current Perspectives in Microbial Ecology,* Klug MJ, Reddy CA, Eds., American Society of Microbiology, Washington, D.C., pp. 540–545.

Bahls LL (1993) Periphyton bioassessment methods for Montana streams. State of Montana Water Quality Bureau, Department of Health and Environmental Sciences, Helena, MT.

Baker JH, Morita RY (1982) A note on the effects of crude oil on microbial activities in stream sediments. *Environ Pollut Ser A* 31:149–157.

Baron J, Norton SA, Beeson DR, Herrman R (1986) Sediment diatom stratigraphy from Rocky Mountain lakes with special reference to atmospheric deposition. *Can J Fish Aquat Sci* 43:1350–1362.

Berkman HE, Rabeni CF (1987) Effects of siltation on stream fish communities. *Environ Biol Fish* 18:285–294.

Besch KW, Roberts-Pichette P (1970) Effects of mining pollution on vascular plants in the Northwest Miramichi River system. *Can J Bot* 48:1647–1656.

Bitton G (1983) Bacterial and biochemical tests for assessing toxicity in the aquatic environment: a review. *CRC Crit Rev Environ Control* 13:51–67.

Boston HL, Hill WR, Stewart AJ (1991) Evaluating direct toxicity and food-chain effects in aquatic systems using natural periphyton communities. In *Plants for Toxicity Assessments*, Vol. 2, Gorsuch JW, Lower WR, Wang W, Lewis MA, Eds., ASTM STP 1115, American Society for Testing and Materials, Philadelphia, PA, pp. 126–145.

Bott TL, Rogenmuser K (1978) Effects of No. 2 fuel oil, Nigerian crude oil, and used crankase oil on attached algal communities: acute and chronic toxicity of water-soluble constituents. *Appl Environ Microbiol* 36:673–682.

Bott TL, Dunn CS, Naiman RJ, Ovink RW, Petersen RC (1985) Benthic metabolism in four temperate stream systems: an interbiome comparison and evaluation of the river continuum concept. *Hydrobiologia* 123:3–45.

Boyle TP (1984) The effect of environmental contaminants on aquatic algae. In *Algae as Ecological Indicators*, Shubert LE, Ed., Academic Press, London, pp. 237–256.

Boyle TP, Smillie GM, Anderson JC, Beeson DR (1990) A sensitivity analysis of nine diversity and seven similarity indices. *Res J Water Pollut Control Fed* 62:749–762.

Brinson MM, Lugo AE, Brown S (1981) Primary productivity, decomposition and consumer activity in freshwater wetlands. *Annu Rev Ecol Syst* 12:123–161.

Burton GA, Lanza GR (1985) Sediment microbial activity tests for the detection of toxicant impacts. In *Aquatic Toxicology and Hazard Assessment*, Caldwell RD, Purdy R, Bahner RC, Eds., ASTM STP 854, American Society for Testing and Materials, Philadelphia, PA, pp. 214–228.

Burton GA, Lanza GR (1987) Aquatic microbial activity and macrofaunal profile of an Oklahoma stream. *Water Res* 21:1173–1182.

Burton GA, Stemmer BL (1988) Evaluation of surrogate tests in toxicant impact assessments. *Toxic Assess* 3:255–269.

Burton GA, Drotar A, Lazorchak JM, Bahls LL (1987) Relationship of microbial activity and *Ceriodaphnia* responses to mining impacts on the Clark Fork River, Montana. *Arch Environ Contam Toxicol* 16:523–530.

Burton MAS (1990) Terrestrial and aquatic bryophytes as monitors of environmental contaminants in urban and industrial habitats. *Bot J Linnean Soc* 104:267–280.

Burton TM, Stanford RM, Allan JW (1985) Acidification effects on stream biota and organic matter processing. *Can J Fish Aquat Sci* 42:669–675.

Cairns J Jr. (1968) The effects of Dieldrin on diatoms. *Mosq News* 28:177–179.

Cairns J Jr. (1990) The genesis of biomonitoring in aquatic ecosystems. *Environ Prof* 12:169–176.

Cairns J Jr., Pratt JR (1986) Developing a sampling strategy. In *Rationale for Sampling and Interpretation of Ecological Data in the Assessment of Freshwater Ecosystems*, Isom BG, Ed., ASTM STP 894, American Society for Testing and Materials, Philadelphia, PA, pp. 168–186.

Carpenter J, Odum WE, Mills A (1983) Leaf litter decomposition in a reservoir affected by acid mine drainage. *Oikos* 41:165–172.

Carrick HJ, Lowe RL, Rotenberry JT (1988) Guilds of benthic algae along nutrient gradients: relationships to algal community diversity. *J N Am Benthol Soc* 7:117–128.

Carter J (1990) A new *Eunotia* and its great morphological variations under stress caused by a habitat loaded with copper salts. *Ouvrage dedie a H. Germain, Koletz* 1990:13–17.

Charles DF (1985) Relationships between surface sediment diatoms assemblages and lakewater characteristics in Adirondack lakes. *Ecology* 66:994–1011.

Christensen ER, Nyholm N (1984) Ecotoxicological assays with algae: Weibull dose-response curves. *Environ Sci Technol* 18:713–718.

Clark JR, Dickson KL, Cairns J Jr. (1979) Estimating aufwuchs biomass. In *Methods and Measurement of Periphyton Communities: A Review,* Weitzel RL, Ed., ASTM STP 690, American Society for Testing and Materials, Philadelphia, PA, pp. 116–141.

Connell JH, Sousa WP (1983) On the evidence needed to judge ecological stability or persistence. *Am Nat* 121:789–824.

Crossey MJ, LaPoint TW (1988) A comparison of periphyton community structural and functional responses to heavy metals. *Hydrobiologia* 162:109–121.

Cummins KW (1988) The study of stream ecosystems: a functional view. In *Concepts of Ecosystem Ecology*, Pomeroy LR, Alberts JJ, Eds., Springer-Verlag, New York, pp. 247–262.

Cummins KW (1991) Establishing biological criteria: functional views of biotic community organization. In *Biological Criteria: Research and Regulation*, EPA-440/5-91-005, Office of Water, U.S. Environmental Protection Agency, pp. 3–8.

Cummins KW, Sedell JR, Swanson FJ, Minshall GW, Fisher SG, Cushing CE, Petersen RC, Vannote RL (1983) Organic matter budgets for stream ecosystems: problems in their evaluation. In *Stream Ecology*, Barnes JR, Minshall GW, Eds., Plenum Press, New York, pp. 299–353.

Dale HM, Miller GE (1978) Changes in aquatic macrophyte flora of Whitewater Lake near Sudbury, Ontario from 1947–1977. *Can Field Nat* 92:264–270.

Dennison WC, Orth RJ, Moore KA, Steveneson JC, Carter V, Kollar S, Bergstrom PW, Batiuk RA (1993) Assessing water quality with submersed aquatic vegetation. *BioScience* 43:86–94.

Descy JP (1979) A new approach to water quality estimation using diatoms. *Nova Hedwigia* 64:305–323.

Dixit SS, Smol JP, Kingston JC, Charles DF (1992) Diatoms: powerful indicators of environmental change. *Environ Sci Technol* 26:22–33.

Dougherty JM (1989) Anaerobic subsurface soil microcosms: the effect of anthropogenic organic compounds on microbial communities. Ph.D. Dissertation. University of Texas, Dallas.

Eisle PJ, Hartung R (1976) The effects of methoxychlor on riffle invertebrate populations and communities. *Trans Am Fish Soc* 105:628–633.

Elwood JW, Newbold JD, O'Neill RV, Stark RW, Singley PT (1981) The role of microbes associated with organic and inorganic substrates in phosphorus spiralling in a woodland stream. *Verh Int Verein Limnol* 21:850–856.

Elwood JW, Newbold JD, O'Neill RV, Van Winkle W (1983) Resource spiralling: an operational paradigm for analyzing lotic ecosystems. In *Dynamics of Lotic Ecosystems*, Fontaine TD III, Bartell SM, Eds., Ann Arbor Science Publishers, Ann Arbor, MI, pp. 3–27.

Federle TW, Livingston RJ, Meeter DA, White DC (1986) A quantitative comparison of microbial community structure of estuarine sediments from microcosms and the field. *Can J Microbiol* 32:319–325.

Fisher SG (1977) Organic matter processing by a stream-segment ecosystem: Fort River, Massachusetts, USA. *Int Rev Gesampten Hydrobiol* 62:701–727.

Fisher SG, Likens GE (1973) Energy flow in Bear Brook, New Hampshire: an integrative approach to stream ecosystem metabolism. *Ecol Monogr* 43:421–439.

Forbes AM, Magnuson JJ (1980) Decomposition and microbial colonization of leaves in a stream modified by coal ash effluent. *Hydrobiologia* 76:263–267.

Franzen WG, McFarlane GA (1980) An analysis of the aquatic macrophyte, *Myriophyllum exalbescens*, as an indicator of metal contamination of aquatic ecosystems near a base metal smelter. *Bull Environ Contam Toxicol* 24:597–605.

Fritz SC (1990) Twentieth-century salinity and water-level fluctuations in Devils Lake, North Dakota: a test of a diatom-based transfer function. *Limnol Oceanogr* 35:1771–1781.

Gammon JR (1976) The fish populations of the middle 340 km of the Wabash River. *Purdue Univ Water Resour Cent Tech Rep* 86:1–73.

Gammon JR, Spacie A, Hamelink JL, Kaesler RL (1981) Role of electrofishing in assessing the environmental quality of the Wabash River. In *Ecological Assessments of Effluent Impacts on Communities of Indigenous Aquatic Organisms*, Bates JM, Weber CI, Eds., ASTM STP 703, American Society for Testing and Materials, Philadelphia, PA, pp. 307–324.

Giesy JP Jr. (1978) Cadmium inhibition of leaf decomposition in an aquatic microcosm. *Chemosphere* 7:467–476.

Gilfillan ES, Page DS, Bass AE, Foster JC, Fickett PM, Ellis WG, Rusk S, Brown C (1989) Use of Na/K ratios in leaf tissues to determine effects of petroleum on salt exclusion in marine haliphytes. *Mar Pollut Bull* 20:272–276.

Gray LJ, Ward JV (1983) Leaf litter breakdown in streams receiving treated and untreated mine drainage. *Environ Int* 9:135–138.

Gregory SV (1978) Phosphorus dynamics on organic and inorganic substrates in streams. *Verh Int Verein Limnol* 20:1340–1346.

Griffiths RP, Morita RY (1981) Study of microbial activity and crude oil-microbial interactions in the waters and sediments of Cook Inlet and the Beaufort Sea. In *Environmental Assessment of the Alaskan Continental Shelf*, National Oceanographic and Atmospheric Administration, Juneau, AL.

Grossman GD, Moyle PB, Whitaker JO Jr. (1982) Stochasticity in structural and functional characteristics of an Indiana stream fish assemblage: a test of community theory. *Am Nat* 120:423–454.

Grossman GD, Dowd JF, Crawford M (1990) Assemblage stability in stream fishes: a review. *Environ Manage* 14:661–671.

Guilizzoni P (1991) The role of heavy metals and toxic materials in the physiological ecology of submersed macrophytes. *Aquat Bot* 41:87–109.

Guthrie RK, Cherry DS, Singleton FL (1978) Responses of heterogeneous bacterial populations to pH changes in coal ash effluent. *Water Resour Bull* 14:803–808.

Guzkowska MAJ, Gasse F (1990) Diatoms as indicators of water quality in some English urban lakes. *Freshwater Biol* 23a:233–250.

Hall CAS (1972) Migration and metabolism in a temperate ecosystem. *Ecology* 53:585–604.

Hall RI, Smol JP (1992) A weighted-averaging regression and calibration model for inferrring total phosphorus concentration from diatoms in British Columbia (Canada) lakes. *Freshwater Biol* 27:417–434.

Harrel RC, Dorris JC (1968) Stream order, morphometry, physico-chemical conditions, and community structure of benthic macroinvertebrates in an intermittent stream system. *Am Midl Nat* 80:220–251.

Harrel RC, Davis BJ, Dorris TC (1967) Stream order and species diversity of fishes in an intermittent Oklahoma stream. *Am Midl Nat* 78:428–437.

Haslam SM (1978) River plants. In *The Macrophytic Vegetation of Watercourses*, Cambridge University Press, London.

Hellawell JM (1977) Change in natural and managed ecosystems: detection, measurement and assessment. *Proc R Acad London B. Biol Sci* 197:31–56.

Herbold B (1984) Structure of an Indiana stream fish association: choosing an appropriate model. *Am Nat* 124:561–572.

Hill BH (1981) Distribution and production of *Justicia americana* in the New River, Virginia. *Castanea* 46:162–169.

Hill BH (1985) The breakdown of macrophytes in a reservoir wetland. *Aquat Bot* 21:23–31.

Hill BH (1987) *Typha* productivity in a Texas pond: implications for energy and nutrient dynamics in freshwater wetlands. *Aquat Bot* 27:385–394.

Hill BH, Gardner TJ (1987) Benthic metabolism in two Texas prairie streams. *Southwest Nat* 32:305–311.

Hill BH, Webster JR (1984) Productivity of *Podostemum ceratophyllum* in the New River, Virginia. *Am J Bot* 71:130–136.

Hill BH, Gardner TJ, Ekisola OF (1988) Breakdown of gallery forest leaf litter in intermittent and perennial prairie streams. *Southwest Nat* 33:323–331.

Hill BH, Gardner TJ, Ekisola OF (1992a) Microbial use of leaf litter in prairie streams. *J N Am Benthol Soc* 11:11–19.

Hill BH, Gardner TJ, Ekisola, OF (1992b) Predictability of streamflow and particulate organic matter concentration as indicators of stability in prairie streams. *Hydrobiologia* 242:7–18.

Hill, BH, Parrish, L, Rodriguez, G (1992c) Periphyton community response to elevated metal concentrations in a Rocky Mountain stream. *Bull N Am Benthol Soc* 9:178.

Hill, BH, McFarland, BH, Parrish, L, Willingham, WT (1993) Introduced tiles versus natural substrates for the assessment of metal impacts on periphyton communities in a Rocky Mountain stream. *Bull N Am Benthol Soc* 10:191.

Hill, BH, Lazorchak, JM, McCormick, FH, Willingham, WT (1997) The effects of elevated metals on benthic community metabolism in a Rocky Mountain stream. *Environ Pollut* (in press).

Hughes JS, Vilkas AG (1983) Toxicity of N,N-dimethylformamide used as a solvent in toxicity tests with the green alga, *Selenastrum capricornutum*. *Bull Environ Contam Toxicol* 31:98–104.

Hunsaker CT, Graham RL, Suter GW III, O'Neill RV, Barnthouse LW, Gardner RH (1990) Assessing ecological risk on a regional scale. *Environ Manage* 14:325–332.

Hutchinson GE (1975) *A Treatise on Limnology, Volume 3. Limnological Botany*, John Wiley & Sons, New York.

Jensen S, van der Maarel E (1980) Numerical approaches to lake classification with special reference to macrophyte communities. *Vegetation* 42:117–128.

Jones TW, Winchel L (1984) Uptake and photosynthetic inhibition by atrazine and its degradation products on four species of submerged vascular plants. *J Environ Qual* 13:243–247.

Jones TW, Kemp WM, Estes PS, Stevenson JC (1986) Atrazine uptake, photosynthetic inhibition and short-term recovery for the submersed vascular plant, *Potamogeton perfoliatus*, L. *Arch Environ Contam Toxicol* 15:277–283.

Judy BM, Lower WR, Miles CD, Thomas MW, Krause GF (1990) Chlorophyll fluorescence of a higher plant as an assay for toxicity assessment of soil and water. In *Plants for Toxicity Assessment*, Wang W, Gorsuch JW, Lower WR, Eds., ASTM STP 1091, American Society for Testing and Materials, Philadelphia, PA, pp. 308–318.

Karr JR (1981) Assessment of biotic integrity using fish communities. *Fisheries* 6:21–27.

Karr JR, Dudley DR (1981) Ecological perspective on water quality goals. *Environ Manage* 5:55–68.

Kingston JC, Birks HJB, Uutala AJ, Cumming BF, Smol JP (1992) Assessing trends in fisheries resources and lake water aluminum from paleolimnological analyses of siliceous algae. *Can J Fish Aquat Sci* 49:116–127.

Klaine SJ, Ward CH (1983) Growth-optimized algal bioassays for toxicity evaluation. *Environ Toxicol Chem* 2:245–249.

Kolkwitz R, Marsson M (1908) Okologie der pflanzlichen saprobien. *Berliner Dtsch Bot Ges (Stuttgart)* 26:505–519.

Lange-Bertalot H (1979) Pollution tolerance of diatoms as a criterion for water quality estimation. *Nova Hedwigia* 64:285–304.

Lanza GR, Cairns J Jr. (1972) Physio-morphological effects of abrupt thermal stress on diatoms. *Trans Am Microsc Soc* 91:276–298.

Leland HV, Carter JL (1985) Effects of copper on production of periphyton, nitrogen fixation, and processing of leaf litter in a Sierra Nevada, California, stream. *Freshwater Biol* 15:155–173.

Lenhard G (1968) A standardization procedure for the dehydrogenase activity in samples from anaerobic treatment systems. *Water Res* 2:161–167.

Leopold A (1941) Lakes in relation to terrestrial life patterns. In *A Symposium on Hydrobiology*, The University of Wisconsin Press, Madison, pp. 17–22.

Letterman RD, Mitsch WJ (1978) Impact of mine drainage on a mountain stream in Pennsylvania. *Environ Pollut* 17:53–73.

Little JE, Sjogren RE, Carbon GR (1979) Measurement of proteolysis in natural waters. *Appl Environ Microbiol* 37:900–908.

Lowe RL (1974) *Environmental Requirements and Pollution Tolerance of Freshwater Diatoms*, EPA-670/4-74-005, U.S. Environmental Protection Agency, Cincinnati, OH.

MacArthur R (1955) Fluctuations of animal populations and a measure of community stability. *Ecology* 36:533–536.

Maki AW, Johnson HE (1976) Evaluation of a toxicant on the metabolism of model stream communities. *J Fish Res Board Can* 33:2740–2746.

Margalef R (1960) Ideas for a synthetic approach to the ecology of running waters. *Int Rev Gesampten Hydrobiol* 45:13–153.

Margalef R (1963) On certain unifying principles in ecology. *Am Nat* 97:357–374.

Margulis L, Chase D, Guerrero R (1986) Microbial communities. *BioScience* 36:160–170.

Marshall JS, Mellinger DL (1980) Dynamics of cadmium-stressed plankton communities. *Can J Fish Aquat Sci* 37:403–414.

Matthews RA, Buikema AL Jr., Cairns J Jr., Rodgers JH Jr. (1982) Biological monitoring: IIA. Receiving system functional methods, relationships, and indices. *Water Res* 16:129–139.

McFarland BH, Hill BH, Willingham WT (1997) Abnormal *Fragilaria* sp. (Bacillariophyceae) in streams impacted by mine drainage. *J Freshwat Ecol* (in press).

Mills MR, Beck G, Brumley J, Call SM, Grubbs J, Houp R, Metzmeier L, Smathers K (1993) *Methods for Assessing Biological Integrity of Surface Waters*, Kentucky Department of Environmental Protection, Frankfort.

Motten AF, Hall CAS (1972) Edaphic factors override a possible gradient of ecological maturity indices in a small stream. *Limnol Oceanogr* 17:922–926.

Mulholland PJ, Newbold JD, Elwood JW, Ferrer LA, Webster JR (1985a) Phosphorus spiralling in a woodland stream: seasonal variations. *Ecology* 66:1012–1023.

Mulholland PJ, Elwood JW, Newbold JD, Ferrer LA (1985b) Effect of a leaf-shredding invertebrate on organic matter dynamics and phosphorus spiralling in heterotrophic laboratory streams. *Oecologia (Berlin)* 66:199–206.

Mulholland PJ, Steinman AD, Elwood JW (1990) Measurement of phosphorus uptake length in streams: comparison of radiotracer and stable PO_4 releases. *Can J Fish Aquat Sci* 47:2351–2357.

Naiman RJ, Sedell JR (1979) Benthic organic matter as a function of stream order in Oregon. *Arch Hydrobiol* 87:404–422.

Newbold JD, Elwood JW, O'Neill RV, Van Winkle W (1981) Measuring nutrient spiralling in streams. *Can J Fish Aquat Sci* 38:860–863.

Newbold JD, O'Neill RV, Elwood JW, Van Winkle W (1982) Nutrient spiralling in streams: implications for nutrient limitation and invertebrate activity. *Am Nat* 120:628–652.

Newbold JD, Elwood JW, O'Neill RV, Sheldon AL (1983) Phosphorus dynamics in a woodland stream ecosystem: a study of nutrient spiralling. *Ecology* 64:1249–1265.

Newman MC, McIntosh AW (1989) Appropriateness of Aufwuchs as a monitor of bioaccumulation. *Environ Pollut* 60:83–100.

Newman RM, Perry JA, Tam E, Crawford RL (1987) Effects of chronic chlorine exposure on litter processing in outdoor experimental streams. *Freshwater Biol* 18:415–428.

Niemi GJ, Detenbeck NE, Perry JA (1993) Comparative analysis of variables to measure recovery rates in streams. *Environ Toxicol Chem* 12:1541–1547.

Odum EP (1969) The strategy of ecosystem development. *Science* 164:262–270.

Odum EP (1985) Trends expected in stressed ecosystems. *BioScience* 35:419–422.

Ohio EPA (1988) *Biological Criteria for the Protection of Aquatic Life. Volume I. The Role of Biological Data in Water Quality Assessment*, Ohio Environmental Protection Agency, Columbus.

O'Neill RV, DeAngelis DL, Waide JB, Allen TFH (1986) A hierarchical concept of ecosystems. In *Monographs in Population Biology*, No. 23, Princeton University Press, New Jersey.

Osborne LL, Davies RW, Linton KJ (1979) Efects of limestone strip mining on benthic macroinvertebrate communities. *Water Res* 13:1285–1290.

Patrick R (1973) Use of algae, especially diatoms, in the assessment of water quality. In *Biological Methods for the Assessment of Water Quality*, Cairns J Jr., Dickson KL, Eds., ASTM STP 528, American Society of Testing and Materials, Philadelphia, PA, pp. 76–95.

Payne AG, Hall RH (1979) A method for measuring algal toxicity and its application to the safety assessment of new chemicals. In *Aquatic Toxicology*, Marking LL, Kimerle RA, Eds., ASTM STP 667, American Society for Testing and Materials, Philadelphia, PA, pp. 171–180.

Perry JA, Troelstrup NH Jr., Newsome M, Shelley B (1987) Whole ecosystem manipulation experiments: the search for generality. *Water Sci Technol* 19:55–71.

Pratt JR, Cairns J Jr. (1985) Functional groups in the Protozoa: roles in differing ecosystems. *J Protozool* 32:409–417.

Puckett KJ (1976) The effect of heavy metals on some aspects of lichen physiology. *Can J Bot* 54:2695–2703.

Puckett KJ, Burton MAS (1981) The effect of trace elements on lower plants. In *Effect of Heavy Metal Pollution on Plants*, Vol. 2, Lepp NW, Ed., Applied Sciences Publishers, London, pp. 213–238.

Ram NM, Plotkin S (1983) Assessing aquatic productivity in the Houstonic River using the algal assay: bottle test. *Water Res* 17:1095–1106.

Ramachandran S, Rajendran N, Nandakumar R, Venugopalan VK (1984) Effect of pesticides on photosunthesis and respiration of marine macrophytes. *Aquat Bot* 19:395–399.

Rapport DJ, Regier HA, Hutchinson TC (1985) Ecosystem behavior under stress. *Am Nat* 125:617–640.

Rich PH, Wetzel RG (1978) Detritus in lake ecosystems. *Am Nat* 112:57–71.

Rodgers JH Jr., Dickson KL, Cairns J Jr. (1979) A review and analysis of some methods used to measure functional aspects of periphyton. In *Methods and Measurement of Periphyton Communities: A Review*, Weitzel RL, Ed., ASTM STP 690, American Society for Testing and Materials, Philadelphia, PA, pp. 142–167.

Sayler GS, Puziss M, Silver M (1979) Alkaline phosphatase assay for freshwater sediments: application to perturbed sediment systems. *Appl Environ Microbiol* 38:922–927.

Sayler GS, Sherrill TW, Perkins RE, Mallory LM, Shiaris MP, Pedersen D (1982) Impact of coal-coking effluent on sediment microbial communities: a multivariate approach. *Appl Environ Microbiol* 44:1118–1129.

Schlosser IJ (1982) Trophic structure, reproductive success and growth rate of fishes in a natural and modified headwater stream. *Can J Fish Aquat Sci* 39:968–978.

Schlosser IJ (1990) Environmental variation, life history attributes and community structure in stream fishes: implications for environmental management and assessment. *Environ Manage* 14:621–628.

Schlosser IJ (1991) Stream fish ecology: a landscape perspective. *BioScience* 41:704–712.

Sculthorpe CD (1967) *The Biology of Aquatic Vascular Plants*, Edward Arnold Ltd., London.

Seddon B (1972) Aquatic macrophytes as limnological indicators. *Freshwater Biol* 2:107–130.

Seidel K (1976) Macrophytes and water purification. In *Biological Control of Water Pollution*, Tourbier J, Pierson RW Jr., Eds., University of Pennsylvania Press, Philadelphia, PA, pp. 109–121.

Shannon CE, Weaver W (1963) *The Mathematical Theory of Communication*, University of Illinois Press, Chicago.

Sheehan PJ (1984a) Effects on community and ecosystem structure and dynamics. In *Effects of Pollutants at the Ecosystem Level*, Sheehan PJ, Miller DR, Butler GC, Bourdeau P, Eds., John Wiley & Sons, New York, pp. 51–99.

Sheehan PJ (1984b) Functional changes in the ecosystem. In *Effects of Pollutants at the Ecosystem Level*, Sheehan PJ, Miller DR, Butler GC, Bourdeau P, Eds., John Wiley & Sons, New York, pp. 101–145.

Sheldon AL (1968) Species diversity and longitudinal succession in stream fishes. *Ecology* 49:333–351.

Sigmon CF, Kania HJ, Beyers RJ (1977) Reductions in biomass and diversity resulting from exposure to mercury in artificial streams. *J Fish Res Board Can* 34:493–500.

Skujins JJ (1976) Extracellular enzymes in soil. *CRC Crit Rev Microbiol* 4:383–421.

Stevenson RJ, Lowe RL (1986) Sampling and interpretation of algal patterns for water quality assessments. In *Rationale for Sampling and Interpretation of Ecological Data in the Assessment of Freshwater Ecosystems*, Isom BG, Ed., ASTM STP 894, American Society for Testing and Materials, Philadelphia, PA, pp. 118–149.

Stream Solute Workshop (1990) Concepts and methods for assessing solute dynamics in stream ecosystems. *J N Am Benthol Soc* 9:95–119.

Sudhakar G, Jyothi B, Venkateswarlu V (1991) Metal pollution and its impact on algae in flowing waters in India. *Arch Environ Contam Toxicol* 21:556–566.

ter Braak CJF, Barendregt LG (1986) Weighted averaging of species indicator values: its efficeincy in environmental calibration. *Math Biosci* 78:57–72.

ter Braak CJF, van Dam H (1989) Inferring pH from diatoms: a comparison of old and new calibration methods. *Hydrobiologia* 178:209–223.

Thomas MLH (1978) Comparison of oiled and unoiled intertidal communities in Chedabucto Bay, Nova Scotia. *J Fish Res Board Can* 35:707–716.

Thompson P-A, Couture P (1990) Aspects of carbon metabolism in the recovery of *Selenastrum capricornutum* populations exposed to cadmium. *Aquat Toxicol* 17:1–14.

Tu CM (1981) Effects of pesticides on activities of enzymes and microorganisms in clay soil. *J Environ Sci Health* B16:179–191.

Tyler G (1976) Heavy metal pollution, phosphatase activity, and mineralization of organic phosphorus in forest soils. *Soil Biol Biochem* 8:327–332.

U.S. EPA (1978) The *Selenastrum capricornutum* Printz algal assay bottle test. EPA-600/9-78-018, U.S. Environmental Protection Agency, Corvallis, OR.

U.S. EPA (1989a) *Short-Term Methods for Estimating the Chronic Toxicity of Effluents and Receiving Waters to Freshwater Organisms*, 2nd ed., EPA/600/4-89/001, U.S. Environmental Protection Agency, Cincinnati, OH.

U.S. EPA (1989b) Rapid bioassessment protocols for use in streams and rivers: benthic macroinvertebrate and fish. EPA/444/4-89-001, U.S. Environmental Protection Agency, Cincinnati, OH.

U.S. EPA (1990) Biological criteria. In *National Program Guidance for Surface Waters*, EPA-440/5-90-004, U.S. Environmental Protection Agency, Washington, D.C.

Vannote RL, Minshall GW, Cummins KW, Sedell JR, Cushing CE (1980) The river continuum concept. *Can J Fish Aquat Sci* 37:130–137.

Van Voris P, O'Neill RV, Emmanuel WR, Shugart HH Jr. (1980) Functional complexity and ecosystem stability. *Ecology* 61:1352–1360.

Waide JB, Webster JR (1975) Engineering systems analysis: applicability to ecosystems. In *Systems Analysis and Simulation in Ecology*, Vol. 4, Patten BC, Ed., Academic Press, New York, pp. 329–371.

Wallace JB, Webster JR, Woodall WR Jr. (1977) The role of filter feeders in flowing waters. *Arch Hydrobiol* 79:506–532.

Walsh GE, Weber DE, Simon TL, Brashers LK (1991) Toxicity tests of effluents with marsh plants in water and sediments. *Environ Toxicol Chem* 10:517–525.

Wang W (1986) Phytotoxicity tests of aquatic pollutants by using common duckweed. *Environ Pollut* 11:1–14.

Wang W (1987) Toxicity of nickel to common duckweed (*Lemna minor*). *Environ Toxicol Chem* 6:961–967.

Wang W (1988) Site-specific barium toxicty to common duckweed, *Lemna minor*. *Aquat Toxicol* 12:203–212.

Wang W (1990) Literature review on duckweed toxicity testing. *Environ Res* 52:7–22.

Weber CI (1973) Biological monitoring of the aquatic environment. In *Biological Methods for the Assessment of Water Quality,* Cairns J Jr., Dickson KL, Eds., ASTM STP 528, American Society of Testing and Materials, Philadelphia, PA, pp. 46–60.

Weber CI, McFarland BH (1981) Effects of copper on the periphyton of a small calcareous stream. In *Ecological Assessments of Effluent Impacts on Communities of Indigenous Aquatic Organisms,* Bates JM, Weber CI, Eds., ASTM STP 730, American Society for Testing and Materials, Philadelphia, PA, pp. 101–131.

Webster JR (1975) Analysis of potassium and calcium dynamics in ecosystems on three southern Appalachian watersheds of contrasting vegetation. Ph.D. Dissertation. University of Georgia, Athens.

Webster JR, Benfield EF (1986) Vascular plant breakdown in freshwater ecosystems. *Annu Rev Ecol Syst* 17:567–594.

Webster JR, Patten BC (1979) Effects of watershed perturbation on stream potassium and calcium dynamics. *Ecol Monogr* 49:51–72.

Webster JR, Waide JB, Patten BC (1975) Nutrient cycling and stability in ecosystems. In *Mineral Cycling in Southeastern Ecosystems,* Howell FG, Gentry JB, Smith MH, Eds., ERDA symposium series, Department of Energy, Washington, D.C., pp. 1–27.

Webster JR, Gurtz ME, Haines JJ, Meyer JL, Swank WT, Waide JB, Wallace JB (1983) Stability of stream ecosystems. In *Stream Ecology,* Barnes JR, Minshall GW, Eds., Plenum Press, New York, pp. 355–395.

Webster JR, D'Angelo DJ, Peters GT (1991) Nitrate and phosphate uptake in streams at Coweeta Hydrologic Laboratory. *Verh Int Verein Limnol* 24:1681–1686.

Weibe WJ (1984) Some potentials for the use of microorganisms in ecological theory. In *Current Perspectives in Microbial Ecology,* Klug MJ, Reddy CA, Eds., American Society of Microbiology, Washington, D.C., pp. 17–21.

Westlake DF (1965) Some basic data for investigation of the production of aquatic macrophytes. *Mem Ist Ital Idrobiol* 18:229–248.

Winner RW, Owen HA (1991) Seasonal variability in the sensitivity of freshwater phytoplankton communities to a chronic copper stress. *Aquat Toxicol* 19:73–88.

Wong PK, Chang L (1991) Effects of copper, chromium and nickel on growth, photosynthesis and chlorophyll *a* synthesis of *Chlorella pyrenoidosa* 251. *Environ Pollut* 72:127–139.

Yant PR, Karr JR, Angermeier PL (1984) Stochasticity in stream fish communities: an alternative interpretation. *Am Nat* 124:573–582.

chapter eleven

Allium *test for screening chemicals; evaluation of cytological parameters*

Geirid Fiskesjö

Introduction

Increasing need for short-term tests

With the increasing interest in the use of cytological investigations in short-term tests (STTs) for environmental monitoring, higher plants have come into focus again. That plants could be used for the screening of chemicals and also for the prediction of human cancer was foreseen by Levan (1951). Kihlman (1966) found "a good correlation between the chromosome-breaking activities of chemicals in plant and animal cells," and in 1979, deSerres claimed that "if it is feasible to extrapolate from *Salmonella* to man, then it certainly should be feasible to extrapolate from higher plants to higher animals, perhaps with even greater confidence since they are both eucaryotic organisms," even if the different metabolism of a certain chemical may cause extrapolation problems also from plant systems (e.g., Nilan and Vig 1976; Higashi 1988). It has, however, been postulated that biotransformations are generally qualitatively similar in plant and animal systems (e.g., Menn 1978; Plewa and Gentile 1982; Sandermann 1988), and Ehrenberg (1989) summed it up: "… a trend is indicated by a growing volume of evidence for simple *in vitro* tests, giving a more reliable basis for risk estimation of cancer initiators than long-term tests with laboratory animals."

In a review on plant toxicity tests, Wang (1992) points to the sensivity of simple root elongation tests and recommends plants especially for toxicity testing of complex mixtures.

Moreover, in a collaborative study on plant test systems, Sandhu et al. (1991) report that "analysis of the Gene-Tox data base have shown that the predictive value of plant systems may be as good as any other bioassay."

The Allium *test*

The now classical test material *Allium cepa* L. (the common onion; 2n = 16) has been used for studies of the effects of chemicals on chromosomes since 1938 (Levan 1938). The present *Allium* test method has been developed during a succession of years (Fiskesjö 1982, 1985a, 1993a). With a growing need for monitoring the quantitative impact of chemicals in living organisms, the *Allium* test has been applied also to the measuring of root growth restrictions (Fiskesjö 1985b, 1988, 1993b,c,

1994, 1995). By means of series of onions, the mean values for growth in a number of treatment concentrations will make growth curves and EC (effect concentration) values obtainable; thus facilitating comparison with other toxicity test systems.

Recently, Ennever et al. (1988) have classified certain plant genotoxicity assays according to their sensitivity and specificity; here, the well-known *A. cepa* was placed among plants giving few false negative results, thus indicating the applicability in a test battery for the prediction of risks from chemicals causing genotoxic effects.

The *Allium* test is used in many laboratories, often in the present form, with macroscopical as well as microscopical studies (e.g., Liu et al. 1992; Rank and Hviid Nielsen 1993).

For a review on *Allium* in use as an environmental test system, see Grant (1982a). When agricultural chemicals were studied by various plant cytogenetics and comparisons were made between animal and plant systems (among them *Allium*), it was shown that "specifically plant systems have shown excellent correlations with mammalian systems" (Grant 1982b).

Materials and methods

Biological material

Equal-sized bulbs of the yellow onion *A. cepa* L. are chosen from commercial batches, ready for planting in the fields. These batches always consist of a population of onions and therefore exhibit some variation in growth. Thus, for the determination of growth retardation after a certain treatment, the application of a series of bulbs is necessary. (To use a cloned material would not be practicable for the large number of bulbs needed.)

The commercial onions may have been treated with herbicides; however, the use of series of onions for treatments and controls at each test occasion will compensate for possible damage from a possible herbicide pretreatment. Moreover, unscaling and rinsing of the bulbs will diminish the risk of damage caused by other factors than the intended treatment itself. Experience has shown that with the present method, very few aberrations occur in the control root tip cells (<1%).

About the chemicals tested

Over the years, various chemicals have been investigated for effects using the present *Allium* test method.

Wastewaters and complex mixtures

The use of series of onions was developed in connection with the test of water from a river where the outlet of industrial water led to the deterioration of plant growth in the surroundings (Fiskesjö 1975,

1985a), followed by tests of the chemicals involved (phenoxyacetic acids and chlorophenols) (Fiskesjö 1981a). The results from the river water tests — the shorter the roots, the nearer to the outlet of waste — were used as evidence in a lawsuit against the factory (Fiskesjö 1985a). Later, wastewater from ten industries were compared for biological damage, and seven complex mixtures were tested in five different organisms: *Salmonella*, fish (*Gasterosteus aceulatus*), Crustacea (*Nitocra spinipes*), unicellular algae (Granmo 1984) and *Allium* (Fiskesjö 1985a). In these and other tests, *A. cepa* was shown to be one of the most sensitive species showing toxicity (Fiskesjö 1985b). The detailed method for the use of the *Allium* test in wastewater monitoring has been described (Fiskesjö 1993b).

Metal salts
Two mercury compounds were selected as test agents for experiments designed to compare four test systems (Fiskesjö 1982). Both metals showed good agreement between the effects in the four systems tested (bacteria, human lymphocytes, the hamster cell line V79, and the plant *A. cepa*), as well as to humans *in vivo* (Fiskesjö 1985b; Gerstner and Huff 1977). Several metal salts have been subjected to the *Allium* test (Fiskesjö 1988), with some producing c-mitotic effects and relatively few causing chromosome breaks.

Benzo(a)pyrene and N-methyl-N-nitro-N-nitrosoguanidine
For the establishment of *A. cepa* as a standard test material, investigation of the classical test chemicals benzo(*a*)pyrene (BP) and *N*-methyl-*N*-nitro-*N*-nitrosoguanidine (MNNG) was obligatory. Data indicate that *Allium* root cells dosed with BP have a set of the important mixed function oxidase (MFO) enzymes, assuring that the *Allium* test has wide applicability. The mutagenic MNNG showed the "expected" high frequency of chromosome breaks (again good correlation to effects in other organisms) (Fiskesjö 1981b).

MEIC chemicals (MEIC — the multicenter evaluation of in vitro cytotoxicity)
The largest portion of test chemicals comprises the MEIC chemicals in a ring test started by the Scandinavian Society of Cell Toxicology for an evaluation of short-term tests against effects in humans (Bondesson et al. 1989). The chemicals on the MEIC list were chosen at random with the only qualification being that some data were available about effects in humans. About 100 laboratories are now participating. The MEIC list is comprised of 50 chemicals.

The results from *Allium* tests with the first ten MEIC chemicals have been compared to results in four other test systems (Fiskesjö and Levan 1993). Twenty-nine MEIC chemicals so far tested in the *Allium* test are marked with MEIC (Tables 1 and 2). Also, within the MEIC study, the results from *Allium* tests were in good agreement with other results.

Method

Unscaled bulbs of *A. cepa* are grown on test tubes containing 15 ml of the test liquid or of control water (tap water or nutrient solution; see Fiskesjö 1993c). Since root setting may be rather poor in some 20% of the onions, 12 should be started for a 10-bulb series, with the two poorest growing discarded after 24 h of growth. Figure 1 demonstrates an experiment with a series of ten onions in each test series with cadmium chloride ($CdCl_2$) concentrations from 10^{-3} to 10^{-5} M.

Test liquids should be changed every 24 h. After 48 h, root tips are cut and fixed, either for storage in a freezer (three parts ethanol and one part acetic acid) or for immediate slide preparation (nine parts 45% acetic acid and one part 1 M hydrochloric acid). After storage in a freezer and also by direct slide preparation, the root tips should be soaked in the 9:1 fixative for 5 min at 50°C (in test tubes in a water bath). Thereafter, the cells are squashed and stained in orcein (2% orcein in 45% acetic acid), and the cover glasses are framed with Krönigs cement.

After 72 h, the root bundles are measured (with a ruler): one measure for each of the ten onions in a series. The mean value is used for a calculation of root growth in the test series in percent of control growth. From a growth curve (growth/treatment), the EC values are obtained (EC10, 50, and 90, giving the concentrations causing 10, 50, and 90% of growth restrictions, respectively). Figure 2 shows a growth curve drawn from the results demonstrated in Figure 1. EC50, the most commonly used effective concentration, is found at the concentration 3.1×10^{-5} M Cd.

A recovery test may be included in the test. After the measurings at 72 h, five of the ten test tubes in a series are emptied and filled with control water. In the other five tubes only the evaporated amount of test liquid is replaced. After 96 h, root bundle lengths are again measured and compared. The roots growing for 24 h in control water may have recovered and increased in length, showing less difference to the control roots.

Prior to fixation, a colchicine treatment may be performed (0.1% colchicine for 2 to 3 h). This treatment will destroy the spindle and spread the chromosomes in a c-metaphase, thus facilitating the study of chromosome or chromatid breaks. However, since colchicine is a strong aneugen, effects of other aneugens will be covered, and, from the "main treatment," only clastogenic effects may be revealed.

A detailed description of the *Allium* test method for toxicity (root growth) and for genotoxicity (cytological aberrations) is published elsewhere (e.g., Fiskesjö 1985b, 1988, 1993a,c, 1994, 1995).

Table 1   C-Mitotic Effects in the *Allium* Test of a Variety of Compounds

Compound	EC50 (growth) (M)	Conc. causing highest effect of C-mitoses (M)	C-mitotic effects (%)	Micronuclei (no. in 9 root tips)	Source
NiCl₂	1.7×10^{-5}	3.3×10^{-2}	100.0	ND	Fiskesjö 1988
Digoxin (MEIC)[a]	2.0×10^{-5}	1.3×10^{-4}	99.3	ND	Fiskesjö and Levan 1993
Methyl-HgCl	9.0×10^{-7}	2.0×10^{-6}	81.9	+[b]	Fiskesjö 1979, 1988
Methanol (MEIC)	4.6×10^{-1}	1.2×10^{-0}	79.0	ND	Fiskesjö and Levan 1993
2,4,5-Trichlorophenol[c]	1.7×10^{-6}	5.0×10^{-5}	60.6	ND	Fiskesjö et al. 1981
CdCl₂	3.1×10^{-5}	3.3×10^{-4}	33.3	ND	Fiskesjö 1988
LiSO₄ (MEIC)	6.0×10^{-3}	3.3×10^{-2}	25.8	21	Fiskesjö unpublished
Diazepam (MEIC)	4.2×10^{-5}	1.7×10^{-3}	24.5	12	Fiskesjö and Levan 1993
ThSO₄ (MEIC)	6.5×10^{-8}	3.3×10^{-6}	15.9	3	Fiskesjö unpublished
MCPA[c]	1.2×10^{-7}	4.5×10^{-5}	15.8	ND	Fiskesjö et al. 1981
BeSO₄[c]	4.8×10^{-4}	1.0×10^{-3}	10.2	ND	Fiskesjö 1988
1,1,1-Trichloroethane (MEIC)	1.0×10^{-2}	2.5×10^{-2}	9.1	14	Fiskesjö unpublished
KCN (MEIC)	2.8×10^{-5}	1.5×10^{-4}	8.6	13	Fiskesjö unpublished
FeSO₄ (MEIC)	6.1×10^{-4}	1.0×10^{-3}	8.3	ND	Fiskesjö and Levan 1993
Ethylene glycol[c] (MEIC)	7.2×10^{-1}	8.9×10^{-1}	7.0	ND	Fiskesjö and Levan 1993
Propanolol-HCl (MEIC)	5.5×10^{-5}	1.7×10^{-4}	5.8	38	Fiskesjö unpublished
Thioridazine-HCl (MEIC)	2.1×10^{-6}	2.4×10^{-6}	5.0	24	Fiskesjö unpublished
Theophyllin[c] (MEIC)	3.1×10^{-4}	1.0×10^{-3}	4.8	13	Fiskesjö unpublished
Acetylsalicylic acid (MEIC)	6.3×10^{-5}	5.5×10^{-4}	4.5	ND	Fiskesjö and Levan 1993
MNNG[c]	1.2×10^{-5}	6.8×10^{-6}	4.1	++[b]	Fiskesjö 1981

Compound					Reference
MnCl$_2$[c]	5.2×10^{-3}	3.3×10^{-2}	3.1	ND	Fiskesjö 1988
CuSO$_4$ (MEIC)	2.7×10^{-6}	5.0×10^{-6}	3.1	ND	Fiskesjö 1988
Amitriptyline (MEIC)	2.1×10^{-5}	3.2×10^{-5}	2.6	ND	Fiskesjö and Levan 1993
Paraquat[c] (MEIC)	1.8×10^{-5}	4.0×10^{-5}	2.4	42	Fiskesjö unpublished
Ethanol (MEIC)	2.4×10^{-1}	3.5×10^{-1}	2.3	ND	Fiskesjö and Levan 1993
Dextropropoxyphene-HCl[c] (MEIC)	4.8×10^{-5}	5.9×10^{-5}	2.0	6	Fiskesjö unpublished
Paracetamol[c] (MEIC)	2.7×10^{-3}	1.3×10^{-3}	1.5	+[b]	Fiskesjö and Levan 1993
AsO$_3$ (MEIC)	3.4×10^{-5}	5.0×10^{-5}	1.4	3	Fiskesjö unpublished
Isopropanol (MEIC)	1.8×10^{-1}	1.3×10^{-1}	1.3	ND	Fiskesjö and Levan 1993
Benzo(*a*)pyrene[c]	3.6×10^{-5}	4.0×10^{-5}	1.3	(+)[b]	Fiskesjö 1981
Phenobarbital[c] (MEIC)	1.1×10^{-3}	9.2×10^{-4}	1.2	1	Fiskesjö unpublished
NaCl (MEIC)	6.5×10^{-2}	8.0×10^{-2}	1.0	ND	Fiskesjö unpublished
HgCl$_2$[c] (MEIC)	3.3×10^{-6}	3.3×10^{-6}	1.0	ND	Fiskesjö 1988
NaF (MEIC)	6.5×10^{-3}	1.0×10^{-2}	0.8	ND	Fiskesjö unpublished
2,4-D[c] (MEIC)	2.7×10^{-7}	4.5×10^{-8}	0.8	ND	Fiskesjö et al. 1981
2,4,5-T[c]	5.0×10^{-7}	3.9×10^{-6}	0.6	ND	Fiskesjö et al. 1981
Xylene[c] (MEIC)	7.0×10^{-4}	4.1×10^{-4}	0.3	0	Fiskesjö unpublished
Dichlorophenol[c]	4.4×10^{-4}	3.1×10^{-5}	0.2	ND	Fiskesjö et al. 1981

Note: The c-mitotic effects comprise c-mitoses + vagrant chromosomes + multipolar anaphases. For comparison, EC50 for toxicity is listed also.

[a] (MEIC): this compound is tested within the project MEIC (multicenter evaluation of *in vitro* cytotoxicity).

[b] (+): few micronuclei; +: some micronuclei; ++: many micronuclei.

[c] This compound also causes chromosome breaks; see Table 2.

Table 2 Chromosome Breaking Effects in the *Allium* Test of a Variety of Compounds

Compound	EC50 (growth) (M)	Conc. causing highest effect of breaks (M)	Chromosome fragments and/or bridges (%)	Micronuclei (no. in 9 root tips)	Source
MNNG[c]	1.2×10^{-5}	6.8×10^{-6}	24.3	++[b]	Fiskesjö 1981
Paracetamol[c] (MEIC)[a]	2.7×10^{-3}	1.3×10^{-3}	20.0	+[b]	Fiskesjö and Levan 1993
Paraquat[c] (MEIC)	1.8×10^{-5}	8.0×10^{-5}	7.7	27	Fiskesjö unpublished
BESO$_4$	4.8×10^{-4}	1.0×10^{-3}	4.1	ND	Fiskesjö 1988
Theophyllin[c] (MEIC)	3.1×10^{-4}	1.0×10^{-2}	4.0	14	Fiskesjö unpublished
2,4,5-T[c]	5.0×10^{-7}	3.9×10^{-6}	3.2	ND	Fiskesjö et al. 1981
2,4,5-Trichlorophenol[c]	1.7×10^{-6}	5.0×10^{-5}	2.5	ND	Fiskesjö et al. 1981
MCPA[c]	1.2×10^{-7}	4.5×10^{-7}	1.9	ND	Fiskesjö et al. 1981
Methyl-HgCl[c]	9.0×10^{-7}	5.0×10^{-8}	1.8	+[b]	Fiskesjö 1988
2,4-D[c] (MEIC)	2.7×10^{-7}	4.5×10^{-9}	1.8	ND	Fiskesjö et al. 1981
MnCl$_2$[c]	5.2×10^{-3}	3.3×10^{-2}	1.6	ND	Fiskesjö 1988
2,4-Dichlorophenol[c]	4.4×10^{-4}	3.1×10^{-5}	0.6	ND	Fiskesjö et al. 1981

Nicotine (MEIC)	1.5×10^{-3}	5.0×10^{-4}	0.6	2	Fiskesjö unpublished
Benzo(a)pyrene[c]	3.6×10^{-5}	4.0×10^{-5}	0.5	(+)[b]	Fiskesjö 1981
HgCl$_2$ (MEIC)	3.3×10^{-6}	1.0×10^{-6}	0.4	ND	Fiskesjö 1988
Dextropropoxyphene-HCl[c] (MEIC)	4.8×10^{-5}	5.9×10^{-5}	0.4	6	Fiskesjö unpublished
Phenobarbital[c] (MEIC)	1.1×10^{-3}	9.2×10^{-4}	0.4	1	Fiskesjö unpublished
Xylene[a] (MEIC)	7.0×10^{-4}	4.1×10^{-4}	0.3	0	Fiskesjö unpublished
Thioridazine-HCl[c] (MEIC)	2.1×10^{-6}	2.4×10^{-6}	0.3	24	Fiskesjö unpublished
Malathion (MEIC)	2.7×10^{-3}	1.5×10^{-3}	0	3	Fiskesjö unpublished

Note: Chromosome-breaking effects are recorded as fragments + fragments and bridges + bridges. For comparison, EC50 for toxicity is listed.

[a] (MEIC): this compound is tested within the project MEIC (multicenter evaluation of *in vitro* cytotoxicity).

[b] (+): few micronuclei; +: some micronuclei; ++: many micronuclei.

[c] This compound also causes c-mitotic effects; see Table 1.

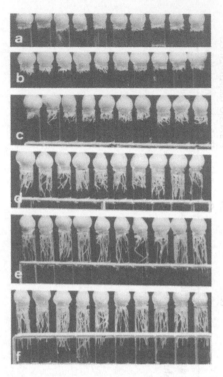

Figure 1 *Allium* test showing series of onions after 3 d of growth in concentrations of Cd (as CdCl$_2$) from 10^{-3} to 10^{-5} M: (a) 1×10^{-3} M, (b) 3×10^{-4} M, (c) 1×10^{-4} M, (d) 3×10^{-5} M, (e) 1×10^{-5} M, (f) control (tap water).

Standard parameters

Microscopical investigations are performed on three slides with each having three root tips for each concentration of a compound tested, using the standard parameters for cytological effects.

In Table 3, effects of thioridazine-HCl (a weak aneugen) treatments exemplify the standard screening with the following column captions:

- Treatment concentration.
- Root growth in percent of control (a macroscopic parameter included in Table 3 for an orientation).
- Mitotic Index (MI) obtained by observation of 1500 cells; in each slide in 500 cells distributed at three to five locations of meristem cells. (MI = number of mitoses per 1000 cells counted.)

The microscopical effects are recorded, if possible, in 300 mitotic cells (only metaphases or anaphases), preferably 100 per slide. These classified cells are assigned to the following columns in Table 3:

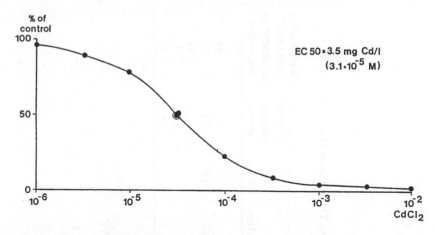

Figure 2 Growth curve of *A. cepa* roots after 3 d treatment with CdCl$_2$. From the curve, EC50 is obtained.

- Normal metaphases
- Normal anaphases
- Early anaphases (the two anaphase groups still not completely apart)
- Sticky chromosomes (a toxic, irreversible effect)
- Fragments and bridges (recorded as fragments, fragments and bridges, and bridges); since these three groups are probably all the result of chromosome breaks, they may also be counted together (an irreversible effect)
- C-mitotic effects, which consist of regular c-mitoses, vagrant (lagging) chromosomes (weak c-mitotic effect), and multipolar anaphases (also weak c-mitotic effects) (may be a reversible effect)
- Micronuclei (MNi): Since MNi may be screened in slides containing no mitoses, the three slides are searched completely (and not only, as for c-mitoses and fragments, until 300 mitoses are classified), thus giving a better base for comparison between the occurrence of MNi and of chromosome aberrations in 3 × 3 root tips from each treatment concentration; instead of the number of MNi, only the number of micronucleated cells (MNC) are recorded

Table 4 displays the results for paraquat (a clastogen) with the same standard parameters. The frequency of MNC, however, shows a correlation different from that in Table 3 (see the following discussion).

Figures 3, 4, and 5 illustrate the most important genotoxic aberrations. Figure 3 demonstrates the c-mitotic parameters: (a) multipolar anaphase, (b) anaphase with vagrant chromosomes, and (c) c-metaphase. Figure 4 shows the parameters for chromosome-breaking events: anaphase with (a) fragment, (b) fragments and bridges, and (c) only

Table 3 Standard Table for Results in the *Allium* Test After Treatments with Thioridazine-HCl

| Treatment | | Root length (% of control) | Mitotic Index (MI) | No. of classified mitoses | Microscopical effects (%) | | | | | | | Micronucleated cells (MNC) (no. of MNC in 9 root tips) |
M	mg/l				Normal meta-phases	Normal ana-phases	Early ana-phases	Sticky chromo-somes	Bridges	C-mitoses	Vagrant chromo-somes	
1.2×10^{-5}	5	13.1	0	ND[a]								
4.9×10^{-6}	2	19.5	5.3	43	(41.9)[b]	(16.3)	(20.9)	(20.9)				54
2.4×10^{-6}	1	42.3	32.0	282	41.1	28.0	16.0	9.6	0.3	1.8	3.2	24
1.2×10^{-6}	0.5	84.2	45.3	300	48.7	30.3	13.0	8.0				13
4.9×10^{-7}	0.2	104.5	43.3	300	41.0	33.3	19.0	5.0			1.7	4
Control		100	45.3	300	46.0	32.7	19.0	1.7			0.6	6

[a] ND = not done.

[b] Less than 100 mitoses recorded.

Note: There were no fragments, bridges and fragments, or multipolar anaphase results for microscopical effects.

Table 4 Standard Table for Results in the *Allium* Test After Treatments with Paraquat

Treatment M	mg/l	Root length (% of control)	Mitotic Index (MI)	No. of classified mitoses	Microscopical effects (%)								Micronucleated cells (MNC) (no. of MNC in 9 root tips)
					Normal meta-phases	Normal ana-phases	Early ana-phases	Sticky chromo-somes	Frag-ments	Bridges and frag-ments	Bridges	Vagrant chromo-somes	
8.0 × 10⁻⁴	250	33.9	4.0	94	(45.8)[a]	(28.7)	(18.1)	(7.4)	3.6	1.5	2.5		2
4.0 × 10⁻⁴	125	37.3	24.0	197	44.7	31.0	16.2	2.5	0.6	3.0	4.1		23
8.0 × 10⁻⁵	25	37.9	28.7	169	50.3	25.4	11.8	3.6		0.6		1.2	27
4.0 × 10⁻⁵	12.5	37.9	20.0	170	58.8	20.6	11.2	3.5	2.9	3.0		2.4	42
8.0 × 10⁻⁶	2.5	71.2	28.7	265	52.4	21.9	15.8	3.0	0.8		0.8	2.3	41
4.0 × 10⁻⁶	1.25	67.2	38.0	287	47.7	33.1	14.6	1.1	1.1	0.7	1.4	0.3	71
Control		100	38.7	300	60.0	26.0	13.7	0.3					0

^{a} Less than 100 mitoses recorded.

Note: There were no c-mitoses or multipolar anaphase results for microscopical effects.

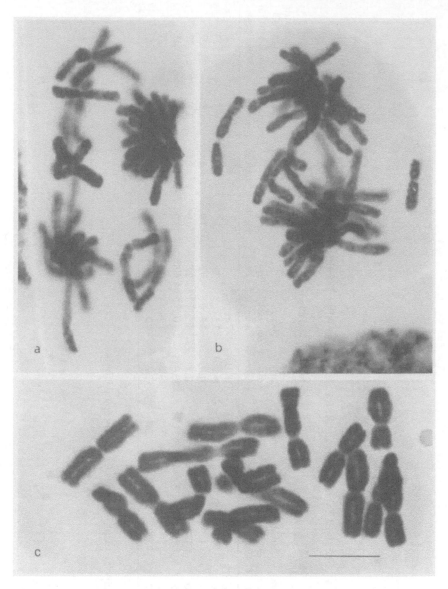

Figure 3 C-mitotic parameters for microscopic screening in the *Allium* test: (a) multipolar anaphase, (b) vagrant chromosomes, (c) c-mitosis (c-metaphase). (a,b) Treatment for 2 d with $5 \times 10^{-7} M$ Hg (as methylmercury chloride) + 6 × $10^{-6} M$ Se (as sodium selenite). (c) Treatment for 2 d with $2 \times 10^{-6} M$ Hg + 2.5 × $10^{-5} M$ Se. Scale = 10 μm.

bridges visible. Figure 5 gives various examples of micronuclei: (a, b) cells with one or two MNi, respectively; (c) a multinucleated cell (the result of a multipolar anaphase); and (d) extracellular micronuclei (ECMN), a parameter discussed but not recorded in the present *Allium* tests.

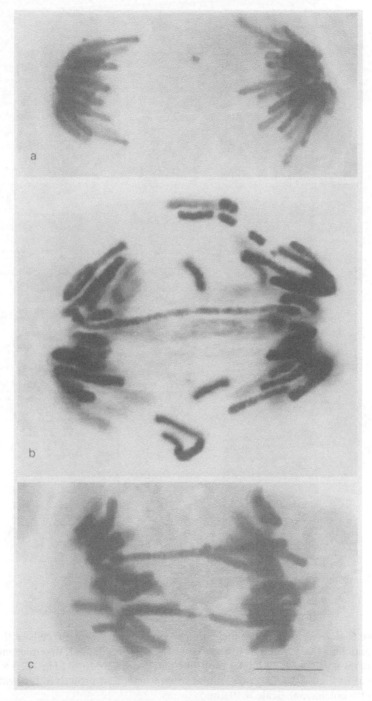

Figure 4 Cytological parameters for chromosome-breaking events in the *Allium* test: (a) fragment, (b) fragments and bridges, (c) bridges. (a, c) Treatment for 2 d with river water contaminated with waste of phenoxyacetic acids and chlorophenols. (b) Treatment for 2 d with paraquat 4×10^{-6} *M*. Scale = 10 μm.

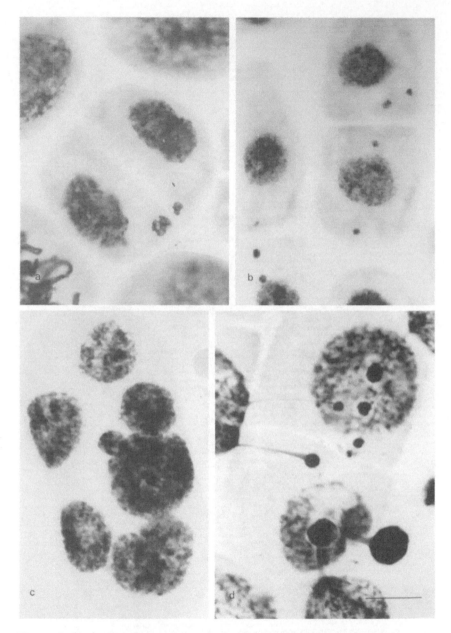

Figure 5 Examples of micronuclei in root tip cells of *A. cepa*: (a) cells with one micronucleus (river water as in Figures 4a,c; 2 d); (b) cells with two micronuclei (paraquat 4×10^{-6} *M*; 2 d); (c) a multinucleated cell (2×10^{-6} *M* Hg + 2.5×10^{-5} *M* Se as in Figure 3c); (d) extracellular micronuclei (2 d growth in tap water + 2 h 0.1% colchicine). Scale = 10 μm.

Results and observations

Among the chemicals tested in the *Allium* test system for toxicity and genotoxicity, 40 are represented in Table 1 (aneugens; chemicals causing c-mitosis) and in Table 2 (clastogens; chemicals causing chromosome breaks). Some of the chemicals tested (marked with a table footnote) induce both these types of genotoxic events, but mostly a compound clearly gives more of one type than of the other type of effect.

In Table 1, aneugens are listed after their c-mitotic efficiency, and Table 2 similarly displays a list of clastogens.

Two concentrations are given for each compound: the EC50 value gives an orientation of the common toxicity, and the "cone causing the highest effect" states the concentration level for the observation of the highest level of c-mitotic events (Table 1) and of chromosome breakage (Table 2). In most cases, this highest effect concentration happens to be the same in both tables for the compounds causing both types of events, with the exception of only four compounds (paraquat, theophyllin, 2,4-D, MCPA). For the following discussion, the occurrence of MNi has been listed (when available; thus, only for a limited number of compounds). Finally, an author reference is given.

The figures for effect are taken from a complete standard table for each compound. For instance, from Table 3, the highest number of c-mitotic events, 1.8 + 3.2% (c-mitoses + vagrant chromosomes), are found at the concentration $2.4 \times 10^{-6} M$, and the sum, 5.0%, is transferred to Table 1 in the column for c-mitotic effects (%). Similarly, in Table 4, the highest frequency of chromosome-breaking effects is found in the concentration $8.0 \times 10^{-5} M$: 0.6% fragments + 3.0% bridges and fragments + 4.1% bridges make 7.7%, which is transferred to the column for chromosome fragments and/or bridges (%) for paraquat in Table 2.

Discussion

Comparison of toxicity and genotoxicity

Chemical compounds display a large variety in the degree of toxicity, which appears clearly from the results presented in Tables 1 and 2, where the EC50 values give information on general toxicity, thus providing a possibility for comparison between compounds.

However, since toxicity is not always strictly correlated to genotoxicity, it is important to distinguish between the two types of toxicity. In the *Allium* test, there usually seems to be a certain correspondence between growth retardation (toxicity) and certain chromosomal deviations (genotoxicity). When chromosome aberrations occur, there are almost always some growth restriction. For example, in Tables 1 and 2, the concentrations for highest effect are most often found to be somewhat higher than the corresponding EC50 value; however, seven of the aneugens and eight of the clastogens tested have a highest effect of genotoxic damage below the EC50 value of the chemical. This shows that the commonly used EC50 for toxicity is a parameter useful only for comparison between compounds and only concerning toxicity; the EC50 values cannot be directly used for information on genotoxicity. Moreover, not even an EC10 for toxicity will be sufficient for the estimation of risks of genotoxic events, especially not the risks of low doses.

In the following, some examples will be given of divergence between the two types of toxicity.

Low toxicity, high level of c-mitoses, high chromosome-breaking capacity

Certain compounds have relatively low toxicity, but nevertheless high genotoxicity, such as $NiCl_2$ and $LiSO_4$ (Table 1) or paracetamol and theophyllin (Table 2). The last two compounds are of importance in medicine and are therefore of special interest. Paracetamol, for instance, is a common analgesic. It has a lower toxicity in the *Allium* test than another widely used analgesic, acetylsalicylic acid, but it is paracetamol that causes chromosome breaks. In *Allium*, these genotoxic aberrations occur in concentrations between 1.0×10^{-4} and 1.0×10^{-3} M (0.02 to 0.2 g/l) paracetamol, and similar doses (milligrams per liter as milligrams per kilogram body weight) are reported to cause chromosome breaks in humans, *in vitro* as well as *in vivo* (Hongslo et al. 1991). It may be mentioned here, that FASS (a handbook on pharmaceutical specialties in Sweden; 1992) states that no risks are known with the use of paracetamol during pregnancy. Since chromosome studies are not routine for medical compounds, however, there may be a hidden risk with paracetamol and other medicines causing chromosome breaks at low doses.

High toxicity, high level of c-mitoses, low chromosome-breaking capacity

When methylmercury (methyl-HgCl or MMC) was investigated in four different test systems, toxicity effects were consistent for the four widely different materials (bacteria, human lymphocytes, the mammalian cell line V79, and *Allium*) (Fiskesjö 1982). In two of the systems where cytological investigations were performed, *Allium* and human lymphocytes, a similar high degree of c-mitotic effects was observed after MMC treatment. The c-mitotic effects, however, did not correspond to the mutagenicity in the V79 cells. Instead, low mutagenicity in these mammalian cells were correlated to a low chromosome-breaking effect observed in the *Allium* cells, also a genotoxic correlation. Given this, it appears that for a full genotoxic analysis both chromosome breaks and c-mitotic events should be recorded, and both are observable in the *Allium* test.

High toxicity, relatively low genotoxicity

As examples of extremely toxic compounds with relatively low geno-toxicity, the phenoxyacetic acids 2,4,5-T (with a highest effect of 3.2% chromosome breaks after treatment in $3.9 \times 10^{-6} M$) and 2,4-D (with a highest effect of 1.8% breaks in $4.5 \times 10^{-9} M$) (Tables 1 and 2) may also be mentioned. The toxicity EC50 values are about the same for both compounds ($5.0 \times 10^{-7} M$ and $2.7 \times 10^{-7} M$, respectively). From these observations, it does not seem reasonable that 2,4,5-T is prohibited, while 2,4-D, which is toxic and genotoxic in much lower doses than 2,4,5-T, is still in use.

These phenoxyacetic acids also cause a certain low frequency of c-mitoses (2,4,5-T: 0.6% in $3.9 \times 10^{-6} M$; 2,4-D: 0.8% in $4.5 \times 10^{-8} M$). However, since chromosome breaks or c-mitotic events may occur up to 1% in control material, it seems that damage frequencies below 1% cannot be considered reliable. Therefore, when a clearer result is needed, an investigation comprising a higher number of cells can probably exhibit a better distinction between low doses and control.

The occurrence of genotoxic aberrations in very low doses also points to the uncertainty of limit values. Variations in time of treatment, sensitivity during the cell cycle, temperature, changes in the form of a chemical, and other conditions may make a suggested limit not reliable.

Clastogens and aneugens

In spite of a comprehensive debate on the predictive value of short-term tests (Mendelsohn 1982; Tennant et al. 1987; Young 1988; Parodi et al. 1988; Clive 1988; Waters et al. 1988), an increasing optimism may be discerned in the numerous reports in this field.

The correlations usually discussed are mainly those between clastogens and cancer; however, by this restriction to certain endpoints one may miss the effects of other mechanisms (e.g., Omenn and Lave 1988). In most test systems, chromosome breaks have long been the main endpoint; the aneugens have not been taken into consideration. For instance, when the frequency of chromosome aberrations is compared to the occurrence of cancer, the endpoint "chromosome aberrations" means direct DNA damage (e.g., Ishidate 1988; Savage 1976; and others). However, in a battery of test systems, the importance of various endpoints has been emphasized (e.g., Brusick 1988; Gatehouse 1990).

Liang (1983) states that there is generally good agreement between clastogens and carcinogens, but "since many cancer cells exhibit abnormal chromosome numbers (aneuploidy) and since several human syndromes (e.g. Down's, Turner's, Klinefelter's) reveal aneuploid constitution, mitotic poisons are one group of environmental agents deserving extensive studies."

In a review on plant test systems, Sharma (1990) displays a list of various markers in plant test systems, including clastogenic *and* aneugenic endpoints.

Parry and Parry (1989), dealing specifically with aneugens, suggest that "such aneuploidy-producing chemicals may represent a significant genetic and/or carcinogenic risk to the human population, and their detection and evaluation requires the development and use of appropriate test systems."

It is now known that all carcinogens are not mutagens or clastogens. Could the aneugens, being a majority among chemicals, constitute part of a missing gap for better correlation of genotoxicity/carcinogenicity?

In the *Allium* test, clastogens as well as aneugens may be detected and recorded in the microscopic screening, and, having a well-documented good correspondence to other test organisms, the *A. cepa* material may well be justified for a place in a test battery for routine screening of biological effects of chemicals.

Micronuclei as a test parameter

The occurrence of micronuclei (MNi) has been used as a single parameter for the recording of genotoxic damage (e.g., Schmid 1975; Fenech 1990; Ma 1990; Ma et al. 1992).

For the estimation of genotoxic damage as a whole, the recording of MNi will suffice. Of importance then is only if MNi occur or not, thus revealing a total of chromosomal aberrations leading to MNi. However, to obtain information about the type of genotoxic damage caused by a certain chemical, a detailed screening of various cytological parameters is necessary, such as chromosome-breaking or c-mitotic events. The screening of MNi is said to be a more rapid method than the screening of chromosomal aberrations. Of course, a certain training is necessary, but, being relatively easy to learn, a more detailed screening will give much more information with about the same efforts.

In the present *Allium* test, MNi have not been recorded until recently, mainly because the cytological conditions provide very good opportunities for chromosome studies. Also, in exceptional cases, the *Allium* material has a varying content of MNi, making the actual batch of onion bulbs less useful for MNi screening. MNi may also occur, although seldom, without any observations of other genotoxic damage. Of course, a few MNi may be the result of earlier herbicide treatment. As an example of an occurrence of MNi, without any breaks or c-mitoses in the same material, the results from malathion treatment are listed at the bottom of Table 2.

Micronuclei correlated to other parameters

In *Allium*, a positive correlation is most often found between treatment dose and chromosome aberrations (c-mitoses as well as breaks); however, the MNi correlations vary.

Positive correlation of MNi to dose and breaks

The positive MNi/dose/breaks correlation demonstrated for a clastogen in Table 4 has been found after microwave treatment of human lymphocytes (Garaj-Vrhovac et al. 1992) and V79 hamster cells (Garaj-Vrhovac et al. 1991) and after ethylene oxide treatment of V79 cells (Zhong et al. 1991). Surralles et al. (1993) observed a nonuniform response for MNi in human lymphocytes for the two clastogens investigated, ethyl methane-sulfonate and mitomycin). Gómez-Arroyo et al. (1986) found different responses for MNi after a number of chemical treatments in *Vicia faba*.

Moreover, Garaj-Vrhovac et al. (1990) found differences in the clastogenic response of a chemical and a physical agent, hinting to a differentiated classification of clastogenic effects and their specific correlations.

Negative correlation of MNi to dose and c-mitoses

The reversed correlation between MNi and dose, as shown for a weak aneugen in Table 3, has been found for other aneugens: in water hyacinth after Cd treatment (Rosas et al. 1984) and in human lymphocytes after colchicine and vincristine treatments (Surralles et al. 1993). However, Panda et al. (1989) reported a positive correlation for MNi/dose/c-mitoses in *Allium* root tips.

Further classification of MNi

These few examples hint to a certain similarity of response in different test systems: positive correlation of MNi/dose for clastogens and the reversed correlation between MNi/dose for certain aneugens (perhaps aneugens should also be classified further). Explanations of the varying results in MNi correlations may be found in the findings that spontaneous elevation of MNi is caused primarily by increased chromosome breakage and only in a few instances by mitotic poisons (Kratochvil et al. 1991). Another plausible explanation is that the establishment of MNi may vary throughout the cell cycle (Xue et al. 1986a). However, more research is needed to understand these relationships. Specific methods, either identifying the centromeres in MNi (Fenech and Morley 1989; Norppa et al. 1993; Ellard et al. 1993; Ventura et al. 1993) or the cytokinesis block method, studying only one cell generation (Fenech 1990), may improve the results and the possibility for evaluation of the relationships between chromosome aberrations and the occurrence of MNi.

MNi variations in Allium

In Tables 3 and 4, the occurrence of MNi in *Allium* root tip cells may be compared to the frequency of chromosome breaks and c-mitosis in the same treatments.

In Table 3, where results from thioridazine-HCl treatment serve as an example of a (weak) c-mitotic compound, the occurrence of MNi correlated to the treatment doses and also weakly to the few c-mitotic events. A reason for this may be connected with the delay in the mitotic cycle caused by a c-mitotic agent by accumulating the metaphases (as c-metaphases), as the MNi are accumulated in interphase cells.

In Table 4, where paraquat states an example of a chromosome-breaking compound, the number of MNi seems to be correlated to the mitotic index, and therefore also to the number of mitoses possible to classify, instead of to the frequency of chromosome-breaking events. The reason for this may be that MNi are accumulated in the interphases during the 2-d treatment, whereas in dividing cells the number of breaks will represent only the effect in one cell generation.

The results in Tables 3 and 4 show an obvious uncertainty in conclusions drawn from the frequency of MNi recorded in the present *Allium* test. However, occurrence of MNi above the control level clearly demonstrates a genotoxic effect and is therefore valuable per se. Thus, to make a reasonable hypothesis for the variation in the occurrence of MNi in *Allium* root tip cells, more experiments have to be done. For comparison to other standard tests using micronuclei as a parameter, it is suggested that screening of MNi be included in the standard routine for the *Allium* test.

Extracellular micronuclei (ECMN)

Certain forms of MNi have a specific appearance; they seem to have been extruded from the main nucleus and are connected to the nucleus by thin threads, possibly consisting of elongated nuclear membrane material. Such MNi may be located either in the same cell as the main nucleus or, as is most often the case, in cytoplasm of another cell. This may be the result of a second cell division after a first misdivision, but even extrusions from one cell into two or more other cells seem to occur (Figure 5d), thus hinting at a more complicated explanation for these specific MNi. This phenomenon has been discussed only rarely in the literature. Vaarama (1941) named similar observations in pollen mother cells of *Sagittaria* "cytomixis." Doronin (1986) observed extrusions of nucleoli-like structures from the nuclei in cells of mouse embryos, and Xue et al. (1986b) observed extracellular MNi (ECMN) in human lymphocytes, suggesting that these ECMN constitute a pathological phenomenon.

However, since in *Allium*, cytomixis or ECMN has been observed also in control root tips (possibly caused by herbicide treatment), the

MNi, observed as extrusions connected to the main nucleus, are not included in the recording of MNi in the present *Allium* tests.

Summary

The *Allium* test has many advantages: (1) the good cytological conditions in *Allium* root tip cells make the observations of various genotoxic parameters possible; (2) by the application of a series of onions in each test concentration, toxicity (EC50 for mean root growth) may be obtained within the same test occasion; (3) good correlation has been shown to other test systems; and (4) in addition to normal laboratory equipment, only a microscope with a camera is needed.

Genotoxicity, as well as toxicity, should be investigated. Since a high level of genotoxic damage has been found in spite of a relatively low toxicity (e.g., paracetamol), the setting of limit values will otherwise be obscured. Also, a chemical may cause both aneugenic and clastogenic effects, and some (e.g., $NiCl_2$ or methyl-HgCl) are weak clastogens but very strong aneugens, showing the importance of screening for both these types of genotoxic parameters. Moreover, of the 40 chemicals submitted to the *Allium* test, the highest proportion (33 compounds) causes c-mitotic effects (in 1% or higher), while only 11 compounds are chromosome-breaking compounds (also in 1% or higher, since aberrations may occur in control material to a degree below 1%). The higher frequency of aneugens among these randomly chosen chemicals further shows the importance of screening both clastogens *and* aneugens.

Micronuclei (MNi) will be produced by both aneugens and clastogens; however, neither the correlations between the MNi and mitotic aberrations nor the relationship between any of these parameters and carcinogenicity are unambiguous or fully understood. Therefore, a continued screening of both c-mitotic events as well as of micronuclei is recommended. Root tip cells of *A. cepa* L. (2n = 16) provide excellent conditions for these studies.

The *Allium* test has a wide applicability in environmental monitoring, as well as in screening, for the biological effects of chemical compounds.

References

Bondesson I, Ekwall B, Hellberg S, Romert L, Stenberg K, Walum E (1989) MEIC — a new international multicenter project to evaluate the relevance to human toxicity of in vitro cytotoxicity tests. *Cell Biol Toxicol* 5(3):311–347.

Brusick D (1988) Evolution of testing strategies for genetic toxicity. *Mutat Res* 205:69–78.

Clive D (1988) Genetic toxicology: can we design predictive in vivo assays? *Mutat Res* 205:313–380.

deSerres FJ (1979) Higher plant systems as monitors of environmental mu-
    tagens. In *Application of Short-Term Bioassays in the Fractionation of Com-*
    plex Environmental Mixtures, Waters MD, Nesnow S, Huisingh JL,
    Sandhu SS, Claxton L, Eds., Plenum Publishing Corporation, New York,
    pp. 101–109.
Doronin Yu K (1986) Chromatin diminution, abnormality of mitoses, and cel-
    lular death in early mouse embryos. *Tsitol Genet* 28(5):405–500.
Ehrenberg L (1989) Risk assessment of genotoxic agents: problems and possible
    solution. In *New Trends in Genetic Risk Assessment*, Jolles G, Cordier A,
    Eds., Academic Press, London, pp. 433–448.
Ellard S, Parry EM, Parry JM (1993) The induction and analysis of micronuclei
    in mammalian cell cultures and human peripheral lymphocytes. *Mutat*
    Res Compl 291(3):231.
Ennever FK, Andreano G, Rosenkranz HS (1988) The ability of plant genotox-
    icity to predict carcinogenicity. *Mutat Res* 205:99–105.
FASS (1992) Farmacevtiska specialiteter i Sverige. LINFO, Stockholm, Sweden.
Fenech M (1990) The cytokinesis-block micronucleus assay in nucleated cells.
    In *Mutation and the Environment*, Part B, Mendelsohn ML, Albertini RJ,
    Eds., Wiley-Liss, New York, pp. 195–206.
Fenech M, Morley AA (1989) Kinetochore detection in micronuclei: an alterna-
    tive method for measuring chromosome loss. *Mutagenesis* 4(2):98–104.
Fiskesjö G (1975) Rapport om ett biologiskt test (Alliumtest) utfört med vatten
    från Braån och Höje å i Skåne. *Vatten* 4:304–316.
Fiskesjö G (1981a) Chlorophenoxyacetic acids and chlorophenols in the mod-
    ified Allium test. *Chem Biol Interact* 34:333–344.
Fiskesjö G (1981b) Benzo(a)pyrene and N-methyl-N-nitro-N-nitrosoguanidin
    in the Allium test. *Hereditas* 95:155–162.
Fiskesjö G (1982) Evaluation of short-term tests for toxicity and mutagenicity
    with special reference to mercury and selenium. Ph.D. Thesis. Depart-
    ment of Genetics, University of Lund, Sweden.
Fiskesjö G (1985a) Allium test or river water from Braån and Saxån before and
    after closure of a chemical factory. *Ambio* 14(2):99–103.
Fiskesjö G (1985b) The Allium test as a standard in environmental monitoring.
    Hereditas 102:99–112.
Fiskesjö G (1988) The Allium test — an alternative in environmental studies:
    the relative toxicity of metal ions. *Mutat Res* 197(2):243–260.
Fiskesjö G (1993a) The Allium test — a potential standard for the assessment
    of environmental toxicity. In *Environmental Toxicology and Risk Assess-*
    ment, Vol. 2, ASTM STP 1216, Gorsuch JW, Dwyer FJ, Ingersoll CG,
    LaPoint TW, Eds., American Society for Testing and Materials, Phila-
    delphia, PA.
Fiskesjö G (1993b) The Allium test in wastewater monitoring. *Environ Toxicol*
    Water Qual 8:291–298.
Fiskesjö G (1993c) Allium test I. A 2-3 days plant test for toxicity assessment
    of various chemicals by measuring the mean root growth of a series of
    onions (*Allium cepa* L.). *Environ Toxicol Water Qual* 8:461–470.
Fiskesjö G (1994) Allium test II. Assessment of a chemical's genotoxicity po-
    tential by recording aberrations in chromosomes and cell divisions in
    root tips of *Allium cepa* L. *Environ Toxicol Water Qual* 9:235–241.

Fiskesjö G (1995) *Allium* test. In *Methods in Molecular Biology, Vol. 43: In Vitro Toxicity Testing Protocols,* O'Hare S, Atterwill CK, Eds., Humana Press, Inc., Totowa, NJ.

Fiskesjö G, Levan A (1993) Evaluation of the first ten MEIC chemicals in the *Allium* test. *ATLA* 21:139–149.

Garaj-Vrhovac V, Fučić A, Horvat D (1990) Comparison of chromosome aberration and micronucleus induction in human lymphocytes after occupational exposure to vinyl chloride monomor and microwave radiation. *Period Biol* 92(4):411–416.

Garaj-Vrhovac V, Horvat D, Koren Z (1991) The relationship between colony-forming ability, chromosome aberrations and incidence of micronuclei in V79 chinese hamster cells exposed to microwave radiation. *Mutat Res* 263(3):143–150.

Garaj-Vrhovac V, Fučić A, Horvat D (1992) The correlation between the frequency of micronuclei and specific chromosome aberrations in human lymphocytes exposed to microwave radiation in vitro. *Mutat Res* 281:181–186.

Gatehouse DG (1990) An industrial and UK perspective on short-term testing. In *Mutation and the Environment,* Part D, Mendelsohn ML, Albertini RJ, Eds., Wiley-Liss, New York, pp. 249–259.

Gerstner HB, Huff JF (1977) Clinical toxicology of mercury. *J Toxicol Environ Health* 2:491–526.

Gómez-Arroyo S, Castillo-Ruíz P, Villalobos-Pietrini R (1986) Chromosomal alterations induced in *Vicia faba* by different industrial solvents: thinner, toluene, benzen, n-hexane, n-heptane and ethyl acetate. *Cytologia* 51:133–142.

Granmo Å (1984) Biological tests from waste waters from Stenungsund. Report to the National Swedish Environmental Protection Board SNV PM 1845 (Swedish, English summary), Stockholm, Sweden, 134 pp.

Grant WF (1982a) Chromosome aberrations assays in Allium. A report of the U.S. Environmental Protection Agency Gene-Tox Program. *Mutat Res* 99:273–291.

Grant WF (1982b) Cytogenetic studies of agricultural chemicals in plants. In *Genetic Toxicology. An Agricultural Perspective,* Fleck RA, Hollaender A, Eds., Plenum Press, New York, pp. 353–378.

Higashi K (1988) Metabolic activation of environmental chemicals by microsomal enzymes of higher plants. *Mutat Res* 197:273–288.

Hongslo JK, Brögger A, Björge C, Holme JA (1991) Increased frequency of sister-chromatid exchange and chromatid breaks in lymphocytes after treatment of human volunteers with therapeutic doses of paracetamol. *Mutat Res* 261:1–8.

Ishidate M Jr. (1988) A proposed battery of tests for the initial evaluation of the mutagenic potential of medicinal and industrial chemicals. *Mutat Res* 205:397–407.

Kihlman BA (1966) *Actions of Chemicals on Dividing Cells,* Prentice-Hall, Englewood Cliffs, NJ.

Kratochvil M, Ruemmelein B, Reimers U, Ehlert U, Weichenthal M, Mensing H, Breitbart EW, Ruediger HW (1991) Constitutively increased micronuclei are predominantly caused by acentric fragments. *Mutat Res* 249(1):223–226.

Levan A (1938) The effect of colchicine on root mitoses in Allium. *Hereditas* 29:381–443.

Levan A (1951) Chemically induced chromosome reactions in *Allium cepa* and *Vicia faba*. *Cold Spring Harbor Symp Quant Biol* 16:233–243.

Liang JC (1983) Cytogenetics and public health — assays for environmental mutagens. *Cancer Bull* 35(3):138–143.

Liu DH, Jiang WS, Li MX (1992) Effects of trivalent and hexavalent chromium on root growth and cell division of *Allium cepa*. *Hereditas* 117:23–29.

Ma T-H (1990) Tradescantia-micronucleus test on clastogens and in situ monitoring. In *Mutation and the Environment*, Part E, Mendelsohn ML, Albertini RJ, Eds., Wiley-Liss, New York, pp. 83–90.

Ma T-H, Xu J, Xia W, Jong X, Sun W, Lin G (1992) Proficiency of the Tradescantia-micronucleus image analysis system for scoring micronucleus frequencies and data analysis. *Mutat Res* 270:39–44.

Mendelsohn ML (1982) Extrapolation of mutagenicity testing to the human. In *Genetic Toxicology. An Agricultural Perspective*, Flack RA, Hollaender A, Eds., Plenum Press, New York, pp. 113–116.

Menn JJ (1978) Comparative aspects of pesticide metabolism in plants and animals. *Environ Health Perspect* 27:113–124.

Nilan RA, Vig BK (1976) Plant test systems for detection of chemical mutagens. In *Chemical Mutagens. Principles and Methods for Their Detection*, Vol. 4, Hollaender A, Ed., Plenum Press, New York, pp. 143–170.

Norppa H, Renzi L, Lindholm C, Autio K (1993) Micronuclei containing whole chromosomes in cytokinesis-blocked human lymphocytes. *Mutat Res Compl* 291(3):260.

Omenn GS, Lave LB (1988) Scientific and cost-effectiveness criteria in selecting batteries for short-term tests. *Mutat Res* 205:41–49.

Panda KK, Lenka M, Panda BB (1989) Allium micronucleus (MNC) assay to assess bioavailability, bioconcentration and genotoxicity of mercury from solid waste deposits of a chloralkali plant, and antagonism of L-cysteine. *Sci Total Environ* 79:25–36.

Parodi S, Taningher M, Santi L (1988) Utilization of the quantitative component of positive and negative results of short-term tests. *Mutat Res* 205:283–294.

Parry JM, Parry EM (1989) Induced chromosome aneuploidy: its role in the assessment of the genetic toxicology of environmental chemicals. In *New Trends in Genetic Risk Assessment*, Jolles G, Cordier A, Eds., Academic Press, London, pp. 261–298.

Plewa MJ, Gentile JM (1982) The activation of chemicals into mutagens by green plants. In *Chemical Mutagens. Principles and Methods for Their Detection*, Vol. 7, deSerres FJ, Hollaender A, Eds., Plenum Press, New York, pp. 401–420.

Rank J, Hviid Nielsen M (1993) A modified Allium test as a tool in the screening of the genotoxicity of complex mixtures. *Hereditas* 118:49–53.

Rosas I, Carbajal ME, Gómez-Arroyo S, Belmont R, Villabolos-Pietrini R (1984) Cytogenetic effects of cadmium accumulation on water hyacinth (*Eichhornia crassipes*). *Environ Res* 33:386–395.

Sandermann H Jr. (1988) Mutagenic activation of xenobiotics by plant enzymes. *Mutat Res* 197:183–194.

Sandhu SS, deSerres FJ, Gopalan HNB, Grant WF, Veleminsky J, Becking GC (1991) Status report of the International Programme on Chemical Safety's Collaborative Study on plant test systems. *Mutat Res* 257:19–25.

Savage JRK (1976) Classification and relationships of induced chromosomal structural changes. *J Med Genet* 13:103–122.

Schmid W (1975) The micronucleus test. *Mutat Res* 31:9–15.

Sharma CBSR (1990) In situ detection of genotoxicity — possibilities and measurements. In *Mutation and the Environment*, Part E, Wiley-Liss, New York, pp. 67–76.

Surralles J, Carbonell E, Xamena N, Creus A, Marcos R, Antoccia A, Degrassi F, Tanzarella C (1993) Influence of cytochalasin-B concentration on the micronucleus frequency in cultured human lymphocytes. *Mutat Res Compl* 291:277.

Tennant RW, Margolin BH, Shelby MD, Zeiger Z, Haseman JK, Spalding J, Caspary W, Resnick M, Stasiewicz S, Anderson B, Minor R (1987) Prediction of chemical carcinogenicity in rodents from in vitro genetic toxicity assays. *Science* 236:933–941.

Vaarama A (1941) Beocachtungen über die cytomixis in Meiotischen pollenmutterzellen von *Sagittaria natans pall. Ann Acad Sci Fenn Ser* A4:1–20.

Wang W (1992) Use of plants for the assessment of environmental contaminants. *Rev Environ Contam Toxicol* 126:87–127.

Waters MD, Bergman HB, Nesnow S (1988) The genetic toxicology of Gene-Tox non-carcinogens. *Mutat Res* 205:139–182.

Ventura L, Renzi L, Russo A, Lewis AG (1993) An approach to the micronucleus assay in Syrian hamster embryo cells by using cytochalasin B and CREST immunostaining. *Mutat Res Compl* 291:282–283.

Xue K, Ji G, Sun Y, Zhou P (1986a) Preliminary studies on the relationship between micronucleus formation and cell cycle. I. Effects of various store and culture intervals of irradiated human whole blood on micronucleus frequency. *Acta Genet Sin* 13(5):397–402.

Xue K, Sun Y, Zhou P, Ji G (1986b) Preliminary studies on the relationship between micronucleus formation and cell cycle II. Studies on micronucleus formation by autoradiography and microscopy. *Acta Genet Sin* 13(6):460–463.

Young SS (1988) Do short-term tests predict carcinogenicity? *Science* 241:1232–1233.

Zhong BZ, Whong WZ, Wallace WE, Ong TM (1991) Comparative study of micronucleus assay and chromosomal aberration analysis in V79 cells exposed to ethylene oxide. *Teratog Carcinog Mutagen* 11(5):227–233.

chapter twelve

The use of vascular plants as "field" biomonitors

Rebecca L. Powell

Introduction

Ecosystems are continually exposed to natural forces such as freezing, fire, flooding, etc., and, as a result, are in a constant state of flux (Sheehan and Loucks 1994; Harwell et al. 1994). When the frequency or nature of these disturbances are altered or anthropogenic stressors such as xenobiotic chemicals or nutrients are introduced, the ecological system can be irreversibly changed (Harwell et al. 1994).

In an effort to better understand the complex nature of ecosystems, the U.S. Environmental Protection Agency (EPA) (1992) recently published a framework for evaluating the ecological effects that anthropogenic stressors may have on the environment. A major component of this risk assessment process revolves around exposure of the biotic entities to chemical and nonchemical stressors. Exposure is regulated not so much by the amount of contamination present, but more importantly by the bioavailability of the compound and the duration in which the organism is in contact with the contaminant. Current analytical methods and laboratory tests, while useful, are not designed to mimic environmental conditions, organism conditions, exposure conditions, or, more importantly, the biological interactions that occur in the field (Sheehan and Loucks 1994).

Analytical determination of chemical toxins is the most direct method for assessing the presence of toxins. These methods, however, fail to provide critical information regarding the relationship between the biotic and abiotic components of the environment (Kovacs and Podani 1986). Analytical methods do not take into account physiochemical parameters such as temperature, pH, and water hardness that may influence the bioavailability of the compound (Cairns and van Der Schalie 1980; Lambou and Williams 1980; Lovett Doust et al. 1994b), nor do they take into account synergistic or antagonistic effects (Lambou and Williams 1980; Kovacs and Podani 1986). Analytical methodologies may also be a limiting factor when substances are toxic below the detection limits or released sporadically or in a pulsed fashion and rapidly dispersed (Cairns and van Der Schalie 1980). Analytical methodology may also be inadequate for dealing with mixtures or effluents when a complete list of pollutants is unavailable (Lambou and Williams 1980; Cairns and van Der Schalie 1980). Finally, the cost of continuous analytical monitoring would be prohibitive (Lambou and Williams 1980).

Laboratory-based, single-species toxicity tests for single toxicants and chemical mixtures provide standardized methods for estimating toxicity to selected organisms. These tests, however, do have their limitations. Results from these types of tests are dependent on several variables, including the test conditions which may vary significantly from natural systems (Lovett Doust et al. 1993; Wang and Freemark 1995). Laboratory tests are conducted under favorable conditions for the test organisms. They are frequently short term (<96 h) and may consist of a single pulse, unlike the more erratic exposures in the field (Sheehan and Loucks 1994). Chronic or sublethal effects are often ignored, as well as interactions among individuals and other species (Kimball and Levin 1985). Pulsed releases which disperse rapidly may not be identified (Lovett Doust et al. 1994b). Finally, only a limited number of species have been successfully cultured in the laboratory and are recommended for use in toxicity tests (Swanson et al. 1991).

Debate is still strong about extrapolating results from laboratory-based tests to the field (Lambou and Williams 1980; Kimball and Levin 1985; Cairns and Niederlehner 1987; Fletcher 1991; Sheehan and Loucks 1994; Chapman 1995). Contradictory results often occur between laboratory and field tests, with laboratory tests generally being more conservative. One proposed theory for this discrepancy is that plants, in their native environment, can become acclimated to sublethal concentrations of toxicants, thereby reducing their sensitivity (Wang and Freemark 1995). Garrod (1989) also lists numerous environmental (quality and quantity of light, temperature, wind) and plant factors (growth stage, genotype) which may be responsible for the variation. It is recommended that single-species toxicity tests be used to compare relative toxicity for establishing national pollutant criteria, not for making direct comparisons with complex systems (Kimball and Levin 1985; Fleming et al. 1991).

Alternate methods are necessary to understand how organisms respond to anthropogenic stressors under natural conditions. Such methods may include field biomonitoring, which allow the test organism to absorb and integrate doses of toxicants over different environmental and climatological conditions (Cairns and van Der Schalie 1980; Weinstein et al. 1990). These methods provide a continuous summation of the organisms response to stressors as influenced by physical, chemical, and biological processes (Wheeler et al. 1992; Slabbokoorn 1993; Lovett Doust et al. 1993). The use of biomonitors in the field could yield valuable information, not only on the presence of anthropogenic stressors, but, more importantly, on the adverse impact the stressors are having on the environment. Clements and Goldsmith (1924), in describing their work with plants, said, "instruments may be used to determine the factors of a habitat, but the significance of those factors in the development or structure of the plant or community must be determined by the plant."

Field biomonitoring, as used in this chapter, refers to the use of test organisms, which are living and growing in the "field," to provide a quantitative assessment of environmental quality. This chapter will focus on the important role that plants could and should play when making management decisions. Particular emphasis will be placed on the value of using vascular plants in the field as "biomonitors" to integrate exposure to stressors with other biological and environmental factors. Although stressors can be any physical, chemical, or biological entity that induces a negative effect (U.S. EPA 1992), the focus in this chapter is on chemical stressors.

The important role of plants in the environment

Food chain transfer and habitat loss are two major concerns facing environmental managers today (Brody et al. 1993; McVey and Macler

1993; Barnthouse and Brown 1994). Because of the significant role that plants play in both of these areas (Hinman and Klaine 1992; Brody et al. 1993; McVey and Macler 1993; Barnthouse and Brown 1994), it is critical to include plants when making assessments regarding the environment (Fletcher 1991).

Plants are the primary producers, supporting almost all other life forms. As such, they play an active role in transferring contaminants to higher trophic levels (Manny et al. 1991; McVey and Macler 1993) and should be considered when conducting an environmental assessment. For instance, although there did not appear to be any exposure-related impacts to the plant community at Rocky Mountain Arsenal, Denver, CO (Fordham and Reagan 1993), some of the chemicals in question can accumulate in plants. Because many animals, including birds and invertebrates, rely heavily on plants as a food source, the plants provided a potential exposure route that needed to be examined. Manny et al. (1991) reported similar concerns following a study of metal contamination in selected macrophytes in the Detroit River. For all metals analyzed, concentrations were 3800 to 161,000 times higher in the plants than in the water (Table 1). Again, food chain exposure must be considered.

Table 1 Mean Metal Concentrations in Drifting Macrophytes (μg g^{-1}) and in Filtered Detroit River Water (μg ml^{-1} × 10^{-6}), Along with Enrichment Ratios for Each Metal in Macrophytes Relative to Water

| Metal | Metal concentration | | Enrichment ratio |
	Macrophytes	Water	
Cd	5.2	60[a]	86,700
Co	3.8	1000[b]	3,800
Cr	10.3	2000[b]	5,200
Cu	31.1	1786[a]	17,400
Hg	0.1	8[a]	12,500
Pb	16.1	100[c]	161,000
Ni	41.4	1924[a]	21,500
Zn	117.8	7078[a]	16,600

[a] U.S. EPA 1988.

[b] From U.S. EPA Storet for station numbers 820414 and 820018 in the Detroit River, 1980–1988.

[c] EC and U.S. EPA 1988.

From Manny BA, et al. (1991) *Hydrobiologia* 219:333. With permission.

Plants also influence habitat by providing cover, stabilizing the soil and sediment, and regulating the temperature and flow of water. The concern that whole plant communities could be irreversibly altered when exposed to anthropogenic stressors is real (Outridge and Noller 1991). If left unchecked, these changes could have a significant, indirect effect on higher organisms, which may persist long after the original stressor has attenuated (Lovett et al. 1993; Peterson and Nyholm 1993; Sheehan and Loucks 1994; Harwell et al. 1994). Brody et al. (1993), using data collected from the Lake Verret Basin in southern Louisiana, modeled potential changes in vegetation in response to altered hydrology caused by the construction of levees. The researchers, using a subsidence rate of 1.0 cm/year, predicted that within 70 years the bottomland hardwood forest would be completely replaced by a swamp community. Significant changes in wildlife habitat would be the end result.

Despite the important role of plants, Benenati (1990) noted that only a small percentage of the environmental data submitted under the Toxic Substances Control Act (TSCA) guidelines for existing chemicals (TSCA Section 4), Premanufacturing Notices (TSCA Section 5), and "substantial risk" chemicals (TSCA Section 8e) was related to plants. Although laboratory and field methods for testing different species vary significantly, making relative comparisons difficult (Hughes 1992), evidence suggests that organisms do respond differently to stressors (Benenati 1990; Cairns and Niederlehner 1987; Mhatre 1991). Species which are sensitive to a particular stressor will demonstrate adverse symptoms prior to those with more resistance (Chapekar 1991; Lovett Doust et al. 1994b).

Plants have been reported to be more sensitive than animals for a number of compounds, such as herbicides (Wang and Freemark 1995). Even within the plant community, species differ in their sensitivity to stressors, with variability directly related to taxonomic classification (Fletcher et al. 1990; Wang and Williams, in press). The most commonly used group of plants for phytotoxicity tests are the algae (Benenati 1990); yet several researchers (Fletcher 1990; Swanson et al. 1991; Wang 1992), conducting database searches for comparative information regarding sensitivity, indicate that algae may not be good predictors for macrophytes. Aquatic macrophytes were found to be more sensitive to toxicants such as herbicides than algae (Thomas et al. 1986; Lovett Doust et al. 1994b), with the rooted macrophytes being the most sensitive (Swanson et al. 1991; Lovett Doust et al. 1994b). With metals, however, macrophytes tended to be one to two orders of magnitude less sensitive than algae (Outridge and Noller 1991). This lack of consistency may be due to physical differences in the tested plants, to the mode of action of the test chemical, or simply reflect a lack of standardized testing procedures that are comparable among the different functional groups (Swanson et al. 1991).

Because of the ecological relevance of plants, the observed differences in species response to stressors, and our own limited knowledge of the integrated relationships among the different taxonomic classes, plants should be considered when making environmental decisions. As Barnthouse and Brown (1994) said, "because of its relative conspicuousness and direct association with particular abiotic factors, vegetation (i.e., primary producers) when studied can provide considerable information for characterizing an ecosystem."

History of plants used as environmental indicators

Although not extensive, the use of plants in the field as "indicators" of environmental quality has a long history. Algal blooms (U.S. EPA 1978) have long been associated with excessive nutrient loading in aquatic systems, while lichens (LeBlanc and DeSloover 1970; Nieboer et al. 1972), bryophytes (Gilbert 1968), and mosses (Parkarinen and Tolonen 1972; Little and Martin 1974) have been used as signals of air quality.

The first documented record of plants being used as instruments for measuring environmental factors dates back to the early 1900s. It was noted that when plants were placed in different environments, they could be used to establish the presence of various conditions and the degree to which the conditions affected plant growth (Clements and Goldsmith 1924; Weinstein et al. 1990). In 1917, McLean conducted the first comprehensive study in which a quantitative relationship between climatic conditions and the growth of plants was established through the use of standard plants. In his report, McLean (1917) states, "the plant is thus regarded as a sort of integrating and recording instrument, and the reading of which is zero at the beginning of each observation period." In 1924, Clements and Goldsmith (1924) coined the term "phytometer" when referring to plants used to measure or assess plant growth and vigor in the presence of contaminants or other environmental stressors.

Advantages of using biomonitors in the field

There are several advantages of using biomonitors in the field to assess conditions. First, bioavailability of toxicants and nutrients is highly dependent on environmental conditions (Lovett Doust et al. 1994b; Outridge and Noller 1991; Tisdale and Nelson 1975). Utilizing biomonitors to measure the response of organisms or communities to a stressor under "natural" conditions, where biotic and abiotic factors are integrated, reduces the need to make assumptions regarding bioavailability of a substance (Lovett Doust et al. 1994b; Chaphekar 1991; Weinstein et al. 1990). For example, metal availability and subsequent plant uptake are highly dependent on the pH and Eh of the soil (Schierup and Larsen 1981; Campbell et al. 1988). Plant roots can change both of these properties within the rhizosphere, thereby altering the amount of metal that is biologically available (Crowder 1991; Outridge and Noller 1991). Bulk soil analysis may not measure conditions at the membrane surface where the changes occur.

Second, there is an inherent variability within ecosystems (Sheehan and Loucks 1994). The use of biomonitors in the field, in conjunction with other tests, could help identify effects caused by anthropogenic stressors vs. those caused by natural forces (Mhatre 1991) by allowing for the separation of covarying parameters (Outridge and Noller 1991). For instance, the interactions of pH and metal uptake by aquatic plants were studied by Sprenger and McIntosh (1989). The researchers concluded that the natural pH of the water was a more important factor in determining metal uptake by *Utricularia purpurea*, a free-floating plant, than the actual concentration of the metals in the water column. In rooted plants, however, sediment concentrations rather than pH influenced metal uptake. This ability to separate natural successional changes from those induced by anthropogenic stressors would also allow for more accurate assessment of "reference" conditions.

Third, trees, used as biomonitors, can provide valuable historical information regarding past environmental conditions. The science of examining and dating tree rings, referred to as dendroecology, allows the scientist to make references regarding environmental conditions such as fires, floods, insect outbreaks, and toxicant exposure (Lepp 1975; Chaphekar 1991). Alterations in these "long-lived" monitors, which have withstood natural variability, would be ecologically significant (Harwell et al. 1994). In a limited study, Kardell and Larsson (1978) showed a marked increase of lead in an oak tree following the introduction of tetraethyl lead as a gasoline additive. Techniques are currently being developed which use lasers beams to identify the presence of contaminants in tree cores (Evans personal communication, 1995).

Fourth, field biomonitoring eliminates the need to collect and transport environmental samples to the laboratory, reducing the possibility of altering the samples (Callahan et al. 1991).

Finally, toxicants can persist long after releases to the environment have ceased (Manny and Kenaga 1991). From a cost perspective view, the use of biomonitors may be substituted for expensive analytical instruments, or at least add to the information that is obtained with the instruments, for monitoring long-term impacts as well as stressor abatement (Weinstein et al. 1990).

In addition to these general advantages of using field biomonitors, plants possess many characteristics which make them a favored test species for many researchers. They are easy to grow, relatively inexpensive to maintain, and the best organisms for detecting phytotoxic compounds such as herbicides (Wang 1988, 1991). They are also subject to all possible exposure routes (Walsh et al. 1991). They can either be started from seed, which has a long storage life; field collected; or cloned (Lovett Doust et al. 1994b). Plants provide an adequate amount of tissue which can be analyzed for stress either destructively or nondestructively for timed monitoring (Walsh et al. 1991; Lovett Doust et al. 1993). Plants can also be archived for later reference (Weinstein et al. 1990).

Limitations of using biomonitors in the field

There are some limitations, however, with using biomonitors in the field (Mhatre 1991). Traditionally, plants have been used more as indicators of environmental conditions, providing qualitative rather than quantitative data. They may be site specific. Similar responses to different stressors may be demonstrated, making it hard to separate causes. Finally, biomonitors may develop genetic resistance to the stressors (Lovett Doust et al. 1994b). Despite these obstacles, the idea of using plants as biomonitors for assimilating natural environmental conditions with the presence of stressors is promising. Increased research in this area could help alleviate some of these issues.

The value of plants in assessing direct and indirect impacts

Plant biomonitors can be used to directly measure adverse impacts to the plant community. If a correlation between residue and effect has been established, then plants, which are capable of accumulating toxicants in their tissue (Chaphekar 1991; Lovett Doust et al. 1994b; Lambou and Williams 1980; Weinstein et al. 1990), can provide the earliest and most relevant assessment of the cumulative effect that a stressor is having on the plant community (Mhatre 1991; Cairns and van Der Schalie 1980). By examining the type and extent of damage to a susceptible plant, an early warning about possible hazards to either natural or cultivated plants can be obtained (Mhatre 1991; Chaphekar 1991). For instance, Lovett Doust et al. (1993) used *Vallisneria americana* to detect the presence of organochlorine contamination in water and sediments. The researchers observed that as concentrations of the organochlorine in the water and sediments increased, so did concentrations in the plant tissues, which resulted in a decrease in several growth parameters. An "impairment index" was calculated to provide the relative degree of contamination at the study site, as compared to a reference site, yielding valuable information for making management decisions (Lovett Doust et al. 1993).

Plants can also be used indirectly for detecting potential problems that may significantly impact higher trophic levels. For instance, it was noted that waterfowl nesting near Kesterson Reservoir, San Joaquin Valley, CA were experiencing poor reproductive success in the early 1980s (McVey and Macler 1993). Selenium in irrigation drainage water was determined to be the cause for this adverse effect. Selenium has a high potential to bioaccumulate, as shown by data from Ohlendorf et

Table 2 Selenium Concentrations (ppm, Dry Weight) in Composite Samples of Plants, Invertebrates, and Mosquitofish Collected at Volta, CA (Reference Site) and Kesterson Wildlife Refuge, CA (Study Site) May 1983

| | Volta | | | Kesterson | | |
Sample	N[a]	Mean[b]	Range	N	Mean	Range
Filamentous algae	0/4	ND[c]		6/6	35.2	12–68
Rooted plants	1/1	0.43		18/18	52	18–79
Net plankton	4/4	2.03	1.4–2.9	7/7	85.4	58–124
Water boatman (*Corixidae*)	5/5	1.91	1.1–2.5	2/2	22.1	20–24
Midge larvae (*Chironomidae*)	3/3	2.09	1.5–3.0	3/3	139	71–200
Dragonfly nymphs (*Anisoptera*)	2/2	1.29	1.2–1.4	6/6	122	66–179
Damselfly nymphs (*Zygoptera*)	2/2	1.45	1.2–1.7	3/3	175	118–218
Mosquitofish (*Gambusia affinis*)	5/5	1.29	1.2–1.4	12/12	170	115–283

[a] Number with measurable concentrations/number analyzed.

[b] Geometric means computed only when selenium was measurable in at least 50% of samples. When only one sample was analyzed, the concentration is shown in this column.

[c] ND = not detected.

From Ohlendorf HM, et al. (1986) *Sci Total Environ* 52:49. With permission.

al. (1986) in Table 2, and can be readily transferred through the food chain (Wu et al. 1995). At Kesterson Reservoir, elevated levels of selenium were measured at several different trophic levels, confirming that the food chain was an important exposure pathway (McVey and Macler 1993; Wu et al. 1995). Serious problems may have been avoided if accumulation and possible impact at the plant community level had been monitored and early action taken.

Types of plant biomonitors

Field biomonitors can be divided into two basic groups. The first group include the indigenous organisms, which are those plants that are already present in the field and act as "passive" biomonitors (Chaphekar 1991; Lovett Doust et al. 1994b). Using native species can provide pertinent information regarding the spatial distribution of bioavailable pollutants. Because the most susceptible organisms may have already been eliminated, however, the extent of the stressor may be underestimated (Lovett Doust et al. 1994b). Exposure to long-term, chronic levels of stressors may also increase the tolerance of endemic biomonitors, providing a false estimation of the impact. This was demonstrated by Cairns and van Der Schalie (1980) with bluegill sunfish

(*Lepomis macrochirus*). The researchers saw only decreased activity when they exposed the fish to 3.0 mg l⁻¹ zinc, following an acclimation period of 29 weeks when the fish were exposed to zinc at 1/100 of the 96-h EC50 (0.075 mg l⁻¹) concentration. Similar acclamations have been reported for algae (Wang 1987). Reference site selection can play a critical role in alleviating some concerns regarding the use of passive biomonitors.

The second group of biomonitors consist of organisms that are either cultured in the laboratory or collected from a noncontaminated reference site and transplanted into "stressed" environments for a known period of time (Lovett Doust et al. 1994b; Chaphekar 1991; Slabbokoorn 1993; Kovacs and Podani 1986). Using this type of system allows for the exposure time to be controlled, so that both temporal and spatial environmental changes can be determined.

Selection of plant biomonitors

Plant species selected as biomonitors should have certain characteristics. They should be endemic to the area of concern and preferably available as seed or clones to eliminate some of the genotypic variability between plants (Outridge and Noller 1991; Lovett Doust et al. 1993, 1994b). Seeds should germinate uniformly and rapidly to produce sturdy seedlings (Al-Farraj et al. 1984; Wheeler et al. 1992). Plants should also have relatively high growth rates to provide rapid, measurable responses. The plants need to be sensitive to the stressor, yet tolerant enough to survive transplantation (Outridge and Noller 1991; Wheeler et al. 1992; Slabbokoorn 1993). Finally, a statistically reliable relationship between the concentration of the toxicant in the plant and the concentration in the environment should be demonstrated (Outridge and Noller 1991).

The author (unpublished data) observed this type of relationship when assessing native wetland vegetation for potential impact due to elevated concentrations of boron. Although there was significant variation among plants species in their ability to accumulate boron, concentrations measured in the leaf tissue, in general, mimicked surface water and sediment concentrations (Table 3). Correlations with boron in the leaf tissue were generally better with surface water than with sediments and again varied with species (Table 4).

The use of a database such as PHYTOTOX, which contains information on over 1500 different vascular species exposed to a variety of test chemicals under both laboratory and field conditions, may be useful in selecting the appropriate plant biomonitors (Royce et al. 1984; Fletcher et al. 1988; Fletcher 1990). Currently, this database focuses on agricultural species (Smith 1991). Data regarding the influence of chemical and nonchemical stressors on noncrop plants would be a valuable addition.

Table 3 Concentrations of Boron (ppm) in Surface Water, Sediments, and Selected Vegetation Along One Transect at Wetland Study Site

Station	Surface water	Sediment	Peltandra virginica	Persea borbonia	Pontedaria cordata	Taxodium distichum
1	11.80	24	135	—[a]	148	—
2	10.50	70	102	—	156	1040
3	14.30	111	150	—	190	797
4	7.33	20	69	323	—	598
5	3.48	—	48	187	—	—
6	2.61	—	30	96	33	—
7	0.60	3	—	42	27	—

[a] Sample not collected.

Table 4 Correlation Matrix of Plant Leaf Tissue
Concentrations with Surface Water and Sediment
Concentrations for All Samples Combined

Leaf tissue concentration	Surface water concentration	Sediment concentration
Peltandra virginica	$R^2 = 0.193$	$R^2 = 0.200$
	$n = 37^a$	$n = 28$
Persea palustris	$R^2 = 0.330$	$R^2 = 0.072$
	$n = 21$	$n = 18$
Pontederia cordata	$R^2 = 0.652$	$R^2 = 0.650$
	$n = 22$	$n = 15$
Taxodium distichum	$R^2 = 0.573$	$R^2 = 0.427$
	$n = 25$	$n = 22$

[a] n = number of samples.

Endpoint selection

It is possible to evaluate the presence of a stressor either directly by measuring the concentration of the toxicant in the plant tissue or indirectly by observing unique or specific responses in an organism or population (Kovacs and Podani 1986; Mhatre 1991; U.S. EPA 1992).

Responses or symptoms can also provide a semiquantitative measurement of the accumulation of a toxicant by the biomonitor, if the intensity of the symptoms is dose related (Horowitz 1976; Weinstein et al. 1990). For example, the gladiolus plant is an ornamental species that has been shown to be extremely sensitive to airborne fluoride. Distinctive symptoms of injury appear as tip necrosis. In a study to evaluate the distribution of airborne fluoride near some industrial sites in Switzerland, Weinstein et al. (1990) visually ranked the degree of tip necrosis observed on gladiolus plants. The researchers noted that the severity of injury in the gladiolus plants correlated with the concentration of fluoride measured in several other plant species during a previous study.

Responses may be measured at several different levels of biological organization, starting at the subcellular level and moving toward whole ecosystems (Kovacs and Podani 1986; Chaphekar 1991; Lovett Doust et al. 1994b; U.S. EPA 1992). At the subcellular level, changes may be biochemical or physiological in nature. Organismal-level alterations may involve anatomical or morphological changes, as well as reduction in reproductive success and shortened lifespan. On a larger scale, whole populations or communities may be altered, while ecosystem structure and function may be impacted (Chaphekar 1991; Kovacs and Podani 1986).

Because the effects at one biological level are often assessed by measuring the significance of the impact at the next higher level (Shee-

Table 5 Simplified Conceptual Model of the Apparent Timescales of
Physiological and Ecological Processes Associated with Plant Community
Response to Chronic Air Pollution[a]

Response variable	Timescale interval
Pollutant uptake	10^{-1}–10^3 min
Reduced photosynthesis; altered membrane permeability	10^1–10^3 min
Reduced labile carbohydrate pool	10^0–10^1 d
Reduced growth of root tips and new leaves	10^1–10^2 d
Decreased leaf area	10^2–10^{25} d
Differences in species growth performance	10^2–10^{25} d
Reduced community canopy cover	10^2–10^3 d
Reduced reproductive capacity	10^2–10^3 d
Shifts in interspecific competitive advantage	10^2–$10^{3.5}$ d
Alteration of community composition	$10^{2.5}$–10^4 d
Change in species diversity	10^3–10^4 d
Change in communtiy structure (physiognomy)	$10^{3.5}$–$10^{4.5}$ d
Functional ecosystem changes (e.g., decline in nutrient cycling efficiency, net productivity)	$10^{3.5}$–$10^{4.5}$ d

[a] Timescales are suggested for a hypothesized forest community exposed to chronic
ozone levels comparable to much of the eastern United States of the 1980s. The
timescale intervals, which are not verified empirically and are not intended to be
associated with a specific site, are suggested as the ranges within which response
symptoms would be clearly detected given current capabilities in pollution effects
research.

From Armentano TV, Bennett J (1992) in *Air Pollution Effects on Biodiversity,* Barker JR,
Tingey TD, Eds., Van Nostrand Reinhold, New York. With permission.

han and Loucks 1994), there is an inherent lag period between the time
of initial exposure and a measured response (Harwell et al 1994; Lovett
Doust et al. 1994b). This period increases as the level of response is
elevated, as was illustrated by Armentano and Bennett (1992) for a
fictional forest exposed to chronic ozone levels (Table 5). As the level
of response increased from the plant to the community to the ecosystem, so did the time frame for detecting a significant impact.

Historically, many environmental concerns have been handled in
a reactive manner, i.e., the near extinction of several "top-of-the-food-
chain" organisms forced researchers and regulators to reexamine the
use of DDT and its potential to bioaccumulate.

The challenge today is to be more proactive and to prevent major
environmental alterations or, at least, minimize their impact. An "early
warning" system should provide rapid, reliable detection of effects at
the lowest practical level (Cairns and van Der Schalie 1980; Lovett
Doust et al. 1994b). Because lower biological levels can be more sensitive to change than higher levels (U.S. EPA 1992), research in the area
of biochemical biomarkers has intensified (Huggett et al. 1992). Caution
is needed, however, because not all "lower"-level alterations result in

a negative impact at the level studied or at higher levels (U.S. EPA 1992). For example, the reduction in a plant's ability to photosynthesize in the presence of a toxicant may not be ecologically significant unless important factors such as the survival or reproductive energy of the plant is diminished. Therefore, it is recommended that a suite of responses or endpoints from different organizational levels be included when making ecological decisions (U.S. EPA 1992).

PHYTOTOX can be a useful resource for selecting measurable plant responses or endpoints. To be relevant, however, a relationship between the selected endpoint and the ecologically significant impact to the plant must first be established, i.e., what is the net result of losing 50% of an algal population? If other algal species replace the lost population and primary productivity remains the same, is there a significant impact?

Plants exposed to environmental and anthropogenic stressors demonstrate numerous, easily recognized, visual symptoms such as lesions, changes in pigmentation, stunting, etc. (Weinstein et al. 1990). Survival, reproductive activity, and clonal growth are other common measures of plant stress (Lovett Doust et al. 1993). Physiological, biochemical, or genetic changes such as DNA/RNA synthesis, peroxidase activity, altered respiration, leaf chlorophyll content, and fluorescence can also be used to demonstrate exposure to stressors (Fletcher 1990; Miles 1990; Judy et al. 1990; Velagaleti et al. 1990). There are several methods currently being developed for measuring these subcellular alterations.

For instance, several researchers (Byl and Klaine 1991; Byl et al. 1994; Fleming personal communication) have noted that, for rooted aquatic vascular plants, increases in peroxidase activity are dose dependent. For several test chemicals, this increase in enzyme activity occurs prior to reductions in vegetative growth (Figure 1), as can be seen in data from Byl et al. (1994), indicating that the use of peroxidase as an indicator of plant stress may be a more sensitive endpoint than traditional endpoints such as growth.

Powell et al. (1996), looking at reductions in chlorophyll content of aquatic plants exposed to selected compounds, have also demonstrated a dose relationship. Research is currently underway to determine the relative sensitivity, as well as the ecological relevance, of several endpoints, including chlorophyll concentration, growth, and residue analysis.

Exposure and uptake

Plants may be exposed to contaminants in numerous ways. Leaf and stem tissue may be exposed to pollutants from the atmosphere either from direct application of chemicals or from deposition of particulate matter. In the case of aquatic plants, leaf and stem tissue may also be exposed to toxicants in water. *Lemna*, which floats on the water surface,

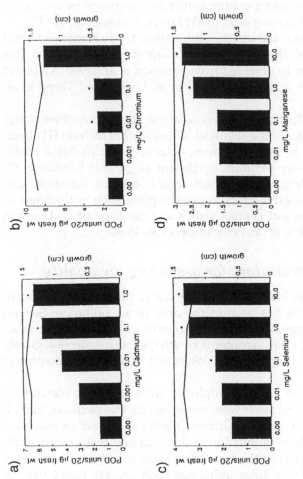

Figure 1 Influence of cadmium (a), chromium (b), selenium (c), and manganese (d) on growth (line) and peroxidase (POD) activity (bars) in *Hydrilla verticillata*. An asterisk indicates significant difference from the control ($p \leq 0.05$). (From Byl TD, et al. [1994] *Environ Toxicol Chem* 13:509. With permission.)

has been shown to be a good monitor for detecting pesticide drift and hydrophobic compounds (Swanson et al. 1991). Foliage may also be exposed to chemicals which have volatilized from the soil (Paterson et al. 1990; Hulster et al. 1994).

Roots can be exposed to contaminants in soil, sediment, or water. In aquatic systems, rooted macrophytes have been shown to have a significant ability to adsorb and accumulate chemicals via their roots and shoots (Lovett Doust et al. 1993), making them extremely useful as biomonitors in detecting contaminants in sediments as well as in the water column (Swanson et al. 1991; Lovett Doust et al. 1993).

A discussion about the uptake and transport of chemicals within the plants is beyond the scope of this chapter. Information on this subject, however, can be obtained from Paterson et al. (1990), Gobas et al. (1991), Hulster et al. (1994), Schroll et al. (1994), and Trapp et al. (1994).

Many factors influence the exposure of the plant to xenobiotics and the subsequent uptake of the chemical. These factors include (1) chemical properties such as vapor pressure, octanol–water partition coefficient (K_{ow}), octanol–air partition coefficient (K_{oa}), and solubility; (2) environmental conditions such as soil organic content, humidity, and pH; and (3) morphological structures of the plant such as thickness of the cuticle and other leaf characteristics and the type of root system (Garrod 1989; Paterson et. al 1990; Trapp et al. 1994).

Use of plant biomonitors for assessing air quality

Most of the work that has focused on the use of plants as indicators of environmental quality has been in the area of air pollution. Several researchers have suggested that certain plants are very sensitive to air pollutants, showing clear symptoms of effects, both morphologically and physiologically (Ruston 1921; Weinstein et al. 1990; Chaphekar 1991).

In England's Tyne Valley, bryophytes were relatively scarce near large towns. Numerous causes were suggested for this decrease, including drought and air pollution. Gilbert (1968) conducted an extensive survey of the bryophyte population in this area and was able to establish a relationship between the number of species and density of bryophytes and the distance from industrial facilities. He noted that, in general, both the number of species and the density increased with increasing distance from industrialized areas. Mapping the bryophyte populations correlated exactly with distributions of the major air pollutant, sulfur dioxide (SO_2). Similar conditions have been observed in Montreal, Canada. While conducting a survey, LeBlanc and DeSloover (1970) noted that the epiphytic mosses and lichen populations increased as the researchers moved from industrialized zones to rural areas. Along with increased densities, fertility of each species also increased. Although air sampling was not conducted by the researchers, air pollution was again the suggested cause for the altered populations.

Based on these earlier studies, it appears that the growth and development of plants can be directly related to air quality. If a statistically sound relationship between plant injury and the quantity and duration of exposure to pollutants could be established, then plants would be valuable biological tools for indirectly measuring air pollution (LeBlanc and DeSloover 1970).

Schroder and Debus (1991) observed such a relationship when spruce trees were exposed for 41 d to Halone 1211 (difluoro-chloro-bromo-methane). An increase in the activity of glutathione-S-transferase, a constitutive detoxification enzyme, was measured. While not specific to Halone 1221 exposure, the researches also observed changes in needle fresh weight, protein content, and pigment pattern. An alteration in the formation of secondary plant metabolites, important in disease and insect resistance, has also been related to air pollution (Coleman et al. 1992). Other researchers (Richardson et al. 1990; Calamari et al. 1994) have also been successful in establishing a relationship between selected air pollutants and plant response.

Plants vary in their sensitivities to air pollutants. Evergreens and legumes appear to be among the most sensitive to air pollution (Ruston 1921). Lichens also appear to be very sensitive to atmospheric SO_2 (Lovett Doust et al. 1994a). Barley, alfalfa, and apples have been shown to develop chronic injury at an SO_2 concentration of 0.10 ppm (LeBlanc and DeSloover 1970). Plant sensitivity to air toxicants can also be influenced by natural factors. For instance, Gilbert (1968) noted that, in the field, bryophyte species exposed to elevated levels of SO_2 demonstrated a preference for growing on high pH substrates. Experimentally, Gilbert confirmed that plants growing on substrates with higher pHs were able to convert sulfur to less toxic forms of $SO_3^=$ and $SO_4^=$, thereby allowing these species to survive.

Plants respond to air pollution in a variety of ways. With bryophytes, symptoms start with a slight loss of color in the tips of the more exposed leaves. With time, this color loss extends down the leaves to the shoots until there is a complete loss of chlorophyll (Gilbert 1968; LeBlanc and DeSloover 1970). Other characteristics of plants exposed to air pollution include brown spotting on leaves, earlier leaf drop in autumn, stunting, reduced intensity of flower color, and fewer roots hairs and fibrous roots (Ruston 1921). The vitality of the plants may also be effected by air pollution. Plants either may not produce seed or produce seed that is smaller and possess less germination energy (Ruston 1921). Biochemical changes can also occur for those plants exposed to air pollution. The amount of CO_2 assimilated by the plant has been inversely related to the amount of air pollution the plant was exposed to (Ruston 1921). Peroxidase activity in plants is enhanced following exposure to air pollutants (Keller 1974; Niemter 1979; Endress et al. 1980). Finally, enzyme activity associated with germination of plants has been shown to be inhibited in the presence of air pollution (Ruston 1921).

Use of plant biomonitors for assessing water quality

Less work has been done using plant biomonitors for aquatic pollutants. Emphasis in aquatic toxicity has been placed on the development of single-species ecotoxicity tests where conditions are well defined (Lovett Doust et al. 1994b). Standard laboratory testing procedures for algae and *Lemna* are currently available (U.S. EPA 1979; OECD 1984; ISO 1989; ASTM 1993a,b), while methods with submerged and emergent macrophytes are under development (Fleming et al. 1991; APHA et al. 1992; ASTM 1997; Nelson personal communication, 1996). These types of tests are necessary and provide relevant information regarding potential toxicity (Lovett Doust et al. 1994b). They also provide a means for comparing the toxicity of compounds and establishing water quality criteria (Kimball and Levin 1985). Unfortunately, standardized laboratory tests fail to provide critical information regarding the biological interactions of the test compound in the "real world" (Lambou and Williams 1980; Kimball and Levin 1985; Lovett Doust et al. 1993).

The availability of nutrients and toxicants to aquatic plants is highly dependent on chemical, physical, and biological conditions (Outridge and Noller 1991; Lovett Doust et al. 1993). For instance, increased rainfall has been shown to enhance metal uptake by plants by increasing the movement of metals into the lakes (Franzin and McFarlane 1980), yet flooding conditions limit uptake by creating reduced conditions (Lee et al. 1982). The presence of other ions such as calcium (Ca) and magnesium (Mg) in the water column also reduces the plants ability to take up metals (Franzin and McFarlane 1980). In rice (*Oryza sativa*), lower temperatures and shading were both shown to reduce translocation of metals from the root system to the shoots (Chino and Baba 1981). Seasonal influences also need to be considered (Outridge and Noller 1991). Metabolic changes within the plant, status of nutrients or toxicants, and microbial action can all affect the response of a plant to the stressor (Outridge and Noller 1991).

Most of the research with aquatic plants has been with heavy metals. Vascular plants, known as hyperaccumulators, can accumulate selected heavy metals in their tissues several thousand times the ambient concentrations (Baker and Brooks 1989; Smith 1991; Lovett Doust et al. 1994b). In general, submerged plants with both roots and leaf tissue in contact with contaminants tend to accumulate higher concentrations of metals than either free-floating plants or emergent species (Outridge and Noller 1991; Lovett Doust et al. 1994b). Greater surface area to volume may account for this difference (Franzin and McFarlane

1980). Heavy metals have been shown to decrease root mass (Mhatre 1991), cause shoot stunting (Mhatre 1991; Foy et al. 1978), reduce the germination and growth of seedlings (Mhatre and Chaphekar 1982), as well as numerous other symptoms (Tiagi and Aery 1985). Outridge and Noller (1991) provided a good review of freshwater vascular species known to accumulate large quantities of heavy metals, which may make good biomonitors in aquatic systems.

Lovett Doust et al. (1994a,b) have been exploring the use of *V. americana*, American wildcelery, as a potential biomonitor for aquatic systems. *V. americana*, a dioecious submerged aquatic macrophyte, has many characteristics considered important in a biomonitor, including the ability to accumulate pollutants and propagate both sexually and through clonal growth, and is a valued component in the food chain. The researchers (Lovett Doust et al. 1994a), examining the use of *V. americana* as a biomonitor of organochlorine contamination, found that leaf production, rate of clonal growth, sexual reproduction, as well as plant survival were all affected with varying levels of organochlorines (penta- and hexachlorobenzene, octachlorostyrene, *trans*-nonachlor, 1,1-dichloro-2,2-*bis*(*p*-chlorophenyl)-ethylene, and a number of poly-chlorinated biphenyls). Examining a number of endpoints allowed the researchers to measure both acute and chronic effects, providing critical information regarding subtle responses to the toxicant.

Use of plant biomonitors for assessing soil fertility

Another use of plants as monitors is in the area of soil fertility (Al-Farraj et al. 1984; Slabbokoorn 1993; Spink et al. 1994). Notenboom (1993), in his evaluation of soil fertility of river marginal wetlands in Europe, noted that many of these areas were heavily disturbed. As a result, species richness, number of rare species, and species composition were reduced. One suggested cause for this diminished value was increased fertility in these wetlands (Moore et al. 1989).

Currently, procedures for assessing soil fertility rely on chemical and physical analysis to estimate the concentration of bioavailable nutrients. Soil chemistry alone, however, may not adequately reflect year-round soil fertility (Wheeler et al. 1992; Slabbokoorn 1993; Notenboom 1993). The nutrient cycle is very complex and is influenced by numerous environmental and anthropogenetic factors (Al-Farraj et al. 1984; Spink et al. 1994). Seasonal alterations in nutrient levels, changes in nutrient forms, flooding, and erosion, as well as direct fertilization, runoff from nearby farms and livestock operations, and liming, can alter the soil fertility (Slabbokoorn 1993). Soil properties such as texture, moisture, and microbial activity also influence the bioavaibility of nutrients (Tisdale and Nelson 1975).

Using plants as biomonitors can provide an accurate assessment of soil fertility by integrating the effects of these factors with the vigor or "health" of the plants (Salbbokoorn 1993; Spink et al. 1994). For example, working with floodplain vegetation, Al-Farraj et al. (1984) noted that there was not a strong relationship between extractable concentrations of nitrogen and phosphorus and the net productivity of the above-ground vegetation. Possible explanations for this lack of correlation include the influence of other environmental factors on the growth of the plants, physiochemical alterations in bioavailability of the nutrients, and inadequate representation of year-round nutrient availability. Standard analytical methods can provide information on the nutrient concentrations in the soil, but fail to account for the other factors which influence the overall growth and vigor of the plants.

Plant biomonitors vs. animal biomonitors

Many animals have been successfully used as biomonitors, including clams and mussels (DeKock 1986; Dorherty and Cherry 1988; V-Brough 1988; Schmitt et al. 1987), fish (Dwyer et al. 1988; Parks et al. 1991), tadpoles (Cooke 1981), birds (Mineau et al. 1984), snapping turtles (Meyers-Schone and Walton 1994; Overmann and Krajicek 1995), and earthworms (Callahan et al. 1991). There are, however, many advantages to using plants instead of animal species as biomonitors.

First, as stated earlier, evidence indicates that plants are often more sensitive to certain types of contaminants than animals (Miller et al. 1985; Smith 1991; Wang and Freemark 1995). In order to be an effective "alarm system," the test organism needs to provide warning of a possible hazard before ecologically significant damage has occurred. Waiting until detrimental effects are demonstrated at the higher trophic levels can result in irreparable damage to the ecosystem. Plants, being at the base of the food chain, may experience the effects of toxicants sooner than higher trophic levels, thereby reducing the lag period between exposure and significant impact (Lovett Doust et al. 1994b).

Plants also have the advantage over most animals in that they are stationary. The use of fish, birds, or other organisms that either migrate or have large habitats create additional variables that must be taken into account when interpreting the results (Lovett Doust et al. 1994b). Animal species such as clams and mussels are less mobile; however, they still must be caged (Lovett Doust et al. 1994b). This could alter the organisms' growth, behavior, and survival (Lambou and Williams 1980).

Another problem associated with clams is their unique ability to close up tight when exposed to unfavorable conditions which might occur during pulsed or intermediate release events (Lovett Doust et al. 1994b). This may mask the presence of the toxicant. Plants, especially those with root systems in contact with the sediments, do not need to be caged, and while their responses may vary among species, plants can not selectively disregard contaminants (Lovett Doust et al. 1994b). Pulsed and/or infrequent exposures can be detected through the use of stationary plants, providing a more accurate account of the contaminant loading at the study site (Clements and Goldsmith 1924; Lovett Doust et al. 1994b).

The stationary nature of plants can also be beneficial when investigating the source of environmental problems (Ray and White 1976). For instance, when a fish kill occurs, there is usually a lag period between exposure and mortality of the fish. Also, the dead fish are often observed downstream from the initial exposure site. Examining plants that have the ability to accumulate toxicants can help identify the exposure location.

Relationship of field biomonitoring to other ecological tests

The use of biomonitors in the field combined with analytical chemistry and laboratory toxicity tests allows for a complete assessment, not only of the amount of contamination present, but, more importantly, the influence the toxicant is having on the organisms of concern. Each method is valuable for answering different parts of the question regarding environmental contamination (Lovett Doust et al. 1993). For instance, in ecological risk assessments, biomonitoring could be useful as a tool to establish if ambient conditions have been or are being altered. If concerns do exist, then additional testing, including single-species toxicity tests, under controlled conditions and analytical analysis could be used to identify the potential cause(s) and better define the magnitude of the problem (Lambou and Williams 1980). Together, these methods can provide the critical information necessary for making management decisions.

As an example, ozone was first identified as a phytotoxicant on grapes and tobacco in 1958 (Heck et al. 1977). Since that time, numerous other crop species have been identified as ozone sensitive, based on the appearance of foliar symptoms including necrotic flecking. Heck et al. (1977) confirmed a casual relationship between ozone and plant injury through the use of controlled greenhouse studies. The next step involved long-term field studies in which economically valuable crop species were exposed to different ozone concentrations. Reductions in yield were measured (Heck et al. 1983). Analytical monitoring of ozone across the United States during a several-year period indicated that ozone was present at concentrations capable of causing damage to sensitive species (Heck et al. 1983).

Data obtained from the monitoring program, the greenhouse studies, and the field biomonitoring studies were used to predict the economic impact that ozone has on crop production. The final results indicated that a 25% reduction in the tropospheric ozone would result a $1.9 billion annual net benefit to the producers and consumers due to increased yields (Heck 1993).

Conclusions

Ecosystems are constantly subject to stress from both natural and anthropogenic sources. Any change in the biota, whether positive or negative, is an indication that a new factor is present in the environment. To be protective of the environment requires the ability to quickly and reliably detect adverse factors before a potentially long-term and cumulative impact occurs.

The EPA's framework (U.S. EPA 1992) for assessing ecological risk emphasizes the need to understand exposure as well as effects. Exposure is controlled by the bioavailability of the contaminant as well as the duration of the contact. Numerous factors affect exposure, including soil properties such as pH, redox potential, and texture and environmental conditions such as flooding. The organism's habit and the nature of the release, i.e., continual vs. pulsed, also influence exposure. Current practices that rely mostly on analytical methodology and single-species toxicity tests to be protective of the whole ecosystem are not designed to account for all the biotic and abiotic factors which influence how an organism will be impacted by anthropogenic stressors.

Additional methods such as field biomonitoring could be used to integrate the previously listed factors with the ambient concentration of the toxicant. Field biomonitoring takes advantage of the fact that some animals and plants accumulate specific toxicants in their tissues and, in response, demonstrate symptoms characteristic of exposure to that stressor. Biomonitors provide a direct measure of the availability of the contaminant as influenced by the organisms' environment (Phillips 1978).

Plants, either growing naturally or translocated into potentially contaminated sites, can be used as biomonitors to provide an indication of significant ecological change. Plants have been successfully used in the field to detect changes in ambient conditions. They can also help separate natural successional factors from anthropogenic stressors, provide historical information, and minimize the need for more expensive testing procedures.

Most research with plant biomonitors has been in the areas of assessing air quality and detecting heavy metals in aquatic systems. There are several advantages to using plants as biomonitors. Plants support almost all other life forms and, therefore, can have a significant influence on higher trophic levels. They also play a key role in defining habitat. Alterations in the plant community may provide the earliest "warning" system of impending disaster.

Because organisms have been shown to vary in their relative sensitivity to stressors, numerous trophic levels, including plants, should be considered when making environmental decisions. Wang and Freemark (1995) list several areas including effluents, pesticides, and waste sites where plant testing is becoming more predominant.

Decisions regarding toxicity of pollutants can be very complicated. No single, multipurpose test is available for assessing total environmental quality. Instead, combining the use of plant biomonitors, which integrate exposure with other environmental and biological factors, with air or water monitoring would allow for direct damage to vegetation, as well as indirect impacts to higher trophic levels to be assessed (Weinstein et al. 1990). Laboratory tests could provide additional insight on plant uptake of contaminants and the significance to the plant, thereby maximizing the value of information obtained from field biomonitoring (Mortimer 1985).

Although there are some obstacles with using biomonitors in the field, the idea of using plants for assimilating natural environmental conditions with the presence of stressors is promising. Increased research in this area could help maximize the benefits associated with using plants as "instruments" for assessing environmental quality.

Acknowledgments

Special appreciation is extended to Richard Kimerle and Donald Grothe for their valued input on this manuscript.

References

Al-Farraj MM, Giller KE, Wheeler BD (1984) Phytometric estimation of fertility of waterlogged rich-fen peats using *Epilobium hirsutum* L. *Plant Soil* 81:283–289.

American Society for Testing and Materials (ASTM) (1993a) Standard guide for conducting static toxicity tests with *Lemna gibba* G3. *ASTM Annual Book of Standards*, Vol. 11.04, ASTM E1415-91, American Society for Testing and Materials, Philadelphia, PA.

American Society for Testing and Materials (ASTM) (1993b) Standard guide for conducting static 96-hour toxicity tests with microalgae. *ASTM Annual Book of Standards*, Vol. 11.04, ASTM E1218-90, American Society for Testing and Materials, Philadelphia, PA.

American Society for Testing and Materials (ASTM) (1997) Test guide for conducting renewal phytotoxicity tests with freshwater emergent macrophytes, D1841-96, American Society of Testing and Materials, Philadelphia, PA.

APHA (American Public Health Association), American Water Works Association, and Water Environment Federation (1992) *Standard Methods for the Examination of Water and Wastewater*, 18th ed., APHA, American Water Works Association, and Water Environment Federation, Washington, D.C.

Armentano TV, Bennett J (1992) Air pollution effects on the diversity and structure of communities. In *Air Pollution Effects on Biodiversity*, Barker JR, Tingey TD, Eds., Van Nostrand Reinhold, New York.

Baker A, Brooks R (1989) Terrestrial higher plants with hyperaccumulate metallic elements — a review of their distribution, ecology and phytochemistry. *Biorecovery* 1:81–126.

Barnthouse L, Brown J (1994) Issue paper on conceptual model development. In *Ecological Risk Assessment Issue Papers*, EPA/630/R-94/009, U.S. Environmental Protection Agency, Washington, D.C

Benenati F (1990) Keynote address: plants — keystone to risk assessment. In *Plants for Toxicity Assessment*, Wang W, Gorsuch JW, Lower WR, Eds., ASTM STP 1091, American Society of Testing and Materials, Philadelphia, PA, pp. 5–13.

Byl TD, Klaine SJ (1991) Peroxidase activity as an indicator of sublethal stress in the aquatic plant *Hydrilla verticullata* (Royle). In *Plants for Toxicity Assessment: Second Volume*, Gorsuch JW, Lower WR, Lewis MA, Wang W, Eds., ASTM STP 1115, American Society of Testing and Materials, Philadelphia, PA, pp. 101–106.

Byl TD, Sutton HD, Klaine SJ (1994) Evaluation of peroxidase as a biochemical indicator of toxic chemical exposure in the aquatic plant *Hydrilla verticillata*, Royle. *Environ Toxicol Chem* 13:509–515.

Brody MS, Valette Y, Troyer ME (1993) Ecological risk assessment case study: modeling future losses of bottomland forest wetlands and changes in wildlife habitat within a Louisiana basin. In *A Review of Ecological Assessment Case Studies from a Risk Assessment Perspective*, EPA/630/R-92/005, U.S. Environmental Protection Agency, Washington, D.C.

Cairns J Jr., van Der Schalie WH (1980) Biological monitoring part I — early warning systems. *Water Res* 14:1179–1196.

Cairns J Jr., Niederlehner BR (1987) Problems associated with selecting the most sensitive species for toxicity testing. *Hydrobiologia* 153:87–94.

Calamari D, Tremolada P, DiGuardo A, Vighi M (1994) Chlorinated hydrocarbons in pine needles in Europe: fingerprint for the past and recent use. *Environ Sci Technol* 28:429–434.

Callahan CA, Menzie CA, Burmaster EE, Wilborn DC, Ernst T (1991) On-site methods for assessing chemical impact on the soil environment using earthworms: a case study at the Baird and McGuire superfund site, Holbrook, Massachusetts. *Environ Toxicol Chem* 10:817–826.

Campbell PGC, Lewis AG, Chapman PM, Crowder AA, Fletcher WK, Imber B, Luoma SN, Stokes PM, Winfrey M (1988) Biologically Available Metals in Sediments. National Research Council of Canada (NRCC) No. 27694. Ottawa, Ontario.

Chaphekar SB (1991) An overview on bio-indicators. *J Environ Biol* 12:163–168.

Chapman PM (1995) Extrapolating laboratory toxicity results to the field: letter to the editor. *Environ Toxicol Chem* 14:927–930.

Chino M, Baba A (1981) The effects of some environmental factors on the partitioning of zinc and cadmium between roots and tops of rice plants. *J Plant Nutr* 3:203–214.

Clements FE, Goldsmith GW (1924) *The Phytometer Method in Ecology: The Plant and Community as Instruments,* Publication Number 356, Garnegie Institution of Washington, Washington, D.C., 106 pp.

Coleman JA, Jones CG, Krischik VA (1992) Phytocentric and exploiter perspectives of phytopathology. In *Advances in Plant Pathology,* Vol. 8, Academic Press, London.

Cooke AS (1981) Tadpoles as indicators of harmful levels of pollution in the field. *Environ Pollut Ser A* 25:123–133.

Crowder A (1991) Acidification, metals and macrophytes. *Environ Pollut* 71:171–203.

DeKock WC (1986) Monitoring bio-available marine contaminants with mussels (*Mytilus edules* L.) in the Netherlands. *Environ Monit Assess* 7:209–220.

Dorherty FG, Cherry DS (1988) Tolerance of Asiatic Clam *Corbicula* spp. to lethal levels of toxic stressors — a review. *Environ Pollut* 51:269–313.

Dwyer F, Schmitt C, Finger S, Mehrle P (1988) Biochemical changes in lonear sunfish, *Lepomis megalotis,* associated with lead, cadmium, and zinc from mine tailings. *J Fish Biol* 33:307–317.

EC (Environment Canada) and U.S. EPA (1988) Upper Great Lakes connecting channels study. Final report. Vol. 2, Great Lakes National Program Office, Chicago, IL, 626 pp.

Endress AG, Suarex SJ, Taylor OC (1980) Peroxidase activity in plant leaves exposed to gaseous HCl or ozone. *Environ Pollut Ser A* 22:47–58.

Fleming WJ (1995) Personal communication.

Fleming WJ, Ailstock MS, Monot JJ, Norman CM (1991) Response of Sago Pondweed, a submerged aquatic macrophyte, to herbicides in three laboratory culture systems. In *Plants for Toxicity Assessment: Second Volume,* Gorsuch JW, Lower WR, Lewis MA, Wang W, Eds., ASTM STP 1115, American Society of Testing and Materials, Philadelphia, PA, pp. 258–266.

Fletcher J (1990) Use of algae versus vascular plants to test for chemical toxicity. In *Plants for Toxicity Assessment,* Wang W, Gorsuch JW, Lower WR, Eds., ASTM STP 1091, American Society of Testing and Materials, Philadelphia, PA, pp. 33–39.

Fletcher J (1991) Keynote speech: a brief overview of plant toxicity testing. In *Plants for Toxicity Assessment: Second Volume,* Gorsuch JW, Lower WR, Wang W, Lewis MA, Eds., ASTM STP 1115, American Society of Testing and Materials, Philadelphia, PA, pp. 5–11.

Fletcher J, Johnson F, McFarlane J (1988) Database assessment of phytotoxicity data published on terrestrial vascular plants. *Environ Toxicol Chem* 7:615–622.

Fletcher JS, Johnson FL, McFarlane JC (1990) Influence of greenhouse versus field testing and taxonomic differences on plant sensitivity to chemical treatment. *Environ Toxicol Chem* 9:769–776.

Fordham CL, Reagan DP (1993) Ecological risk assessment case study: assessing ecological risk at Rocky Mountain Arsenal. In *A Review of Ecological Assessment Case Studies from a Risk Assessment Perspective*, EPA/630/R-92/005, U.S. Environmental Protection Agency, Washington, D.C.

Foy CD, Chaney RL, White MC (1978) The physiology of metal toxicity in plants. *Annu Rev Plant Physiol* 29:511–566.

Franzin WG, McFarlane GA (1980) An analysis of aquatic macrophyte, *Myriophyllum exalbescens*, as an indicator of metal contamination of aquatic ecosystems near a base metal smelter. *Bull Environ Contam Toxicol* 24:597–605.

Garrod JF (1989) Comparative responses of laboratory and field grown test plants to herbicides. *Aspects Appl Biol* 21:51–56.

Gilbert OL (1968) Bryophytes as indicators of air pollution in the Tyne Valley. *New Phytol* 67:15–30.

Gobas FAPC, McNeil EJ, Lovett-Doust L, Haffner GD (1991) Bioconcentration of chlorinated aromatic hydrocarbons in aquatic macrophytes. *Environ Sci Technol* 25:924–929.

Harwell M, Gentile J, Norton B, Cooper W (1994) Issue paper on ecological significance. In *Ecological Risk Assessment Issue Papers*, EPA/630/R-94/009, U.S. Environmental Protection Agency, Washington, D.C.

Heck WW (1993) Ecological risk assessment case study: the national crop loss assessment network. In *A Review of Ecological Assessment Case Studies from a Risk Assessment Perspective*, EPA/630/R-92/005, U.S. Environmental Protection Agency, Washington, D.C.

Heck WW, Mudd JB, Miller PR (1977) Plants and microorganisms. In *Ozone and Other Photochemical Oxidants*, National Academy of Sciences, Washington, D.C., pp. 437–585.

Heck WW, Adams RM, Cure WW, Heagle AS, Heggestad HE, Kohut RJ, Kress LW, Rawlings JO, Taylor OC (1983) A reassessment of crop loss from ozone. *Environ Sci Technol* 17:573A.

Hinman ML, Klaine SJ (1992) Uptake and translocation of selected organic pesticides by the rooted aquatic plant *Hydrilla verticillata* Royle. *Environ Sci Technol* 26:609–613.

Horowitz M (1976) Application of bioassay techniques to herbicide investigations. *Weed Res.* 16:209–215.

Hulster A, Muller JF, Marschner H (1994) Soil-plant transfer of polychlorinated dibenzo-*p*-dioxins and dibenzofurans to vegetables of the cucumber family (*Cucurbitaceae*). *Environ Sci Technol* 28:1110–1115.

Huggett RJ, Kimerle RS, Mehrle PM Jr., Bergman HL, Eds. (1992) *Biomarkers: Biochemical, Physiological, and Histological Markers of Anthropogenic Stress*, Lewis Publishers, Chelsea, MI.

Hughes JS (1992) The use of aquatic plant toxicity tests in biomonitoring programs. Canadian Technical Report of Fisheries and Aquatic Sciences. No. 1863, Department of Fisheries and Oceans, Ottawa, Ontario.

International Standards Organization (ISO) (1989) Water quality — fresh water algal growth inhibition test with *Scenedesmus subspicatus* and *Selenastrum capricornutum*. ISO 8692. Nederlands Normalisatie Instituut, Delft, the Netherlands.

Judy BM, Lower WR, Miles CD, Thomas MW, Krause GF (1990) Chlorophyll fluorescence of a higher plant as an assay for toxicity assessment of soil and water. In *Plants for Toxicity Assessment*, Wang W, Gorsuch JW, Lower WR, Eds., ASTM STP 1091, American Society of Testing and Materials, Philadelphia, PA, pp. 308–318.

Kardell L, Larsson J (1978) Lead and cadmium in Oak tree rings (*Quercus robur* L.). *Ambio* 7:117–121.

Keller T (1974) The use of peroxidase activity for monitoring and mapping air pollution areas. *Eur J For Pathol* 4:11–19.

Kimball KD, Levin SA (1985) Limitations of laboratory bioassays: the need for ecosystem-level testing. *Bioscience* 35:165–171.

Kovacs M, Podani J (1986) Bioindication: a short review on the use of plants as indicators of heavy metals. *Acta Biol Hung* 37(1):19–29.

Lambou VW, Williams LR (1980) Biological Monitoring of Hazardous Wastes in Aquatic Systems. Second Interagency Workshop on *In-Situ* Water Sensing: Biological Sensors, Pensacola Beach, FL, pp. 11–18.

LeBlanc F, DeSloover J (1970) Relation between industrialization and the distribution and growth of epephytic lichens and mosses in Montreal. *Can J Bot* 48:1485–1496.

Lee CR, Folsom BR Jr., Engler RM (1982) Availability and plant uptake of heavy metals from contaminated dredged material placed in flooded and upland disposal environments. *Environ Int* 7:65–71.

Lepp NW (1975) The potential of tree-ring analysis for monitoring heavy metal pollution patterns. *Environ Pollut* 9:49–61.

Little P, Martin MH (1974) Biological monitoring of heavy metal pollution. *Environ Pollut* 6:1–19.

Lovett Doust L, Lovett Doust J, Schmidt M (1993) In praise of plants as biomonitors-send in the clones. *Function Ecol* 7:754–758.

Lovett Doust L, Lovett Doust J, Biernacki M (1994a) American wildcelery, *Vallisneria americana*, as a biomonitor of organic contaminants in aquatic ecosystems. *J Great Lakes Res* 20:333–354.

Lovett Doust J, Schmidt M, Lovett Doust L (1994b) Biological assessment of aquatic pollution: a review, with emphasis on plants as biomonitors. *Biol Rev* 69:147–186.

Manny BA, Kenaga D (1991) The Detroit River: effects of contaminants and human activities on aquatic plants and animals and their habitats. *Hydrobiologia* 219:269–279.

Manny BA, Nichols SJ, Schloesser DW (1991) Heavy metals in aquatic macrophytes drifting in a large river. *Hydrobiologia* 219:333–344.

McLean FT (1917) A preliminary study of climatic conditions in Maryland, as related to plant growth. *Physiol Res* 2:129.

McVey M, Macler BA (1993) Ecological risk assessment case study: selenium effects at Kesterson Reservoir. In *A Review of Ecological Assessment Case Studies from a Risk Assessment Perspective*, EPA/630/R-92/005, U.S. Environmental Protection Agency, Washington, D.C.

Meyers-Schone L, Walton B (1994) Turtles as monitors of chemical contaminants in the environment. *Rev Environ Contam Toxicol* 135:93–153.

Mhatre GN (1991) Bioindicators and biomonitoring of heavy metals. *J Environ Biol* 12:201–209.

Mhatre GN, Chaphekar SB (1982) Effect of heavy metals on seed germination and early growth. *J Environ Biol* 3(2):53–64.

Miles D (1990) The role of chlorophyll fluorescence as a bioassay for assessment of toxicity in plants. In *Plants for Toxicity Assessment*, Wang W, Gorsuch JW, Lower WR, Eds., ASTM STP 1091, American Society of Testing and Materials, Philadelphia, PA, pp. 297–307.

Miller W, Peterson S, Green J, Callahan C (1985) Comparative toxicology of laboratory organisms for assessing hazardous waste sites. *J Environ Qual* 14:569–574.

Mineau P, Fox GA, Nostrom RJ, Welseloh DV, Hallett DJ, Ellenton JA (1984) Using the herring gull to monitor levels and effects on organochlorine contamination in the Canadian Great Lakes. *Adv Environ Sci Technol* 14:426–452.

Moore DRJ, Keedy PA, Gaudet CL, Wisheu IC (1989) Conservation of wetlands: do infertile wetlands deserve a higher priority? *Biol Conserv* 47:203–217.

Mortimer DC (1985) Freshwater aquatic macrophytes as heavy metal monitors — the Ottawa River experience. *Environ Monit Assess* 5:311–323.

Nelson MK (1996) Personal communication.

Nieboer E, Ahmed HM, Puckett KJ, Richardson DHS (1972) Heavy metal content of lichens in relation to distance from a nickel smelter in Sudbury, Ontario. *Lichenologist* 5:292–304.

Niemter S (1979) Influence of zinc smelter emissions on peroxidase activity in Scots pine needle of various families. *Eur J For Pathol* 9:142–147.

Notenboom E (1993) A phytometric assessment of soil fertility of river marginal wetlands in France, England, and Ireland. Report Number 931102, Dept. of Plant Ecology and Evolutional Biology, Utrecht University, the Netherlands.

Ohlendorf HM, Hoffman DJ, Saiki MK, Aldrich TW (1986) Embryonic mortality and abnormalities of aquatic birds: apparent impacts of selenium from irrigation drainwater. *Sci Total Environ* 52:49–63.

Organization for Economic Cooperation and Development (OECD) (1984) Alga, Growth Inhibition Test. OECD Guideline for Testing of Chemicals, Guideline 201. OECD, Paris.

Outridge PM, Noller BN (1991) Accumulation of toxic trace elements by freshwater vascular plants. *Rev Environ Contam Toxicol* 121:1–63.

Overmann SR, Krajicek JJ (1995) Snapping turtles (*Chelydra serpentina*) as biomonitors of lead contamination of the Big River in Missouri's old lead belt. *Environ Toxicol Chem* 14:689–695.

Parkarinen P, Tolonen K (1972) Regional survey of heavy metals in peat mosses (*Sphagum*). *Ambio* 5(1):38–40.

Parks JW, Curry C, Romani D, Russell DD (1991) Young northern pike, yellow perch and crayfish as biomonitors in a mercury contaminated watercourse. *Environ Monit Assess* 16:39–73.

Paterson S, Mackay D, Tam D, Shiu WY (1990) Uptake of organic chemicals by plants: a review of processes, correlations and models. *Chemosphere* 21:297–331.

Peterson HG, Nyholm N (1993) Algal bioassays for metal toxicity identification. *Water Pollut Res J Can* 28:129–153.

Phillips DJ (1978) Use of biological indicator organisms to quantitate organo-chlorine pollutants in aquatic environments — a review. *Environ Pollut* 16:167–229.

Powell RL, Kimerle RA, Moser EM (1996) Development of a plant bioassay to assess toxicity of chemical stressors to emergent macrophytes. *Environ Toxicol Chem* 15:1570–1576.

Ray S, White W (1976) Selected aquatic plants as indicators species for heavy metal pollution. *J Environ Sci Health* A11(12):717–725.

Richardson CJ, Sasek TW, DiGiulio RT (1990) Use of physiological and bio-chemical markers for assessing air pollution stress in trees. In *Plants for Toxicity Assessment*, Wang W, Gorsuch JW, Lower WR, Eds., ASTM STP 1091, American Society of Testing and Materials, Philadelphia, PA, pp. 143–155.

Royce C, Fletcher J, Risser P, Benenati F (1984) PHYTOTOX: a database dealing with the effects of organic chemical on terrestrial vascular plants. *J Chem Info Comput Sci* 24:7–10.

Ruston AG (1921) The plant as an index of smoke pollution. *Ann Appl Biol* 7:390–403.

Schierup H-H, Larsen VJ (1981) Macrophyte cycling of zinc, copper, lead and cadmium in the littoral zone of a polluted and a non-polluted lake. I. Availability, uptake and translocation of heavy metals in *Phragmites australis* (Cav.) Trin. *Aquat Bot* 11:197–210.

Schmitt C, Finger S, May T, Kaiser M (1987) Bioavailability of lead and cadmium from mine tailings to the pocket-book mussel (*Lampsilis ventricosa*). In *Die-Offs of Freshwater Mussels in the United States,* Neves R, Ed., U.S. Fish and Wildlife Service, Washington, D.C., pp. 115–145.

Schroder P, Debus R (1991) Response of spruce trees (*Picea Abies.* L. KARST) to fumigation with Halone 1211 — first results of a pilot study. In *Plants for Toxicity Assessment: Second Volume*, Gorsuch JW, Lower WR, Wang W, Lewis MA, Eds., ASTM STP 1115, American Society of Testing and Materials, Philadelphia, PA, pp. 258–266.

Schroll R, Bierling B, Cao G, Dorfler U, Lahaniati M, Langenbach T, Scheunert I, Winkler R (1994) Uptake pathways of organic chemicals from soil by agricultural plants. *Chemosphere* 28:297–303.

Sheehan PJ, Loucks OL (1994) Issue paper on effects characterization. In *Ecological Risk Assessment Issue Papers*, EPA/630/R-94/009, U.S. Environmental Protection Agency, Washington, D.C.

Slabbokoorn H (1993) A fertilization experiment using *Epilobium hirsutum* L. as a phytometer. Report Number 930614, Dept. of Plant Ecology and Evolutional Biology, Utrecht University, the Netherlands.

Smith BM (1991) An inter- and intra-agency survey of the use of plans for toxicity assessment. In *Plants for Toxicity Assessment: Second Volume,* Gorsuch JW, Lower WR, Wang W, Lewis MA, Eds., ASTM STP 1115, American Society of Testing and Materials, Philadelphia, PA, pp. 41–59.

Spink AJ, de Bruin JW, Notenboom E, Van Oorschot M, van der Peijl M, Slabbekoorn H, Vergoeven JTA (1994) Phytometric estimation of fertility. Proceedings of the 4th workshop of the Functional Analysis of European Wetland Ecosystems Project. Miriaflores de la Sierra, Madrid. Personal communication.

Sprenger M, McIntosh A (1989) Relationship between concentrations of aluminum, cadmium, lead and zinc in water, sediments and aquatic macrophytes in sex acidic lakes. *Arch Environ Contam Toxicol* 18:225–231.

Swanson SM, Richard CP, Freemark KE, MacQuarrie P (1991) Testing for pesticide toxicity to aquatic plants: recommendations for test species. In *Plants for Toxicity Assessment: Second Volume*, Gorsuch JW, Lower WR, Wang W, Lewis MA, Eds., ASTM STP 1115, American Society of Testing and Materials, Philadelphia, PA, pp. 77–97.

Thomas JM, Skalski JR, Cline JF, McShabe MC, Miller WE, Peterson SA, Callahan CA, Greene JC (1986) Characterization of chemical waste site contamination and determination of its extent using bioassays. *Environ Toxicol Chem* 5:487–501.

Tiagi YD, Aery NC (1985) Plant indicators of heavy metals. In *Biological Monitoring of the State of Environment: Bioindicators*, fir I.C.S.U. by I.R.L. pp. 207–222.

Tisdale SL, Nelson WL (1975) *Soil Fertility and Fertilizers*, Macmillan Publishing Co., New York, 694 pp.

Trapp S, McFarlane C, Matthies M (1994) Model for uptake of xenobiotics into plants: validation with bromacil experiments. *Environ Toxicol Chem* 13:413–422.

U.S. Environmental Protection Agency (1978) The *Selenastrum capricornutum* Printz algal assay bottle test. EPA-600/9-78-018, U.S. Environmental Protection Agency, Washington, D.C.

U.S. Environmental Protection Agency (1979) Algal, *Selenastrum capricornutum* growth test. EPA-600/4-89-001, U.S. Environmental Protection Agency, Washington, D.C.

U.S. Environmental Protection Agency (1988) Input-output mass loading studies of toxic and conventional pollutants in the Trenton Channel, Detroit River. EPA/600/3-88/004, U.S. Environmental Protection Agency, Large Lakes Res. Stat., Grosse Ile, MI.

U.S. Environmental Protection Agency (1992) Framework for ecological risk assessment. EPA/630/R-92/001, U.S. Environmental Protection Agency, Washington, D.C.

V-Brough K (1988) Heavy metal pollution from a point source demonstrated by mussel (*Unio pictorum*) at the Lake Balaton, Hungary. *Bull Environ Contam Toxicol* 41:910–914.

Velagaleti RR, Kramer D, Marsh SS, Reichenbach NG, Fleischman DE (1990) Some approaches to rapid and pre-symptom diagnosis of chemical stress in plants. In *Plants for Toxicity Assessment*, Wang W, Gorsuch JW, Lower WR, Eds., ASTM STP 1091, American Society of Testing and Materials, Philadelphia, PA, pp. 333–345.

Walsh GE, Weber ED, Simon TL, Brashers LK, Moore JC (1991) Use of marsh plants for toxicity testing of water and sediment. In *Plants for Toxicity Assessment: Second Volume*, Gorsuch JW, Lower WR, Lewis MA, Wang W, Eds., ASTM STP 1115, American Society of Testing and Materials, Philadelphia, pp. 341–354.

Wang W (1987) Factors affecting metal toxicity to (and accumulation by) aquatic organisms (overview). *Environ Int* 13:437–457.

Wang W (1991) Higher plants (common duckweed, lettuce and rice) for effluent toxicity assessment. In *Plants for Toxicity Assessment: Second Volume,* Gorsuch JW, Lower WR, Lewis MA, Wang W, Eds., ASTM STP 1115, American Society of Testing and Materials, Philadelphia, PA, pp. 68–76.

Wang W (1992) Use of plants for the assessment for environmental contaminants. *Rev Environ Contam Toxicol* 126:87–127.

Wang W, Freemark K (1995) The use of plants for environmental monitoring and assessment. *Ecotoxicol Environ Saf* 30:289–301.

Wang W, Williams JM (1988) Screening and biomonitoring of industrial effluents using phytotoxicity tests. *Environ Toxicol Chem* 7:645–652.

Weinstein LH, Laurence JA, Mansl RH, Walti K (1990) The use of native and cultivated plants as bioindicators and biomonitors of pollution damage. In *Plants for Toxicity Assessment*, Wang W, Gorsuch JW, Lower WR, Eds., ASTM STP 1091, American Society of Testing and Materials, Philadelphia, PA, pp. 117–126.

Wheeler BD, Shaw SC, Cook RED (1992) Phytometric assessment of the fertility of undrained rich-fen soils. *J Appl Ecol* 29:466–475.

Wu L, Chen J, Ganji KK, Banuelos GS (1995) Distribution and biomagnification of selenium in a restored upland grassland contaminated by selenium from agricultural drain water. *Environ Toxicol Chem* 14:733–742.

chapter thirteen

Metal accumulation by aquatic macrophytes

Wuncheng Wang and Michael A. Lewis

Introduction

Aquatic macrophytes, typically classified as submerged, floating, or emergent plant species, are widely distributed in various aquatic environments, from fresh to salt water, as well as the littoral through pelagic zones. They have several characteristics favorable for metal accumulation. First, in terms of biomass, aquatic and wetland macrophytes are the predominant organisms in the highly productive, littoral ecosystems, such as wetlands and seabeds (Brix and Schierup 1989). Second, leaves and epiphytes provide an expanded area to trap particulate matter, sorb metal ions, and accumulate and sequester pollutants (Ward

1987; Bishop and DeWaters 1988; Levine et al. 1990). Third, rooted species can absorb metals through their roots and rhizomes as well as through their leaves (Welsh and Denny 1980; Heisey and Damman 1982). Fourth, aquatic macrophytes are stationary and constantly exposed to (and absorbing) contaminants such as metals (Say et al. 1981).

In this chapter, the discussion will focus primarily on the ability of vascular plants and mosses, both fresh- and saltwater species, to accumulate toxic heavy metals or metalloids. Macroalgae, microalgae, and other species will be mentioned for comparison only. This chapter serves as an overview of studies on the topic since 1980 and provides basic, synthesized information. A few major works prior to 1980 also are cited. Many earlier and other uncited publications are included in the Additional Literature section at the end of this chapter.

Selected factors affecting metal accumulation

Several biological and nonbiological factors are known to affect metal accumulation by aquatic macrophytes. The biological factors include species, age, generation, and route and mode of metal accumulation. The nonbiological factors can be divided into environmental and chemical factors. Examples are temperature, light, metal concentration, and exposure time. Table 1 summarizes the important factors mentioned in the literature.

Table 1   Major Factors Affecting Metal Accumulation by Aquatic Plants

	Nonbiological factors	
Biological factors	Environmental factors	Chemical factors
Species	Temperature	Concentration
Age	Light	Duration of exposure
Generation	Season	Disturbance or mixing
Translocation	pH	Competing ions
Route of uptake (leaf or root)	Salinity	Complexing agents
Mode of uptake (live or dead)	Plant nutrients	

Plant species

Species type is important, as evidenced in Table 2. Several studies compared relative uptake rates of different plant species; among them, filamentous alga (*Cladophora* sp.) was reported to accumulate large concentrations of metals (Abo-Rady 1980; Capone et al. 1983; Wells et al. 1980). For example, among macrophytes collected from the River

Leine in Germany, *C. glomerata* accumulated more copper (Cu), mercury (Hg), and nickel (Ni) than pondweed (*Potamogeton pectinatus*) and horned pondweed (*Zannichellia palustris*), whereas *Zannichellia* accumulated more cadmium (Cd) and zinc (Zn) than *Potamogeton* and *Cladophora* (Abo-Rady 1980). Similarly, *Cladophora* was found to accumulate larger metal concentrations (15 metals) from Lake Huron in the United States when compared with 21 other species of macrophytes (Wells et al. 1980). *Cladophora* accumulated two- to fourfold more chromium (Cr) and lead (Pb) than ditch grass (*Ruppia* sp.) (Capone et al. 1983).

Aquatic macrophytes, compared with other plant and animal species, have been reported to have a large or comparable capacity for metal accumulation. Cherry et al. (1984) found that in a coal-ash settling basin, duckweed (*Lemna perpusilla*) and invertebrates (aquatic insects, crayfish, and snails) accumulated approximately the same concentrations of metals (arsenic [As], Cd, Cr, Cu, selenium [Se], and Zn), whereas mosquitofish accumulated 35% less than duckweed and the invertebrates. Guthrie and Cherry (1979) reported that among biota, green alga (*Hydrodictyon* sp.) and *L. perpusilla* accumulated predominantly aluminum (Al) and iron (Fe), and other macrophytes were the major accumulators of barium (Ba) and manganese (Mn). In a baseline study of a relatively nonindustrialized watershed, both aquatic macrophytes and mollusks in the Kennebec River in northern Maine contained larger concentrations of Cr, Cu, Pb, and Zn than fish (Friant 1979). Genin-Durbet et al. (1988) also reported that waterweed (*Elodea* sp.) had a large capacity to accumulate metals, especially Cd and Hg, in a microcosm study that included bacteria and algae. Jana (1988) reported the decreasing order of Cr accumulation in certain plant species as water hyacinth (floating macrophyte), *Hydrilla verticillata* (submerged macrophyte), and *Oedogonium aerolatum* (alga); the decreasing order of Hg accumulation was *Hydrilla*, *Oedogonium*, and water hyacinth.

Metal accumulations in macrophytes may be different from those in fish in the same body of water. Ajmal et al. (1987) reported the decreasing order of metal concentrations in water hyacinth as Fe, Mn, Zn, Ni, Cu, Cr, Pb, cobalt (Co), and Cd; whereas in fish (*Heteropnuestes fossilis*), the decreasing order of concentration was Fe, Zn, Mn, Pb, Ni, Co, Cu, Cd, and Cr. In general, Fe had the largest concentration in water and sediment and also was the metal with the largest accumulations in these plants and fish.

Among submerged, floating, and emergent macrophytes, submerged species are likely to accumulate the largest metal concentration (Franzin and McFarlane 1980; Aulio and Salim 1982; Heisey and Damman 1982; van der Werff and Pruyt 1982; Ozimek 1985; Schuler et al. 1990). Of the five species of pondweed (*Potamogeton* spp.) examined by Aulio and Salim (1982), *P. perfoliatus*, *P. obtusifolius*, and *P. praelongus*

Table 2 Metal Accumulation by Aquatic Macrophytes as Affected by Plant Species

Metals	Comments	Ref.
Cd, Cu, Hg, Ni, Pb, Zn	*Cladophora* was especially useful, compared with other species, as a biomonitor of metals.	Abo-Rady 1980
Cu, Zn	Smaller concentrations in floating plants than in submerged plants.	Aulio and Salim 1982
Cr, Pb	*Cladophora* accumulated two- to fourfold more metals than *Ruppia*.	Capone et al. 1983
As, Cd, Cr, Cu, Se, Zn	Macrophytes accumulated 11.1 ppm (mean value for six metals), whereas invertebrates and mosquitofish accumulated 11.8 and 7.5 ppm, respectively.	Cherry et al. 1984
As, Cu, Mn, Pb, Zn	*Myriophyllum* accumulated more metals than *Calla, Nupher, Sparganium,* and *Utricularia*	Franzin and McFarlane 1980
Cr, Cu, Pb, Zn	Macrophytes accumulated metals equal to or more than fish and mollusks from area of paper-mill preoperation survey	Friant 1979
B	Common duckweed contained 10 to 45 times more B than *Ceratophyllum* in the same body of water.	Glandon and McNabb 1978
Cu	Smaller concentration in emergent species than in floating and submerged species.	Heisey and Damman 1982
Cr	Decreasing order of accumulation was water hyacinth, *Hydrilla,* and *Oedogonium.*	Jana 1988
Hg	Decreasing order of accumulation was *Hydrilla, Oedogonium,* and water hyacinth.	
Mo	Above-ground organs of cattail and reed were comparable.	
Pb	Young, above-ground organs of reed and cattail contained similar metal content in early June; thereafter, reed contained six times more Pb than cattail.	Kufel 1991

Metal	Observation	Reference
As	Concentrations in 10 macrophytes collected from 22 pools varied from 2- to 20-fold differences.	Lee et al. 1991
As, Cd, Cu, Hg, Pb	Water hyacinth collected from one area had three- to sevenfold larger metal content than from another area of the same lake	Lin and Zhang 1990
Cd, Cu, Pb, Zn	*Posidonia* accumulated more than eelgrass.	Malea and Haritonidis 1989
Hg	Submerged species accumulation rate *Utricularia* > *Certophyllum* = *Nitella* > *Najas*.	Mortimer 1985
Hg	*Elodea canadensis* accumulated two times more than *E. densa*.	Salim 1988
Ni	At pH ≤ 5.0, plant leaves removed Ni in the order: pine > cinchona > oak > cyprus.	Schuler et al. 1990
Se	Emergent species contained smaller concentrations than submerged species; cattail accumulated larger concentration than alkali bulrush.	van der Merwe et al. 1990
Mn, Zn	*Typha* accumulated more than *Arundo*.	van der Werff and Pruyt 1982
Ni	*Arundo* accumulated more than *Typha*.	
Cd, Cu, Pb, Zn	Submerged and partly submerged species accumulated larger metal content than floating species (*Lemna* and *Spirodela*).	Wells et al. 1980
15 metals[a]	Among 71 specimens (of 22 species) tested, *Cladophora* and *Typha* accumulated the largest metal concentrations.	
Cu, Pb	*Potamogeton crispus* accumulated more than twofold larger metal concentrations in shoots and roots than *P. perfoliatus*.	Welsh and Denny 1980

[a] Ag, As, Ba, Cd, Ce, Co, Cr, Cs, Ni, Pb, Sb, Se, Th, U, and Zn were determined using neutron activation analysis.

were submerged taxa; *P. gramineus* had submerged and floating leaves; and *P. natans* had only floating leaves. Submerged species accumulated the largest Cu, Fe, Mn, and Zn concentrations, possibly due to foliar uptake. Glandon and McNabb (1978), on the contrary, reported that common duckweed (*L. minor*) accumulated 10 to 45 times more boron than the submerged species of hornwort (*Ceratophyllum* sp.). Among five macrophytes, Franzin and McFarlane (1980) selected water milfoil (*Myriophyllum exalbescens*) as the indicator species for study of an area potentially affected by proximity to a base-metal smelter. Heisey and Damman (1982) reported that Pb accumulation was primarily through sediment, so that differences among accumulations in emergent, floating, and submerged plants were negligible. In the same study, they indicated that Cu accumulations by *Potamogeton* and other macrophytes were indicative of Cu concentrations in the water column.

Metal accumulation also can vary among submerged species of macrophytes. Welsh and Denny (1980) reported that *P. crispus* accumulated more than two times as much Cu and Pb in both shoots and roots than *P. perfoliatus*. Mortimer (1985) reported that *Elodea canadensis* accumulated more than two times as much inorganic Hg than *E. densa* accumulated in a 10-d flow-through experiment. The Hg accumulation rates of other submerged species were in the decreasing order of *Utricularia, Ceratophyllum,* stonewort (*Nitella* sp.), and *Najas* (Mortimer 1985). Everard and Denny (1985) observed that in 5 min of exposure to an initial Pb concentration of 10 mg/l, the leaves of several different species contained similar concentrations; after 24 h, *E. canadensis* and a freshwater diatom, *Navicula* sp., accumulated twice as much Pb as *P. pectinatus* or the moss *Fontinalis antipyretica*.

Among marine macrophytes, *Posidonia oceanica* accumulated larger concentrations of Cd, Cu, Pb, and Zn than eelgrass, *Zostera marina* (Brix and Lyngby 1982; Malea and Haritonidis 1989). The mean Hg concentrations in seaweed showed considerable variation between sampling sites and species; seaweed (*Gracilaria verrucosa*) was found to accumulate larger Hg concentrations when compared with *Fucus vesiculosus* and *Ulva lactuca* (Ferreira 1991). The larger accumulation was probably due to the greater productivity of this species and its longer period of immersion through tidal cycles. In a salt marsh on the west coast of India, Joshi et al. (1987) reported that among five saltmarsh species, *Atriplex griffithii* and glasswort (*Salicornia brachiata*) accumulated large concentrations of Fe and Mn, whereas rush (*Juncus maritimus*) accumulated large concentrations of Cu.

Both reed (*Phragmites australis*) and cattail (*Typha augustifolia*) are common in littoral zones and have a dominant role in nutrient and metal cycling. Kufel (1991) found that molybdenum (Mo) concentrations in the above-ground organs of reed and cattail were comparable. On June 2, 1980, young shoots of both species contained the same concentrations of Pb (0.20 µg/g). On June 22, 1980 (and thereafter), reed shoots contained larger Pb concentrations than cattail shoots, 2.8 compared to 0.50 µg/g.

Aquatic and marsh macrophytes accounted for 15 to 20% of the water surface of the Okefenokee Swamp along the border between Georgia and Florida (Bosserman 1985). Bosserman reported that among eight plant species in the swamp, stems of yellow water lily (*Nuphar advena*) and leaves of bladderwort (*Utricularia* sp.) and redroot (*Lachnanthes caroliniana*) had the largest concentrations of Al, Cd, Co, Cr, Cu, Ni, Pb, and Zn. Other species used for comparison in the study were white water lily (*Nymphaea odorata*), pickerel weed (*Pontederia cordata*), yellow-eyed grass (*Xyris smalliana*), sphagnum (*Sphagnum cuspidatum*), and beakrush (*Rhynchospora inundata*).

In a streambed community survey of various ecosystem components, Brisbin et al. (1989) reported that the best indicator species for radioactive cesium (Cs) contamination along Steel Creek, SC were cattail (*Typha* sp.) and buckwheat (*Polygonum* sp.).

Season

Several studies indicated that metal accumulation declined during the summer in many species: pickerel weed (*Pontederia cordata*), *A. graffithii*, *M. exalbescens*, *Z. marina*, and *Ponsidonia oceanica* (Heisey and Damman 1982; Kimball and Baker 1982; Lyngby and Brix 1982; Joshi et al. 1987; Malea and Haritonidis 1989). The seasonal patterns, however, are quite complex and depend on the type of metal speciation, plant species, and location. For example, Heisey and Damman (1982) found that Cu concentrations in the shoots of *Pontederia* were smallest in late August and early September, Cu in *Potamogeton* did not exhibit a seasonal pattern, and Pb concentrations in shoots of *Pontederia* declined slowly throughout the growing season. Kimball and Baker (1982) reported that *Myriophyllum* sp. in Lake Winnipesaukee, NH appeared to have a late winter-spring concentration maximum for Fe and Mn, a summer concentration minimum for Zn, and no detectable pattern for Cu.

Larsen and Schierup (1981) reported that Cu concentrations in stems of ditch reed (*Phragmites australis*) and Zn concentrations in stems and leaves reached a maximum during the growing season and decreased thereafter. Summer maxima of Mn, Ni, and Zn in emergent plants (*T. capensis* and *Arundo donax*) were reported in areas affected by gold mines and industrial effluents and corresponded with the periods when stream water became more alkaline (van der Merwe et al. 1990).

In *Z. marina*, the metal concentrations in the above-ground parts of the plant were largest during winter-early spring, declined throughout the summer, and increased during fall (Lyngby and Brix 1982). Ward (1987) reported that the concentrations of Cd in the seagrass varied 27-fold when the seagrass specimens were collected during

October 1980 and February 1991 in an area contaminated by a lead smelter in South Australia. He suggested that metal biomonitoring using seagrass leaves at different locations should be done at the same time of year to avoid temporal variation. Joshi et al. (1987) reported that Fe, Mn, Ni, and Zn concentrations in five species of succulent halophytes in Indian salt marshes increased during the winter, but not during the monsoon season.

Kufel and Kufel (1985) reported that there were seasonal variations in the uptake of different metals in *Typha* and *Phragmites* sp. in Lake Gardynskie, Poland. First, Mo concentrations in plant shoots showed an early summer maximum followed by a slow decrease during the remainder of the growing season. Second, Cd, Co, and Pb concentrations were nearly constant throughout the year. Third, Cu and Mn concentrations showed a late summer maximum. Metal accumulation by aquatic mosses such as *Fontinalis antipyretica* was reported to lack seasonal variation and thus has a strong advantage over many other macrophytes in documenting environmental changes (Kelly et al. 1987).

pH

The pH of a solution can affect metal speciation and metal uptake (Table 3). Nasu et al. (1983) reported that within a pH range of 3.6 to 5.1, a pH increase resulted in the increase of Cd and Cu concentrations in duckweed (*Lemna paucicostata*) and decreased duckweed growth as expressed in wet weight. O'Keeffe et al. (1984) observed that the Cd uptake rate by water hyacinth (*Eichhornia crassipes*) in Hoagland culture medium was very sensitive in the pH range of 2.0 to 4.0 and less sensitive above pH 4.0. At pH 2.0, the roots of the plants were bleached, and the plants did not survive. When pH increased to 4.0, the uptake rate also increased. O'Keeffe et al. (1984) suggested that the dissociated form of a charged group (such as the carboxyl group) in the plant tissues increased as pH increased from 2.0 to 4.0, facilitating metal uptake. In contrast, another report showed that Cu uptake by *E. crassipes* was not significantly different throughout a pH range of 3.0 to 10 (Lee and Hardy 1987). Likewise, in a laboratory study using the 2-cm tips of moss, Wehr et al. (1987) reported that pH had only a slight effect on Zn accumulation.

The interactive relation between pH and temperature also can have an effect on metal accumulation. Chawla et al. (1991) found that at 20°C the effect of pH on Cd uptake in common duckweed (*L. minor*) was less prominent than at lower temperatures. Apparently, under optimum conditions for plant growth, the plants grew more vigorously and were able to sequester Cd more effectively until the plants succumbed to the effects of metal toxicity.

In an 8-year study of a coal-ash settling basin, Cherry et al. (1984) found that As, Cd, and Se accumulations in duckweed (*L. perpusilla*) were larger at pH 7.4 than at pH 5.4 or 6.5. Cr and Cu accumulations by the same plant were larger at pH 5.4 than at pH 6.5 or 7.4. Zn accumulation by duckweed was larger at pH 6.5 than at pH 5.4 and 7.4.

In an investigation of aquatic liverwort (*Scapania undulata*) in rivers and streams in England, France, Germany, and Ireland, Whitton et al. (1982) found that metal concentrations in plants varied spatially by as much as three orders of magnitude; Fe and Cu ranged from 1120 to 89,000 µg/g and 9 to 1700 µg/g, respectively. Statistical analyses of the data indicated that a pH increase resulted in increased accumulation of Cd and Zn, but had only a moderate effect on Pb accumulation.

The effect of pH on metal accumulation by plants can be demonstrated by using characteristics of different metals. Comparing thallium (Tl) and Cd accumulation by *L. minor*, Smith and Kwan (1989) showed that Tl was a "mobile" metal with little affinity to organic ligands. Thus, the plant had minor Tl accumulation. Cd, on the other hand, had a strong affinity to organic ligands, which, being amphoteric, are strongly affected by pH. As a result, Cd accumulation by the plant is strongly pH dependent.

Other factors

There are many other nonbiological factors affecting the amount of metal accumulation by plants (Table 3). In general, metal accumulation by plants increases as temperature increases (Dieckmann 1982; Maury-Brachet et al. 1990; Chawla et al. 1991). Maury-Brachet et al. (1990) studied Hg accumulation by *Elodea densa* from a sediment source and found that temperature (range 18 to 24°C) and photoperiod (range 8 to 16 h) separately and in combination had a positive effect on methylmercury accumulation. The synergistic effects of these two nonbiological factors varied according to which plant organs were analyzed for metal accumulation.

Metal accumulations by macrophytes can be affected by metal concentrations in water and sediments. Lin and Zhang (1990) reported that As, Cd, Cu, Hg, and Pb concentrations in water hyacinth (*Eichhornia crassipes*) varied from three- to sevenfold when comparing specimens collected from different locations in Lake Dianchi, China, indicating that even in the same lake water and sediment qualities vary. Lee et al. (1991) reported that As concentrations in 10 macrophytes collected from 22 pools affected by tin-mining activities in Malaysia varied 2- to 20-fold. In a laboratory study, Lee et al. (1991) reported that *Hydrilla verticillata* accumulated As as a function of the initial concentration in water.

The amount of metal accumulation by aquatic macrophytes can be affected by the presence of chelating agents or other ions in solution. Nasu et al. (1983) reported that only 30 µM of EDTA (ethylenediaminetetraacetic acid) were sufficient to prevent Cu uptake by *L. paucicostata*

Table 3   Metal Accumulation by Aquatic Macrophytes as Affected by Selected Environmental Factors

Metals	Aquatic macrophytes	Comments	Ref.
		A. Season	
Cu	*Pontederia*	Cu concentrations in shoots decreased in late summer.	Heisey and Damman 1982
Pb	*Pontederia*	Pb concentrations in shoots decreased slowly throughout the growing season.	
Fe, Mn, Ni, Zn	*Atriplex griffithii*	Metal accumulation increased during winter.	Joshi et al. 1987
Cu, Fe, Mn	Water milfoil	Late winter maximum.	Kimball and Baker 1982
Zn	Water milfoil	No significant seasonal variation.	
Cu	Ditch reed	Accumulation in stems increased throughout the growing season.	Larsen and Schierup 1981
Cd, Cu, Pb, Zn	Eelgrass	Metal concentrations in the above-ground organs were largest during winter-early spring, decreased throughout the summer, and increased during fall.	Lyngby and Brix 1982
Cu, Pb, Zn	*Posidonia*	Metal concentrations were largest in winter and early spring and smallest in summer.	Malea and Haritonidis 1989
Cd	*Posidonia*	Metal concentrations were largest during summer and smallest in spring.	
Hg	5 species[a]	Metal concentrations changed considerably within a month and also from month to month.	Mortimer 1985
Se	Widgeongrass	Metal concentrations were substantially larger in August and December than in May; concentrations in seeds were larger in December than in August.	Schuler et al. 1990
Ba, Cd, Zn	Moss	No seasonal pattern of metal accumulation.	Wehr and Whitton 1983a

B. pH

Metal	Species	Description	Reference
Al, Mn, Zn	Aquatic liverwort	Metal accumulation decreased as water pH decreased.	Caines et al. 1985
Cd	Duckweed	Metal accumulation depended on pH-temperature interactions; at 30, 20, and 10°C, the largest uptake was at pH 6.5, 7.5, and 7.0.	Chawla et al. 1991
As, Cd, Cr, Cu, Se, Zn	7 species[b]	Metal accumulation depended on pH and metal species.	Cherry et al. 1984
Cu	Water hyacinth	pH over the range of 3.0 to 10 did not significantly affect uptake.	Lee and Hardy 1987
Cd	Duckweed	Metal accumulations increased as the initial pH increased (between 3.6 to 5.1).	Nasu et al. 1983
Cu	Duckweed	Metal accumulation did not change when the initial pH was between 3.6 to 5.1.	
Cd	Water hyacinth	Uptake rate increased as pH increased from 2.0 to 4.0; at higher pH, the trend reversed.	O'Keeffe et al. 1984
Cd	Moss	Cd concentrations in moss collected from England and Europe (52 sites) were significantly correlated with pH ($p < 0.01$).	Say and Whitton 1983
Pb, Zn	Aquatic liverwort	Pb accumulation increased moderately as pH increased; Zn accumulation increased as pH increased.	Whitton et al. 1982

C. Other factors

Metal	Species	Description	Reference
Cd	Duckweed	Metal accumulation depended on pH-temperature interaction.	Chawla et al. 1991
Cd	Seaweed	Salinity and light affected metal accumulation.	Dieckmamm 1982
Pb	Moss	Rainfall increased metal accumulation.	Everard and Denny 1985

Table 3 (continued)

Metals	Aquatic macrophytes	Comments	Ref.
As, Cd, Cu, Fe, Mn, Pb, Zn	*Myriophyllum*	Greater rainfall possibly caused plants to accumulate up to eightfold more metals.	Franzin and McFarlane 1980
Pb, Zn	Duckweed, water velvet	The presence of one metal inhibited the uptake of another metal.	Jain et al. 1990
Cu	Water hyacinth	Complexing agents reduced Cu uptake; volume increased and Cu uptake increased. Stirring of solution increased Cu uptake. Competing ions Cd, Co, Fe, Hg, Mg, Ni, and Zn decreased Cu uptake.	Lee and Hardy 1987
Hg	*Elodea*	Increased temperature and photoperiod increased accumulation separately and in combination.	Maury-Brachet et al. 1990
Cu	*Bacopa*	Cu accumulation was stimulated by Cd.	Sinha and Chandra 1990
Cd	*Bacopa*	Cd accumulation was inhibited by Cu.	
Zn	Moss	Nutrient-rich solution reduced metal accumulation.	Wehr et al. 1987
Zn	Moss	Ca, Mg, Mn, and chelating agent reduced metal accumulation.	
Cd, Zn	Aquatic liverwort	Metal accumulation increased as Ca concentration increased.	Whitton et al. 1982

[a] Macrophytes were *Elodea*, *Phalaris*, *Sagittaria*, and *Sparganium* (*S. angustifolium* and *S. eurycarpum*).

[b] Mean values of *Andropogon*, *Cyperus*, *Hydrodictyon*, *Lemna*, *Oscillatoria*, *Pontederia*, and *Typha*.

at concentrations of 5.0 to 10 µM, whereas 400 µM of EDTA were necessary to prevent Cd uptake at the same concentrations of 5.0 to 10 µM. Nasu et al. (1983) noted that the ammonium ion suppressed Cu absorption, whereas Cd absorption was only slightly suppressed. Lee et al. (1991) reported that the presence of a phosphate concentration of 5 mg/l did not affect accumulation of As (0.4 mg/l initial concentration) by *H. verticillata*. The accumulation, however, stopped when the phosphate concentration reached 160 mg/l. Transplanting the moss *Rhynchostegium riparioides* between two streams, Wehr et al. (1987) found that Zn accumulation was less in a nutrient-rich (large nitrate and phosphate concentrations) than in a nutrient-poor stream. Other nonbiological factors such as large concentrations of NaCl, metal ligands, and surfactants (both anionic and nonionic) were reported to have little effect on Cu absorption by *Iris pseudacorus* (Piccardi and Clauser 1983).

Metal accumulation by aquatic macrophytes also can be affected by competing metal ions. Jain et al. (1990) compared Pb and Zn uptake by water fern (*Azolla* sp.) and *Lemna* sp. and found that the presence of one metal ion in solution decreased the uptake rate of the other, indicating competitive absorption of these two metals (probably the same absorption site in plant tissues). Nasu et al. (1984) reported that Cd and Cu were both extremely toxic to *Lemna* sp.; Cd inhibited frond multiplication, whereas Cu inhibited both frond multiplication and frond growth (i.e., biomass increase of each frond). They demonstrated that the presence of Cd did not affect the absorption of Cu, whereas the absorption of Cd decreased considerably with increasing concentrations of Cu in the medium. Mo et al. (1988) reported that the uptake of Cu ions by *Lemna* sp. was suppressed significantly by the presence of Pb ions, and the presence of Al ions increased the suppressing effects. Franzin and McFarlane (1980) compared interlake variation of metal accumulation and noted that Ca concentrations in lake waters might have a modifying effect on plant uptake of all metals. In a lake highly contaminated by aerial deposition of toxic metals (Cd, Cu, and Zn), Ca appeared to inhibit metal accumulation by *Myriophyllum*. In laboratory experiments, elements decreasing Zn accumulation included Ca, Mg, and Mn (Wehr et al. 1987). Lee et al. (1991) reported that the amount of As accumulated by *H. verticillata* varied depending on whether deionized-and-distilled water or an ambient water sample was used.

In an investigation of metal accumulation by *M. exalbescens* in contaminated lakes near a base-metal smelter in Flin Flon, Canada, Franzin and McFarlane (1980) reported that metal concentrations in water milfoil were much larger in 1976 than in 1975: 12,000 compared to 1600 µg/g for Zn, 350 compared to 66 µg/g for Cu, 31 compared to 6 µg/g for Cd, and 240 compared to 27 µg/g for Pb. The large interyear variation was attributed to increased rainfall in 1976, thus increasing the transport of metals into the lakes. Another possibility for the increase,

suggested in other studies (see the following), was the greater disturbance of lake sediments due to the rainfall.

Storms and associated disturbances can resuspend sediment as a
secondary source of contamination. The sediments of Lake Ullswater,
Great Britain were enriched with Pb from past mining activities. Everard and Denny (1985) reported that by placing moss monitors (*Sphagnum papillosum*) in the lake water, Pb was taken up by the moss at the
rate of 0.1 mg of Pb per gram dry weight in 72 h during stormy weather.
In the same lake, there was no Pb uptake during periods of calm
weather. Relevant results also were reported using laboratory experiments. Lee and Hardy (1987) found that Cu uptake by *E. crassipes*
increased during stirring of a test solution in the laboratory experiment,
suggesting that Cu removal was partially diffusion limited.

Metal accumulation factor

Because of their fibrous root systems with large contact areas, aquatic
plants generally have the ability to accumulate larger metal concentrations in plant organs than those present in the surrounding water. The
accumulation factor (AF) is defined as the metal concentration in plant
tissue (micrograms per gram) divided by the metal concentration in
water (micrograms per milliliter); for practical purposes, AF is dimensionless. AF also has been known as a concentration factor (Charpentier
et al. 1987; Jain et al. 1990; Lin and Zhang 1990), a bioconcentration
factor (Cherry et al. 1984; Smith and Kwan 1989), an enrichment ratio
(Satake et al. 1989), a concentration ratio (Twining 1988), and a bioconcentration ratio (Knowlton et al. 1983).

Other expressions also are used to express the amount of accumulation. Mo et al. (1988) expressed the proportion of metal uptake as the
weight percentage in plants of the original metal solution. De Wet et
al. (1990) defined a bioaccumulation factor using metal contents in
plants vs. metal contents in sediment. Reported results of AF studies
are listed in Table 4 for laboratory and field studies.

In laboratory studies, the metal concentration in water can be
defined precisely, and the AF can be determined. Several studies have
indicated that the AF for a variety of metals and aquatic macrophytes
can be large. Faraday and Churchill (1979) reported that the Cd AF for
Z. marina was dependent on plant organs, exposure time, and Cd
concentration in the medium. When the Cd concentration in water was
1 mg/l and exposure time was 3 d, AF values in the root rhizome and
the leaves were 48 and 94, respectively. Mortimer (1985) used 0.7 µg/l
of radioactive inorganic Hg and found that *Elodea densa* and *E. canadensis* gradually increased Hg accumulation from test day 2 to test day 10;
the AF increased from 200 to 580, respectively. In a similar experiment
using 0.28 µg/l of inorganic Hg, four submerged macrophytes (*Cerato-*

phyllum, Najas, Nitella, and *Utricularia*) showed substantially larger Hg AFs, from 900 to 3300, compared with *Elodea* species (Mortimer 1985).

In general, when the metal concentration in water increases, the amount of metal accumulation in plants increases, whereas the AF values decrease. For example, Nakada et al. (1979) reported AFs of Cd in *Elodea* were 1700, 1100, 1100, and 820 when the initial Cd concentrations were 0.025, 0.05, 0.10, 0.50 mg/l, respectively. Likewise, AF values of Cu in a prolific, partially submerged macrophyte, *Bacopa monnieri,* were 1200, 1100, 900, and 450 when the initial Cu concentrations were 0.5, 1.0, 2.0, and 5.0 mg/l, respectively (Sinha and Chandra 1990).

In uncontrolled field studies where moderate concentrations of metal ions were present, the AF can be larger than 1 million (Table 4). Satake et al. (1989) collected aquatic liverwort (*Scapania undulata*) specimens from a tributary of Glenridding Beck, England that received drainage water from an abandoned lead mine. They reported that Pb in the shoots ranged from 0.7 to 2.4% on the basis of dry weight, giving AFs of 350,000 to 1.2 million in the shoot tip and shoot base, respectively. They found that Pb was localized in the cell wall and not in the nucleus and other components.

Other, extremely large AFs also were reported, including Al, Mn, and Zn in streams where the H^+ ion content was 1 µeq/l (microequivalents per liter); Cu in *Myriophyllum*; Hg in marine macroalgae; As in tin-mining regions in Malaysia; and As, Cd, Cu, Hg, and Pb in Lake Dianchi, China (Caines et al. 1985; Janauer 1985; Lin and Zhang 1990; Ferreira 1991; Lee et al. 1991). Some of these AF values far exceeded 10,000. Janauer (1985) reported that the decreasing order of AF by plant species was *Fontinalis, Elodea, Ranunculus, Myriophyllum, Potamogeton,* and *Lemna.* Among 20 different species of aquatic macrophytes, Janauer (1985) also reported that the decreasing order of AF for metal species was Mn, Fe/Zn, Cu, Cd, and Ni.

Reported studies on metal accumulation by aquatic macrophytes

Laboratory studies

Many laboratory studies have been conducted using various metal ions, aquatic macrophytes, and experimental conditions (Table 5). Results given in Table 5 reflect the complexities of using plants for environmental impact assessment because there is no consensus test methodology. Nevertheless, important findings of these studies are summarized as follows.

Several studies indicated that the rate of metal accumulation by aquatic macrophytes is rapid. Sen and Mondal (1990) found that when 1 l of water containing 0.75 or 1.2 mg/l of Cu was treated with 20 g of floating moss *Salvinia,* 90% of the Cu was removed in 1 d. In another

Table 4 Metal Accumulation Factor, Dimensionless, Expressed as [(Concentration in Plant, µg/g)/Concentration in Water, µg/mL)], All on Dry-Weight Basis, Except as Noted by (w), Wet-Weight Basis

Metals	Plants (organs)		Accumulation factor	Experimental conditions	Ref.
			A. Laboratory studies		
Cd	Duckweed		650	0.1 mg/l for 4 d	Charpentier et al. 1987
			640	1.0 mg/l for 4 d	
Hg	*Salvinia*	(root)	210(w)	3.0 mg/l for 2 d	De et al. 1985
		(shoot)	70(w)		
		(root)	340(w)	6.0 mg/l for 2 d	
		(shoot)	170(w)		
		(root)	1,100(w)	60 mg/l for 2 d	
		(shoot)	560(w)		
Cd	Eelgrass	(rhizome)	48	1.0 mg/l for 3 d	Faraday and Churchill 1979
		(leaf)	94		
Pb	Duckweed		68	1.0 mg/l for 14 d	Jain et al. 1990
			64	2.0 mg/l for 14 d	
			59	4.0 mg/l for 14 d	
			51	8.0 mg/l for 14 d	
Zn	Duckweed		44	1.0 mg/l for 14 d	
			39	2.0 mg/l for 14 d	
			35	4.0 mg/l for 14 d	
			28	8.0 mg/l for 14 d	
Pb	Water velvet		70	1.0 mg/l for 14 d	
			66	2.0 mg/l for 14 d	
			62	4.0 mg/l for 14 d	
			55	8.0 mg/l for 14 d	

Metal	Plant		Value	Concentration/duration	Reference
Zn	Water velvet		47	1.0 mg/l for 14 d	Jain et al. 1990
			42	2.0 mg/l for 14 d	
			38	4.0 mg/l for 14 d	
			31	8.0 mg/l for 14 d	
Hg	*Ceratophyllum*		1,300	0.28 µg/l for 14 d	Mortimer 1985
	Najas		900	0.28 µg/l for 14 d	
	Nitella		1,300	0.28 µg/l for 14 d	
	Utricularia		3,300	0.28 µg/l for 14 d	
Hg	*Elodea canadensis*		580	0.7 µg/l for 10 d	
	E. densa		200	0.7 µg/l for 10 d	
Cd	*E. densa*		900	10 µg/l for 12 d	
Ni			200	10 µg/l for 12 d	
Pb			1,400	10 µg/l for 12 d	
Cd	Water hyacinth	(AG)	44	1.0 mg/L for 16 d	Muramoto and Oki 1983
		(BG)	100	1.0 mg/l for 16 d	
		(AG)	47	4.0 mg/l for 16 d	
		(BG)	47	4.0 mg/l for 16 d	
		(AG)	35	8.0 mg/l for 16 d	
		(BG)	29	8.0 mg/l for 16 d	
Hg	Water hyacinth	(AG)	0.044	0.5 mg/l for 16 d	
		(BG)	440	0.5 mg/l for 16 d	
		(AG)	0.032	1.0 mg/l for 16 d	
		(BG)	490	1.0 mg/l for 16 d	
		(AG)	0.048	2.0 mg/l for 16 d	
		(BG)	340	2.0 mg/l for 16 d	
Pb	Water hyacinth	(AG)	4.2	1.0 mg/l for 16 d	
		(BG)	300	1.0 mg/l for 16 d	
		(AG)	19	4.0 mg/l for 16 d	

Table 4 (continued)

Metals	Plants (organs)		Accumulation factor	Experimental conditions	Ref.
Pb	Water hyacinth	(BG)	220	4.0 mg/l for 16 d	Muramoto and Oki 1983
		(AG)	20	8.0 mg/l for 16 d	
		(BG)	127	8.0 mg/l for 16 d	
Cd	Elodea		1,700	0.025 mg/l for 30 d	Nakada et al. 1979
			1,100	0.05 mg/l for 30 d	
			1,100	0.10 mg/l for 30 d	
			820	0.50 mg/l for 30 d	
Cu	Elodea		1,400	0.025 mg/l for 30 d	
			1,000	0.05 mg/l for 30 d	
			610	0.10 mg/l for 30 d	
Pb	Elodea		6,200	0.025 mg/l for 30 d	
			4,300	0.05 mg/l for 30 d	
			2,800	0.10 mg/l for 30 d	
			680	0.50 mg/l for 30 d	
Zn	Elodea		3,000	0.10 mg/l for 30 d	
			710	0.50 mg/l for 30 d	
Cd	Bacopa		530	0.50 mg/l for 7 d	Sinha and Chandra 1990
			680	1.0 mg/l for 7 d	
			660	2.0 mg/l for 7 d	
			390	5.0 mg/l for 7 d	
Cu	Bacopa		1,200	0.50 mg/l for 7 d	
			1,100	1.0 mg/l for 7 d	
			900	2.0 mg/l for 7 d	
			450	5.0 mg/l for 7 d	

B. Field studies

Cd	3 species[a]	300	River Leine, Germany	Abo-Rady 1980
Cu		340	River Leine, Germany	
Hg	3 species[a]	760	River Leine, Germany	Abo-Rady 1980
Ni		590	River Leine, Germany	
Pb		2,400	River Leine, Germany	
Zn		2,200	River Leine, Germany	
Al	Aquatic liverwort	16,000	I group, H ion 58 µeq/l	Caines et al. 1985
		87,000	II group, H ion 10 µeq/l	
		356,000	III group, H ion 1 µeq/l	
Mn	Aquatic liverwort	700	I group, H ion 58 µeq/l	
		116,000	II group, H ion 10 µeq/l	
		2,300,000	III group, H ion 1 µeq/l	
Zn	Aquatic liverwort	3,700	I group, H ion 58 µeq/l	
		8,800	II group, H ion 10 µeq/l	
		67,000	III group, H ion 1 µeq/l	
As	Water hyacinth (root)	4,800	Tin-mining drainage regions, Malaysia	Lee et al. 1991
	(leaf)	650	Tin-mining drainage region, Malaysia	
	Water lettuce (root)	8,600	Tin-mining drainage regions, Malaysia	
	(leaf)	2,300	Tin-mining drainage region, Malaysia	

Table 4 (continued)

Metals	Plants (organs)		Accumulation factor	Experimental conditions	Ref.
	Lotus		460	Tin-mining drainage regions, Malaysia	
	Water gentian		11,000	Tin-mining drainage regions, Malaysia	
	Kenkung		1,300	Tin-mining drainage regions, Malaysia	
	Hydrilla		7,100	Tin-mining drainage regions, Malaysia	
	Chara		1,500	Tin-mining drainage regions, Malaysia	
	Ultricularia		9,000	Tin-mining drainage regions, Malaysia	
	Nettlia		21,000	Tin-mining drainage regions, Malaysia	
	Asiatic eelgrass		2,000	Tin-mining drainage regions, Malaysia	
As	Water hyacinth	(root)	2,200	Lake Dianchi, China	Lin and Zhang 1990
Cd			14,000	Lake Dianchi, China	
Cu			11,000	Lake Dianchi, China	
Hg	Water hyacinth		7,000	Lake Dianchi, China	Lin and Zhang 1990
Pb			16,000		

	Duckweed	Coal-ash drainage system, South Carolina	Rodgers et al. 1978
As	111		
Cd	29		
Cr	68		
Cu	29		
Hg	26		
Mn	2,800		
Se	24		
Zn	17		

Aquatic liverwort (shoot tip)	350,000	Lead-mine drainage, England	Satake et al. 1989
(shoot base)	1,200,000		
Pb			

Note: AG and BG are above-ground and below-ground organs, respectively.

[a] Mean values for *Cladophora, Potamogeton,* and *Zannichellia.*

Table 5 Metal Accumulation by Aquatic Macrophytes in Laboratory Studies, All Expressed in µg/g Dry-Weight Basis, Except as Noted by (w), Wet-Weight Basis

Metals	Plants (organs)	Accumulation factor	Experimental conditions	Ref.
Cd	Duckweed (root)	350	0.50 mg/l for 14 d	Charpentier et al. 1987
Cd	Duckweed (frond)	110	0.50 mg/l for 14 d	
Cd	Duckweed	0.1–13	Accumulation dependent on concentration, pH, temperature	Chawla et al. 1991
As	Water hyacinth	340	10 mg/l for 2 d	Chigbo et al. 1982
Cd		570	10 mg/l for 2 d	
Hg		510	10 mg/l for 2 d	
Pb		480	10 mg/l for 2 d	
Hg	*Salvinia* (root)	210(w)	3 mg/l for 2 d	De et al. 1985
	(shoot)	70(w)	3 mg/l for 2 d	
	(root)	340(w)	6 mg/l for 2 d	
	(shoot)	170(w)	6 mg/l for 2 d	
	(root)	1,100(w)	60 mg/l for 2 d	
	(shoot)	560(w)	60 mg/l for 2 d	
Pb	Duckweed	110	1.0 mg/l for 1 h	Everard and Denny 1985
Cr	*Ceratophyllum*	75	0.005 mg/l for 2 d	Garg and Chandra 1990
		120	0.05 mg/l for 2 d	
		160	0.10 mg/l for 2 d	
		220	1.0 mg/l for 2 d	
		580	2.0 mg/l for 2 d	
Pb	Duckweed	1,100	1.0 mg/l for 14 d	Jain et al. 1990
Zn		720	1.0 mg/l for 14 d	

Metal	Species	Tissue		Exposure	Reference
Pb	Water velvet		1,200	1.0 mg/l for 14 d	Jana 1988
Zn			830	1.0 mg/l for 14 d	
Cr	Hydrilla		470	1.0 mg/l for 28 d	
Hg	Hydrilla		840	1.0 mg/l for 28 d	
Cr	Oedogonium		310	1.0 mg/l for 28 d	
Hg			670	1.0 mg/l for 28 d	
Cr	Water hyacinth	(root)	980	1.0 mg/l for 28 d	
		(shoot)	78	1.0 mg/l for 28 d	
Hg		(root)	820	1.0 mg/l for 28 d	
		(shoot)	300	1.0 mg/l for 28 d	
Zn	Eelgrass	(root)	700	1.0 mg/l for 7 d	Lyngby et al. 1982
		(rhizome)	210	1.0 mg/l for 7 d	
Cd	Water hyacinth	(root)	2,650	1.0 mg/l for 16 d	Muramoto and Oki 1983
		(shoot)	460	1.0 mg/l for 16 d	
		(root)	7,600	4.0 mg/l for 16 d	
		(shoot)	1,700	4.0 mg/l for 16 d	
		(root)	11,000	8.0 mg/l for 16 d	
		(shoot)	2,300	8.0 mg/l for 16 d	
Hg		(root)	240	0.5 mg/l for 16 d	
			580	1.0 mg/l for 16 d	
			680	2.0 mg/l for 16 d	
Pb		(root)	6,200	1.0 mg/l for 16 d	
		(shoot)	490	1.0 mg/l for 16 d	
		(root)	23,000	4.0 mg/l for 16 d	
		(shoot)	820	4.0 mg/l for 16 d	
		(root)	26,000	8.0 mg/l for 16 d	
		(shoot)	1,800	8.0 mg/l for 16 d	

Table 5 (continued)

Metals	Plants (organs)		Accumulation	Experimental conditions	Ref.
Cd	Water hyacinth	(root)	620	0.4 mg/l for 21 d, in mother plants	Nir et al. 1990
		(shoot)	56		
Pb		(root)	1,100	0.4 mg/l for 21 d, in daughter plants	
		(shoot)	110		
Cd	*Azolla*	(root)	24,000	5–20 mg/l for 3–7 d	Sela et al. 1989
		(frond)	9,600	5–20 mg/l for 3–7 d	
Cr		(root)	10,000	5–20 mg/l for 3–7 d	
		(frond)	1,400	5–20 mg/l for 3–7 d	
Cu		(root)	44,000	5–20 mg/l for 3–7 d	
		(frond)	6,100	5–20 mg/l for 3–7 d	
Ni		(root)	20,000	5–20 mg/l for 3–7 d	
		(frond)	7,100	5–20 mg/l for 3–7 d	
Zn		(root)	12,000	5–20 mg/l for 3–7 d	
Hg	*Salvinia*		33(w)	0.5 mg/l for 3 d	Sen and Mondal 1987
			67(w)	1.0 mg/l for 3 d	
			190(w)	3.0 mg/l for 3 d	
			300(w)	5.0 mg/l for 3 d	
Cu	*Salvinia*		67(w)	1.0 mg/l for 5 d	Sen and Mondal 1990
			670(w)	10 mg/l for 5 d	
			1,300(w)	20 mg/l for 5 d	
Cr	*Pistia*	(root)	60(w)	5 mg/l for 5 d	Sen et al. 1987
		(shoot)	1(w)	5 mg/l for 5 d	
		(root)	110(w)	10 mg/l for 5 d	
		(shoot)	2(w)	10 mg/l for 5 d	

Metal	Plant		Accumulation	Exposure	Reference
Cd	Bacopa	(root)	200(w)	50 mg/l for 5 d	Sinha and Chandra 1990
		(shoot)	7(w)	50 mg/l for 5 d	
		(root)	270(w)	100 mg/l for 5 d	
		(shoot)	12(w)	100 mg/l for 5 d	
Cu			3,400	5 mg/l for 7 d	
			3,800	5 mg/l for 7 d	

study using the same species, 0.5 to 1 mg/l of Hg were removed completely in 3 d (Sen and Mondal 1987). O'Keeffe et al. (1984) and Lee and Hardy (1987) both reported that Cd and Cu uptakes by water hyacinth reflected a biphasic rate. The fast phase lasted about 4 h and was followed by a slow phase of about 48 to 72 h exposure. Other reports also indicated the maximum uptake of Cd, Cr, and Hg occurs in 1 to 2 d (De et al. 1985; Garg and Chandra 1990; Chawla et al. 1991). In contrast, long metal uptake periods have been reported. Muramoto and Oki (1983) found Cd concentrations in roots of *Eichhornia crassipes* increased exponentially during the first 12 d of exposure. Gotoh and Iriye (1989), using the solution renewal method every 10 d and metal concentrations of sublethal ranges, reported that uptake of Cd, Cr, Cu, Mn, and Zn by *Egeria densa* increased for 30 d.

Decaying, senescent macrophytes were found to accumulate more Pb than live plants, 250 to 360 µg/g compared to 5 to 90 µg/g (Knowlton et al. 1983). Similarly, Sela et al. (1989) reported that the dead tissues contained three to seven times larger metal concentrations than the living tissues. It was speculated that during senescence one or both of the following processes occurred: the partial breakdown of macrophyte tissues resulted in increased surface areas for increased metal accumulation or the increased epiphyte organism populations accompanying tissue degradation increased metal absorption. Mo et al. (1988) proposed an uptake mechanism of metals by *Lemna* in two stages. In the first stage, metal ions are absorbed by live plants. Metal ions are then continually absorbed by injured or dead plants in the second stage. Possibly, during the second stage, epiphyte organisms play a major role in continuing metal uptake.

Metal accumulation by aquatic macrophytes, under controlled laboratory conditions, is dependent on the metal concentration in water. Typically, the amount of metal accumulation increases proportionally as the metal concentration in water increases. This relation has been observed for As, Cd, Cr, Cu, and Hg (Faraday and Churchill 1979; Nakada et al. 1979; Mortimer 1985; Gotoh and Iriye 1989; Nir et al. 1990; Lee et al. 1991).

Many reports indicate that macrophyte roots accumulate larger concentrations of Ba, Cd, Cr, Cu, Fe, Hg, Ni, Pb, and Zn than other plant organs (Lyngby et al. 1982; Piccardi and Clauser 1983; Jana 1988; Maury et al. 1988; Samecka-Cymerman 1988; Gries and Garbe 1989; Sela et al. 1989). The amounts of Cr and Hg accumulated in the roots of *Eichhornia crassipes* were 2- to 12-fold larger than in shoots (Jana 1988), whereas the amounts of Cd, Cr, Cu, Ni, and Zn accumulated in *A. filiculoides* were 2- to 5-fold larger in roots than in shoots (Sela et al. 1989). Similarly, Piccardi and Clauser (1983) reported that Cu concentrations in roots of *I. pseudacorus* were larger than in rhizome and leaf organs. In *Z. marina*, Lyngby et al. (1982) found that all parts of the marine species had the ability to absorb Zn, whereas in the case of

Elodea densa, Maury et al. (1988) found that root absorption of methylmercury was the dominant route of uptake and that the leaves were the principal organ for storage.

The amount of metal accumulations in different generations of plants varied in a chronic, sublethal experiment. Nir et al. (1990) reported that Cd accumulations in root and leaf organs of the mother plants of *Eichhornia crassipes* were 620 and 56 μg/g, respectively, whereas in the daughter plants, accumulations were 1000 and 110 μg/g, respectively.

An interactive effect can occur when a water solution contains more than one metal ion as the ions compete for uptake. Testing *L. paucicostata* and *B. monnieri*, Nasu et al. (1984) and Sinha and Chandra (1990) both reported that the absorption of Cu was not affected (or stimulated) by the presence of Cd. Both studies also noted, however, that absorption of Cd was reduced considerably with increasing Cu concentrations in the solution. The copresence of Pb and Zn resulted in the decreased metal uptake of the other (Jain et al. 1990).

An interactive effect can occur when a water solution contains more than one aquatic macrophyte since different species accumulate ions from the solution at different rates. Everard and Denny (1985) reported that in a mixed culture where *Navicula* covered the surface of *L. trisula*, Pb uptake by *Lemna* was approximately 27% of that which occurred with the *Lemna* species alone. Ward (1987) reported that epiphytes contained some metals (Mn and Ni) in larger concentrations than in the leaves of seagrass, *Ponsidonia australis*. These results suggest that epiphytes possibly are competitors for metal uptake into the foliar tissues.

Field studies

Many field studies have been conducted to compare the amount of metal accumulation by aquatic macrophytes at contaminated and uncontaminated sites (Table 6). Three examples are described in this section.

A base-metal processing plant in Pori, Finland discharged Cu as part of an effluent into a river (Aulio 1980). The dominant macrophyte in the river was yellow water lilies, *Nuphar lutea*. Plant specimens were collected from various locations along the river at an interval of 300 m. The specimens were washed, sorted according to plant organs, and oven dried at 60°C. One-gram samples were dry ashed at 450°C, dissolved in hydrochloric acid, filtered, and determined for Cu concentrations using an atomic absorption spectrophotometer. Total Cu accumulations in various plant organs are presented in Figure 1.

The amount of Cu contained in the plant material was markedly greater at the sampling sites near the metal processing plant, and the Cu content decreased with increasing distance upstream or down-

Table 6   Metal Concentrations (μg//g Dry Weight) of Aquatic Macrophytes in Referenced and Contaminated Sites

Aquatic macrophytes (organs)		Metals	Metal concentrations		Sources of contamination	Ref.
			Referenced	Contaminated		
Yellow water lily	(petioles)	Cu	30	110	Base-metal processing	Aulio 1980
	(leaves)		30	90		
	(rhizomes)		20	50		
	(flowers)		10	20		
Yellow water lily	(rhizome)	Cu	2.4	50	Mining and smelting	Campbell et al. 1985
	(stem)		3	12		
	(rhizome)	Zn	20	93		
	(stem)		8	50		
Rice	(root)	Cu	5.7	100	Pyrite mining	Fernandes and Henriques 1990
	(shoot)		2.0	5.7		
	(grain)		1.9	2.6		
	(root)	Fe	15,000	34,000		
	(shoot)		360	300		
	(grain)		100	62		
	(root)	Mn	440	1,100		
	(shoot)		160	700		
	(grain)		58	47		
	(root)	Pb	16	20		
	(shoot)		9	13		
	(grain)		19	14		
	(root)	Zn	400	26		
	(shoot)	Zn	48	15		
	(grain)		21	17		

Species	Metal			Source	Reference
Myriophyllum	Cd	<0.1	6	Base-metal processing	Franzin and McFarlane 1980
	Cu	32	66		
	Fe	1,390	1,460		
	Mn	92	203		
	Pb	<0.6	27		
	Zn	78	1,640		
Sparganium	Cd	<0.2	5		
	Cu	38	63		
	Fe	2,400	3,000		
	Mn	190	590		
	Pb	<0.9	27		
	Zn	54	1,500		
Water hyacinth	Co	8.5	16	Urban and industrial effluents	Gonzalez et al. 1989
	Cr	38	38		
	Cu	15	55		
	Fe	6,000	6,600		
	Mn	1,900	12,000		
	Pb	16	51		
	Zn	78	150		
Pontederia (shoot)	Cu	3	24	Printed-circuit factory and municipal effluents	Heisey and Damman 1982
(rhizome)		5	81		
(shoot)	Pb	4	22		
(rhizome)		11	88		
Potamogeton	Cu	13	220		
	Pb	9	18		

Table 6 (continued)

Aquatic macrophytes (organs)		Metals	Metal concentrations		Sources of contamination	Ref.
			Referenced	Contaminated		
Ditch reed	(leaf)	Cd	0.56	0.39	Sewage	Larsen and Schierup 1981
		Cu	17	6.1		
		Pb	14	13		
		Zn	140	160		
Potamogeton						
Pipewort	(root)	Cd	0.66	45	Electroplating plant	McIntosh et al. 1978
	(shoot)	Cd	8.2	2.7	Metal processing plant	Miller et al. 1983
			3.0	2.2		
	(root)	Cu	13	460		
	(shoot)		12	300		
	(root)	Ni	11	290		
	(shoot)		7	310		
	(root)	Pb	23	100		
	(shoot)		11	51		
	(root)	Zn	100	39		
	(shoot)		51	46		
Potamogeton	(AG)	Cd	0.65	4.9	Base-metal mining	Ray and White 1976
	(BG)		1.3	6.7		
	(AG)	Cu	2.8	170		
	(BG)		8.2	200		
	(AG)	Pb	2.5	4.8		
	(BG)		3.7	13		
	(AG)	Zn	110	2,900		
	(BG)		100	1,800		

Species	Organ	Metal			Source	Description
Phragmites	(root)	Cd	1.2	0.58	Schierup and Larsen 1981	Industrial and domestic effluents
		Cu	18	15		
		Pb	9.3	3.5		
		Zn	245	158		
Chlorophyta		Se	0.27	31	Schuler et al. 1990	Surface agricultural drainage
Ruppia			0.59	74		
Burlrush	(rhizome)		2.0	170		
Cattail	(rhizome)		0.60	140		
Cattail	(leaf)		0.15	37		
Cattail		Mn	350	3,700	van der Merwe et al. 1990	Acid-mine drainage
		Ni	2.5	82		
		Zn	77	520		
Arundo donax		Mn	120	4,000		
		Ni	4.6	580		
		Zn	69	390		
Seagrass		Cd	2.8	541	Ward 1987	Pb smelter

Note: AG and BG are above-ground and below-ground organs, respectively.

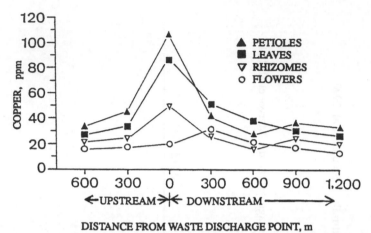

DISTANCE FROM WASTE DISCHARGE POINT, m

Figure 1 Cu accumulation in yellow water lily (*Nuphar lutea*) sampled at different distances from the waste discharge point of a metal-processing plant in Finland. (Modified from Aulio K [1980] *Bull Environ Contam Toxicol* 25:716. With permission.)

stream from the Cu waste discharge point. In addition, the amount of Cu accumulation in petioles, leaves, and rhizomes of *N. lutea* correlated significantly ($p < 0.001$) with the Cu concentrations in the streambed sediments.

Palestine Lake is a small, eutrophic lake located in Indiana (McIntosh et al. 1978). An electroplating plant was located approximately 3.3 km upstream from the lake on Williamson Ditch, and the inflow from Williamson Ditch contained large concentrations of Cd and Zn. The aquatic macrophyte *Potamogeton crispus* filled much of the lake in mid-April 1976. Plant samples were collected from various locations in the lake and determined for Cd content.

The Cd concentrations in *Potamogeton* samples clearly delineated zones of metal contamination in the lake, ranging from extremely contaminated to uncontaminated (Figure 2). Region I (most contaminated) was located in the north-central part of the lake and was the area receiving the influx of Cd-laden water. Region II (moderately contaminated) comprised the areas immediately adjacent to (north and south of) Region I. Region III (little contamination) was the remaining part of the western half of the lake. Region IV (no contamination) was the eastern half of the lake. The results of metal zonation in plants were correlated with the hydrology of the lake. The general flow path in the lake was from the eastern to the western part, ultimately discharging from the northwestern part of the lake.

The Natchaug River in Connecticut flows through a predominantly forested watershed with little urbanization (Heisey and Damman 1982). The Willimantic River also originates in a pristine area, but most of its watershed is moderately populated and receives effluents from a

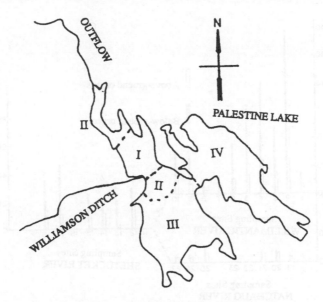

Figure 2 Cd accumulation in pondweed (*Potamogeton crispus*) in Palestine Lake, Indiana. I: most contaminated; mean Cd concentration 45 µg/g dry weight. II: moderately contaminated; mean Cd concentration 17 µg/g dry weight. III: little contaminated; mean Cd concentration 3.1 µg/g dry weight. IV: no contamination; mean Cd concentration 0.66 µg/g dry weight. ----: approximate boundary of regions defined on basis of contamination. (Modified from McIntosh AW, et al. [1978] *J Environ Qual* 7:304. With permission.)

printed-circuit factory. The Shetucket River is formed at the confluence of the Natchaug River and the Willimantic River. Aquatic macrophytes *Pontederia cordata* and *Potamogeton epihydrus* were abundant in the study area and were sampled at intervals along the three rivers. Cu concentrations of *Pontederia* sp. at various sampling sites are presented in Figure 3.

Roots and rhizomes of *Pontederia* spp. had larger Cu concentrations than the shoots. In the relatively unpolluted Natchaug River, the difference was small, whereas in the Willimantic River Cu concentrations in the below-ground organs were two to four times larger than those in the shoots. Cu concentrations in rhizomes and roots were on the average four times larger in the Shetucket River and six times larger in the Willimantic River than in the Natchaug River.

The results of these examples suggest that aquatic macrophytes can have a large capacity for metal accumulation, can tolerate large metal concentrations in water, and can be ideal *in situ* biomonitors for metal contamination. It is noteworthy that aquatic macrophytes are stationary and can reflect intermittent fluctuations of metal concentrations in an ambient environment. The selection of plant species for biomonitoring, however, is dependent on local conditions and the availability of aquatic macrophytes.

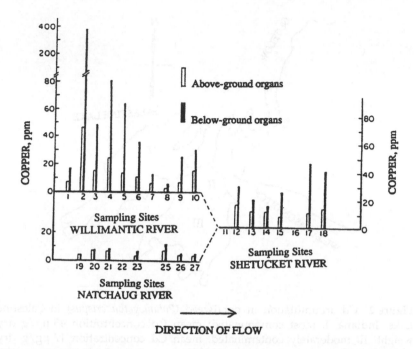

Figure 3 Cu accumulation in pickerel weed (*Pontederia cordata*) collected at stations (indicated by number) in streams in Connecticut. (Modified from Heisey RM, Damman AWH [1982] *Aquat Bot* 14:218. With permission.)

Many other ecological studies also indicate metal accumulation by aquatic macrophytes. These studies are summarized in Table 7. Among plant organs, the below-ground organs (roots and rhizomes) generally accumulate larger metal concentrations than the above-ground organs. This relation has been reported in studies using species of *Typha latifolia*, *Spartina alterniflora*, and *Z. marina* (Taylor and Crowder 1983; Kraus 1988; Lyngby and Brix 1989). Other studies, however, indicated that the above-ground organs of *Z. marina* contained significantly larger concentrations of Cd, Cu, and Zn than the below-ground organs (Lyngby and Brix 1982; Brix et al. 1983; Brix and Lyngby 1984).

The metal accumulation in different plant organs can be source specific. Campbell et al. (1985) reported that the Cu accumulated in stem and rhizome probably originated from the sediments, whereas Zn was largely from the water column. Welsh and Denny (1980) indicated that Cu accumulation was mainly through root absorption from sediments, whereas Pb accumulation in the shoots was through adsorption from the water solution.

In an extensive survey of aquatic mosses in England and Europe, Whitton et al. (1982) reported that Zn concentrations determined in samples from the 2-cm tip of mosses correlated significantly ($p < 0.001$) with Zn concentrations in water. Pb concentrations in the mosses also correlated significantly ($p < 0.01$) with Pb concentrations in water (Say et al. 1981; Say and Whitton 1983; Wehr and Whitton 1983a). For streams that had no indigenous moss population, Kelly et al. (1987) placed "moss bags" and moss on natural boulders in streams and reported that both methods were effective and there was no significant difference between these methods on the uptake of Zn.

Summary

Aquatic macrophytes are useful in environmental studies due to their significant potential for metal accumulation. This chapter reviews the use of these plants for studies of metal accumulation; summarizes factors affecting the uptake process; and discusses the metal accumulation factor (AF), which is the ratio of metal concentration in plant tissue to metal concentration in water.

Metal accumulation by aquatic macrophytes can be affected by many factors. These factors can be classified as biological and nonbiological; both factors can be classified further. Examples include plant species and different organs, season, pH, metal concentration, and exposure time. Submerged species are likely to accumulate larger metal concentrations than floating species.

Seasonal effects on metal accumulation have been reported, but seasonal patterns are complex. Several studies indicated that the amounts of metal accumulation by certain plants declined during summer, although the seasonal patterns of uptake depended on types and species of metal ions, plant species and organs, and location. Other investigations found that the amount of metal accumulation had a summer maximum, winter-early spring maximum, early fall maximum, or was nearly constant throughout the year.

The pH of a solution can affect metal speciation and metal uptake. In a pH range of 3.6 to 5.1, a report indicated that larger Cd and Cu concentrations in duckweed plants occur than in water. However, another study found that metal uptake by water hyacinth was inhibited at a pH range of 2.0 to 4.0, but was much less affected above pH 4.0.

Generally, the amounts of metal accumulation increase as temperature increases. Results from studies indicate that temperature and photoperiod have a strong effect on metal accumulation, either as separate or concurrent effects. The amount of metal accumulation by macrophytes also can be affected by the presence of competing metals in solution, as well as by other coexisting macrophytes.

Several studies indicate that the metal AF of aquatic macrophytes was large; some AF values of more than 1 million are reported in the literature. The rate of metal accumulation by certain aquatic macrophytes was rapid and reported to be biphasic in some cases. Metal accumulation increased as metal concentration in water increased, whereas AF decreased as metal concentration in water increased. The roots and rhizomes of aquatic macrophytes are generally the primary areas of metal accumulation.

Table 7 Metal Accumulation in Ambient Plant Samples, in μg/g Dry-Weight Basis, Except as Noted (w), Wet-Weight Basis

Plants (organs)	Metals	Accumulation	Comments	Ref.
Potamogeton	Cu	26–84	General survey in Pennsylvania; ranges of 8 species	Adams et al. 1973
	Fe	1,100		
	Mn	1,500–4,000		
	Zn	52–180		
Myriophyllum	Cu	20–41	General survey in Pennsylvania; ranges of 2 species	
	Fe	1,100		
	Mn	1,800–4,200		
	Zn	120–320		
Pondweed	Cu	40	Mean values of 5 species, Finland	Aulio and Salim 1982
	Fe	4,600		
	Mn	4,800		
	Zn	240		
Eelgrass (AG)	Cd	0.62	Mean value of 40 specimens, Denmark	Brix et al. 1983
(BG)		0.30		
(AG)	Cu	4.8	Mean value of 40 specimens	
(BG)		3.3		
(AG)	Pb	1.1	Mean value of 36 specimens	
(BG)		1.0		
(AG)	Zn	78	Mean value of 40 specimens	
(BG)		55		
Ruppia (leaf)	Cr	3.5(w)	Urban and petrochemical plant, Italy	Capone et al. 1983
(BG)		2.1(w)		
(leaf)	Pb	13(w)		
(BG)		19(w)		

		Metal	Value	Location	Reference
Elodea	(whole)	Pb	200	Near Ullswater, England	Everard and Denny 1985
	(leaf)		170		
	(stem)		120		
Typha	(root)	Cu	26	Pyrite mining in Portugal	Fernandes and Henriques 1990
	(shoot)		7		
	(root)	Fe	5,700		
	(shoot)		810		
	(root)	Mn	290		
	(shoot)		370		
	(root)	Pb	40		
	(shoot)		10		
	(root)	Zn	190		
	(shoot)		34		
Phragmites	(shoot)	Cu	7.4		
		Fe	524		
		Mn	550		
		Pb	14		
		Zn	28		
Duckweed		B	3,200	Sewage oxidation ponds, Michigan	Glandon and McNabb 1978
Ceratophyllum		B	137		
Scapania		As	710	Mineral deposits, North Wales	Jones 1985
Moss		Cd	11	Former Pb mining areas, England	Kelly et al. 1987
		Pb	460		
		Zn	3,500		
Water milfoil	(leaf)	Fe	290–12,000	Tributary and nontributary sites	Kimball and Baker 1982
		Mn	130–6,400		
Potamogeton	(shoot)	Pb	1,480	Root uptake from spiked soil	Knowlton et al. 1983
Reed	(shoot)	Mo	0.1	An eutrophic lake, Poland	Kufel 1991
	(rhizome)		0.9		

Table 7 (continued)

Plants (organs)	Metals	Accumulation	Comments	Ref.
(root)	Pb	0.6		
(shoot)		2.1		
(rhizome)		1.7		
Cattail (shoot)	Mo	0.5		
(rhizome)		0.8		
(root)		1.3		
(shoot)	Pb	0.04		
(rhizome)		0.1		
(root)		0.5		Lyngby and Brix 1982
Eelgrass (AG)	Cd	1.4	Limfjord, Denmark	
(BG)		0.6		
(AG)	Cu	40		
(BG)		20		
(AG)	Pb	22		
(BG)		25		
(AG)	Zn	160		
(BG)		90		
Smooth cordgrass (root)	Cd	3.1	Near an urban estuary,	Kraus 1988
	Cr	29	New Jersey	
	Cu	160		
	Ni	25		
	Pb	65		
Water hyacinth (AG)	As	55	Dianchi Lake, China	Lin and Zhang 1990
(BG)		240		
(AG)	Cd	14		

Species	Fraction	Metal	Value	Site	Reference
Victoria	(BG)	Cu	340	The Amazon Basin affected by gold mining	Martinelli et al. 1988
Water hyacinth	(AG)		100		
	(BG)		240		
	(AG)	Hg	0.12		
	(BG)		0.32		
	(AG)	Pb	15		
	(BG)		130		
Elodea	(leaf)	Hg	0.91	Unpolluted section of a river, Japan	Nakada et al. 1979
	(root)	Hg	1.04		
	(leaf)	Cd	0.9		
	(root)		14		
	(leaf)	Cu	11		
	(root)		24		
	(leaf)	Fe	860		
	(root)		1,800		
	(leaf)	Mn	390		
	(root)		900		
	(leaf)	Pb	12		
	(root)		320		
	(leaf)	Zn	110		
	(root)		330		
Rice		Fe	1,088	Mean values of seven locations during peak flood, India	Rother and Whitton 1988
		Mn	490		
		Zn	79		
Scapania		Ba	5,000	Sudetic Mountain streams, Poland	Samecka-Cymerman 1988
Aquatic liverwort		B	520	European streams	Samecka-Cymerman et al. 1991
		Ba	420		
		Cd	16		
		Co	180		

Table 7 (continued)

Plants (organs)		Metals	Accumulation	Comments	Ref.
		Cr	120		
		Cu	290		
		Li	11		
		Mn	11,000		
		Mo	690		
		Ni	240		
		Pb	460		
		Sr	960		
		V	120		
		Zn	210		
Water hyacinth	(leaf)	Cr	3.8	A stream in August, Venezuela	Schorin et al. 1991
	(root)		7.8		
	(leaf)	Ni	2.2		
	(root)		2.8		
	(leaf)	Pb	3.2		
	(root)		5.5		
	(leaf)	Zn	31		
	(root)		100		
Moss		Cd	65	Maximum concentration in 105 sites in northern England	Wehr and Whitton 1983b
		Pb	8,700		
		Zn	8,800		

			Welsh and Denny 1980	Oligotrophic-mesotrophic lakes in England
Potamogeton	(root)	Cu		27
	(shoot)			27
	(root)	Pb		1,200
	(shoot)			210
Myriophyllum	(root)	Cu		48
	(shoot)			36
	(root)	Pb		830
	(shoot)			210

Note: AG and BG are above-ground and below-ground organs, respectively.

Appendix 1   Common and Latin Names of Plants Mentioned in This
Chapter

Common name	Latin name
Beakrush	*Rhynchospora inundata*
Bladderwort	*Utricularia* sp.
Cattail	*Typha* spp. (*T. augustifolia,* *T. latifolia*)
Diatom	*Navicula* sp.
Ditch grass	*Ruppia* sp.
Ditch reed	*Phragmites australis*
Duckweed	*Lemna* spp. (*L. perpusilla,* *L. minor, L. paucicostata*)
Eelgrass	*Zostera marina*
Floating moss	*Salvinia* sp.
Glasswort	*Salicornia brachiata*
Green alga	*Hydrodictyon* sp.
Horned pondweed	*Zannichellia pallustris*
Hornwort	*Ceratophyllum* sp.
Liverwort	*Scapania undulata*
Moss	*Fontinalis antipyretica* *Rhynchostegium riparioides* *Sphagnum papillosum*
Pickerel weed	*Pontederia cordata*
Pondweed	*Potamogeton* spp. (*P. pectinatus,* *P. crispus, P. perfoliatus*)
Redroot	*Lachnanthes caroliniana*
Rush	*Juncus maritimus*
Seaweed	*Gracilaria verrucosa*
Smooth cordgrass, salt grass	*Spartina alterniflora*
Sphagnum	*Sphagnum cuspidatum*
Stonewort	*Nitella* sp.
Water fern	*Azolla filiculoides*
Water hyacinth	*Eichhornia crassipes*
Water milfoil	*Myriophyllum exalbescens*
Waterweed	*Elodea* spp. (*E. canadensis, E. densa*)
White water lily	*Nymphaea odorata*
Yellow-eyed grass	*Xyris smalliana*
Yellow water lily	*Nuphar* spp. (*N. advena, N. lutea*)

References

Abo-Rady MDK (1980) Aquatic macrophytes as indicator for heavy metal pollution in the River Leine (West Germany). *Arch Hydrobiol* 89:387–404.

Adams FS, Cole H, Massie LB (1973) Element constitution of selected aquatic vascular plants from Pennsylvania: submersed and floating leaved species and rooted emergent species. *Environ Pollut* 5:117–147.

Ajmal M, Raziuddin, Khan R, Khan AU (1987) Heavy metals in water, sediments, fish and plants of River Hindon, U.P., India. *Hydrobiologia* 148:151–157.

Aulio K (1980) Accumulation of copper in fluvial sediments and yellow water lilies (*Nuphar lutea*) at varying distances from a metal processing plant. *Bull Environ Contam Toxicol* 25:713–717.

Aulio K, Salim M (1982) Enrichment of copper, zinc, manganese, and iron in five species of pondweeds (*Potamogeton* spp.). *Bull Environ Contam Toxicol* 29:320–325.

Bishop PL, DeWaters J (1988) *Biotechnology for Degradation of Toxic Chemicals in Hazardous Wastes*, Noyes Data Corp., Park Ridge, NJ.

Bosserman RW (1985) Distribution of heavy metals in aquatic macrophytes from Okefenokee Swamp. In *Heavy Metals in Water Organisms*, Symposia Biologica Hungarica, Vol. 29, Akademiai Kiado, Budapest, Hungary, pp. 173–182.

Brisbin IL, Breshears DD, Brown KL, Ladd M, Smith MH (1989) Relationships between levels of radiocaesium in components of terrestrial and aquatic food webs of a contaminated streambed and floodplain community. *J Appl Ecol* 26:173–182.

Brix H, Lyngby JE (1982) The distribution of cadmium, copper, lead, and zinc in eelgrass (*Zostera marina* L.). *Sci Total Environ* 24:51–63.

Brix H, Lyngby JE (1984) A survey of the metallic composition of *Zostera marina* (L.) in the Limfjord, Denmark. *Arch Hydrobiol* 99:347–359.

Brix H, Schierup HH (1989) Use of aquatic macrophytes in water-pollution control. *Ambio* 18:100–107.

Brix H, Lyngby JE, Schierup HH (1983) Eelgrass (*Zostera marina* L.) as an indicator organism of trace metals in the Limfjord, Denmark. *Mar Environ Res* 8:165–181.

Caines LA, Watt AW, Wells DE (1985) The uptake and release of some trace metals by aquatic bryophytes in acidified waters in Scotland. *Environ Pollut Ser B* 10:1–18.

Campbell PGC, Tessier A, Bisson M, Bougie R (1985) Accumulation of copper and zinc in the yellow water lily, *Nuphar variegatum*: relationships to metal partitioning in the adjacent lake sediments. *Can J Fish Aquat Sci* 42:23–32.

Capone W, Mascia C, Porcu M, Masala MLT (1983) Uptake of lead and chromium by primary producers and consumers in a polluted lagoon. *Mar Pollut Bull* 14:97–102.

Charpentier S, Garnier J, Flaugnatti R (1987) Toxicity and bioaccumulation of cadmium in experimental cultures of duckweed, *Lemna polyrrhiza* L. *Bull Environ Contam Toxicol* 38:1055–1061.

Chawla G, Singh J, Viswanathan PN (1991) Effect of pH and temperature on the uptake of cadmium by *Lemna minor* L. *Bull Environ Contam Toxicol* 47:84–90.

Cherry DS, Guthrie RK, Davis EM, Harvey RS (1984) Coal ash basin effects (particulates, metals, acidic pH) upon aquatic biota: an eight-year evaluation. *Water Resour Bull* 20:535–544.

Chigbo FE, Smith RW, Shore FL (1982) Uptake of arsenic, cadmium, lead, and mercury from polluted waters by the water hyacinth *Eichhornia crassipes*. *Environ Pollut Ser A* 27:31–36.

De AF, Sen AK, Modak DP, Jana S (1985) Studies of toxic effects of Hg (II) on *Pistia stratiotes*. *Water Air Soil Pollut* 24:351–360.

De Wet LPD, Schoonbee HJ, Pretorius J, Bezuidenhout LM (1990) Bioaccumulation of selected heavy metals by the water fern, *Azolla filiculoides* Lam. in a wetland ecosystem affected by sewage, mine and industrial pollution. *Water SA* 16:281–286.

Dieckmann GS (1982) Seasonal variation in uptake and retention of cadmium in the seaweed *Zostera marina* L. from the Kiel Fjord (Western Baltic Sea). NTIS, Springfield, VA (abstract only).

Everard M, Denny P (1985) Flux of lead in submerged plants and its relevance to freshwater system. *Aquat Bot* 21:181–193.

Faraday WE, Churchill AC (1979) Uptake of cadmium by the seaweed *Zostera marina*. *Mar Biol* 53:293–298.

Fernandes JC, Henriques FS (1990) Heavy metal contents of paddy fields of Alcacer Do Sal, Portugal. *Sci Total Environ* 90:89–97.

Ferreira JG (1991) Factors governing mercury accumulation in three species of marine macroalgae. *Aquat Bot* 39:335–343.

Franzin WG, McFarlane GA (1980) An analysis of the aquatic macrophytes, *Myriophyllum exalbescens*, as an indicator of metal contamination of aquatic ecosystems near a base metal smelter. *Bull Environ Contam Toxicol* 24:597–605.

Friant SL (1979) Trace metal concentrations in selected biological, sediment, and water column samples in a northern New England river. *Water Air Soil Pollut* 11:455–465.

Garg P, Chandra P (1990) Toxicity and accumulation of chromium in *Ceratophyllum demersum* L. *Bull Environ Contam Toxicol* 44:473–478.

Genin-Durbet C, Champetier Y, Blazy P (1988) Metallic bioaccumulation in aquatic plants. *CR Acad Sci Ser 2* 18:1875–1877 (abstract only).

Glandon RT, McNabb CD (1978) The uptake of boron by *Lemna minor*. *Aquat Bot* 4:53–64.

Gonzalez H, Lodenius M, Otero M (1989) Water hyacinth as indicator of heavy metal pollution in the tropics. *Bull Environ Contam Toxicol* 6:910–914.

Gotoh T, Iriye T (1989) The behavior of heavy metallic elements in plants. (1) The uptake of heavy metallic elements by aquatic plants. *Jpn J Limnol* 50:321–331.

Gries C, Garbe D (1989) Biomass an nitrogen, phosphorus, and heavy metal content of *Phragmites australis* during the third growing season in a root zone waste water treatment. *Arch Hydrobiol* 117:91–107.

Guthrie RK, Cherry DS (1979) The uptake of chemical elements from coal ash and settling basin effluent by primary producers. I. Relative concentrations in predominant plants. *Sci Total Environ* 12:217–222.

Heisey RM, Damman AWH (1982) Copper and lead uptake by aquatic macrophytes in eastern Connecticut, U.S.A. *Aquat Bot* 14:213–229.

Jain SK, Vasudevan P, Jha N (1990) *Azolla pinnata* R.Br. and *Lemna minor* L. for removal of lead and zinc from polluted water. *Water Res* 24:177–183.

Jana S (1988) Accumulation of Hg and Cr by three aquatic species and subsequent changes in several physiological and biochemical plant parameters. *Water Air Soil Pollut* 38:105–109.

Janauer GA (1985) Heavy metal accumulation and physiological effects on Austrian macrophytes. In *Heavy Metals in Water Organism*, Symposia Biologica Hungarica, Vol. 29, Akademiai Kiado, Budapest, Hungary.

Jones KC (1985) Gold, silver and other elements in aquatic bryophytes from a mineralized area of North Wales. *UK J Geochem Explor* 24:237–246.

Joshi AJ, Engenhart M, Wickern M, Breckle SW (1987) Seasonal changes in the trace metals in salt marsh angiosperms. *J Plant Physiol* 128:173–177.

Kelly MG, Girton C, Whitton BA (1987) Use of moss-bags for monitoring heavy metals in rivers. *Water Res* 21:1429–1435.

Kimball KD, Baker AL (1982) Variations in the mineral content of *Myriophyllum heterophyllum* Michx related to site and season. *Aquat Bot* 14:139–149.

Knowlton WF, Boyle TP, Jones JR (1983) Uptake of lead from aquatic sediment by submerged macrophytes and crayfish. *Arch Environ Contam Toxicol* 12:535–541.

Kraus ML (1988) Accumulation and excretion of five heavy metals by the saltmarsh cordgrass *Spartina alterniflora. Bull NJ Acad Sci* 33:39–43.

Kufel I (1991) Lead and molybdenum in reed and cattail — open versus closed type of metal cycling. *Aquat Bot* 40:275–288.

Kufel I, Kufel L (1985) Heavy metals and mineral nutrient budget in *Phragmites australis* and *Typha angustifolia*. In *Heavy Metals in Water Organisms.*, Symposia Biologica Hungarica, Vol. 29, Akademiai Kiado, Budapest, Hungary, pp. 61–66.

Larsen VJ, Schierup H-H (1981) Macrophyte cycling of zinc, copper, lead and cadmium in the littoral zone of a polluted and non-polluted lake. II. Seasonal changes in heavy metal content of above-ground biomass and decomposing leaves of *Phragmites australis* (Cav.) Trin. *Aquat Bot* 11:211–230.

Lee CK, Low KS, Hew NS (1991) Accumulation of arsenic by aquatic plants. *Sci Total Environ* 103:215–227.

Lee TA, Hardy JK (1987) Copper uptake by the water hyacinth. *J Environ Sci Health* A22:141–160.

Levine SN, Rudnick DT, Kelly JR, Morton RD, Buttel LA (1990) Pollution dynamics as influenced by seagrass beds: experiments with tributyltin in *Thalassia* microcosms. *Mar Environ Res* 30:297–322.

Lin YX, Zhang XM (1990) Accumulation of heavy metals and the variation of amino acids and protein in *Eichhornia crassipes* (Mart.) solms in the Dianchi Lake. *Oceanol Limnol Sinica* 21:179–184 (in Chinese).

Lyngby JE, Brix H (1982) Seasonal and environmental variation in cadmium, copper, lead and zinc concentrations in eelgrass (*Zostera marina* L.) in the Limfjord, Denmark. *Aquat Bot* 14:59–74.

Lyngby JE, Brix H (1989) Heavy metals in eelgrass (*Zostera marina* L.) during growth and decomposition. *Hydrobiologia* 176/177:189–196.

Lyngby JE, Brix H, Schierup H-H (1982) Absorption and translocation of zinc in eelgrass (*Zostera marina* L.) *J Exp Mar Biol Ecol* 58:259–270.

Malea P, Haritonidis S (1989) Uptake of Cu, Cd, Zn and Pb in *Posidonia oceanica* (Linnaeus) from Antikyra Gulf, Greece: preliminary note. *Mar Environ Res* 28:495–498.

Martinelli LA, Ferreira JR, Forsberg BR, Victoria RL (1988) Mercury contamination in the Amazon: a gold rush consequence. *Ambio* 17:252–254.

Maury R, Boudou A, Ribeyre F, Engrand P (1988) Experimental study of mercury transfer between artificially contaminated sediment (CH₃HgCl) and macrophytes (*Elodea densa*). *Aquat Toxicol* 12:213–228.

Maury-Brachet R, Ribeyre F, Boudou A (1990) Action and interactions of temperature and photoperiod on mercury accumulation by *Elodea densa* from sediment sources. *Ecotoxicol Environ Saf* 20:141–155.

McIntosh AW, Shephard BK, Mayes RA, Atchison GJ, Nelson DW (1978) Some aspects of sediment distribution and macrophyte cycling of heavy metals in a contaminated lake. *J Environ Qual* 7:301–305.

Miller GE, Wile I, Hitchin GG (1983) Patterns of accumulation of selected metals in numbers of the soft-water macrophyte flora of central Ontario lakes. *Aquat Bot* 15:53–64.

Mo SC, Choi DS, Robinson JW (1988) A study of the uptake by duckweed of aluminum, copper, and lead from aqueous solution. *J Environ Sci Health* A23:139–156.

Mortimer DC (1985) Freshwater aquatic macrophytes as heavy metal monitors — the Ottawa River experience. *Environ Monit Assess* 5:311–323.

Muramoto S, Oki Y (1983) Removal of some heavy metals from polluted water by water hyacinth (*Eichhornia crassipes*). *Bull Environ Contam Toxicol* 30:170–177.

Nakada M, Fukaya K, Takeshita S, Wada Y (1979) The accumulation of heavy metals in the submerged plant (*Elodea nuttallii*). *Bull Environ Contam Toxicol* 22:21–27.

Nasu Y, Kugimoto M, Tanaka O, Takimoto A (1983) Comparative studies on the absorption of cadmium and copper in *Lemna paucicostata*. *Environ Pollut Ser A* 32:201–209.

Nasu Y, Kugimoto M, Tanaka O, Yanase D, Takimoto A (1984) Effects of cadmium and copper co-existing in the medium on the growth and flowering of *Lemna paucicostata* in relation to their absorption. *Environ Pollut Ser A* 33:267–274.

Nir R, Gasith A, Perry AS (1990) Cadmium uptake and toxicity to water hyacinth: effect of repeated exposures under controlled conditions. *Bull Environ Contam Toxicol* 44:149–157.

O'Keeffe DH, Hardy JK, Rao RA (1984) Cadmium uptake by the water hyacinth: effects of solution factors. *Environ Pollut Ser A* 34:133–147.

Ozimek T (1985) Heavy metal content in macrophytes from ponds supplied with post-sewage water. In *Heavy Metals in Water Organisms,* Symposia Biologica Hungarica, Vol. 29, Akademiai Kiado, Budapest, Hungary, pp. 41–49.

Piccardi EB, Clauser M (1983) Absorption of copper by *Iris pseudacorus*. *Water Air Soil Pollut* 19:185–192.

Ray SN, White WJ (1976) Selected aquatic plants as indicator species for heavy metal pollution. *J Environ Sci Health* A11:717–725.

Rodgers JH, Cherry DS, Guthrie RK (1978) Cycling of elements in duckweed (*Lemna perpusilla*) in an ash settling basin and swamp drainage system. *Water Res* 12:765–770.

Rother JA, Whitton BA (1988) Mineral composition of *Azolla pinnata* in relation to composition of floodwaters in Bangladesh. *Arch Hydrobiol* 113:371–380.

Salim R (1988) Removal of nickel (II) from water using decaying leaves — effects of pH and type of leaves. *J Environ Sci Health* A23:183–197.

Samecka-Cymerman A (1988) Tolerance of *Scapania undulata* to barium present in the environment. *Pol Arch Hydrobiol* 35:33–44.

Samecka-Cymerman A, Kempers AJ, Bodelier PL (1991) Preliminary investigations into the background levels of various metals and boron in the aquatic liverwort *Scapania uliginosa* (Sw) Dum. *Aquat Bot* 39:345–352.

Satake K, Takamatsu T, Soma M, Shibata K, Nishikawa M, Say PJ, Whitton BA (1989) Lead accumulation and location in the shoots of the aquatic liverwort *Scapania undulata* (L.) Dum. in stream water at Greenside mine, England. *Aquat Bot* 33:111–122.

Say PJ, Whitton BA (1983) Accumulation of heavy metals by aquatic mosses. I. *Fontinalis antipyretica* Hedw. *Hydrobiologia* 100:245–260.

Say PJ, Harding JPC, Whitton BA (1981) Aquatic mosses as monitor of heavy metal contamination in the River Etherow, England. *Environ Pollut Ser B* 2:295–307.

Schierup H-H, Larsen VJ (1981) Macrophyte cycling of zinc, copper, lead and cadmium in the littoral zone of a polluted and nonpolluted lake. I. Availability, uptake and translocation of heavy metals in *Phragmites australis* (Cav.) Trin. *Aquat Bot* 11:197–210.

Schorin H, de Benzo ZA, Bastidas C, Velosa M, Marcano E (1991) Use of water hyacinths to determine trace metal concentrations in the tropical Morichal Largo River, Venezuela. *Appl Geochem* 6:195–200.

Schuler CA, Anthony RG, Ohlendorf HM (1990) Selenium in wetlands and water fowl foods at Kesterson Reservoir, California, 1984. *Arch Environ Contam Toxicol* 19:845–853.

Sela M, Garty J, Tel-Or E (1989) Accumulation and the effect of heavy metals on the water fern *Azolla filiculoides*. *New Phytol* 112:7–12.

Sen AK, Mondal NG (1987) *Salvinia natans* — as a scavenger of Hg(II). *Water Air Soil Pollut* 34:439–446.

Sen AK, Mondal NG (1990) Removal and uptake of copper(II) by *Salvinia natans* from waste water. *Water Air Soil Pollut* 49:1–6.

Sen AK, Mondal NG, Mandal S (1987) Studies of uptake and toxic effects of Cr(VI) on *Pistia stratiotes*. *Water Sci Techol* 19:119–127.

Sinha S, Chandra P (1990) Removal of Cu and Cd from water by *Bacopa monnieri* L. *Water Air Soil Pollut* 51:271–276.

Smith S, Kwan MKH (1989) Use of aquatic macrophytes as a bioassay method to assess relative toxicity, uptake kinetics and accumulated forms of trace metals. *Hydrobiology* 188/189:345–351.

Taylor GJ, Crowder AA (1983) Uptake and accumulation of heavy metals by *Typha latifolia* in wetlands of the Sudbury, Ontario region. *Can J Bot* 61:63–73.

Twining JR (1988) Radium accumulation from water by foliage of the water lily, *Nymphaea violacea*. *Verh Int Verein Limnol* 23:1954–1962.

van der Merwe GG, Schoonbee HJ, Pretorius J (1990) Observations on concentrations of the heavy metals zinc, manganese, nickel and iron in water, in the sediments and two aquatic macrophytes, *Typha capensis* (Rohrb.) N.E. Br. and *Arundo donax* L., of a stream affected by goldmine and industrial effluents. *Water SA* 16:119–124.

van der Werff M, Pruyt MJ (1982) Long-term effects of heavy metals on aquatic plants. *Chemosphere* 11:727–739.

Ward TJ (1987) Temporal variation of metals in the seagrass *Posidonia australis* and its potential as a sentinel accumulator near a lead smelter. *Mar Biol* 95:315–321.

Wehr JD, Whitton BA (1983a) Accumulation of heavy metals by aquatic mosses. 3: Seasonal changes. *Hydrobiology* 100:285–291.

Wehr JD, Whitton BA (1983b) Accumulation of heavy metals by aquatic mosses. 2: *Rhynchostegium riparioides*. *Hydrobiology* 100:261–284.

Wehr JD, Kelly MG, Whitton BA (1987) Factors affecting accumulation and loss of zinc by the aquatic moss *Rhynchostegium riparioides* (Hedw.) C. Jens. *Aquat Bot* 29:261–274.

Wells JR, Kaufman PB, Jones JD (1980) Heavy metal contents in some macrophytes from Saignaw Bay (Lake Huron, U.S.A.). *Aquat Bot* 9:185–193.

Welsh RPF, Denny P (1980) The uptake of lead and copper by submerged aquatic macrophytes in two English lakes. *J Ecol* 68:443–455.

Whitton BA, Say PJ, Jupp BP (1982) Accumulation of zinc, cadmium and lead by the aquatic liverwort *Scapania*. *Environ Pollut Ser B* 3:299–316.

Additional literature

Beckett PHT, Davis RD (1977) Upper critical levels of toxic elements in plants. *New Phytol* 79:95–105.

Bojanowski R (1973) The occurrence of major and minor chemical elements in the more common Baltic seaweed. *Oceanology* 2:81–152.

Boyd CE (1970) Chemical analyses of some vascular aquatic plants. *Arch Hydrobiol* 67:78–85.

Burton MAS, Peterson PJ (1979) Metal accumulation by aquatic bryophytes from polluted mine streams. *Environ Pollut* 19:39–46.

Clark JR, VanHassel JH, Nicholson RB, Cherry DS, Cairns J (1981) Accumulation and depuration of metals by duckweed (*Lemna perpusilla*). *Ecotoxicol Environ Saf* 5:87–96.

Cooley TN, Martin DF (1979) Cadmium in naturally occurring water hyacinth. *Chemosphere* 2:75–78.

Cowgill UM (1974) The hydrogeochemistry of Linsley Pond, North Branford, Connecticut. II. The chemical composition of the aquatic macrophytes. *Arch Hydrobiol* Suppl 45:1.

Cowgill UM (1975) Mineralogical composition of submerged aquatic macrophytes from Connecticut. *Verh Int Verein Limnol* 19:2749–2757.

DeMarte JA, Hartman RT (1974) Studies on absorption of ^{32}P, ^{59}P, and ^{45}Ca by water-milfoil (*Myriophyllum exalbescens* Fernald). *Ecology* 55:188–195.

Dietz F (1973) The enrichment of heavy metals in submerged plants. *Adv Water Pollut Res* 6:53–62.

Dollar SG, Keeney DR, Chesters G (1971) Mercury accumulation by *Myriophyllum spicatum* L. *Environ Lett* 1:191–198.

Eastbrook GF, Burk DW, Inman DR, Kaufman PB, Wells JR, Jones JD, Ghosheh N (1985) Comparison of heavy metals in aquatic plants on Charity Island, Saginaw Bay, Lake Huron, U.S.A., Saginaw Bay. *Am J Bot* 72:209–216.

Eriksson C, Mortimer DC (1975) Mercury uptake in rooted higher aquatic plants: laboratory studies. *Verh Int Verein Limnol* 19:2087–2093.

Ernst WHO, van der Werff MM (1978) Aquatic angiosperms as indicator of copper contamination. *Arch Hydrobiol* 83:354–366.

Gleason ML, Drifmeyer JE, Zieman JC (1979) Seasonal and environmental variation in Mn, Fe, Cu, and Zn content of *Spartina alterniflora*. *Aquat Bot* 7:385–392.

Harding JPC, Whitton BA (1978) Zinc, cadmium and lead in water, sediments and submerged plants of the Derwent Reservoir, Northern England. *Water Res* 12:307–316.

Harrison PG, Mann KH (1975) Chemical changes during the seasonal cycle of growth and decay in eelgrass (*Zostera marina*) on the Atlantic coast of Canada. *J Fish Res Board Can* 32:615–621.

Heaton C, Frame J, Hardy JK (1986) Lead uptake by *Eichhornia crassipes*. *Toxicol Environ Chem* 11:125–135.

Henriques FS, Fernandes JC (1988) Relationship of estuarine plant contaminants to existing data bases. NTIS, Springfield, VA.

Hutchinson TC, Czyrska H (1975) Heavy metal toxicity and synergism to floating aquatic weeds. *Verh Int Verein Limnol* 19:2102–2111.

Jones KC, Peterson PJ, Davies BE (1985) Silver and other metals in some aquatic bryophytes from streams in the lead mining district of mid-Wales, Great Britain. *Water Air Soil Pollut* 24:329–338.

Jones LHP, Clement CR, Hopper MJ (1973) Lead uptake from solution by perennial *Ryegrass* and its transport from roots to shoots. *Plants Soil* 38:403–414.

Kwan KHM, Smith S (1988) The effect of thallium on the growth of *Lemna minor* and plant tissue concentrations in relation to both exposure and toxicity. *Environ Pollut* 52:203–219.

Mayes RA, McIntosh AW, Anderson VL (1977) Uptake of cadmium and lead by rooted aquatic macrophyte (*Elodea canadensis*). *Ecology* 58:1176–1180.

McLean RO, Jones AK (1975) Studies of tolerance to heavy metals in the flora of the rivers Ystwyth and Cdarach, Wales. *Freshwater Biol* 5:431–444.

McNaughton SJ, Folsom TC, Lee T, Park F, Price C, Roeder D, Schmitz J, Stockwell C (1974) Heavy metal tolerance in *Typha latifolia* without the evolution of tolerant species. *Ecology* 55:1163–1165.

Mortimer DC, Kudo A (1975) Interaction between aquatic plants and bed sediments in mercury uptake from flowing water. *J Environ Qual* 4:491–495.

Rawlence DJ, Whitton JS (1977) Elements in aquatic macrophytes, water, plankton, and sediments surveyed in three North Island lakes. *NZ J Mar Freshwater Res* 11:73–93.

Ray SN, White WJ (1979) Equesetum arvense, and aquatic vascular plant as a biological monitor for heavy metal pollution. *Chemosphere* 8:125–128.

Reiniger P (1977) Concentrations of cadmium in aquatic plants and algal-mass in flooded rice culture. *Environ Pollut* 14:297–301.

Sutton DL, Blackburn RD (1971) Uptake of copper in *Hydrilla. Weed Res* 11:47–53.
Tokunagawa K, Furuta N, Morimoto M (1976) Accumulation of cadmium in *Eichhornia crassipes. J Hyg Chem* 22:24–239.
van der Werff M, Ernst WHO (1979) Kinetics of copper and zinc uptake by leaves and roots of an aquatic plant, *Elodea nuttallii. Z Pflanzenphysiol* 92:1–10.
Wolverton BC, McDonald RC (1978) Bioaccumulation and detection of trace levels of cadmium in aquatic systems by *Eichhornia crassipes. Environ Health Perspect* 27:161–164.

chapter fourteen

Bioaccumulation of xenobiotic organic chemicals by terrestrial plants

Otto J. Schwarz and Lawrence W. Jones

Introduction

Since the initiation of the industrial age with its parallel reliance on the utilization of the hydrocarbon reserves on earth, the biosphere has been increasingly exposed to an ever expanding variety of chemicals of anthropogenic origin (Jury et al. 1983; Lendzian and Kerstiens 1991; Sims and Overcash 1983). According to Gunther and Pestemer (1990), "environmental chemicals" represent a broad class of substances that find their way into the environment as a result of anthropogenic activity and may be hazardous to biota, including humans. These hazards were defined as being influenced by acute or chronic exposure, synergism, or as the result of accumulation and subsequent metabolism by the host organism.

    Many of these chemicals are xenobiotic in nature. Connell (1990) describes xenobiotic substances as "chemicals foreign to life, which are derived synthetically or from an abiotic process." Etymologically, "xenobiotic" is borrowed from the Greek words "xenos," meaning alien or strange, and "bios," meaning life. Interest in this class of chemicals has historically been high because of their wide use as biocides in agriculture (Crisp 1972; Wain and Carter 1967). The magnitude of this chemical invasion has been described as "numbering in the thousands" (O'Connor et al. 1991), as 15,000 to 20,000 manmade chemicals (Jacobs et al. 1987) (both of them in reference to contaminants found in municipal sludge), and as an inventory of 100,106 chemicals (both naturally occurring and xenobiotic) found to have been on the market in the European community over a 10-year period between 1971 and 1981 (Greim et al. 1993). Schmidt-Bleek and Haberlandt (1980) estimated that an additional 2000 chemicals enter the market each year. The number and structural variety represented in the current xenobiotic mix present a formidable challenge to those who wish to understand the individual and collective impacts of the chemicals on the biosphere.

The relationship between the importance of the potential impact that these chemicals might have on the plant biota and our current level of effort to define these effects was underscored by Fletcher (1991): "Plants, the most abundant life form and primary producer for all other life forms, always seem to get least recognition and finish last in environmental matters." In addition, Fletcher defined two major issues of environmental concern with respect to the plant biota: (1) the toxicity of a chemical to plants and (2) the bioaccumulation of these chemicals by plants. This chapter will focus on two aspects of the latter issue: specifically, on progress made in defining the mechanism of entry (active vs. passive) and route of entry (soil vs. air: root vs. foliar) of these chemicals into terrestrial plants.

Bioaccumulation

Bioaccumulation, in its broadest sense, has been defined as "the uptake and retention of pollutants from the environment by organisms via any mechanism or pathway" (Connell and Miller 1984) or, similarly, "the extent to which a living organism accumulates a compound from its surrounding environment by all processes" (Isensee 1991). Bioconcentration in aquatic systems occurs under circumstances that result in xenobiotic accumulation to a higher concentration within the organism than in its surrounding environment (Connell 1990). The ratio of the concentration of the chemical in an organism to the concentration in its surrounding environment defines the bioconcentration factor (BCF) (Isensee 1991). When applied to animals, bioaccumulation includes the possibility of acquiring the xenobiotic via ingestion of a contaminated food organism or sediment residues (Amdur et al. 1991). Ingestion of contaminated food opens the possibility of pollutant transfer from one trophic level to another (Connell and Miller 1984) and, therefore, possible biomagnification of the pollutant.

Because of the trophic position of green plants as the primary producer in the food chain, the process of bioaccumulation starts with the transfer of a chemical from the abiotic environment to the plant. The options for contamination are somewhat restricted when compared to those available to organisms represented in the higher trophic levels (i.e., grazing, predation). The initial accumulation process is restricted to mechanisms that facilitate transfer of the chemical directly from the external, nonliving compartment into various plant structures, cells, and/or tissues, which may or may not be considered living. Examples of nonliving structures may include the plant apoplast, as in the initial rapid uptake into root free space (Briggs and Robertson 1957; Krstich and Schwarz 1989; Shone and Wood 1974), or perhaps just the aerial contamination of the cuticle, which is the primary air–foliage boundary structure (Schonherr and Riederer 1989). As previously mentioned, the propensity of plants and other biota to bioconcentrate xenobiotics is measured through the determination of BCFs (Briggs 1981; Eiceman et al. 1993) under steady-state conditions (Esser 1986). Determining the value of the BCF associated with possible environmental risk for a plant contaminant is a complex undertaking (Esser 1986). For some plant-chemical contaminants, a BCF as low as 1 or less is thought to pose greater than minimum risk to humans (Eiceman et al. 1993). These authors further state that BCF values of less than 0.01 are considered to "represent negligible risk" by most researchers.

Bioaccumulation in plants may be confined to a single process of xenobiotic transfer from the surrounding environment to the root and/or shoot compartment with little subsequent movement within the plant or metabolic transformation (McCrady et al. 1990). If the chemical is metabolically stable and mobile, it may be transferred via apoplast or symplast compartments, or both, throughout most of the plant as parent compound, where it may be found from trace to highly bioconcentrated levels (McFarlane et al. 1987a,b). When the site of its residence is foliar, the possibility exists for chemical loss over time due to photochemical processes resulting in destruction of the parent compound (Huang et al. 1993; McCrady and Maggard 1993). For chemicals that are accumulated directly from soil and are translocated to the leaves, an additional route of parent compound loss may occur by chemical volatilization to the surrounding atmosphere from foliar tissues (Krstich and Schwarz 1990), with the final tissue concentration being determined by equilibrium plant–air partitioning (Bacci et al. 1990b). The atmosphere presents several opportunities for xenobiotic contamination from chemical vapor : chemical sorption to particulates and solubilization in water droplets.

Metabolic interactions with xenobiotics

The possibility of direct plant–chemical metabolic interaction presents a major problem to those concerned with determining the extent of environmental risk of a chemical (Plewa and Gentile 1982; Plewa et al. 1993). The metabolites may be of less, equal, or greater potential risk to the environment (Gentile et al. 1991). Relatively little is known of the metabolic fate and potential biological activity of most xenobiotic metabolites with respect to the total plant biota. Most of the information regarding xenobiotic–plant metabolic interaction has been obtained from the investigation of herbicide metabolism by target (pest) weeds and economically related crop species (Crafts 1964; Freed and Montgomery 1963; Geissbuhler et al. 1963). Metabolism of xenobiotic organics has been generally thought to result in endproducts that are predominantly polar conjugates and insoluble residues (Lamoureux and Rusness 1981). Highly lipophilic compounds may be oxidized to phenols and alcohols, increasing their water solubility and permitting conjugation via glycosidic bond formation (Cole 1983). An important detoxifying step in herbicide metabolism is conjugation with glutathione (Shimabukuro and Walsh 1979). Hydrolysis and photoconversion may also play an important role as mechanisms of plant xenobiotic transformation (Cole 1983). A summary of chemical–plant metabolic interaction is presented in a review by Bell (1992).

Plant cell cultures have been suggested as *in vitro* test systems for screening environmental chemicals for their metabolic breakdown and toxic effects (Ebing et al. 1984; Harms 1973; Harms and Langebartels 1986; Langebartels and Harms 1986; Sandermann et al. 1984; Schuphan et al. 1984). Advantages accrued by using *in vitro* plant cultures included the relatively low maintenance cost and year-round availability, ease of standardization and sterility, and ability to manipulate the physiological status of the cells (Sandermann et al. 1984). *In vitro* systems are also well suited for rapid screening for the metabolism of xenobiotics in plants because of the lack of interfering microorganisms, the rapid sorption of test chemicals, and their high rate of metabolite formation (Harms and Langebartels 1986). Mumma and Davidonis (1983) suggested that it might be difficult to "effectively extrapolate" *in vitro*-generated results to whole plants. However, several other researchers have found little difference between whole-plant metabolic interactions and *in vitro* cell culture interactions on the chemicals tested (Harms 1973; Langebartels and Harms 1986; Schuphan et al. 1984). Ebing et al. (1984) pointed out that *in vitro* testing would not produce results identical to experimental data obtained from whole plants grown under natural environmental conditions because of "additional abiotic influences" and the microbial activity inherent in the nonsterile environment. A recent review focusing on plant–microbial–toxicant interactions present in the soil environment describes their interactions with respect to toxicant degradation (Anderson et al. 1993).

Mode of chemical uptake

The process of bioaccumulation of xenobiotic organic chemicals in plants is a complex blend of the physicochemical nature of the substance and its interaction with the plant biota. The actual level of bioconcentration achieved is, in the final analysis, the summation of all input and loss processes that govern the plant–chemical interaction (Ryan et al. 1988). Of primary interest are the major routes to plant contamination (via soil and air) and the accumulation mechanism(s) controlling chemical transfer that operates in each of these environmental compartments (Figure 1). The chemical may be present in one

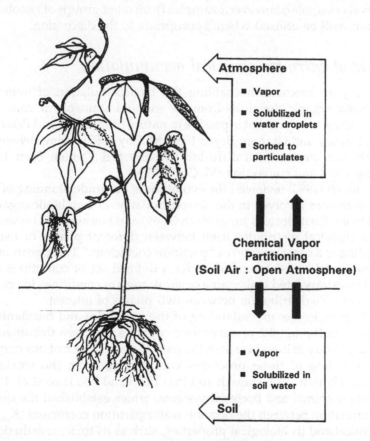

Figure 1 Major entry routes for plant contamination by xenobiotic organic chemicals.

or the other or both environmental compartments, but it is only a candidate for plant uptake and possible bioaccumulation if it is bioavailable to the resident organism (O'Connor et al. 1991). Chemical transfer between the soil and atmospheric compartments may also modify the final site and extent of chemical–plant interaction (Bacci and Gaggi 1985, 1986, 1987; Bacci et al. 1988, 1992; Gaggi and Bacci 1985; Gaggi et al. 1985; McCrady et al. 1990; Muller et al. 1993).

The bioavailability of a chemical for roots or any other subterranean plant organ (Hance 1988) may be thought of as the amount of the substance contained in the soil–water (Gunther and Pestemer 1990) and also in the soil–vapor phase (Bell 1992; Jacobs et al. 1987). For shoots, bioavailability of a chemical may be thought of as the amount of the substance present as vapor, dissolved in water droplets, or sorbed to atmospheric particulates (Grover 1991). Because of the enormous structural diversity found in xenobiotic chemicals, the discussion will be focused primarily on those xenobiotic organics that are nonionic and relatively nonpolar; however, examples from other groups of xenobiotic organics will be utilized where appropriate to the discussion.

Route of entry for chemical accumulation

The primary mechanism enabling the bioaccumulation of nonionic xenobiotic organic chemicals from the soil and atmospheric compartments seems to be one that is passive in nature (Edgington and Peterson 1977; Hartley and Graham-Bryce 1980; McCoy 1987), being driven by the physical processes of diffusion (Devine and Vanden Born 1991; Goring 1967) and convection (McCoy 1987).

Connell (1990) reviewed the evolution of our understanding of the basic processes involved in the bioaccumulation of xenobiotic organics in all biota. Berthelot and Jungfleisch (1869) and Nernst (1891) revealed that a chemical distributes itself between different phases of matter according to a constant ratio or "partition coefficient." This partitioning behavior could be determined under a defined set of conditions and could be extrapolated to the same or similar sets of conditions to predict the chemical's distribution between two phases of interest.

Progress in the understanding of the principles and mechanisms involved in the uptake of xenobiotic organics has been documented by a large body of literature over the past 30 years. Much of our current understanding of these processes was derived from the works of Hansch (Hansch 1973; Hansch and Leo 1979) and Leo (Leo et al. 1969) in various animal and bacterial systems, which established the use of the correlation between the octanol–water partition coefficient (K_{ow}) of a chemical and its biological properties, such as its toxic, anesthetic, or carcinogenic activity. In a comparative study of 70 organic chemicals and 4 different biological systems, Leo et al. (1969) found that out of 6 parameters tested, log K_{ow} correlated best with biological activity. Their work contributed to the use of quantitative structure activity relationships (QSAR) for the prediction of the potential biological properties of compounds. At the same time, other researchers (Dearden and Townsend 1976) demonstrated that the relationship between the log of the partition coefficient (log P) (log K_{ow} if partitioned between octanol and water) and biological activity was a complex one, and log P, alone, could not always predict biological effectiveness.

Recently, more sophisticated approaches to describing chemical–plant interaction based upon mass balance and an expanded use of physicochemical properties of the organic chemical have been presented (Bacci et al. 1990b; Boersma et al. 1988; Calamari et al. 1987; McCoy 1987; Paterson and Mackay 1989; Paterson et al. 1990; Ryan et al. 1988; Schramm et al. 1987). These models have allowed the evaluation of bioaccumulation characteristics, such as the potential for uptake and the translocation and eventual loss as vapor from foliage. Nonetheless, the simple and useful log K_{ow} has found continued use in predicting plant xenobiotic organic interaction (Bacci et al. 1990b; Briggs et al. 1982, 1983; Kerler and Schonherr 1988a,b). In addition, plant researchers have nearly always found log K_{ow} to be correlated with the mobility of the xenobiotic chemical within plant tissues after root sorption (Briggs et al. 1976, 1982; Crowdy 1973; Leroux and Gredt 1977; Martin and Edgington 1981; Shone et al. 1974; Shone and Wood 1974; Small et al. 1977; Tyree et al. 1979). Bell (1992) summarized plant uptake and transport results concerning a broad range of organic pollutants from soils. He concluded that chemicals that were taken up by a passive process, were not degraded by the plant, and possessed a high log K_{ow} (e.g., dioxin [6.14], PCBs [4.12–6.11], PAHs [4.07–7.66]) were "most likely to be accumulated by or in the root and not to be translocated out of it." Chemicals sorbed to the root that are less lipophilic and more water soluble were thought to be translocatable with the possibility of being accumulated in the leaves.

The current understanding of the initial processes of bioaccumulation of xenobiotic organics is based on early works that were designed to determine the biological activity of a broad series of chemicals possessing biocidal properties. Unfortunately, there is a dearth of this sort of information concerning the potential for bioaccumulation in plants for many of the xenobiotic organic chemicals now considered priority pollutants (Bell 1992; Fletcher et al. 1988). The information that is available has been derived from both field observation and laboratory-based experiments. The laboratory-based experiments which were designed to mimic natural systems produce data more easily related to actual environmental situations (Bacci and Gaggi 1985; Bacci et al. 1990a; Scheunert et al. 1986), but make deductions with reference to mechanism more difficult. Other laboratory experiments attempted to simplify the chemical–plant interaction through the use of dissected plant organs (i.e., isolated roots) (Krstich and Schwarz 1989, 1990; McFarlane et al. 1987a; McFarlane and Wickliff 1985; Wickliff et al. 1984; see also references to early literature contained in Table 1). Highly defined, computer controlled, and monitored environments have been designed to measure whole-plant chemical response (McFarlane and Pfleeger 1986, 1987). Relating these data to field observations (Hulzebos et al.

Table 1   Selected Examples of Studies Using Excised Root Systems to Measure the Uptake of Xenobiotic Organic Chemicals

Common name	Species	Age of tissue	Chemical[a]	Uptake mechanism	Ref.
Barley	Hordeum vulgare L.	5 d	2,4-D; monuron (herbicide)	2,4-D-metabolic energy required monuron-passive	Donaldson et al. 1973
Barley	H. vulgare L.	6 d	Simazine (herbicide)	Passive	Shone and Wood 1972
Barley	H. vulgare L.	8 d	Siduron (herbicide)	Initial rates of absorption were linear and found to be a function of external concentration	Splittstoesser 1968
Barley	H. vulgare L.	4 d	Triarimol (fungicide)	Passive	Wallerstein et al. 1976
Barley	H. vulgare L.	8 d	Bromacil (herbicide)	Passive	Wickliff et al. 1984
Barley	H. vulgare L.	7 d	Atrazine (herbicide)	Passive (absorption of living and dead roots equal)	Zhirmunskaya and Kol'tsova 1973
Bean	Phaseolus vulgaris L.	6–8 d	Amitrole (herbicide)	Passive	Lichtner 1983
Corn	Zea mays L.	5 d	Napropamide (herbicide)	Passive diffusion followed by binding into residual fraction	Barrett and Ashton 1983
Corn	Z. mays L.	4 d — then protoplasts	Atrazine 2,4-D (herbicide)	Passive Passive/probably active	Darmstadt et al. 1983

Corn	*Z. mays* L.	Carbendazim (fungicide)	7 d	Passive	Leroux and Gredt 1975
Corn	*Z. mays* L.	Benzimidazole derivatives: carbendazium EBC fuberidazole thiabendazole BSX benomyl cypendazole BIX HOE27591 thiophanates: thiophanate-ethyl thiophanate-methyl thiophamine TX (fungicide)	7 d	All passive	Leroux and Gredt 1977
Corn	*Z. mays* L.	Oryzalin (herbicide)	2 d	Passive	Upadhyaya and Nooden 1980
Foxtails	*Setaria* taxa	Atrazine Simazine, propazine, cyanazine (herbicides)	6–9 d	Passive Passive/active —	Orwick et al. 1976

Table 1 (continued)

Common name	Species	Age of tissue	Chemical[a]	Uptake mechanism	Ref.
Peanuts	*Arachis hypogaea*	10 d	Trifuralin, methzole (herbicides)	— —	Hawxby et al. 1972
Soybean	*Glycine max* L. Merr.	14 d	Amiben Atrazine Chlorpropham EPTC Linuron (herbicide)	Suggest uptake dominated by physical processes	Moody et al. 1970b
Soybean	*G. max* L. Merr.	14 days	(As above) (herbicide)	Efflux after 1 h uptake was a result of release from free space only	Moody et al. 1970a
Velvet bean	*Stizdobium deeringianum* Bort	Root 10–15 min cultured for up to 3 months	CMU (herbicide)	—	Muzik et al. 1954
Velvetleaf	*Abutilon theophrasti*	5–6 d	Atrazine (herbicide)	Initial absorption passive long-term metabolically dependent	Price and Balke 1983
Velvetleaf and Morning glory	*A. theophrasti medic Ipomoea hederaceae* L. Jacq	3 d	Amiben (herbicide)	Initial rates of absorption linear to external amiben concentration	Stoller 1969

Wheat and oat	*Triticum* spp. *Avena* spp.	6 d	Amitrole diclofop-methyl (herbicide)	Jacobson and Shimabukuro (1980)

[a] Amiben, 3-amino-2,5-dichlorobenzoic acid; Amitrole, 3-amino-1,2,4-triazole; Atrazine, [2-chloro-4-ethylamino-7-isopropylamino-s-triazine]; Benomyl, methyl-1-(butylcarbamoyl)-2-benamidazole carbamate; Bromacil, 5-bromo-3-*sec*-butyl-6-methyluracil; Carbendazim, methyl benzimidazole-2-yl carbamate; Chloropropham, isopropyl-*m*-chlorocarbanilate; CMU, 3-(*p*-chlorophenyl)-1,1-dimethylurea; Cyanazine,2-chloro-4-(1-cyano-1-methylamino)-6-ethylamino-1,3,5-triazine; Cypendazole, methyl 1-(5-cyanopentyl carbamoyl)benzimidazol-2-yl carbamate; 2,4-D, 2,4-dichlorophenoxyacetic acid; Diclofop-methyl, methyl 2-[4-(2',4'-dichlorophenoxy)phenoxy] propanoate; EPTC, S-ethyl dipropylthiocarbamate; Fuberidazole, 2-(2-furyl)benzimidazole; HOE27591, 1-methoxycarbonyl-3-hexoendonorbornylmethyl-5-hexahydrotriazinobenzimidazole; Linuron, 3-(3,4-dichlorophenyl)-1-methoxy-1-methylurea; Methazole, 2-(3,4-dichlorophenyl)-4-methyl-1,2,4-oxadiazolidine-3,5-dione methazole; Monuron, 3-(*p*-chlorophenyl)-1,1-dimethylurea; Napropamide, [2-(α-naphthoxy)-N,N-diethyl]propionamide]; Oryzalin, 3,5-dinitro-N^4,N^4-dipropylsulfanilamide; Propazine, 2-chloro-4,6-*bis*-isopropylamino-1,3,5-triazine; Siduron, 1-(2-methylcyclohexyl)-3-phenylurea; Simazine, 2-chloro-4,6-*bis*-ethylamino-1,3,5-triazine; Thiabendazole, 2-(4'-thiazoyl)benzimidazole; Thiophanate-ethyl, diethyl[(1,2-phenylene)*bis*-(imino carbonothioyl)*bis*-(carbamate)]; Thiophanate-methyl, dimethyl[(1,2-phenylene)*bis*-(imino carbonothioyl)*bis*-(carbamate)]; Triarimol, α-(2,4-dichlorophenyl)-α-phenyl-5-pyrimidine-methanol; Trifluralin, α,α,α-trifluoro-2,6-dinitro-N,N-dipropyl-*p*-toluidine; Benzimidazole derivatives: BIX, BSX, EBC, Thiophamine, TX (structures not readily available).

1993) or using information derived in the form of uptake or transfer coefficients or chemical bioconcentration factors in the assessment of environmental risk has posed some difficulties (Ryan et al. 1988).

Soil to root transfer

The need to establish an increased degree of rigor in determining the mechanism that enables the primary contamination of a terrestrial plant was recognized by Donaldson et al. (1973). These authors listed six criteria for determining whether the accumulation of a xenobiotic organic chemical was active or passive (Table 2). These were adapted from the criteria used to establish the mechanism of uptake of inorganic ions (Epstein 1972; Pitman 1976a,b) and represent an appropriate starting point for an analysis of the process of bioaccumulation via soil to root transfer. These criteria can be used to define the accumulation process being monitored as primarily active, as primarily passive, or as some combination of the two (Krstich and Schwarz 1989).

Table 2   Criteria for Establishing an Active Mechanism of Bioaccumulation
of Xenobiotic Organics Via Roots

---

1.   The temperature coefficient (Q_{10}) of sorption should be at least 2.
2.   Sorption should be oxygen dependent; anaerobiosis should drastically limit uptake.
3.   The rate of sorption should be hyperbolic, indicating saturation of a carrier-mediated process.
4.   Structurally similar compounds should compete for sorption, again indicating the presence of a substrate-specific carrier site.
5.   The sorption of the chemical should be inhibited by enzyme/metabolic poisons.
6.   The chemical should be accumulated against a concentration gradient and should reach the cellular cytoplasm in an unchanged form.

Data from Donaldson TW, et al. (1973) *Plant Physiol* 52:638.

Plant test systems readily amenable to application of all six of these criteria are somewhat limited. For this reason, many researchers chose isolated root tissue as a model system. Table 1 lists examples of studies that used isolated root tissue as a model system to characterize the transfer of xenobiotics (all of which are biocides) from the soil compartment into the plant root. As can be seen, only a very few plant species were utilized for this type of investigation (i.e., economically important crops and weeds). In most of the studies, a statement as to the mechanism of uptake was made based upon one or more of the criteria listed by Donaldson et al. (1973). These chemicals were found to be absorbed in a two-phase process: an initial rapid phase, followed by a second, much less rapid and longer accumulation phase. The initial rapid sorption phase is passive, diffusion controlled, and represents the entrance of the chemical into root apparent free space (AFS). Depending upon the chemical (Darmstadt et al. 1984; Price and Balke 1983), the sorption process may continue as passive and allow the chemical to permeate the entire volume of the exposed tissue. In a few instances, however, the second phase of sorption has been shown to be metabolically dependent. Notably, 2,4-dinitrophenoxyacetic acid (2,4-D) was said to utilize some sort of carrier-mediated, metabolically dependent uptake mechanism (Donaldson et al. 1973). Metabolic dependence for long-term sorption has been described for atrazine (2-chloro-4[ethylamino]-6-[isopropylamino]-s-triazine) (Price and Balke 1983), napropamide (2-[α-naphthoxy]-*N,N*-diethylpropiona-mide) (Barrett and Ashton 1983), and picloram (4-amino-3,5,6-trichlo-ropicolinic acid) (Morrison and Vanden Born 1975).

Although no single example of xenobiotic chemical uptake illustrates all of the complexities of chemical–root interaction which may be encountered, the work of Leroux and Gredt (1975, 1977) serves to illustrate the kind of information that can be obtained. These authors described the influx and efflux characteristics of the systemic fungicide carbendazim (methyl benzimidazol-2-yl carbamate), using the excised primary roots of 7-d-old corn (*Zea mays* L.). Carbendazim is a benzim-idazole derivative that can exist in solution in a number of molecular forms, depending upon the pH of the solution (i.e., cationic, anionic, or nonionized forms). Carbendazim total uptake was measured at pH values between 2 and 12 over a 2-h period. Maximal sorption occurred between pH 5.5 and 8.5. In this pH region, the carbendazim exists in solution mainly in the nonionized (molecular) form. Time course uptake experiments over a 6-h period showed an initial rapid sorption lasting approximately 15 m, followed by a slower accumulation that reached maximum sorption after 1 to 2 h. The initial phase of uptake was assumed to follow first order kinetics governed by the equation:

$$\frac{dQ_{at}}{d_t} = K_a[Q_m - Q_{at}] \tag{1}$$

where
  K_a = absorption rate constant
  Q_{at} = amount of chemical retained in the roots after exposure
      time t
  Q_m = maximal amount of chemical retained in the root
After integrating Equation 1 and obtaining its reciprocal, Equation 2 was obtained:

$$\text{Log}\frac{Q_m}{Q_m - Q_{at}} = K_a t \tag{2}$$

The curve obtained by plotting log $[Q_m/Q_m - Q_{at}]$ vs. "t" indicated that the initial uptake process could be separated into two components. The excised corn roots accumulated carbendazim to levels that exceeded the concentration of the ambient solution. The root concentration factor (RCF) (Shone et al. 1974),

$$RCF = \frac{\mu g \text{ chemical}/g \text{ fresh weight root}}{\mu g \text{ chemical}/ml \text{ ambient solution}} \tag{3}$$

was calculated for carbendazim after 6 h of exposure using two different ambient pH values (i.e., pH 6.4 and 3.8). The RCF values attained were 0.62 (pH 3.8) and 1.25 (pH 6.4). The efflux of carbendazim was measured and analyzed using the methods outlined by Macklon and Higinbotham (1970). The roots contained three compartments at pH 6.4 and two compartments at pH 3.8. The derived rate constants for the first compartment at either pH described the highest rate of loss and were said to be associated with the AFS of the root tissue (Briggs and Robertson 1957):

$$AFS = \frac{\begin{array}{c}\text{quantity of readily diffusible}\\ \text{solute}/g \text{ fresh weight root}\end{array}}{\text{quantity of solute}/ml \text{ ambient solution}} \tag{4}$$

The authors calculated the AFS value for the first compartment and found that its size correlated reasonably well with estimates found in the literature (Crowdy 1972; Shone et al. 1974). The second, much larger compartment available to the chemical at both pH values was said to represent the passive diffusion across the protoplasts of the root cortical cells (Donaldson et al. 1973; Shone et al. 1974). It was concluded that carbendazim was passively sorbed because (1) its rate of sorption was a linear function of the concentration of the ambient solution; (2) metabolic inhibitors and anaerobic conditions had no significant effect on uptake; (3) the temperature coefficient (Q_{10}) of sorption was low (i.e., 1.0 to 1.2); and, (4) in general, compounds of similar structure had little effect on carbendazim sorption — however, 2-methylbenzimidazole did slightly inhibit uptake (20% inhibition). The authors concluded that the uptake of benzimidazole derivatives, i.e., carbendazim at the two pH values studied and benomyl (methyl 1-[butylcarbamyoyl]-2-benzimidazolecarbamate) (data not presented), was directly related to their partition coefficients in an octanol–water system (K_{ow}). This conclusion is in agreement with that of Shone and Wood (1974), who noted that the initial rate of uptake for a number of organic herbicides and a systemic fungicide was positively correlated with their partition coefficients in olive oil–water or *n*-dodecane–water systems. Leroux and

Gredt (1977) surveyed nine benzimidazole derivatives and four thiophanates to determine their mechanism of sorption (i.e., passive/active) by excised corn roots. The parameters surveyed were identical to those presented in the work just discussed (Leroux and Gredt 1975). All chemicals were shown to have a passive diffusion mechanism for uptake into the corn roots. They found that the rate of root sorption was greater for the more lipophilic than it was for the lipophobic chemicals and was positively correlated (e.g., for benzimidazole derivatives [r =0.89] and for thiophanates [r =0.99]) to their partition coefficient in an octanol–water system at pH 6.5.

A great majority of environmental organic compounds exhibit a passive uptake mechanism and, therefore, are assumed to follow a process of successive movement into the various plant compartments governed by the physicochemical process of partitioning (Ryan et al. 1988). Unfortunately, analytical data describing the physicochemical characteristics of the root as well as the foliar accumulation processes are not available for the vast majority of xenobiotic organic chemicals, and the data that are available have been gathered for only a very few species of the plant kingdom. Baseline data for bioaccumulation processes are necessary for the development and validation of models designed to track the environmental fate of structurally heterogeneous chemicals possessing greatly differing physicochemical properties. In a review of the uptake of xenobiotic organic chemicals by plants, Paterson et al. (1990) compiled a listing of 88 terrestrial plant species and corresponding reference citations describing the fate of 69 organic chemical contaminants. Estimates place the number of identified plant species at 270,000 (May 1992) and the total number at more than 300,000 (Cronquist 1988). It is not reasonable to assume that baseline information will ever be available for more than a fraction of the total number of plant species.

Only with improved understanding of the basic principles that control the mechanism(s) of transfer of xenobiotic organic chemicals into plants can increasingly accurate predictions be made as to their fate and potential risk. Intertwined with the problem is the possibility for both singular and interactive toxicant effects because of the simultaneous presence of large numbers of chemical contaminants (i.e., both organic and inorganic) in the environment (Amdur et al. 1991; McCarty and MacKay 1993). In a recent review, Breeze (1993) included a discussion of the effects of the simultaneous presence of multiple pollutants. He concluded that "other atmospheric pollutants may therefore affect the phytotoxicity of herbicide vapor" and that the extent of these effects on natural plant communities was basically not known. Breeze speculated that shifts in the species composition of wild plant communities could occur which would affect plants that receive no direct chemical damage.

As previously mentioned, good correlations have been found between chemical sorption into roots and a compound's octanol–water partition coefficient. More recently, other related physicochemical properties have been studied. Two recent reviews present numerous examples of correlations between physicochemical properties of xenobiotics as they relate to root as well as to whole-plant uptake (Bell 1992; Paterson et al. 1990). Briggs et al. (1982) tested a series of two chemical families (o-methylcarbamoyloximes and substituted phenylureas) possessing log K_{ow} values ranging from –0.57 to 4.6. The RCFs for barley (*Hordeum vulgare*) roots were positively correlated (r =0.917) with the octanol–water partition coefficient of the combined series of chemicals tested. Translocation from roots to shoots in barley, as expressed by the transpiration stream concentration factor (TSCF) (Shone and Wood 1974), was found to be a passive process (i.e., all TSCFs were reported to be less than unity) that occurred maximally for chemicals possessing log K_{ow} values in an intermediate range ($\approx$1.8). It is important to note that both the more lipophilic and the more polar chemicals were less efficiently transferred into the transpiration stream. These results suggested the presence of a selective barrier to compounds possessing log K_{ow} values outside the intermediate range. Briggs et al. (1982) suggested that these results could not be explained "in terms of a series of simple partitions within the root." Compounds that are highly lipophilic and consequently highly hydrophobic would predictably be accumulated in the roots and would not be readily transported to the shoot. These predictions have been confirmed for members of the polychlorinated biphenyls (PCBs) (Bacci et al. 1988; Bacci and Gaggi 1985; Hani 1990; Iwata and Gunther 1976; Iwata et al. 1974; Strek and Weber 1982), polybrominated biphenyls (PBBs) (Chou et al. 1978), and polychlorinated dibenzo-p-dioxins (PCDDs) (Bacci et al. 1992; McCrady et al. 1990; Muller et al. 1993; Reischl et al. 1989b).

Both Trapp et al. (1990) and Topp et al. (1986) extended experimental observations to soil-based test systems. Topp et al. found a negative correlation (r =–0.838) between barley RCFs (expressed as the concentration in roots divided by the concentration in soil) and adsorption coefficients based on soil organic carbon. However, RCFs calculated on the basis of the chemical concentration in the roots divided by the concentration in the soil liquid were positively correlated (r =0.896) to the log K_{ow} of the test compounds. An exception to this pattern was the RCF for p,p'-DDT (1,1,1-trichloro-2,2-bis[p-chlorophenyl]ethane) (log P_{ow} = 6.2). Total uptake (i.e., both root and foliar) as measured by whole-plant accumulation divided by the chemical concentration in air-dried soil (expressed as the "concentration factor" [CF; see Equation 5]) was found to be linearly correlated (r = –0.9449) with molecular weight when plotted on a double-logarithmic scale.

$$CF = \frac{\text{concentration in plants (based on fresh weight)}}{\text{concentration in soil (based on air-dried weight)}} \quad (5)$$

Of 16 ^{14}C-labeled organic chemicals tested, ranging in molecular weight from 78 to approximately 750, only 2 did not fit the correlation. The two chemicals, benzene and pentachlorophenol, were found to be rapidly mineralized during the course of the experiment, complicating their CF determination. These authors concluded that the molecular weight of the chemical was a more useful property for predicting plant uptake than its octanol–water partition coefficient. They also suggested that their results could not be applied to plants with "higher lipid content or with oil channels." Examples given for such plants were carrot (*Daucus carota*), cress (*Lepidium sativum*), and parsnip (*Pastinaca sativa*), where uptake and transport are apparently facilitated by oil cells. Whole-plant CF values were shown to be higher for cress than barley throughout a long-term outdoor experiment. It was concluded, therefore, that correlations found to hold for chemical–barley interactions could not be applied to cress and, through inference, to other higher lipid content plants.

Trapp et al. (1990) used a simple closed-model ecosystem to generate data on the "material balance of chemicals" in a soil–water–plant–air system. These data were used to test a fugacity-based (Mackay and Paterson 1981) simulation model. These authors (Trapp et al. 1990) found that bioconcentration predictions obtained with the use of a fugacity-based model, log K_{ow}, as well as other physicochemical properties of the chemicals were in good agreement with their laboratory-generated data. Their model predicted and agreed with the results generated by their laboratory test system that soil-to-plant bioconcentration factors were best modeled by a combination of fugacity-based transfer coefficients and the ratio of the absolute values of octanol–water and octanol–soil–carbon partition coefficients.

The experimental approaches and questions posed have evolved from highly focused experimental designs, which were concerned with defining parameters of uptake into isolated root systems of closely related chemicals, to more complex model ecosystems, which are designed to track whole-plant bioaccumulation processes of chemicals of both related and widely differing physicochemical properties. Perhaps one of the most sophisticated and elaborate test systems assembled to date was designed to study the uptake and phytotoxicity of toxic, radiolabeled chemicals (McFarlane and Pfleeger 1986, 1987) in mature terrestrial plants under a contained and controlled environment. McFarlane et al. (1987a) used an isolated root uptake test (IRUT) and the whole-plant test system to study the uptake, distribution, and metabolism of several environmental chemicals in soybean (*Glycine max* L.) plants and barley roots. Two of these chemicals were herbicides

(bromacil [5-bromo-3-sec-butyl-6-methyluracil]; 2,6-dichlorobenzoni-
trile [DCNB]) which were known to be absorbed through roots and
then translocated and two were "highly toxic" industrial chemicals
(nitrobenzene [NB], 1,3-dinitrobenzene [DNB]). One of their objectives
was to obtain uptake rate constants using both the IRUT and the whole-
plant test system to assess the utility of the isolated root test. The
patterns of uptake rate constants obtained from both tests were found
to be similar, except for DNB; the reason for the deviation was not
determined. Parent compound analysis of three of the chemicals tested
(bromacil, DCNB, and DNB) indicated very different fates for each.
Bromacil was essentially not metabolized in either roots or leaves,
DCNB was metabolized in the leaves but not in the roots, and DNB
was primarily metabolized in the roots and not in the leaves. Both of
the "toxic" industrial chemicals remained mainly in the roots, while,
as expected, the two systemic herbicides moved throughout the plant.
The authors concluded that a "formal kinetic description" was needed
to interpret the data to allow comparisons "between chemicals and
plant species." This objective was accomplished through mathematical
development (Boersma et al. 1988, 1990) of a compartmentalized model
ultimately aimed at describing the uptake of xenobiotic chemicals by
plants from soil solution.

Air to foliage transfer

The opportunity for both regional and worldwide plant exposure to
atmospheric pollutants (Atlas and Giam 1981) such as biocides (Grover
1991), the organochlorines (Bidleman et al. 1976; Calamari et al. 1991;
Harvey and Steinhauer 1974; Larson et al. 1990; Muir et al. 1993),
polyaromatic hydrocarbons (Matzner 1984; Thomas et al. 1984), and
other persistent organics (Voogt and Jansson 1993) has been well documented. In a recent review, Voogt and Jansson (1993) suggested that
plants may play an important part in both the collection and dispersion
of persistent organics via the atmosphere and that the air-to-leaf route
is the major pathway for accumulation of semivolatile organic chemicals. The probable significance of the air-to-vegetation pathway as an
"important source of human exposure to some organic chemicals" was
pointed out by Travis and Hattmeyer-Frey (1988). The mechanisms
involved in the global distribution of organic chemicals have been
reviewed recently (Bidleman 1988, 1990; Grosjean 1991; Schnoor 1992).
The principle routes to contamination of foliage for biocides and environmental chemicals include the direct sorption of pollutant vapor from
the surrounding atmosphere, deposition on plant surfaces of particulate matter and aerosols, and direct application of chemicals either
intentional or accidental (Paterson et al. 1990).

    The principle barrier between the aerial portion of a plant and the
chemical challenges provided by its atmospheric environment is the

cuticle (Schonherr and Riederer 1989). Although a formidable barrier, the cuticle is permeable to a wide range of chemicals, gases, and water (Bukovac and Petracek 1993; Lendzian and Kerstiens 1991). The cuticle is a "thin continuous extracellular membrane," and its discovery dates back nearly 150 years (Holloway 1982). The cuticular membrane is composed mainly of two types of lipids: the insoluble polymeric cutins that provide the structural framework for the membrane and two groups of soluble waxes, one found on the surface (epicuticular wax) and the other embedded in the cutin matrix (cuticular wax) (Holloway 1982). According to Holloway (1982), it is "inadvisable" to use a generalized model for the cuticular membrane structure because of its dynamic composition. Both its structural and its chemical composition can vary with species and stage of development (Kirkwood 1987). The cuticle may contain carbohydrate fibers, which extend from the underlying epidermal cell wall (Kirkwood 1987; Paterson et al. 1990), and enzymatic activity (Price 1982). This protective covering is occasionally breached by small pores called stomata, which provide direct atmospheric entrance into the foliar interior. Chemicals that are present in the atmosphere as vapor could theoretically enter plant tissues via the cuticle or stomata; however, those compounds that are nonvolatile must penetrate the cuticle. The contribution of the cuticle to the accumulation of environmental chemicals via the atmosphere is thought to be of major importance (Breeze 1993; Schonherr and Riederer 1989; Thompson 1983).

Sabljic et al. (1990) suggest that the main route for uptake of "apolar airborne compounds" is via the leaf surface. They listed three reasons for this route's importance: (1) the total leaf surface area of a typical plant is approximately 20 times greater than the ground area over which it is growing; (2) the mobility of apolar chemicals in the soil and in the transport systems of plants is low because of low water solubility, strong sorption, and, consequentially, retention by lipophilic compartments in the plant; and (3) plant surfaces have a high permeability to apolar compounds. Therefore, the probability for uptake of apolar chemicals by plant foliage rather than by roots was thought to be high. Barrows et al. (1969) conducted greenhouse and field experiments in which corn was grown in dieldrin (1,2,3,4,10,10-hexachloro-6,7-epoxy-1,4,4a,5,6,7,8,8a-octahydro-1,4-endo-exo-5,8-dimenthanonaphthalene)-contaminated soil. Greenhouse-grown plants were "protected from aerial contamination." Aerial contamination was found to be a significant contributor to the total amount of dieldrin accumulated by the foliage. Beall and Nash (1971) reported a series of experiments designed to separate root uptake and translocation to shoots from vapor phase uptake of several biocides. Their results indicated that volatilization of the chemicals from treated soil may have been at least as important as root sorption as a mechanism of foliar contamination. Uptake and translocation experiments from soils contaminated with PCBs (Bacci

and Gaggi 1985) and a group of chlorinated pesticides (hexachlorobenzene, α- and τ-hexachlorocyclohexane, dichlorodiphenyltrichloro-
ethane [*p,p'*DDT] and 1,1-dichloro-2,2-*bis*-[4-chlorophenyl]-ethylene
[*p,p'*DDE]) (Bacci and Gaggi 1986) indicated a relative lack of translocation from soil to root for a variety of vegetable species. The foliar
contamination found was thought to result mainly from vaporization
of the xenobiotics from the polluted soil and sorption by the leaves.

Volatilization from the soil has been found to be a function of a
chemical's vapor pressure, which is determined in part by the ambient
temperature, the chemical's water solubility, and the physicochemical
properties of the soil (Ryan et al. 1988). A model using the organic
carbon coefficient (K_{oc}) (Trapp et al. 1990), Henry's constant, and a first-
order degradation rate coefficient (or chemical half-life) was constructed to describe the transport and loss of trace organics from soils
(Jury et al. 1983, 1984a,b,c). The model was able to successfully predict
the behavior of a diverse group of chemicals, and it may be useful in
screening for compounds that might follow a soil–air–foliage contamination route. Topp et al. (1986) also found that uptake of organic
chemicals was "strongly positively correlated to volatilization" of the
compounds from soil. Additional data supporting the importance of
an aerial route to foliar contamination is presented by McCrady et al.
(1990) and Hulster and Marschner (1993). Both of these references
found that a major route to foliar contamination for PCDD and polychlorinated dibenzofurans (PCDF) in potato (*Solanum tuberosum*), lettuce (*Lactuca sativa*), and "hay" (Hulster and Marschner 1993) and for
tetrachlorodibenzo-*p*-dioxin (TCDD) in corn and soybean (McCrady et
al. 1990) was via airborne deposition and not by root contamination
followed by translocation to the foliage.

Hulster and Marschner (1993) found that under certain circumstances direct foliar contamination by soil particles became important.
Schroll and Scheunert (1993) state that foliar contamination by octachlo-
rodibenzo-*p*-dioxin (OCDD) results from a combination of the chemical
evaporated from the soil plus the amount of chemical present as background atmospheric contamination. They found no evidence either for
the transport to shoots of OCDD sorbed to carrot roots or for shoot-to-
root transfer. They reported that the uptake pathways for OCDD in
carrots paralleled those reported for hexachlorobenzene (Schroll and
Scheunert 1992).

As for the entrance of xenobiotics to the root compartment, aerial
entrance to plant foliage is controlled by the physicochemical properties of the pollutant as they relate to similar properties presented by
the cuticular membrane. Aerial bioavailability is dictated in part by the
atmospheric vapor concentration of the chemical (see preceding paragraph). Uptake of the chemical has been shown to be controlled by
simple partitioning from either the vapor phase (Bacci et al. 1990a,b;

Riederer 1990) or the chemical–water solution (Riederer and Schonherr 1985) into the highly lipophilic environment of the cuticular membrane (Kerler and Schonherr 1988a,b). The possibility of chemical transfer directly to the cuticle from a solid particle by "contact exchange" may be possible; however, no quantitative data are available (Schonherr and Riederer 1989). It has been implicated that the cuticle itself is a major sorption compartment in plants for lipophilic environmental chemicals (Kerler and Schonherr 1988b; Riederer and Schonherr 1984). According to Trapp et al. (1990), chemical candidates for foliar uptake possess "a high K_{ow} and a Henry's law constant high enough for volatilization." A number of chemicals have been found to be accumulated via the aerial–foliar route: PCBs (Bacci and Gaggi 1985; Buckley 1982); hexachlorobenzene (HCB), hexachlorocyclohexane (HCH), DDT, and DDE (Bacci and Gaggi 1986); and PCDD/PCDF (Muller et al. 1993).

Movement through the cuticle is by simple diffusion (Riederer and Schonherr 1984). The accumulation of lipophilic chemicals in plant cuticles can be accurately predicted from their octanol–water partition coefficients (Kerler and Schonherr 1988a) and molecular volumes (Kerler and Schonherr 1988b). Chemical movement from the cuticle to the interior foliar tissues is passively controlled, again by simple diffusion, and is ultimately related to the cuticle water partition coefficient as presented at the inner membrane cell wall interface (apoplast). Schonherr and Riederer (1988) have demonstrated that desorption of chemicals from the inner and outer cuticular surface is unequal. They found that only 2 to 3% of the preloaded 2,4-D was lost via the outer membrane surface; the remaining 2,4-D (i.e., 86 to 92%) was desorbed from the inner surface. Desorption rates were 50 to 80 times higher at the inner surface. This asymmetric desorption was said to be the result of the presence of soluble lipids in the outer surface elements of the cuticle, resulting in a high transport resistance and a low partition coefficient to the 2,4-D when compared to the inner portion of the cuticular membrane. These results suggested that the outer portion of the cuticular membrane presented the rate-limiting barrier to chemical accumulation for many compounds, rather than the subsequent transfer into cells and organelles. Based on these observations, these authors suggested that the use of cuticles as biomonitors by determining the amounts of pollutants sorbed would only be applicable to those chemicals that were nonmetabolized. They reasoned that all other chemicals would not be found or would be in very low concentration because the rate of entry would be much less than the rate of efflux into the metabolic sink. Those compounds that are found to accumulate in the cuticle should be concentrated within the inner volume element. The subsequent uptake, release, and metabolic fate of the chemical has not been well characterized for the majority of environmental chemicals present in the aerial compartment once the xenobiotic has partitioned itself into and eventually across the cuticular membrane.

With the realization that plant foliage could take up atmospheric organic chemicals having low water solubility, high lipoaffinity, and a sufficiently high Henry's law constant to allow vaporization to the troposphere came the use of plant foliage as a biomonitor for the presence of atmospheric pollutants (Bacci et al. 1988; Bacci and Gaggi 1987; Buckley 1982; Calamari et al. 1991; Gaggi and Bacci 1985; Gaggi et al. 1985; Reischl et al. 1988, 1989a,b). Bacci et al. (1990b) state that the bioconcentration of aerial contaminants can be quantified by the equilibrium leaf–air BCF, where BCF is equal to

$$BCF = \frac{(\text{mass of contaminant}/\text{volume of leaf})}{(\text{mass of contaminant}/\text{volume of air})} \tag{6}$$

The BCF values of a total of 14 organic chemicals in this study and 11 in a subsequent study (Bacci et al. 1990a) were correlated with their air–water (K_{aw}) and 1-octanol–water (K_{ow}) partition coefficients. When these data were plotted on a log-log plot as log BCF $\times K_{aw}$ vs. log K_{ow}, a slope approaching one (Equation 1) resulted. BCF could then be expressed as

$$BCF = 0.022\, K_{ow}/K_{aw} \tag{7}$$

The proportionality coefficient (0.022) was thought to represent the lipid fractional content (L) (volume per volume,v/v) of the tissue being measured. The ratio of partition coefficients reduces to the solubility in octanol/solubility in air, both being expressed in moles per cubic meter. This ratio is equivalent to the 1-octanol–air partition coefficient (K_{oa}) defined by Paterson et al. (1990). Equation 7 could now be rewritten as

$$BCF = L\, K_{oa} \tag{8}$$

Although these authors found their predictive equation to be similar to the correlative expression (leaf–air BCF correlated to K_{ow} and Henry's law constant) of Reischl et al. (1989b), differences were noted. Bacci et al. (1990b) concluded that the pollutant chemical's solubility in 1-octanol, its vapor pressure, and the lipid content of the leaf cuticle were of central importance in determining the leaf–air equilibrium of nonpolar and nonreactive chemicals. In a more recent study, Bacci et al. (1992) determined both the uptake and clearance kinetics for a polychlorinated dibenzodioxin (1,2,3,4-tetrachlorodibenzo-*p*-dioxin [1,2,3,4-TCDD]) in azalea (*Azalea indica*) leaves in order to determine the leaf–air BCF. The analytically obtained BCF for 1,2,3,4-TCDD closely approached its calculated theoretical value obtained through the application of Equation 7 (i.e., 5.76×10^7 vs. 9.11×10^7, respectively).

Calamari et al. (1991) suggested that the global plant biomass plays a "significant role in the partitioning of certain environmental chemicals." Pivotal to this role is the air-to-foliar transfer of gaseous organics of low water solubility. They also suggest that, in terrestrial ecosystems, plant foliage can be used to monitor the aerial contamination of persistent organic chemicals such as the chlorinated hydrocarbons. In addition, plant biomass may play an important role in "the circulation and bioaccumulation phenomena of these chemicals" in the biosphere.

Conclusion

The process of bioaccumulation encompasses all aspects of the chemical–plant interaction. It is the summation of all gains and losses of chemical that result from these interactions. The extent of plant uptake and loss, transport, and metabolism for the vast majority of xenobiotic organic chemicals is not known. This chapter has presented a sampling of the plant research that has focused on describing soil–plant and aerial–plant transfer of xenobiotic organic chemicals. As a result of these research efforts, increased awareness of the importance of the plant biota as a major participant in determining the environmental fate of xenobiotic organic chemicals has been achieved (Calamari et al. 1987; Morosini et al. 1993; Riederer 1990).

Our understanding of the predominantly passive mechanism involved in soil–plant and air–plant transfer of environmentally persistent xenobiotics of low water and high lipid solubility has implicated the importance of the latter transfer route in the global contamination of the plant biota. In an excellent review titled "Plant uptake of nonionic organic chemicals from soils," Ryan et al. (1988) state that the accumulation of chemicals by plants may involve both "compound specific active processes, and/or a passive process in which the chemical accompanies the transpiration water through the plant." They suggested that if active processes predominate, then a "rigorous relationship" between the chemical's physicochemical characteristics and plant uptake may be weak or may not exist. However, if passive processes control uptake, these relationships should be highly correlated. As presented herein, research efforts over the past decade have focused on the second mechanism, the elucidation of its plant–chemical correlations. This time period also saw the concomitant development of models to predict chemical fate within the plant (Bacci et al. 1990b; Boersma et al. 1988, 1990; Calamari et al. 1987; Cawlfield et al. 1991; Kleier 1988; McCoy 1987; Paterson and Mackay 1989; Riederer 1990; Schramm et al. 1987; Trapp et al. 1990, 1994). Characteristics and attributes of several recently proposed models are discussed by Paterson et al. (1990). They point out a pressing need for the development of general plant models that would be designed (1) to evaluate a broad range of physical and chemical parameters presented by different species of plants and (2) to accept parametric input from a broad range of chemicals. The goal would be to have the ability to more accurately predict chemical–plant interactions such as "concentrations and persistence in, and transport through, various plant tissues."

In an extensive review concerning the accumulation of organic pollutants from soils, Bell (1992) pointed out that a good deal of the earlier work concerning xenobiotic chemical–plant uptake and transport was designed to maximize the efficiency of herbicides. Bell stated that data of a similar nature was urgently needed for "those organic chemicals that currently exist as pollutants." These data were said to be needed to predict the environmental behavior of these new environmental chemical pollutants. McFarlane (1991), in his dicussion of the factors affecting "the fate of a chemical in the plant environment," concluded that plant "uptake and translocation are similar between species, but that metabolism and phytotoxicity are species specific." As understanding of the interactions of these complex systems continues to mature, knowledge of the basic principles governing plant–chemical interaction promises an increased ability to predict the fate of xenobiotic organic chemicals.

References

Amdur MO, Doull J, Klassen CD, Eds. (1991) *Casarett and Doull's Toxicology — The Basic Science of Poisons*, 4th Ed., Pergamon Press, New York, 1033 pp.

Anderson TA, Guthrie EA, Walton BT (1993) Bioremediation in the rhizosphere. *Environ Sci Technol* 27:2630–2636.

Atlas E, Giam CS (1981) Global transport of organic pollutants: ambient concentrations in the remote marine atmosphere. *Science* 211:163–165.

Bacci E, Gaggi C (1985) Polychlorinated biphenyls in plant foliage: translocation or volatilization from contaminated soils. *Bull Environ Contam Toxicol* 35:673–681.

Bacci E, Gaggi C (1986) Chlorinated pesticides and plant foliage: translocation experiments. *Bull Environ Contam Toxicol* 37:850–857.

Bacci E, Gaggi C (1987) Chlorinated hydrocarbon vapours and plant foliage: kinetics and applications. *Chemosphere* 16:2515–2522.

Bacci E, Calamari D, Gaggi C, Biney C, Focardi S, Morosini M (1988) Organochlorine pesticide and PCB residues in plant foliage. *Chemosphere* 17:693–702.

Bacci E, Calamari D, Gaggi C, Vighi M (1990a) Bioconcentration of organic chemical vapors in plant leaves: experimental measurements and correlation. *Environ Sci Technol* 24:885–889.

Bacci E, Cerejeira MJ, Gaggi C, Chemello G, Calamari D, Vighi M (1990b) Bioconcentration of organic chemical vapours in plant leaves: the azalea model. *Chemosphere* 21:525–535.

Bacci E, Cerejeira MJ, Gaggi C, Chemello G, Calamari D, Vighi M (1992) Chlorinated dioxins: volatilization from soils and bioconcentration in plant leaves. *Bull Environ Contam Toxicol* 48:401–408.

Barrett M, Ashton FM (1983) Napropamide fluxes in corn (*Zea mays*) root tissue. *Weed Sci* 31:43–48.

Barrows HL, Card JH, Armiger WH, Edwards WM (1969) Contribution of aerial contamination to the accumulation of dieldrin by mature corn plants. *Environ Sci Technol* 3:262–263.

Beall ML Jr., Nash RG (1971) Organochlorine insecticide residues in soybean plant tops: root vs. vapor sorption. *Agron J* 63:460–464.

Bell RM (1992) Higher plant accumulation of organic pollutants from soil. EPA/600/R-92/138, U.S. Department of Commerce, National Technical Information Service, Springfield, VA.

Berthelot M, Jungfleisch E (1869) Sur les lois qui president ou partage d'un corps entre deux dissolvants. *CR* 69:338–342.

Bidleman TF (1988) Atmospheric processes. *Environ Sci Technol* 22:361–367.

Bidleman TF (1990) The longrange transport of organic compounds. In *The Longrange Atmospheric Transport of Natural and Contaminant Substances*, Knap AH, Ed., Kluwer Academic, Dordrecht, pp. 259–301.

Bidleman TF, Rice CP, Onley CE (1976) High molecular weight chlorinated hydrocarbons in the air and sea: rates and mechanisms of air/sea transfer. In *Marine Pollutant Transfer*, Windom HL, Duce RA, Eds., Lexington Books, Lexington, MA, pp. 323–351.

Boersma L, Lindstrom FT, McFarlane C, McCoy EL (1988) Uptake of organic chemicals by plants: a theoretical model. *Soil Sci* 146:403–417.

Boersma L, Lindstrom FT, McFarlane C (1990) Model for uptake of organic chemicals by plants. Agricultural Experiment Station, Oregon State University, Station Bulletin 677, Corvallis, OR, 104 pp.

Breeze VG (1993) Phytotoxicity of herbicide vapor. *Rev Environ Contam Toxicol* 132:29–54.

Briggs GE, Robertson RN (1957) Apparent free space. *Annu Rev Plant Physiol* 8:11–30.

Briggs GG (1981) Theoretical and experimental relationships between soil adsorption, octanol-water partition coefficients, water solubilities, bioconcentration factors, and parachor. *J Agric Food Chem* 29:1050–1059.

Briggs GG, Bromilow RH, Edmondson R, Johnston M (1976) Distribution coefficients and systemic activity. *Chem Soc Spec Publ* 29:129–134.

Briggs GG, Bromilow RH, Evans AA (1982) Relationships between lipophilicity and root uptake and translocation of non-ionized chemicals by barley. *Pestic Sci* 13:495–504.

Briggs GG, Bromilow RH, Evans AA, Williams M (1983) Relationships between lipophilicity and the distribution of non-ionized chemicals in barley shoots following uptake by roots. *Pestic Sci* 14:492–500.

Buckley EH (1982) Accumulation of airborne polychlorinated biphenyls in foliage. *Science* 216:520–522.

Bukovac MJ, Petracek PD (1993) Characterizing pesticide and surfactant penetration with isolated plant cuticles. *Pestic Sci* 37:179–194.

Calamari D, Vighi V, Bacci E (1987) The use of terrestrial plant biomass as a parameter in the fugacity model. *Chemosphere* 16:2359–2364.

Calamari D, Bacci E, Focardi S, Gaggi C, Morosini M, Vighl M (1991) Role of plant biomass in the global environmental partitioning of chlorinated hydrocarbons. *Environ Sci Technol* 25:1489–1495.

Cawlfield DE, Lindstrom FT, Boersma L (1991) UTAB: mathematical model for the uptake, transport, and accumulation of inorganic and organic chemicals by plants. Agricultural Experiment Station, Oregon State University, Special Report 868, Corvallis, OR.

Chou SF, Jacobs LW, Penner D, Tiedje JM (1978) Absence of plant uptake and translocation of polybrominated biphenyls (PBBs). *Environ Health Perspect* 23:9–12.

Cole D (1983) Oxidation of xenobiotics in plants. *Prog Pertic Biochem Toxicol* 3:199–243.

Connell DW (1990) *Bioaccumulation of Xenobiotic Compounds*, CRC Press, Boca Raton, FL, 219 pp.

Connell DW, Miller GJ, Eds. (1984) *Chemistry and Ecotoxicology of Pollution*, John Wiley & Son, New York, 444 pp.

Crafts AS (1964) Herbicide behaviour in the plant. In *The Physiology and Biochemistry of Herbicides*, Audus LJ, Ed., Academic Press, London, pp. 75–110.

Crisp CE (1972) Insecticides. In *Proceedings of the 2nd International IUPAC Congress on Pesticide Chemistry*, Vol 1, Tel Aviv, Israel, p. 211.

Cronquist A (1988) *The Evolution and Classification of Flowering Plants — Second Edition*, The New York Botanical Garden, Bronx, NY, 555 pp.

Crowdy SH (1972) Translocation. In *Systemic Fungicides*, Marsh RW, Ed., Longman Group, London, pp. 92–114.

Crowdy SH (1973) Patterns and processes of movement of chemicals in higher plants. Proc. 7th Br. Insect. Fung. Conf., Nottingham, U.K., pp. 831–839.

Darmstadt GL, Balke NE, Schrader LE (1983) Use of corn root protoplasts in herbicide absorption studies. *Pestic Biochem Physiol* 19:172–183.

Darmstadt GL, Balke NE, Price TP (1984) Triazine absorption by excised corn root tissue and isolated corn root protoplasts. *Pestic Biochem Physiol* 21:10–21.

Dearden JC, Townsend MS (1976) A theoretical approach to structure-activity relationships — some implications for the concept of optimal lipophilicity. *Chem Soc Spec Publ* 29:135–141.

Devine MD, Vanden Born WH (1991) Absorption and transport in plants. In *Environmental Chemistry of Herbicides*, Vol. 2, Grover R, Cessna AJ, Eds., CRC Press, Boca Raton, FL, pp. 119–140.

Donaldson TW, Bayer DE, Leonard OA (1973) Absorption of 2,4-dichlorophenoxyacetic acid and 3-(p-chlorophenyl)-1,1-dimethylurea (Monuron) by barley roots. *Plant Physiol* 52:638–645.

Ebing W, Haque A, Schuphan I, Harms C, Langebartels C, Scheel D, Trenck TvD, Sandermann H (1984) Ecochemical assessment of environmental chemicals: draft guideline of the test procedure to evaluate metabolism and degradation of chemicals by plant cell cultures. *Chemosphere* 13:947–957.

Edgington LV, Peterson CA (1977) Systemic fungicides: theory, uptake, and translocation. In *Antifungal Compounds*, Vol. 2, Siegal MR, Sisler HD, Eds., Marcel Dekker, New York, pp. 51–89.

Eiceman GA, Urquhart NS, O'Connor GA (1993) Logistic and economic principles in gas chromatography-mass spectrometry use for plant uptake investigations. *J Environ Qual* 22:167–173.

Epstein E (1972) *Mineral Nutrition of Plants: Principles and Perspectives,* John Wiley & Sons, New York, 412 pp.

Esser HO (1986) A review of correlation between physiochemical properties and bioaccumulation. *Pestic Sci* 17:265–276.

Fletcher J (1991) Keynote speech: a brief overview of plant toxicity testing. In *Plants for Toxicity Assessment,* Vol. 2, Gorsuch JW, Lower WR, Lewis MA, Wang W, Eds., ASTM STP 1115, American Society for Testing and Materials, Philadelphia, PA, pp. 5–11.

Fletcher JS, Johnson FL, McFarlane JC (1988) Database assessment of phytotoxicity data published on terrestrial vascular plants. *Environ Toxicol Chem* 7:615–622.

Freed VH, Montgomery ML (1963) The metabolism of herbicides by plants and soils. *Residue Rev* 3:1–17.

Gaggi C, Bacci E (1985) Accumulation of chlorinated hydrocarbon vapours. *Chemosphere* 14:451–456.

Gaggi C, Bacci E, Calamari D, Fanelli R (1985) Chlorinated hydrocarbons in plant foliage: an indication of the troposheric contamination level. *Chemosphere* 14:1673–1686.

Geissbuhler H, Haselback C, Abei H, Ebner L (1963) The fate of N'-(4-chlorophenoxy)-phenyl NN-dimethylurea (C-1983) in soils and plants. *Weed Res* 3:277–297.

Gentile JM, Johnson P, Robbins S (1991) Activation of aflatoxin B1 and benzo(a)pyrene by tobacco cells in the plant cell/microbe coincubation assay. In *Plants for Toxicity Assessment,* Vol. 2, Gorsuch JW, Lower WR, Lewis MA, Wang W, Eds., American Society for Testing and Materials, Philadelphia, PA, pp. 318–325.

Goring CAI (1967) Physical aspects of soil in relation to the action of soil fungicides. *Annu Rev Phytopathol* 5:285–318.

Greim H, Ahlers J, Bias R, Broecker B, Gamer AO, Gelbke H-P, Haltrich WG, Klimisch H-J, Mangelsdorf I, Schon N, Stropp G, Vogel R, Welter G, Bayer E (1993) Priority setting for the evaluation of existing chemicals — the approach of the German advisory committee on existing chemicals of environmental relevance. *Chemosphere* 26:1653–1666.

Grosjean D (1991) Atmospheric fate of toxic aromatic compounds. *Sci Total Environ* 100:367–414.

Grover R (1991) Nature, transport, and fate of airborne residues. In *Environmental Chemistry of Herbicides,* Grover I, Raj R, Eds., CRC Press, Boca Raton, FL, pp. 89–117.

Gunther P, Pestemer W (1990) Risk assessment for selected xenobiotics by bioassay methods with higher plants. *Environ Manage* 14:381–388.

Hance RJ (1988) Absorption and bioavailability. In *Environmental Chemistry of Herbicides,* Vol. 1, Grover R, Ed., CRC Press, Boca Raton, FL, pp. 1–19.

Hani H (1990) The analysis of inorganic and organic pollutants in soil with special regard to their bioavailability. *Int J Environ Chem* 39:197–208.

Hansch C (1973) A computerized approach to quantitative biochemical structure-activity relationships. *Adv Chem Ser* 114:20–40.

Hansch C, Leo A (1979) *Substituent Constant for Correlation Analysis in Chemistry and Biology,* Wiley-Interscience, New York, 339 pp.

Harms H (1973) Pflanzliche Zellsuspensionskulturen — ihr Leistungsvermogen fur Stoffwechseluntersuchungen. Ein Vergleich uber den Metabolismus aromatischer Sauren in Weizen-Keimpflanzen und Weizen-Zellsuspensionskulturen Landbauforschung. _Volkenrode_ 23:127–132.

Harms H, Langebartels C (1986) Standardized plant cell suspension test systems for an ecotoxicologic evaluation of the metabolic fate of xenobiotics. _Plant Sci_ 45:157–165.

Hartley GS, Graham-Bryce IJ (1980) _Physical Principles of Pesticide Behaviour: The Dynamics of Applied Pesticides in the Local Environment in Relation to Biological Response,_ Vol. 2, Academic Press, London, pp. 596–611.

Harvey GR, Steinbauer W (1974) Atmospheric transport of polychlorobiphenyls to the North Atlantic. _Atmos Environ_ 8:777–782.

Hawxby K, Basler E, Santelmann PW (1972) Temperature effects on absorption and translocation of trifluralin and methazole in peanuts. _Weed Sci_ 20:285–289.

Holloway PJ (1982) Structure and histochemistry of plant cuticular membranes: an overview. In _The Plant Cuticle,_ Cutler DF, Alvin KL, Price CE, Eds., Linnean Society Symposium Series Number 10, Academic Press, New York, pp. 1–32.

Huang X-D, Dixon DG, Greenberg BM (1993) Impacts of uv radiation and photomodification on the toxicity of PAHs to the higher plant _Lemna gibba_ (Duckweed). _Environ Toxicol Chem_ 12:1067–1077.

Hulster A, Marschner H (1993) Transfer of PCDD/PCDF from contaminated soils to food and fodder crop plants. _Chemosphere_ 27:439–446.

Hulzebos EM, Adema DMM, Breemen EMD-V, Henzen L, Dis WAV, Herbold HA, Hoekstra JA, Baerselman R, Gestelm CAMV (1993) Phytotoxicity studies with _Lactuca sativa_ in soil and nutrient solution. _Environ Toxicol Chem_ 12:1079–1094.

Isensee AR (1991) Bioaccumulation and food chain accumulation. In _Environmental Chemistry of Herbicides,_ Grover I, Raj R, Eds., CRC Press, Boca Raton, FL, pp. 187–197.

Iwata Y, Gunther FA (1976) Translocation of the polychlorinated biphenyl Aroclor 1254 from soil into carrots under field conditions. _Arch Environ Contam Toxicol_ 4:44–59.

Iwata Y, Gunter FA, Westlake WE (1974) Uptake of a PCB (Aroclor 1254) from soil by carrots under field conditions. _Bull Environ Contam Toxicol_ II:523–528.

Jacobs LW, O'Connor GA, Overcash MA, Zabik MJ, Rygiewicz P (1987) Effects of trace organics in sewage sludges on soil-plant systems and assessing their risk to humans. In _Land Application of Sludge — Food Chain Implications,_ Page AL, Logan TJ, Ryan JA, Eds., Lewis Publishers, Chelsea, MI, pp. 101–143.

Jacobson A, Shimabukuro RH (1980) The absorption and translocation of amitrole and diclofop-methyl by wheat and oat roots. _Plant Physiol_ 65:59 (abstract).

Jury WA, Spencer WF, Farmer WJ (1983) Behavior assessment model for trace organics in soil. I. Model description. _J Environ Qual_ 12:558–564.

Jury WA, Farmer WJ, Spencer WF (1984a) Behavior assessment model for trace organics in soil. II. Chemical classification and parameter sensitivity. _J Environ Qual_ 13:567–572.

Jury WA, Spencer WF, Farmer WJ (1984b) Behavior assessment model for trace organics in soil. III. Application of screening model. *J Environ Qual* 13:573–579.

Jury WA, Spencer WF, Farmer WJ (1984c) Behavior assessment model for trace organics in soil: IV. Review of experimental evidence. *J Environ Qual* 13:580–586.

Kerler F, Schonherr J (1988a) Accumulation of lipophilic chemicals in plant cuticles: prediction from octanol/water partition coefficients. *Arch Environ Contam Toxicol* 17:1–6.

Kerler F, Schonherr J (1988b) Permeation of lipophilic chemicals across plant cuticles: prediction from partition coefficients and molar volumes. *Arch Environ Contam Toxicol* 17:7–12.

Kirkwood RC (1987) Uptake and movement of herbicides from plant surfaces and the effects of formulation and environment upon them. In *Pesticides on Plant Surfaces*, Cottrell, HJ, Ed., John Wiley & Sons, New York, pp. 1–25.

Kleier DA (1988) Phloem mobility of xenobiotics. I. Mathematical model unifying the weak acid and intermediate permeability theories. *Plant Physiol* 86:803–810.

Krstich MA, Schwarz OJ (1989) Characterization of ^{14}C-naphthol uptake in excised root segments of clover (*Trifolium pratense* L.) and fescue (*Festuca arundinaceae* Screb.) *Environ Monit Assess* 13:35–44.

Krstich MA, Schwarz OJ (1990) Characterization of xenobiotic uptake utilizing an isolated root uptake test (IRUT) and a whole plant uptake test (WPUT). In *Plants for Toxicity Assessment*, Wang W, Gorsuch JW, Lower WR, Eds., ASTM STP 1091, American Society for Testing and Materials, Philadelphia, PA, pp. 87–96.

Lamoureux GL, Rusness DG (1981) Catabolism of glutathione conjugates of pesticides in higher plants. In *Sulfur in Pesticide Action and Metabolism*, Rosen JD, Magee PS, Casida JE, Eds., ACS Symp Ser Vol. 158, American Chemical Society, Washington, D.C., pp. 133–164.

Langebartels C, Harms H (1986) Plant cell suspension cultures as test systems for an ecotoxicological evaluation of chemicals — growth inhibition effects and comparison with the metabolic fate in intact plants. *Ange Bot* 60:113–123.

Larson P, Okla L, Woin P (1990) Atmospheric transport of persistent pollutants governs uptake by terrestrial biota. *Environ Sci Technol* 24:1599–1601.

Lendzian KJ, Kerstiens G (1991) Sorption and transport of gases and vapors in plant cuticles. *Rev Environ Contam Toxicol* 121:65–128.

Leo A, Hansch C, Church C (1969) Comparison of parameters currently used in the study of structure-activity relationships. *J Med Chem* 12:766–771.

Leroux P, Gredt M (1975) Absorption of methyl benzimidazol-2-yl carbamate (carbendazim) by corn roots. *Pestic Biochem Physiol* 5:507–514.

Leroux P, Gredt M (1977) Uptake of systemic fungicides by maize roots. *Neth J Plant Pathol* 83:51–61.

Lichtner FT (1983) Amitrole absorption by bean (*Phaseolus vulgaris* L. cv 'Red Kidney') roots. *Plant Physiol* 71:307–312.

Mackay D, Paterson S (1981) Calculating fugacity. *Environ Sci Technol* 15:1006–1014.

Macklon AES, Higinbotham N (1970) Active and passive transport of potassium in cells of excised pea epicotyls. *Plant Physiol* 45:133–138.

Martin RA, Edgington LV (1981) Comparative systemic translocation of several xenobiotics and sucrose. *Pestic Biochem Physiol* 16:87–96.

Matzner E (1984) Annual rates of deposition of polycyclic aromatic hydrocarbons in different forest ecosystems. *Water Air Soil Pollut* 21:425–434.

May RM (1992) *An Appraisal of Taxonomy in the 1990s — Taxonomy Research and Its Applications Problems and Priorities*, The Linnean Society, Burlington House, Piccadilly, London, 49 pp.

McCarty LS, MacKay D (1993) Enhancing ecotoxicological modeling and assessment. *Environ Sci Technol* 27:1719–1728.

McCoy EL (1987) Plant uptake and accumulation of soil applied trace organic compounds: theoretical development. *Agric Syst* 25:177–197.

McCrady JK, Maggard SP (1993) Uptake and photodegradation of 2,3,7,8-tetrachlorodibenzo-p-dioxin sorbed to grass foliage. *Environ Sci Technol* 27:343–350.

McCrady JK, McFarlane C, Gander LK (1990) The transport and fate of 2,3,7,8-TCDD in soybean and corn. *Chemosphere* 21:359–376.

McFarlane C (1991) Uptake of organic contaminants by plants. In *Municipal Waste Incineration Risk Assessment*, Travis CC, Ed., Plenum Press, New York, pp. 151–164.

McFarlane C, Wickliff C (1985) Excised barley root uptake of several ^{14}C labeled organic compounds. *Environ Monit Assess* 5:385–391.

McFarlane C, Nolt C, Wickliff C, Pfleeger T, Shimabuku R, McDowell M (1987a) The uptake, distribution and metabolism of four organic chemicals by soybean plants and barley roots. *Environ Toxicol Chem* 6:847–856.

McFarlane JC, Pfleeger T (1986) Plant exposure laboratory and chambers. EPA/600/3-86-007a,b, U.S. Environmental Protection Agency, Corvallis, OR.

McFarlane JC, Pfleeger T (1987) Plant exposure chambers for study of toxic chemical/plant interactions. *J Environ Qual* 16:361–371.

McFarlane JC, Pfleeger T, Fletcher J (1987b) Transpiration effect on the uptake and distribution of bromacil, nitrobenzene, and phenol in soybean plants. *J Environ Qual* 16:372–376.

Moody K, Kust CA, Buchholtz KP (1970a) Release of herbicides by soybean roots in culture solutions. *Weed Sci* 18:214–218.

Moody K, Kust CA, Buchholtz DP (1970b) Uptake of herbicides by soybean roots in culture solutions. *Weed Sci* 18:642–647.

Morosini M, Schreitmuller J, Reuter U, Ballschmiter K (1993) Correlation between C-6/C-14 chlorinated hydrocarbon levels in the vegetation and in the boundary layer of the troposphere. *Environ Sci Technol* 27:1517–1523.

Morrison IN, Vanden Born WH (1975) Uptake of picloram by roots of alfalfa and barley. *Can J Bot* 53:1774–1785.

Muir DCG, Segstro MD, Welbourn PM, Toom D, Elsenreich SJ, Macdonald CR, Whelpdale DM (1993) Patterns of accumulation of airborne organochlorine contaminants in lichens from the upper great lakes region of Ontario. *Environ Sci Technol* 27:1201–1210.

Muller JF, Hulster A, Papke O, Ball M, Marschner H (1993) Transfer pathways of PCDD/PCDF to fruits. *Chemosphere* 27:195–201.

Mumma RO, Davidonis GH (1983) Plant tissue culture and pesticide metabolism. In *Progress in Pesticide Biochemistry*, Vol. 3, Hutson DH, Roberts TR, Eds., Wiley, Chichester, UK, pp. 255–278.

Muzik TJ, Cruzado JH, Loustalot AJ (1954) Studies on the absorption, translocation and action of CMU. *Bot Gaz (Chicago)* 116:65–73.

Nernst W (1891) Vertherlungeines Stoffes zwischen zwei Losungsnitteln und zwischer Losungsmittel und Dampfraum. *Z Phys Chim* 8:110–139.

O'Connor GA, Chaney RL, Ryan JA (1991) Bioavailability to plants of sludge-borne toxic organics. *Rev Environ Contam Toxicol* 121:129–155.

Orwick PL, Schreiber MM, Hodges TK (1976) Absorption and efflux of chloro-s-triazines by SETARIA roots. *Weed Res* 16:139–144.

Paterson S, Mackay D (1989) Modeling the uptake and distribution of organic chemicals in plants. In *Intermedia Pollutant Transport: Modeling and Field Measurements*, Allen DT, Kaplan IR, Eds., Plenum Press, New York, pp. 269–282.

Paterson S, Mackay D, Tam D, Shiu WY (1990) Uptake of organic chemicals by plants: a review of processes, correlations and models. *Chemosphere* 21:297–331.

Pitman MG (1976a) Ion uptake by plant roots. In *Encyclopedia of Plant Physiology, New Series Vol. 2, Plant Cells and Tissues*, Luttge U, Pitman MG, Eds., Springer-Verlag, Berlin, pp. 95–128.

Pitman MG (1976b) Nutrient uptake by roots and transport to the xylem: uptake processes. In *Transport and Transfer Processes in Plants*, Wardlaw IF, Passioura JB, Eds., Academic Press, New York, pp. 85–100.

Plewa MJ, Gentile JM (1982) The activation of chemicals into mutagens by green plants. In *Chemical Mutagens: Principles and Methods for Their Detection*, Vol. 8, de Serres J, Hollaender A, Eds., Plenum, New York, pp. 401–420.

Plewa MJ, Gichner T, Xin H, Seo K-Y, Smith SR, Wagner ED (1993) Biochemical and mutagenic characterization of plant-activated aromatic amines. *Environ Toxicol Chem* 12:1353–1363.

Price CE (1982) A review of factors influencing the penetration of pesticides through plant leaves. In *The Plant Cuticle*, Culter CF, Alvin KL, Price CE, Eds., Academic Press, London, pp. 237–252.

Price TP, Balke NE (1983) Characterization of atrazine accumulation by excised Velvetleaf (*Abutilon theophrasti*) roots. *Weed Sci* 31:14–19.

Reischl A, Thoma H, Reissinger M, Hutzinger O (1988) Accumulation of organic air constituents by plant surfaces — spruce needles for monitoring airborne chlorinated hydrocarbons. *Biomed Environ Sci* 1:304–307.

Reischl A, Reissinger M, Thoma H, Hutzinger O (1989a) Accumulation of organic air constituents by plant surfaces. Part IV. Plant surfaces: a sampling system for atmospheric polychlorodibenzo-p-dioxin (PCDD) and polychlorodibenzo-p-furan (PCDF). *Chemosphere* 18:561–568.

Reischl A, Reissinger M, Thoma H, Hutzinger O (1989b) Uptake and accumulation of PCDD/F in terrestrial plants: basic considerations. *Chemosphere* 19:467–474.

Riederer M (1990) Estimating partitioning and transport of organic chemicals in the foliage/atmosphere system: discussion of a fugacity-based model. *Environ Sci Technol* 24:829–837.

Riederer M, Schonherr J (1984) Accumulation and transport of (2,4-dichlo-rophenoxy)acetic acid in plant cuticles. I. Sorption in the cuticular membrane and its components. *Ecotoxicol Environ Saf* 8:236–247.

Riederer M, Schonherr J (1985) Accumulation and transport of (2,4-dichlo-rophenoxy)acetic acid in plant cuticles. II. Permeability of the cuticular membrane. *Ecotoxicol Environ Saf* 9:196–208.

Ryan JA, Bell RM, Davidson JM, O'Connor GA (1988) Plant uptake of non-ionic organic chemicals from soils. *Chemosphere* 17:2299–2323.

Sabljic A, Gusten H, Schonherr J, Riederer M (1990) Modeling plant uptake of airborne organic chemicals. 1. Plant cuticle/water partitioning and molecular connectivity. *Environ Sci Technol* 24:1321–1326.

Sandermann H, Scheel D, Trenck TvD (1984) Use of plant cell cultures to study the metabolism of environmental chemicals. *Ecotoxicol Environ Saf* 8:167–182.

Scheunert I, Qiao Z, Korte F (1986) Comparative studies of the fate of atrazine-^{14}C and pentachlorophenol-^{14}C in various laboratory and outdoor soil-plant systems. *J Environ Sci Health* B21:457–485.

Schmidt-Bleek F, Haberlandt W (1980) The yardstick concept for the hazard evaluation of substances. *Ecotoxicol Environ Saf* 4:445–465.

Schnoor JL (1992) Chemical fate and transport in the environment. In *Fate of Pesticides and Chemicals in the Environment,* Schnoor JL, Ed., Wiley, New York, pp. 1–24.

Schonherr J, Riederer M (1988) Desorption of chemicals from plant cuticles: evidence for asymmetry. *Arch Environ Contam Toxicol* 17:13–19.

Schonherr J, Riederer M (1989) Foliar penetration and accumulation of organic chemicals in plant cuticles. *Rev Environ Contam Toxicol* 108:1–70.

Schramm KW, Reischl A, Hutzinger O (1987) UNITTree — a multimedia compartment model to estimate the fate of lipophilic compounds in plants. *Chemosphere* 16:2653–2663.

Schroll R, Scheunert I (1992) A laboratory system to determine separately the uptake of organic chemicals from soil by plant roots and by leaves after vaporization. *Chemosphere* 24:97–108.

Schroll R, Scheunert I (1993) Uptake pathways of octachlorodibenzo-p-dioxin from soil by carrots. *Chemosphere* 26:1631–1640.

Schuphan I, Haque A, Ebing W (1984) Ecochemical assessment of environmental chemicals. Part I. Standard screening procedure to evaluate chemicals in plant cell cultures. *Chemosphere* 13:301–313.

Shimabukuro RH, Walsh WC (1979) Xenobiotic metabolism in plants: in vitro tissue, organ, and isolated cell techniques. In *Xenobiotic Metabolism: In Vitro Methods,* Paulson GD, Frear DS, Marks EP, Eds., ACS Symposium Series Vol. 97, American Chemical Society, Washington, D.C., pp. 3–34.

Shone MGT, Wood AV (1972) Factors responsible for the tolerance of blackcurrants to simazine. *Weed Res* 12:337–347.

Shone MGT, Wood AV (1974) A comparison of the uptake and translocation of some organic herbicides and a systemic fungicide by barley: I. Absorption in relation to physico-chemical properties. *J Exp Bot* 25:390–400.

Shone MGT, Bartlett BO, Wood AV (1974) A comparison of the uptake and translocation of some organic herbicides and systemic fungicide by barley: II. Relationship between uptake by roots and translocation to shoots. *J Exp Bot* 25:401–409.

Sims RC, Overcash MR (1983) Fate of polynuclear aromatic compounds (PNAs) in soil-plant systems. *Residue Rev* 88:2–68.

Small LW, Martin RA, Edgington LV (1977) A comparison of translocation within plants of pyroxychlor and a 6-amino analogue. *Neth J Plant Pathol* 83:63–70.

Splittstoesser WE (1968) The absorption of potassium and several organic compounds by barely roots: effect of siduron. *Plant Cell Physiol* 9:307–314.

Stoller EW (1969) The kinetics of amiben absorption and metabolism as related to species sensitivity. *Plant Physiol* 44:854–860.

Strek HJ, Weber JB (1982) Behaviour of polychlorinated biphenyls (PCB's) in soils and plants. *Environ Pollut Ser A* 28:291–312.

Thomas W, Ruhling A, Simon H (1984) Accumulation of airbourne pollutants (PAH, chlorinated hydrocarbons, heavy metals) in various plant species and humus. *Environ Pollut, Ser A* 36:295–310.

Thompson N (1983) Diffusion and uptake of chemical vapour volatilizing from a sprayed target area. *Pestic Sci* 14:33–39.

Topp E, Scheunert I, Attar A, Korte F (1986) Factors affecting the uptake of ^{14}C-labeled organic chemicals by plants from soil. *Ecotoxicol Environ Saf* 11:219–228.

Trapp S, Matthies M, Scheunert I, Topp EM (1990) Modeling the bioconcentration of organic chemicals in plants. *Environ Sci Technol* 24:1246–1252.

Trapp S, McFarlane C, Matthies M (1994) Model for uptake of xenobiotics into plants: validation with bromacil experiments. *Environ Toxicol Chem* 13:413–422.

Travis CC, Hattmeyer-Frey HA (1988) Uptake of organics by aerial plant parts: a call for research. *Chemosphere* 17:277–283.

Tyree MT, Peterson CA, Edgington LV (1979) A simple theory regarding ambimobility of xenobiotics with special reference to the nematicide, oxamyl. *Plant Physiol* 63:367–374.

Upadhyaya MK, Nooden LD (1980) Mode of dinitroaniline herbicide action. II. Characterization of carbon-14-oryzalin uptake and binding. *Plant Physiol* 66:1048–1052.

Voogt P de, Jansson B (1993) Verticle and long-range transport of persistent organics in the atmosphere. *Rev Environ Contam Toxicol* 132:1–28.

Wain RL, Carter GA (1967) Uptake, translocation and transformations by higher plants. In *Fungicides. An Advanced Treatise*, Vol. 1, Torgeson DG, Ed., Academic Press, New York, pp. 561–612.

Wallerstein IS, Jacoby B, Dinoor A (1976) Absorption, retention and translocation of the systemic fungicide triarimol in plants. *Pestic Biochem Physiol* 6:530–537.

Wickliff C, McFarlane JC, Ratsch H (1984) Uptake of bromacil by isolated barley roots. *Environ Monit Assess* 4:43–51.

Zhirmunskaya NM, Kol'tsova SS (1973) Investigation of permeability of barley root cells in relation to the herbicide atrazine. *Sov Plant Physiol* 20:123–127.

chapter fifteen

Uptake of polycyclic aromatic hydrocarbons by vegetation: a review of experimental methods*

Todd A. Anderson, Anne M. Hoylman, Nelson T. Edwards, and Barbara T. Walton

Introduction

Polycyclic aromatic hydrocarbons (PAHs) are organic compounds made up of fused benzene rings. Although PAHs occur naturally, their largest environmental source is from the utilization of fossil fuels for energy (Edwards 1983). The combustion, storage, and disposal of fossil fuels are all activities that contribute to the ubiquitous presence of PAHs. Many of the PAHs released to the environment become associ-

---

ated with soil. Subsequently, these soils may become secondary sources for long-term PAH releases to groundwater, surface water, and air.

Many types of sites owned by the electric utility industry, including those at manufactured gas plants (MGPs), contain elevated quantities of PAHs in soil (Linz et al. 1991; Walton and Hoylman 1992). The movement of PAHs from soil to vegetation is a potentially important pathway that may reduce the residence time of PAHs in soils and increase the exposure of animals to PAHs in food. To conduct studies on the uptake of PAHs from soil by vegetation and to evaluate research of others on root uptake of these compounds, it is important to know (1) the types of experimental equipment and methods that have been used to measure plant uptake of PAHs and (2) the advantages and disadvantages of these methods.

Analysis of the published literature shows that there are very few studies in which PAH uptake from soil by plants has been measured with a high degree of precision (Edwards 1983; Walton and Hoylman 1992). Of the existing studies, PAH losses due to volatilization from soil or transpiration from vegetation are rarely reported, yet these parameters are critically important to data interpretation and mass balance determinations. Similarly, the distribution of the parent compound and its metabolites in roots, stems, and leaves is rarely reported for experiments on PAH uptake. Although PAH uptake from soil by vegetation is likely to be minimal when low concentrations of PAHs are present in soil of high organic content, as when municipal sewage sludges are applied to agricultural soil (O'Connor et al. 1991), coal tar disposal at MGP sites presents a very different situation. That is, high concentrations of PAHs may exist in soils low in organic matter such that PAH sorption to soil is minimal and PAH availability for plant uptake is favored (Walton and Hoylman 1992). Thus, existing studies of PAH uptake by vegetation do not fully cover the range of chemical concentrations and environmental variables that exist in the field. Moreover, additional experimental data may be needed for site-specific assessments of the potential for PAH uptake by vegetation from soil.

Because PAHs are a diverse group of chemicals with a wide range of physicochemical properties, not all methods for studying plant uptake of chemicals are applicable to this group of compounds. For example, highly lipophilic PAHs (high log K_{ow}) and those with high vapor pressures may be especially problematic and require special conditions to obtain a mass balance. As a result, special care must be taken in PAH-uptake studies to analyze all matrices — soil, vegetation, and air — to account for the test compound. In order to do this, care must be given to how the plant and test chemical are contained during the experiment. The impact of the experimental system on the physiology of the plant must also be considered. If the plant is confined in an air-tight chamber, the needs of the plant for nutrients, water, carbon dioxide (CO_2), and oxygen (O_2) must still be met. Excess heat and moisture may need to be removed. In addition, analytical methods for extraction, identification, and quantification of the PAHs must be validated.

The purpose of this chapter is to review the experimental apparatuses and methods that have been used to measure the uptake of PAHs by plant roots. A few apparatuses readily adaptable to PAH studies are also included.

The choice of an experimental apparatus to measure PAH uptake must be made on the basis of the specific questions to be answered by the study. Initial considerations will include the plant species, rooting medium, test compounds (including use of radioisotopes as tracers), and the analytic methods to be used. The duration of each experiment, numbers of replicates, sampling frequency, and safety and health considerations for handling PAHs (and radiotracers) must also be considered. Percent recovery of the PAHs from various media (plant, soil/water, and air) must be verified.

The discussion of exposure equipment used for PAH-uptake studies is grouped on the basis of the rooting medium provided for the test plant, that is, liquid rooting media (i.e., hydroponic systems) and solid rooting media (i.e., sand and soil systems). Extraction methods for recovering PAHs from plant tissue and soil are summarized after the review of experimental equipment.

Methods for characterizing uptake of PAHs by plants

Hydroponic systems

Hydroponic systems utilize nutrient solutions as the growing medium for plants in uptake studies. The use of an aqueous rooting medium simplifies the experimental design by reducing competing mechanisms such as microbial degradation and sorption of the test compound to soil. In hydroponic systems, mixing is often employed to ensure that roots are supplied with nutrients and O_2, as well as to provide a constant PAH exposure to the roots. If mixing is not employed, a depletion zone may be created around the roots so that the PAH exposure concentration is not uniform. Under these circumstances, uptake rates can be underestimated; however, mixing can also increase volatilization losses of the PAH.

One of the simplest experimental designs to measure root uptake of organic compounds consists of a modified Erlenmeyer flask (Figure 1). This technique, which was originally described by Hsu and Bartha (1979) for determining microbial degradation of ^{14}C-labeled pesticides in the rhizosphere, is readily adaptable to measuring root uptake of PAHs. A single key feature, the isolation of the roots from the upper stem and leaves with a nonreactive silicone rubber sealant (Stotzky et al. 1961), enables the researcher to restrict entry of an organic com-

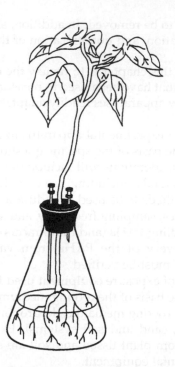

Figure 1 The Erlenmeyer flask is a whole-plant hydroponic apparatus for assessing microbial degradation and root uptake of radiolabeled compounds. Isolation of above-ground plant tissue from roots and spiked solution provides a clear route of uptake. Sampling portals in the flask stopper allow flushing and trapping of ^{14}C-volatile organics and $^{14}CO_2$. (From Hsu TS, Bartha R [1979] *Appl Environ Microbiol* 37:36. With permission.)

pound into the plant to that portion of the plant below the sealant. In addition to isolation of the roots and rooting medium from the shoots, sampling portals permit flushing the flask to help maintain aerobic conditions and permit sampling of volatile compounds, including $^{14}CO_2$, in the headspace above the soil. These measurements are important for mass balance determinations; however, continuous removal of CO_2 from the headspace may be detrimental to normal root development.

A potential problem with this experimental design is the loss of organic compounds through reactions with the rubber stoppers used to cap the flasks. This is a pervasive problem that can thwart all experimental attempts to measure PAH uptake by vegetation, whether the system is hydroponic or not. Edwards et al. (1982) used a hydroponic technique to measure uptake of anthracene by soybean (*Glycine max*), but anchored the plants with polyurethane foam plugs to sorb anthracene that volatilized from solution. The anthracene was extracted from the foam plugs for analysis at the end of the experiment for mass balance quantification. Edwards (1986) further modified the technique

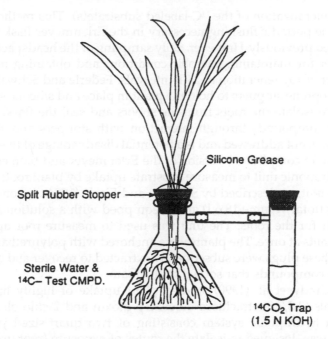

Silicone Grease

Split Rubber Stopper

Sterile Water & ¹⁴C– Test CMPD.

¹⁴CO₂ Trap
(1.5 N KOH)

Figure 2 The biometer flask is a whole-plant hydroponic apparatus for assessing microbial degradation and root uptake of radiolabeled compounds. Isolation of above-ground plant tissue from roots and spiked solution provides a clear route of uptake. The potassium hydroxide solution traps ¹⁴CO₂ generated during mineralization of the test compound. (From Federle TW, Schwab BS [1989] *Appl Environ Microbiol* 55:2092. With permission.)

to measure the uptake, translocation, and metabolism of anthracene in bush bean (*Phaseolus vulgaris* L.). The sample containers were bottles rather than flasks and were considerably larger (2 l). A side arm with a glass-fritted stopper located near the top of the bottle allowed access for taking solution samples, adding PAHs, and replacing water lost by evapotranspiration. These techniques provide an unambiguous route of exposure and permit sampling of multiple environmental compartments (i.e., rooting solution, headspace above the rooting solution, and plant tissue) for PAH analysis. The materials needed are simple and relatively inexpensive, yet this experimental method has been proven to be effective.

Use of a biometer flask (Figure 2) provides information on plant uptake of PAHs similar to that obtained by the use of Erlenmeyer flasks. The presence of a side arm facilitates the monitoring of microbial degradation of ¹⁴C-labeled substrates, as was done in studies by Federle and Schwab (1989) on surfactant degradation by the rhizosphere microbiota associated with aquatic plants. The biometer flask is also adaptable for monitoring root uptake of PAHs. The side arm is filled with a trapping solution (usually KOH) to continuously trap ¹⁴CO₂ produced from mineralization of the ^{14}C-labeled substrate(s). This method eliminates the periodic flushing necessary in the Erlenmeyer flask method described previously. However, daily sampling of the headspace would be better for maintaining aerobic conditions and obtaining measurements of $^{14}CO_2$ generation at 24-h intervals. Federle and Schwab (1989) used neoprene stoppers to hold the plant in place and silicone stopcock grease to isolate the roots from the shoots and seal the flask. Loss of organic compounds through interaction with stoppers and stopcock grease was not addressed and is a potential disadvantage of this design.

A more complicated version of the Erlenmeyer and biometer flask is a hydroponic unit to measure substrate uptake by plant roots (Figure 3). This unit, as described by Schwarz and Eisele (1984), is constructed of glass (total volume 18.6 l) and is equipped with a solution aeration chamber for the roots. The unit was used to measure root uptake in three plants at once. The plants were anchored with polyurethane foam plugs; these plugs were subsequently extracted to recover and quantify volatile compounds that sorbed to the foam.

McCrady et al. (1990) studied plant uptake of highly lipophilic chemicals (2,3,7,8-tetrachloro-dibenzo-*p*-dioxin and 2-chlorobiphenyl) using a hydroponic system consisting of two quart-sized jars. The system was designed to isolate the routes of exposure (root uptake vs. volatilization and subsequent deposition on above-ground foliage) and also to determine the magnitude of translocation via the transpiration stream in corn and soybeans. Mass balance for dioxin was greater than 85%; however, only 32 and 63% of the applied radioactivity was recovered during two experiments with 2-chlorobiphenyl.

An elaborate plant exposure chamber and laboratory described by McFarlane and Pfleeger (1987) (Figures 4 to 7) features a fully automated system for monitoring physiological responses of plants to toxic chemicals. This system was designed to allow exposure of mature plants in an environment where the contaminant is contained and physiologic functions of the plant are normal. The system is equipped with two types of exposure chambers: one for measuring the foliar uptake of volatile compounds in the gas phase and one for monitoring root uptake of test compounds from either a hydroponic solution or soil. All instruments and control functions of the system, including data collection, are regulated by a microcomputer, thus greatly reducing the work required for each experiment and increasing quality control.

Soil/sand systems

Although soil/sand systems add to the complexity of possible interactions occurring between a PAH and the solid matrix, these uptake conditions reflect field conditions more accurately than do hydroponic systems. Importantly, rhizosphere microbial communities of terrestrial

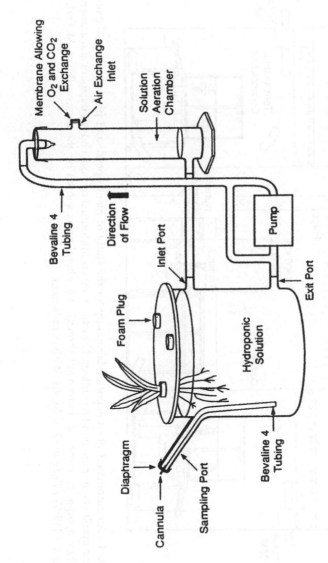

Figure 3 Glass hydroponic root uptake testing unit isolates above-ground plant tissue from root tissue exposed to the test compound. Three plants held in place by foam plugs can be treated at one time. Ambient oxygen levels are maintained through continuous circulation of the nutrient solution. Samples of nutrient solution for analyses are obtained through the sampling port. (From Schwarz OJ, Eisele GR [1984] *Synthetic Fossil Fuel Technologies: Results of Health and Environmental Studies*, Cowser KE, Ed., Butterworth Press, Boston, p. 441. With permission.)

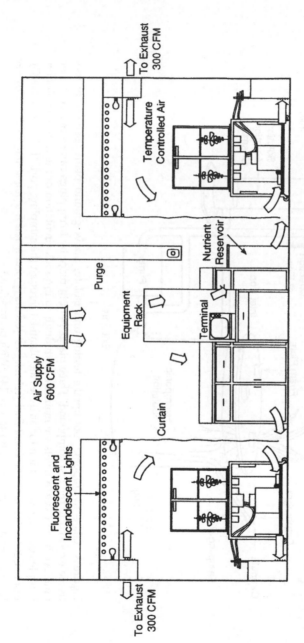

Figure 4   An automated plant exposure laboratory for the study of plant uptake and phytotoxicity of xenobiotics under controlled environmental conditions. Growth chambers are located on both ends of the room and are separated by white opaque curtains. Air movement through the laboratory is shown by the arrows. (From McFarlane JC, Pfleeger T [1987] *J Environ Qual* 16:361. With permission.)

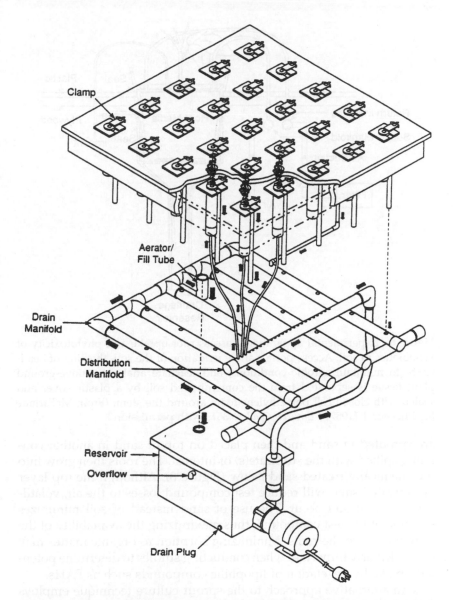

Clamp

Aerator/
Fill Tube

Drain
Manifold

Distribution
Manifold

Reservoir

Overflow

Drain Plug

Figure 5 Hydroponic plant nursery for maturation of plants for uptake studies. Nursery root compartment dimensions are identical to that of exposure chamber. (From McFarlane JC, Pfleeger T [1987] *J Environ Qual* 16:361. With permission.)

plant species are more likely to be cultured in soil/sand matrices than in hydroponic systems.

Perhaps the most straightforward approach for monitoring root uptake from a solid matrix is the sprout culture used by Suzuki et al. (1977) to measure uptake of polychlorinated biphenyls (PCBs). Plants

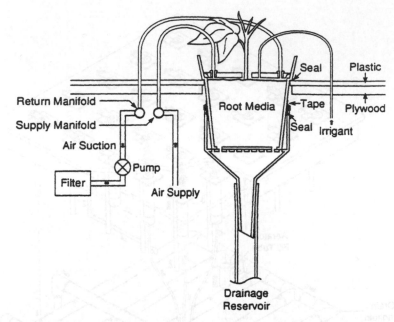

Figure 6 Experimental apparatus to assess root uptake and phytotoxicity of xenobiotics in soil. Access portals allow irrigation of the soil, flushing of head-space to maintain aerobic conditions, and leachate analysis. Above-ground plant tissue is separated from the contaminated soil by a plastic cover and sealed with closed-cell polyethylene foam around the stem. (From McFarlane JC, Pfleeger T [1987] *J Environ Qual* 16:361. With permission.)

are sprouted in sand and then placed on top of sand in another container spiked with the substrate(s) of interest. The roots then grow into the chemically treated sand below (Figure 8). Although the top layer of untreated sand will reduce test compound losses to the air, volatilization could still occur. The use of sand instead of soil minimizes sorption of the test compound, thus maximizing the availability of the contaminant to the roots. Minimizing sorption to organic matter may be particularly important when conducting studies to determine potential uptake by vegetation of lipophilic compounds such as PAHs.

An alternative approach to the sprout culture technique employs soil columns. Webber and Goodin (1990) (Figure 9) used stainless steel and glass columns to monitor the fate of PAHs in sewage sludge applied to soil in which plants are grown. The column is equipped with sampling ports for headspace analysis of volatilized organics and CO_2. These soil columns permit an approximation of critical field soil conditions such as soil density, porosity, water flow, and root growth. When instrumentation to monitor critical soil conditions is included, valuable data can be obtained to quantify the influence of these parameters on the plant uptake process. Other studies have employed alternative column material such as polyvinyl chloride (PVC) (Aprill and Sims

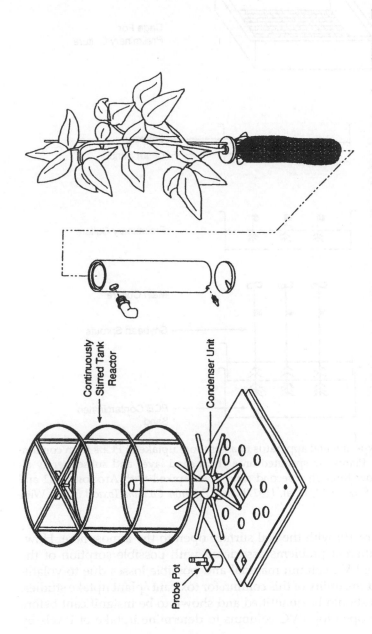

Figure 7 Shoot compartment and hydroponic plant support system for studying root uptake and phytotoxicity of xenobiotics. The shoot compartment on the left consists of an aluminum framework with Teflon™ coating and a closed-cell polyethylene foam gasket. A variable speed fan at the top of the compartment mixes the air in the chamber, and atmospheric moisture condenses on the cold water condenser and is drained away. The hydroponic plant support system on the right separates above-ground tissue from root tissue by a plastic plant support disc and closed-cell polyethylene foam. (From McFarlane JC, Pfleeger T [1987] *J Environ Qual* 16:361. With permission.)

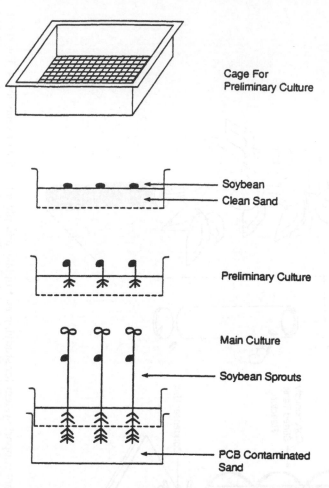

Cage For
Preliminary Culture

Soybean
Clean Sand

Preliminary Culture

Main Culture

Soybean Sprouts

PCB Contaminated
Sand

Figure 8 Experimental apparatus to assess root uptake of PCBs from contaminated sand. Plants are sprouted on a clean sand layer and subsequently allowed to grow through a second sand layer spiked with Aroclors 1242 and 1245. (From Suzuki M, et al. [1977] *Arch Environ Contam Toxicol* 5:343. With permission.)

1990) (Figure 10) with the soil surface open to the atmosphere. However, the inherent problems associated with possible sorption of the PAHs to the PVC column material and possible losses due to volatilization limit the utility of this column for toxicant/plant uptake studies. Such losses should be quantified and shown to be insignificant before using these open top PVC columns to determine uptake of PAHs by vegetation.

Kloskowski et al. (1981) and Scheunert et al. (1983) developed a plant exposure system to address the fate of chemicals in plant–soil systems (Figure 11). Plants are grown in soil treated with ^{14}C-labeled compounds in a 10-l glass desiccator. Air is drawn continuously

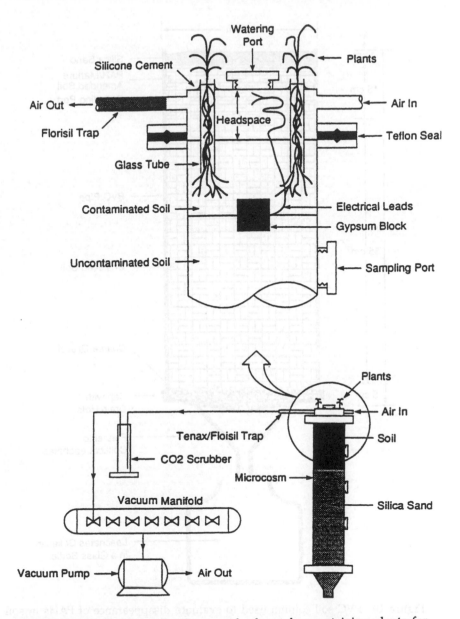

Figure 9 Stainless steel soil column with glass tubes containing plants for monitoring the fate of PAHs in vegetated soil with sewage sludge application. The apparatus has sampling portals for headspace analysis of volatile organics and CO_2, soil moisture measurements, and addition of water and nutrients. Plant roots grow down the glass tube into the spiked soil which is layered over clean silica sand. Plant shoots are separated from below-ground tissue and contaminated soil by silicone cement. (From Webber MD, Goodin JD [1990] Proceedings of the European Communities Land Application of Sludge Working Group, Braunschweig Germany. With permission.)

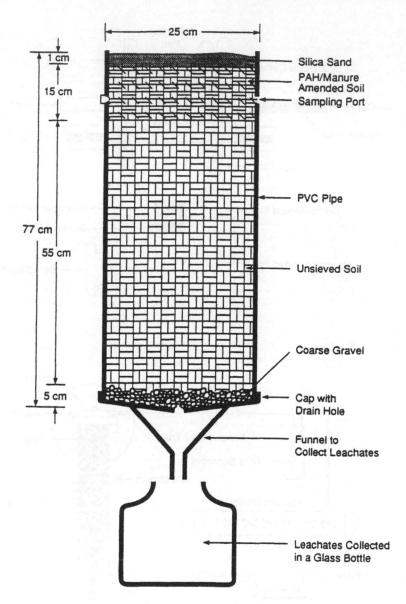

Figure 10 PVC soil column used to evaluate disappearance of PAHs in soil using prairie grasses. Sampling portals for soil analysis and leachate collection are included in the design. Spiked soil is layered between a clean silica sand layer and unsieved soil. Grass seeds are sprouted on the top layer of sand and grow down into the spiked soil layer. (From Aprill W, Sims RC [1990] *Chemosphere* 20:253. With permission.)

through the desiccator (10 ml/min) with exit air passing through consecutive traps to collect and quantify ^{14}C-volatile organics and $^{14}CO_2$. Although the distribution of the ^{14}C residues in plant tissue, soil, $^{14}CO_2$,

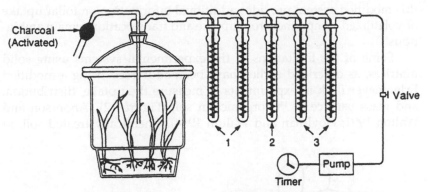

1. Ethyleneglycol-monomethylether
2. Empty bottle
3. Phenethylamine-solution

Figure 11   Experimental design to determine the fate of chemicals in plant–soil systems. Seeds are sown in soil dosed with ^{14}C-labeled compound, and air is drawn through the desiccator. The apparatus has a differential trapping system for ^{14}C-volatile organics and ^{14}CO$_2$. Above-ground plant tissue is not separated from the dosed soil; thus, the route of plant uptake could occur through vapor deposition onto plant surfaces or root uptake and translocation. (Modified from Kloskowski R, et al. [1981] *Chemosphere* 10:1089 and Scheunert I, et al. [1983] *Ecotoxicol Environ Saf* 7:390.)

and ^{14}C-volatile traps can be determined, this apparatus does not provide a clear route of uptake by the plants because the above-ground vegetation is not separated from the soil surface. Volatilization of the ^{14}C-PAH, deposition onto the leaf surfaces, and potential plant uptake of ^{14}CO$_2$ released from the soil would all confound interpretation of plant tissue concentrations. Furthermore, a potential disadvantage of an experimental apparatus with continuous air flow is the opportunity to strip the test compound from the soil at the air–soil interface because of air turbulence and a concentration differential between soil and moving air that would favor movement from the soil to the air. The net effect of continuous air flow may be an overestimation of volatilization rates and an underestimation of potential uptake from the soil by plants growing in that soil.

Schroll and Scheunert (1992) modified the desiccator apparatus to distinguish between specific routes of plant uptake. Teflon™* barriers held in place by clamps are used to separate plant roots from shoots. Root uptake and translocation of ^{14}C-labeled compounds to shoot tissue from soil is determined indirectly as the difference between shoot tissue concentrations in plants exposed to two routes of uptake, via the root and foliar deposition, and shoot tissue concentrations in plants exposed only to foliar deposition of volatilized ^{14}C-labeled compounds. While

* Registered trademark of E. I. du Pont de Nemours and Company, Inc., Wilmington, DE.

this modified system provides additional information on foliar uptake of volatilized compounds, root uptake and translocation remain ambiguous.

Some of the limitations of the experimental systems using solid matrices, as described earlier, have been addressed using a modified Erlenmeyer flask in experiments to measure the uptake, distribution, and mass balance of ^{14}C organics in soil (Figure 12) (Anderson and Walton 1991; Hoylman and Walton 1994). Chemically treated soil, in

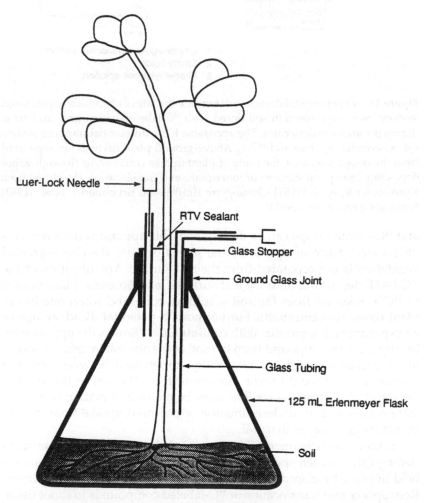

Figure 12 Experimental apparatus to determine plant uptake and fate of ^{14}C-trichloroethylene applied to soil. Above-ground plant tissue is separated from roots and dosed soil by a ground-glass top and silicone sealant. Sampling portals allow analysis of headspace for ^{14}C-volatile organics and $^{14}CO_2$. (From Anderson TA, Walton BT [1991] *Environ Chem Div* 31:197. With permission.)

which a plant is grown, is isolated from above-ground plant tissue, and volatilized ^{14}C compounds are trapped and quantified by flushing the headspace above the soil. Exit air is directed through trapping solutions to collect and quantify $^{14}CO_2$ and volatilized ^{14}C-organic compounds (parent compound plus metabolites) released from the soil into the headspace. This experimental apparatus is a closed system that produces an unambiguous exposure of plant roots to the test compound in the soil. Additionally, the flasks were housed in glass exposure chambers under negative pressure (Jen et al. 1995); thus, radiolabeled compounds translocated from the soil via the transpirational stream and released from the foliage to the ambient air were also trapped and quantified. Hoylman and Walton (1994) have used this modified flask apparatus and glass exposure chamber to measure the uptake and translocation of four ^{14}C-PAHs (naphthalene, phenanthrene, pyrene, and fluoranthene) by vegetation under conditions selected to optimize uptake of the individual PAHs from soil by the plant roots.

Extraction of PAHs from vegetation and soil

The extraction of PAHs from environmental matrices such as vegetation and soil is influenced by salient physicochemical properties such as sorption coefficient, water solubility, and vapor pressure. In addition to the extraction efficiency, the time required for extraction and availability of specialized equipment are important considerations that may influence the selection of a suitable extraction technique.

Analysis of plant tissue

Although there are only a few basic techniques to extract organic chemicals from plant tissues, slight variations in the protocols for each technique provide a multitude of methods for removing chemicals from plants. The goals of plant extraction are to remove the analyte from the plant tissue with minimal loss, degradation, or contamination and to quantify analyte recovery from the plant tissue. The use of ^{14}C-PAHs can be extremely useful when quantification of both PAH parent compounds and metabolites is needed, but the identity of individual metabolites is not critical. Combustion of the tissue to $^{14}CO_2$ and H_2O, followed by liquid scintillation counting of radiocarbon, provides a highly sensitive quantification method that eliminates the need for tissue extraction. Samples of leaf, stem, and root tissue can be oxidized and analyzed quickly and easily. This technique can also be applied to ^3H-PAHs with subsequent combustion of the tissue to CO_2 and ^3H_2O. An important limitation of sample oxidation, however, is that the information obtained is simply the uptake and translocation of radiolabel. The relative distribution of ^{14}C or ^3H among parent compound and metabolic products is not obtained.

Another technique for determining ^{14}C-chemical uptake in plants which has been applied to PAHs is whole-plant autoradiography (Harms 1981). As is the case with oxidation, autoradiography permits only determination of total ^{14}C uptake into the plant, not the identity of ^{14}C material (parent compound or metabolites).

Analysis of nonradiolabeled PAHs in plant tissues requires somewhat complex isolation and cleanup procedures before analysis by gas chromatography (GC) or high performance liquid chromatography (HPLC) (Griest and Caton 1983). Techniques used to solubilize and remove organic chemicals from plant tissues include digestion, sonication, homogenization, and solvent extraction. Digestion involves treatment of the sample with strong alkaline or acidic solutions before extracting with an organic solvent(s). Digesting substances can be prepared (e.g., ethanolic KOH) or purchased from commercial vendors (e.g., Tissue Solubilizer™, Amersham Inc., Arlington Heights, IL).

Both sonication and homogenization are used to disrupt plant cells before solvent extraction (Budde and Eichelberger 1979). Ultrasonic extraction methods use high-intensity ultrasonic vibrations to lyse cells rapidly and increase solvent contact with the sample (Griest and Caton 1983). A disadvantage of sonication, however, is the increase in temperature from the sonication probe. Elevated temperatures can lead to PAH losses from volatilization and sample degradation. Thus, it is extremely important to establish the percent recovery of PAHs by using PAH-spiked controls.

Homogenization or high-intensity mechanical extraction methods employ tissue grinders or homogenizers to disrupt the sample and increase contact with the solvent. Homogenization is a relatively simple, quick, and efficient method for removal of PAHs from solid matrices, including plant tissue. As with sonication, the generation of elevated temperatures is possible, and measures may be necessary to cool the samples during the procedure (e.g., the use of ice baths). In addition, more vigorous extraction techniques may be needed for removal of PAHs from samples with high organic carbon content or prolonged PAH exposure (the latter are sometimes referred to as "aged" samples). Solvent extraction of plant tissues with an organic solvent can be carried out in a separatory funnel or a soxhlet apparatus. A brief description of the various extraction techniques is presented in Table 1.

Digestion extractions have been used successfully by several researchers for the removal of PAHs from plant tissue (Schwarz and Eisele 1984; Larsson 1985; Edwards 1986). Schwarz and Eisele (1984) used 1 ml of Unisol™ (Isolab, Inc.), a commercially available tissue solubilizer, to treat 100-mg samples of plant tissue from experiments on root uptake of ^{14}C-PAHs (i.e., naphthalene, α-naphthol, and methylbenz[c]acridine). Samples were digested in scintillation vials placed in a 37°C water bath. Unisol-complement (a scintillator) was added upon completion of digestion, and the radioactivity in the samples was

Table 1   Methods for the Extraction of PAHs from Plant Tissues

Method	Solvent	Time	Recovery	Ref.
Soxhlet	CH_3COCH_3	24 h	No mass balance	Thomas et al. 1984
Soxhlet	Acetonitrile	24 h	20–75%	Coates et al. 1986
Sonication	Acetonitrile	2 min	47–90%	Coates et al. 1986
Digestion	Tissue Solubilizer	—	—	Schwarz and Eisele 1984
Digestion	Methanolic KOH	3–4	—	Larsson 1985
Digestion	Ethanolic KOH	2 h	—	Edwards 1986
Partitioning	Acetonitrile	3×2 min	32–70%	Coates et al. 1986
Partitioning	CH_3COCH_3	10 min	—	Edwards et al. 1982

determined. The authors presented results of the uptake experiments as a percentage of the total ^{14}C recovered, but the extraction efficiency was not reported for recovery of the PAHs from the three plant species (*Pisum sativum* [pea], *Allium sepa* [onion],and *Lactuca sativa* [lettuce]). Recovery of radiolabeled compounds from the exposure unit (no plants) was determined and indicated approximately 100% recovery of α-naphthol after 100 h, 40% recovery of naphthalene after 100 h, and 45% recovery of methylbenz[c]acridine after 70 h. The major shortcoming of this approach is that percent recovery in the absence of plants may be considerably higher than recovery of the test compound when plants are present. Additional experiments are needed to establish the ability to recover ^{14}C from plant tissue samples spiked with ^{14}C-PAHs.

Larsson (1985) and Edwards (1986) used solutions of methanolic and ethanolic KOH, respectively, to extract PAHs from plant tissues. Larsson (1985) measured the PAH content of lettuce (*L. sativa*) and rye growing at different distances from a roadside. The extraction procedure consisted of treating a 60-g sample with boiling methanolic KOH for 3 to 4 h after addition of an internal standard (ββ-binaphthyl) to the sample. Although extraction efficiencies of the PAHs were not reported, PAH concentrations were adjusted with respect to recovery of the internal standard.

In studies by Edwards (1986) on root uptake of ^{14}C-anthracene by *Phaseolus vulgaris* (bush bean), plant tissues were treated with ethanolic KOH for 2 h in an 80°C water bath. Although the actual extraction efficiency for the digestion procedure was not reported, all but 16% of the total ^{14}C added was accounted for after 30 d; thus, an overall recovery of 84% was achieved

Soxhlet extraction of PAHs from plant tissues has been reported (Thomas et al. 1984; Coates et al. 1986). In a field study near an industrial area in Sweden, Thomas and co-workers (1984) extracted various plant species with acetone for 24 h in a soxhlet apparatus to remove

PAHs. A variety of plant tissues were extracted, including leaves, needles, bark, stems, and roots, but individual extraction efficiencies were not reported for each tissue type.

One of the most thorough studies on the extraction of PAHs from plant tissues was reported by Coates et al. (1986), who compared soxhlet, sonication, and solvent partitioning for the removal of PAHs from spiked sorghum mulch. Nonradiolabeled PAHs (i.e., naphthalene, acenaphthylene, acenaphthene, fluorene, phenanthrene, anthracene, fluoranthene, pyrene, benzo[*a*]anthracene, chrysene, benzo[*b*]fluoranthene, benzo[*k*]fluoranthene, benzo[*a*]pyrene, ideno[1,2,3-*c*,*d*]pyrene, dibenz[*a*,*h*]anthracene, and benzo[*g*,*h*,*i*]perylene) were added to plant mulch, held for 24 h, then extracted. Soxhlet extractions with acetonitrile were conducted for 24 h; sonication with acetonitrile for 2 min, and three partitioning extractions for 2 min each. Percent recoveries for soxhlet extractions ranged from 20% for dibenz[*a*,*h*]anthracene to 75% for pyrene and chrysene. Percent recoveries for sonication extractions ranged from 45% for dibenz[*a*,*h*]anthracene to 90% for phenanthrene, whereas percent recoveries for partitioning ranged from 32% for naphthalene to 70% for phenanthrene. Addition of the PAHs to dead plant material and the limited exposure time of 24 h may have facilitated recovery of the compounds. In addition, the use of mulch probably reduced the amount of plant pigments present in the samples and diminished the need for sample cleanup before analysis.

Grimmer and Böhnke (1975) provide a detailed description of a complex and time-consuming technique that has been used successfully to extract even relatively small concentrations of PAHs from plant materials. The method is especially useful for analysis when it is not possible to use radiolabeled PAHs. The technique involves grinding and refluxing the plant material with 2N KOH in methanol and water for at least 4 h followed by solvent partitioning and back extraction procedures using specified volumes of water, *N*,*N*′dimethylformamide, and cyclohexane. This is followed by drying over sodium sulfate, column (e.g., silica gel) cleanup, and concentration (e.g., evaporation under nitrogen) before HPLC or GC analysis. Using this extraction/cleanup technique, Vahl and Beck (1982) reported a 69% average recovery of 11 PAHs from fresh kale. Analyses were performed by either GC or HPLC with fluorescence detection. Edwards (1991) used a modification of the technique to demonstrate differences in concentrations of 15 PAHs in vegetation collected adjacent to a tar pit near a coal gasification facility and downwind of a coal-fired steam plant.

Analysis of soil

Mass balances are crucial to any study of the uptake of PAHs from soil by vegetation. An important component of the mass balance is the determination of the amount of chemical remaining in the soil or nutrient solution at the conclusion of the uptake experiment. Regardless of whether chromatography or liquid scintillation spectroscopy is used, the objective of the extraction procedure is to quantitatively remove the compound from the sample matrix for further isolation, characterization, and quantification (Griest and Caton 1983). Descriptions of methods used for the extraction of organic compounds from soils are shown in Table 2 and are described in the following.

Soxhlet extraction is the most common method used for removal of PAHs from solid matrices (Griest and Caton 1983) and is recommended by both the American Society for Testing and Materials (ASTM) and the U.S. Environmental Protection Agency (EPA) in standard methods for removal of nonvolatile and semivolatile organics, which include most PAHs, from soils and sludges. However, there are numerous variables that can affect the efficiency of PAH extractions using the soxhlet method, such as solvent composition and cycle time. Unfortunately, critical conditions are often not reported (Griest and Caton 1983).

In a field study at an industrial area in Sweden, Thomas and co-workers (1984) used acetone in a soxhlet extractor to remove PAHs from raw humus materials. Because the extractions were conducted on field samples, neither mass balances nor extraction efficiencies could be determined. Nonetheless, the removal of PAHs from the humus samples after 24 h of extraction demonstrates that soxhlet extraction can be effective. Lee et al. (1987) report no significant difference in the amounts of 14 PAHs extracted from a sediment reference material using various soxhlet and sonication techniques; however, no data on extraction efficiencies are provided.

Sullivan and Mix (1985) used soxhlet extraction to determine PAH concentrations in soil and litter samples from slash and burn sites in Oregon. The authors determined the efficiency of their extraction protocol using a two-phase extraction. Methanol was the initial extracting solvent, followed by a second extraction with 2:3 methanol:benzene for a total extraction time of 48 h. The recovery of spiked ^{3}H-benzo[a]pyrene was 57%, and this extraction efficiency was subsequently factored into the field data. Although the extraction efficiency of the protocol was reported, the authors failed to document additional parameters important to data interpretation, such as the percent organic carbon (% OC) of the soil and the time elapsed between the PAH spike to the soil and the recovery analyses.

More recently, Jones and co-workers (1989a,b) conducted field studies similar to the experiments described previously. Archived soil samples collected over the last century at the Rothamsted Experiment Station in England were extracted by soxhlet with methylene chloride for 4 h. Extraction efficiencies for spiked soils ranged from 30 to 70%, with lower efficiencies obtained for the lower molecular weight PAHs. Although the authors did report the organic carbon content of the soil

Table 2 Methods for the Extraction of PAHs from Soils

Method	Solvent	Time	Recovery	Ref.
Soxhlet	CH_3COCH_3	24 h	No mass balance	Thomas et al. 1984
Soxhlet	CH_3OH Methanol:benzene	48 h	57%	Sullivan and Mix 1985
Soxhlet	Hexane:acetone	16–18 h	61–79%	Fowlie and Bulman 1986
Soxhlet	EPA Method 3540		46–118%	Coover et al. 1987
Soxhlet	CH_2Cl_2	4 h	30–70%	Jones et al. 1989a,b
Soxhlet	CH_2Cl_2	3 h	40% naphthalene 80–100% for other PAHs	Wild et al. 1990a,b
Soxhlet	Hexane:acetone	14–16h	Data but no mass balance	Brilis and Marsden 1990
Soxhlet	CH_2Cl_2:acetone	24 h	92–100%	Aprill and Sims 1990
Homogenization	Cyclohexane	3 min	10–70%	McIntyre et al. 1981
Homogenization	CH_3COCH_3	2 min	50–72%	Fowlie and Bulman 1986
Homogenization	CH_2Cl_2	<1 min	68–104%	Coover et al. 1987
Homogenization	CH_2Cl_2	<1 min	60–70%	Coover and Sims 1987a,b
Homogenization	CH_2Cl_2	3 min	54–65%	Aprill and Sims 1990
Sonication	CH_2Cl_2:acetone	No time reported	Data but no mass balance	Brilis and Marsden 1986
Mechanical	Cyclohexane	No time reported	10–70%	McIntyre et al. 1981

samples (0.9 to 1.1 % OC), details such as the length of contact time before extraction for the spike-recovery tests were omitted. Thus, it is questionable whether similar efficiencies could be obtained for PAHs that have aged with the soil. The data illustrate a common trend observed during extraction of PAHs from soil and vegetation samples: the lower molecular weight PAHs usually yield lower percent recoveries than higher molecular weight PAHs. Volatilization losses of PAHs may occur during spike-recovery tests or PAHs may undergo rapid reactions upon contact with soil constituents that render the PAHs less extractable.

Failure to report key parameters is a common theme among papers describing PAH extraction methods, even in research designed specifically to compare extraction techniques and protocols. A property such as % OC of the soil can be very useful in evaluating data acquired from a specific extraction method or choosing an appropriate extraction method based on the characteristics of the sample. Fowlie and Bulman (1986) compared the extraction efficiencies of soxhlet and homogenization techniques for Donnybrook sandy loam (% OC not reported) spiked with ^{14}C-benzo[a]pyrene and ^{14}C-anthracene at two different concentrations. These spike-recovery tests were extracted following 3- and 5-month incubations of the PAHs. These studies were designed, in part, to permit interactions of the test compounds with the soil matrix. Using 1:1 hexane:acetone, extraction efficiencies for the soxhlet technique ranged from 61 to 79% following a 14 to 16 h extraction, with higher efficiencies for ^{14}C-benzo[a]pyrene than for ^{14}C-anthracene in the nonsterile soil. Extraction efficiencies for homogenization (methanol, 2 min) ranged from 51 to 72%. Higher efficiencies were reported for ^{14}C-benzo[a]pyrene than for ^{14}C-anthracene. The authors did not speculate as to whether microbial degradation of the ^{14}C-anthracene or humification processes could affect the recoveries. Additional spike-recovery experiments under both "sterile" (treated with $HgCl_2$) and nonsterile conditions revealed that extraction efficiencies increased with increasing PAH concentrations. Overall, the soxhlet extraction method gave better percent recoveries than did homogenization, although similar recoveries were obtained by homogenization in considerably less time.

Coover and co-workers (1987) also compared the extraction efficiencies of homogenization and soxhlet techniques. Two soils, a clay loam (3% OC) and a sandy loam (0.5% OC), were spiked with 15 nonradiolabeled PAHs at concentrations from 1 to 1000 µg/g soil and allowed to stand for 2 d. Soxhlet extractions were performed according to EPA Method 3540 (U.S. EPA 1986), in which methylene chloride is used for a 16-h extraction. Homogenization extractions were conducted for 45 s in methylene chloride. Percent recoveries for the soxhlet-extracted clay loam ranged from 49% for benzo[a]pyrene to 118% for dibenz[a,h]anthracene, compared to a range of 46% for benzo[a]pyrene to 115% for dibenz[a,h]anthracene for the soxhlet-extracted sandy loam.

Extraction efficiencies for homogenized clay loam ranged from 70% for benzo[*b*]fluoranthene and benzo[*a*]pyrene to 103% for phenanthrene and benzo[*k*]fluoranthene, compared to 71% for acenaphthylene and benzo[*b*]fluoranthene to 104% for benzo[*k*]fluoranthene for the homogenized sandy loam. Although the soils differed in % OC content, there was little difference in the extraction efficiencies for the various PAHs using the two methods. In addition, equal recoveries were obtained for both low molecular weight and high molecular weight PAHs.

Work by Aprill and Sims (1990) illustrates the high variability that is sometimes observed using similar techniques on different samples. As part of a study to examine disappearance of PAHs in the rhizospheres of prairie grasses, soxhlet extraction was compared with homogenization for four different PAHs (benz[*a*]anthracene, chrysene, benzo[*a*]pyrene, and dibenz[*a,h*]anthracene) at time 0 and after a 219-d incubation. Soxhlet extractions were conducted for 24 h using 1:1 acetone:methylene chloride (U.S. EPA 1986). Homogenization extractions were conducted for 3 min using a Tissumizer™ with methylene chloride as the solvent. Percent recoveries for the soxhlet technique ranged from 92 to 103% compared with 53 to 65% for homogenization, representing at least a 49% decrease in extraction efficiency for the latter technique. Coover and Sims (1987a) used the homogenization technique to recover benzo[*a*]pyrene from clay loam soil (2.9% OC) in microbial degradation experiments. A Tissumizer homogenization system and methylene chloride were used to extract BaP in 45 s. Spike-recovery tests indicated that percent recovery of BaP from the soil ranged from 60 to 70%. Coover and Sims (1987b) also used the Tissumizer system with a variety of PAHs in a sandy loam soil (0.5% OC) to determine PAH disappearance rates at three temperature regimes, but percent recoveries were not reported. Additional studies by Aprill and Sims (1990) also indicated that percent extraction efficiency of PAHs was lower from soils in which prairie grasses were grown than in nonvegetated soils after 219-d incubations; that is, a consistently higher recovery of PAHs was obtained from nonvegetated soils. Recent studies by Walton and co-workers (1994) show that rhizosphere microbial communities may enhance the association of PAHs with the combined humic and fulvic acid fraction of soil. Such associations could contribute to the lower recovery of PAHs from vegetated soils as observed by Aprill and Sims (1990).

Brilis and Marsden (1990) have reviewed extraction methods for PAHs in soil. Comparisons of soxhlet and sonication extraction methods using different solvents indicated that soxhlet extractions gave consistently higher percent recoveries of PAHs than did sonication. Soxhlet extractions were conducted with 1:1 acetone:hexane for 16 h. Sonication extractions were conducted in 1:1 methylene chloride:acetone according to EPA Method 3550. Soil samples from a wood-preserving site were extracted using both techniques, and the concentrations of the extracted PAHs (i.e., naphthalene, acenaphthylene, acenaphthene, fluorene, phenanthrene, anthracene, fluoranthene, pyrene, benzo[a]anthracene, chrysene, benzo[b]fluoranthene, benzo[k]fluoranthene, benzo[a]pyrene, ideno[1,2,3-c,d]pyrene, dibenz[a,h]anthracene, and benzo[g,h,i]perylene) were determined and compared. Analysis of the potential influence of ring number on extraction technique indicated that soxhlet extractions gave equal or greater recovery of all PAHs tested (2-,3-,4-,5-, and 6-aromatic rings). Although extensive analysis of the PAH molecular structure, extraction technique, and recovery were conducted, critical properties of the solid matrix, such as % OC, were not reported.

McIntyre and co-workers (1981) also compared homogenization and low-intensity mechanical shaking for extraction of three different types of sewage sludges containing PAHs. Cyclohexane served as the extraction solvent for both techniques. In addition, both the shaking technique and the homogenization technique were conducted for 3 min. Percent recoveries for the 6 PAHs (i.e., fluoranthene, benzo[k]fluoranthene, benzo[a]pyrene, benzo[g,h,i]perylene, benzo[b]fluoranthene, and ideno[1,2,3-c,d]pyrene) ranged from 10 to 60% using the shaking method for mixed primary sludge, mixed digested sludge, and surplus activated sludge. Recoveries of PAHs were somewhat higher for the homogenization technique, ranging from 20 to 70% for mixed primary sludge, mixed digested sludge, and surplus activated sludge. Although the authors divided the sludge into three categories, the basis for designating the sludge categories was not disclosed. Thus, discussion of the percent recovery data in reference to sludge type is of little predictive value for extrapolating to other solid matrices.

The PAH content of sewage sludge and agricultural soils to which the sewage sludge was applied has been examined by Wild et al. (1990a,b) using soxhlet extraction techniques. Both soil and sewage sludge were extracted with methylene chloride for 3 h. Extraction efficiencies for the soil and sludge were determined by spike-recovery (incubation time not reported). Percent recoveries for naphthalene (40%) were the lowest of the PAHs tested. Recovery of the other PAHs ranged from 80 to 100%. The authors did not report the % OC of the soils and sludges; thus, the importance of soil organic matter cannot be evaluated. The relatively high vapor pressure of naphthalene may also contribute to the lower percent recovery for this compound compared to other PAHs. In addition, microbial activity may contribute to the low recovery of naphthalene from the spiked soils and sludges if the matrices were not sterilized before the spike-recovery tests. Recent studies (Walton et al. 1994) show the association of selected ^{14}C-PAHs with fulvic and humic fractions of soil after a 5-d incubation of ^{14}C-naphthalene, ^{14}C-phenanthrene, and ^{14}C-fluoranthene. The percentage of ^{14}C associated with the fulvic and humic fractions increased in the presence of white sweetclover (*Melilotus alba*) growing in the soil. These

findings are consistent with the Aprill and Sims (1990) observation that
PAH recovery decreased in vegetated soils compared to nonvegetated
soils and further implicate the role of soil microbiota in reducing the
recovery of low molecular weight PAHs from soil.

Conclusions

A variety of experimental apparatuses and methods have been used
to quantify the uptake of PAHs from soil by vegetation; however, most
studies lack a high degree of precision. The experimental designs
reviewed are readily applicable for detailed quantitative analyses of
PAH uptake by vegetation, although many of the systems yield ambiguous data on route of uptake and do not distinguish uptake of PAH
parent compound from uptake of PAH metabolites formed in the soil.
Comparative evaluation of the experimental designs is limited by the
lack of experimental details in many studies. Nonetheless, a generalized experimental design can be formulated to measure plant uptake
of PAHs from soil. Such an experimental design should provide for the
following: (1) an unequivocal route of PAH transport from soil to plant
tissue; (2) actual measurement of PAH residues in air, plant tissue, soil,
and soil headspace; (3) minimize PAH sorption to reactive surfaces
(e.g., neoprene stoppers, PVC columns, and silicone stopcock grease
should be avoided); (4) characterize the rooting matrix as completely
as possible, including % OC; (5) differentiate between abiotic (volatilization) and biotic (microbial degradation, microbial humification)
losses; and (6) determine the mass balance of PAHs.

Three basic extraction techniques — soxhlet, sonication, and
homogenization — have been used to remove PAHs from plant tissues,
nutrient solutions, and sand/soil substrates in which plants are grown.
Extraction solvent(s) and extraction times also constitute important
variables for sample extraction. No standard procedure has emerged
for PAH extractions in these experiments, but the percent efficiency of
extraction procedures used and the characteristics of the samples analyzed (plant species, soil properties, aging of analyte with the soil, etc.)
must be reported for each experiment.

In general, soxhlet extractions yield higher percent recovery of
PAHs than sonication and homogenization, especially for aged PAHs,
soils with high OC content, and high molecular weight PAHs. The
primary disadvantage of soxhlet extraction is the time required to
exhaustively extract the sample. Sonication and homogenization are
quicker, but extraction efficiencies for these two methods are not as
good as that achieved by soxhlet extraction. Differences in extraction
efficiencies for the three techniques are less pronounced for 2- and 3-
ringed PAHs in soils with low (<1%) OC content. Thus, soxhlet extraction is the method of choice for PAHs with 4 or more rings in soils with
greater than 1 % OC or when PAHs have aged with the soil.

Any assessment of PAH uptake by plants must consider the strengths and limitations of the experimental design and analytical methods employed. Selecting experimental conditions to favor plant uptake of PAHs (e.g., low organic carbon soil, high root-to-soil ratio, freshly applied PAHs, and rapidly growing plants) would provide optimum uptake data by which plant uptake under field conditions could be compared. Analytical techniques that distinguish between volatilized organics and CO_2 efflux from the rooting medium and distinguish between parent compound and metabolites enhance the overall understanding of the fate of PAHs in plant–soil systems. The primary question of plant uptake of PAHs is best addressed when both biotic and abiotic pathways for PAH removal are accounted for in the experimental design.

Acknowledgments

The authors gratefully acknowledge helpful contributions to this work by E.F. Neuhauser, Niagara Mohawk Power Corporation, Syracuse, NY; J.W. Goodrich-Mahoney, Electric Power Research Institute (EPRI), Washington, D.C.; and I.P. Murarka, EPRI, Palo Alto, CA. This chapter also benefited from the thoughtful comments of two anonymous reviewers. Research sponsored by EPRI, Research Project 2879-10 to the Oak Ridge National Laboratory, Oak Ridge, TN, managed by Martin Marietta Energy Systems, Inc. under contract DE-ACO5-84OR21400 with the U.S. Department of Energy, Publication No. 4153, Environmental Sciences Division, ORNL.

References

Anderson TA, Walton BT (1991) Fate of ^{14}C-trichloroethylene in soil-plant systems. Abstracts, 202nd National Meeting of the American Chemical Society. *Environ Chem Div* 31:197–200.

Aprill W, Sims RC (1990) Evaluation of the use of prairie grasses for stimulating polycyclic aromatic hydrocarbon treatment in soil. *Chemosphere* 20:253–265.

Brilis GM, Marsden PJ (1990) Comparative evaluation of soxhlet and sonication extraction in the determination of polynuclear aromatic hydrocarbons in soil. *Chemosphere* 21:91–98.

Budde WL, Eichelberger JW (1979) *An EPA Manual for Organics Analysis Using Gas Chromatography-Mass Spectrometry,* EPA-600/8-79-006, U.S. Environmental Protection Agency, Cincinnati, OH.

Coates JT, Elzerman AW, Garrison AW (1986) Extraction and determination of selected polycyclic aromatic hydrocarbons in plant tissues. *J Assoc Off Anal Chem* 69:110–114.

Coover MP, Sims RC (1987a) The rate of benzo(a)pyrene apparent loss in a natural and manure amended clay loam soil. *Hazard Waste Hazard Mater* 4:151–158.

Coover MP, Sims, RC (1987b) The effect of temperature on polycyclic aromatic hydrocarbon persistence in an unacclimated agricultural soil. *Hazard Waste Hazard Matter* 4:69–82.

Coover MP, Sims RC, Doucette W (1987) Extraction of polycyclic aromatic hydrocarbons from spiked soil. *J Assoc Off Anal Chem* 70:1018–1020.

Edwards NT (1983) Polycyclic aromatic hydrocarbons (PAH's) in the terrestrial environment: a review. *J Environ Qual* 12:427–441.

Edwards NT (1986) Uptake, translocation and metabolism of anthracene in bush bean (*Phaseolus vulgaris* L.). *Environ Toxicol Chem* 5:659–665.

Edwards NT (1991) Fate and effects of PAHs in the terrestrial environment: an overview. In *Ecological Exposure and Effects of Airborne Toxic Chemicals: An Overview*, Moser TJ, Barker JR, Tingey DT, Eds., EPA-600/3-91/001, U.S. EPA Environmental Research Laboratory, Corvallis, OR, pp. 48–59.

Edwards NT, Ross-Todd BM, Garver EG (1982) Uptake and metabolism of ^{14}C anthracene by soybean (*Glycine max*). *Environ Exp Bot* 22:349–357.

Federle TW, Schwab BS (1989) Mineralization of surfactants by microbiota of aquatic plants. *Appl Environ Microbiol* 55:2092–2094.

Fowlie PJA, Bulman TL (1986) Extraction of anthracene and benzo[a]pyrene from soil. *Anal Chem* 58:721–723.

Griest WH, Caton JE (1983) Extraction of polycyclic aromatic hydrocarbons for quantitative analysis. In *Handbook of Polycyclic Aromatic Hydrocarbons*, Bjørseth A, Ed., Marcel Dekker, New York, pp. 95–148.

Grimmer G, Böhnke H (1975) Polycyclic aromatic hydrocarbon profile analysis of high-protein foods, oils and fats by gas chromatography. *J Assoc Off Anal Chem* 58(4):725–733.

Harms H (1981) Aufnahme und Metabolismus polycyclischer aromatischer kohlenwasserstoffe (PCKs) in aseptisch kultivierten nahrungspflanzen un Zellsuspensionskulturen. *Landbauforsch Völkenrode* 31:1–6.

Hoylman AM, Walton BT (1994) Fate of polycyclic aromatic hydrocarbons in plant-soil systems: plant responses to a chemical stress in root zone. ORNL/TM-12650. Oak Ridge National Laboratory, Oak Ridge, TN, 122 pp.

Hsu TS, Bartha R (1979) Accelerated mineralization of two organophosphate insecticides in the rhizosphere. *Appl Environ Microbiol* 37:36–41.

Jen MS, Hoylman AM, Edwards NT, Walton BT (1995) Experimental method to measure gaseous uptake of ^{14}C-toluene by foliage. *Environ Exp Bot* 35:389–398.

Jones KC, Stratford JA, Tidridge P, Waterhouse KA, Johnston AE (1989a) Polynuclear aromatic hydrocarbons in an agricultural soil: long-term changes in profile distribution. *Environ Pollut* 56:337–351.

Jones KC, Stratford JA, Waterhouse KA, Vogt NB (1989b) Organic contaminants in Welsh soils: polynuclear aromatic hydrocarbons. *Environ Sci Technol* 23:540–550.

Kloskowski R, Scheunert I, Klein W, Korte F (1981) Laboratory screening of distribution, conversion, and mineralization of chemicals in the soil-plant-system and comparison to outdoor experimental data. *Chemosphere* 10:1089–1100.

Larsson BK (1985) Polycyclic aromatic hydrocarbons and lead in roadside lettuce and rye grain. *J Sci Food Agric* 36:463–470.

Lee H-B, Dookhran G, Chau ASY (1987) Analytical Reference Materials. Part VI. Development and certification of a sediment reference material for selected polynuclear aromatic hydrocarbons. *Analyst* 112:31–35.

Linz DG, Neuhauser EF, Middleton, AC (1991) Perspectives on bioremediation in the gas industry. In *Environmental Biotechnology for Waste Treatment*, Sayler GS, Fox R, Blackburn JW, Eds., Plenum Press, New York, pp. 25–36.

McCrady JK, McFarlane JC, Gander, LK (1990) The transport and fate of 2,3,7,8-TCDD in soybean and corn. *Chemosphere* 21:359–376.

McFarlane JC, Pfleeger T (1987) Plant exposure chambers for the study of toxic chemical-plant interactions. *J Environ Qual* 16:361–371.

McIntyre AE, Perry R, Lester JN (1981) Analysis of polynuclear aromatic hydrocarbons in sewage sludge. *Anal Lett* 14:291–309.

O'Connor GA, Chaney RL, Ryan JA (1991) Bioavailability to plants of sludge-borne toxic organics. *Rev Environ Contam Toxicol* 121:129–155.

Scheunert I, Vockel D, Schmitzer J, Viswanathan R, Klein W, Korte F (1983) Fate of chemicals in plant-soil systems: comparison of laboratory test data with results of open air long-term experiments. *Ecotoxicol Environ Saf* 7:390–399.

Schroll R, Scheunert I (1992) A laboratory system to determine separately the uptake of organic chemicals from soil by plant roots and by leaves after vaporization. *Chemosphere* 24:97–108.

Schwarz OJ, Eisele GR (1984) Food chain transport of synfuels: experimental approaches for acquisition of baseline data. In *Synthetic Fossil Fuel Technologies: Results of Health and Environmental Studies*, Cowser KE, Ed., Butterworth Press, Boston, pp. 441–461.

Stotzky G, Culbreth W, Mish LB (1961) A sealing compound for use in biological work. *Nature* 191:410.

Sullivan TJ, Mix MC (1985) Persistence and fate of polynuclear aromatic hydrocarbons deposited on slash burn sites in the Cascade Mountains and coast range of Oregon. *Arch Environ Contam Toxicol* 14:187–192.

Suzuki M, Aizawa N, Okano G, Takahashi T (1977) Translocation of polychlorobiphenyls in soil into plants: a study by a method of culture of soybean sprouts. *Arch Environ Contam Toxicol* 5:343–352.

Thomas W, Rühling A, Simon H (1984) Accumulation of airborne pollutants (PAH, chlorinated hydrocarbons, heavy metals) in various plant species and humus. *Environ Pollut Ser A* 36:295–310.

U.S. Environmental Protection Agency (1986) Waste/soil treatability studies for four complex wastes: methodologies and results. EPA-600/6-86-001, U.S. Environmental Protection Agency, Washington, D.C.

Vahl M, Beck J (1982) Nordic PAH Project Report No. 15. Polycyclic aromatic hydrocarbons in Danish leaf crops — Part II. Central Institute for Industrial Research, Forskningsvn, 1, Oslo 3, Norway, pp. 23–36.

Walton BT, Hoylman AM (1992) Uptake, translocation, and accumulation of polycyclic aromatic hydrocarbons in vegetation. EPRI TR-101651, Electric Power Research Institute, Palo Alto, CA.

Walton BT, Hoylman AM, Perez MM, Anderson TA, Johnson TR, Guthrie EA, Christman RF (1994) Rhizosphere microbial communities as a plant defense against toxic substances in soils. In *Bioremediation Through Rhizosphere Technology*, Anderson TA, Coats JR, Eds. American Chemical Society, Washington, D.C., pp. 82–92.

Webber MD, Goodin JD (1990) Studies on the fate of organic contaminants in
    sludge treated soils. In Proceedings of the European Communities Land
    Application of Sludge Working Group, Braunschweig Germany.
Wild SR, McGrath SP, Jones KC (1990a) The polynuclear aromatic hydrocarbon
    (PAH) content of archived sewage sludges. *Chemosphere* 20:703–716.
Wild SR, Waterhouse KS, McGrath SP, Jones KC (1990b) Organic contaminants
    in an agricultural soil with a known history of sewage sludge amend-
    ments: polynuclear aromatic hydrocarbons. *Environ Sci Technol*
    24:1706–1711.

chapter sixteen

Plant uptake and metabolism of polychlorinated biphenyls (PCBs)

Ravi K. Puri, Ye Qiuping, Shubender Kapila, William R. Lower, and Vivek Puri

Introduction

Plants play significant roles to keep our environment clean by trapping, absorbing, translocating, and metabolizing organic xenobiotics from soil, water, and air. One of the most notorious, ubiquitous, and persistent environmental contaminants plants are combating have been called polychlorinated biphenyls (PCBs).

PCBs, as a class of synthetic halogenated aromatic hydrocarbons, were introduced into the world market in 1930 as dielectric fluids for high voltage transformers and large electrical capacitors. PCBs gained rapid and widespread industrial use as electrical insulators, lubricants, hydraulic fluids, diffusion pump oils, and plasticizers. PCBs found their way into day-to-day life as flame retardants, polyvinyl chloride (PVC) toys, molded containers, protectants in rubber, in weatherproof coatings, stucco, waxes, varnishes, paints, inks, duplicating fluids, and pesticides formulations, to name just a few.

PCB residues first appeared in 1949 and the level increased progressively through 1965 (Jensen 1966). Later on, PCB residues were detected in a variety of birds, fish, and marine life around the world, indicating pervasive contamination of environmental media and entry into the food chain (Risebrough et al. 1966). Subsequent studies were undertaken to determine their widespread persistence in the environment and to evaluate the toxicological and physiological effects on the biota. Their harmful biological effects, linked to mutagenesis and teratogenesis, resulted in the ban of their application and manufacturing in July 1979 in the United States under the Toxic Substance Control Act (TSCA). The problems of environmental contamination resulting from PCBs were widely publicized and well documented (Winston and Gerstner 1978). Today, PCBs are known as the most ubiquitous and persistent contaminants on the entire planet.

Almost two decades after the restriction imposed on their application and manufacturing, these compounds are still found in the environment. Monitoring programs within several countries revealed significant global environmental contamination from PCBs. Several investigations have indicated the existence of significant quantities of PCBs found in soil, sludges, sediments, water, plants, fish, wildlife, human blood, semen, milk, and biological tissues (Safe et al. 1985; Bush et al. 1986a, b; Jacobson et al. 1989). This is attributed to their environmental stability, global air transport in the form of vapor and particulate, and improper disposal in landfills and dump sites. Thus, PCBs continue to contaminate the food chain.

Extensive studies *in vivo* and *in vitro* have been conducted on the toxicological effects of these compounds in different mammalian systems (Safe 1990; Tilson et al. 1990; Golub et al. 1991). Their fate in the biotic and abiotic system has also been reviewed by Hooper et al. (1990). One of the areas that has received little attention is the uptake, translocation, and transformation of PCBs in terrestrial plants. It is very important to understand the metabolism of these pollutants through vegetation so that the fate and effects of PCBs in the environment can be assessed and remedied.

Limited information on PCB uptake by plants has been reported; these reports have been confined mostly to single chlorinated biphenyl congeners or a single PCB formulation (Iwata and Gunther 1976; Moza et al. 1976, 1979). Many isomers, especially highly chlorinated PCBs, have not been studied with respect to their uptake and metabolism in plants. Moreover, some of the published data are contradictory. Some of the investigations show that there is little or no active transport, and uptake of PCBs by plants is primarily through vapor sorption (Iwata and Gunther 1976; Babish et al. 1979; Fries and Marrow 1981; Buckley 1987), while others point toward an active uptake of PCBs (Wallnöfer et al. 1975; Moza et al. 1976, 1979; Suzuki et al. 1977; Sawhney and Hankin 1984).

Thus, it is very pertinent to comprehend the lifecycle of PCBs through the food chain as part of assessing their fate and distribution in the environment. This chapter is concerned with the uptake and metabolism of PCBs within vegetation. The systemic investigations in our laboratory include the uptake and translocation of PCBs, plant uptake by vapor, and the metabolism of the PCBs in plants.

Review of literature

Chemistry of PCBs

To investigate the mobility of PCBs in any phase of the environment, it is very essential to understand the physical and chemical properties of PCBs. The chemistry of PCBs was reviewed by Erickson (1986) and Hutizinger et al. (1992). PCBs are a mixture of chlorinated biphenyl compounds, comprising 209 individual compounds known as congeners. PCB congeners with an equal number of chlorine substitutions but with differing chlorine locations in the molecule are called isomers. Each isomer consists of one or more different congeners. The number of possible congeners within each of ten possible sets of isomers of PCBs and their toxicity are shown in Table 1.

Chemical and physical properties of PCB congeners differ markedly from one set of isomers to another and, to a lesser degree, among congeners within a given set of isomers. The highly chlorinated congeners, i.e., tetra to deca, are very stable and have very high partition coefficients (i.e., K_{ow} and K_{oc} values). The vapor pressure of congeners decreases with increasing chlorination substitutions. So, low molecular

Table 1 PCB Congeners Isomers and Their Toxicity

No. of chlorine substitutions	IUPAC no.	No. of possible congeners (209)	Congeners of highest toxicity #IUPAC[a]
1	1–3	3	—
2	4–15	12	—
3	16–39	24	—
4	40–81	42	77
5	82–127	46	87, 99, 101, 105, 118, 126
6	128–169	42	128, 138, 153, 156, 169
7	170–193	24	170, 180, 183
8	194–205	12	194, 201
9	206–208	3	—
10	209	1	—

[a] International Union of Pure and Applied Chemistry.

weight PCB congeners tend to volatilize more rapidly than those of high molecular weight. Aqueous solutions of PCB congeners show an analogous trend: solubility decreases with increasing chlorine substitutions. The *n*-octanol–water partition coefficient (K_{ow}) and bioaccumulation factor of PCB congeners also increase with increases in chlorine substitution. Volatilization, solubility, and the bioaccumulation factor are important parameters in controlling distribution and fate of PCB congeners in the environment as well as uptake and mobility in plants.

Commercial PCB mixtures have been sold in many countries under a variety of tradenames, but their PCB content is very similar to an Aroclor series: Aroclor (United States and United Kingdom), Clophen (Germany), Fenchlor (Italy), Kanechlor and Santotherm (Japan), Phenochlor and Pyralene (France), and Sovol (former USSR). Common mixtures of PCBs are Aroclors 1221, 1232, 1242, 1248, 1254, 1260, and 1262, which contained 21, 32, 42, 48, 54, 60, and 62% of chlorine by weight, respectively. Aroclors are designed by a four-digit number, the first two digits of which are 12, representing the 12 carbon of the biphenyl skeleton, and the second two digits as the weight percentage chlorine of the mixture. General physical and toxicological properties of various Aroclors have been reviewed by Cairns et al. (1986) and Sawhney (1986).

Plant uptake and translocation of PCBs

Several investigations have shown that PCBs can be translocated from soil to various parts of the plants. Furthermore, the PCBs can accumulate in particular tissues in higher concentration than others. For example, Walsh et al. (1974) reported the uptake of Aroclor 1242 by *Rhizophora mangle* L. (Red Mangrove seedlings). The authors found that most of the PCBs were accumulated in the hypocotyl and leaves, and no significant accumulation was detected in the stem. Wallnöfer et al. (1975) studied PCB uptake by sugar beets from a soil containing 0.3 ppm of Aroclor 1254. During a 60- to 70-d period, PCB accumulation was 0.05 ppm in the peel of the beet root, 0.01 to 0.15 ppm in the whole plant, and 0.003 ppm in the leaves. The uptake and translocation of PCBs by rice plants were reported by Nakanishi et al. (1975). Moza et al. (1976) studied the uptake of PCBs in carrots and sugar beets. Carrot uptake of dichlorobiphenyl exceeded that of sugar beet, and the authors concluded that carrots can take up lipophilic xenobiotics more rapidly than sugar beets. These investigators also reported that chlorinated biphenyls tended to adhere to the root surface and accumulate in the epidermal region of the cuticle. Further translocation of these contaminants in the plant was not reported. Uptake of up to 0.15 ppm of PCBs by soybean sprout growth in soil containing 100 ppm of Aroclors 1242 and 1254 has been reported by Suzuki et al. (1977). Lawrence and Tosine (1977) studied the uptake of PCBs in corn plants from a soil sludge mixture. They reported that the corn absorbed PCBs readily, especially Aroclor 1260. The concentration in leaves was reported in the range of 45 to 81 µg/kg.

Uptake of PCBs by *Brassica oleracea* var. *capitata* (cabbage) grown on sewage sludge has been reported by Babish et al. (1979). Nadeau et al. (1982) reported that uptake of PCBs by marsh plants was fairly high, indicating that the pollutant could be accumulated and transferred in the food chain.

Sawhney and Hankin (1984) investigated the uptake of PCBs in beets (*Beta vulgaris* L.), turnips (*Brassica rapa* L.), and beans (*Phaseolus vulgaris*). They found that beet and turnip leaves accumulated a larger concentration of PCBs than the roots. In beans, the leaves and pods contained higher concentrations than the stems, while low concentrations were detected in the seeds. The authors also reported that the lower-chlorinated PCB isomers were more abundant in plants than the higher-chlorinated isomers.

The accumulation of PCBs in the foliage of *Edera helix* (ivy), *Quercus robus* (oak), *Castanea sativa* (chestnut), *Juniperus communis* (juniper), and *Pinus pinaster* (pine needles) has been reported by Gaggi et al. (1985). Gosselin et al. (1986) investigated the uptake of PCBs by potato plants grown on sludge-contaminated soils and concluded that PCBs were below detectable levels in potato leaves and pulp, but the peel, which was in direct contact with the soil, contained 0.04 ppm of PCBs.

Wegmann et al. (1987) observed the uptake of PCBs and reported that the maximum concentration of pollutant was found in roots of ray grass and rape plants.

O'Connor et al. (1990) studied plant uptake of sludge-borne PCBs in a greenhouse by two food chain crops and grass species (*Festuca*). They reported that *Daucus carota* (carrots) were contaminated with PCBs and contamination was restricted to carrot peels only. No active translocation was found in any other plants. Gan and Berthouex (1994) investigated the uptake of PCBs from sludge-amended farmland. They detected 79 PCB congeners in the surface soil, corn stover, and corn grain. Results indicated that the rate constant for the disappearance of each individual PCB congener in the soil was independent of the initial PCBs sludge concentration, the sludge loading rate, and sludge application pattern. Most of the 2-, 3-, and 4-chlorinated PCB congeners showed significant decreases in soil, with half-lives ranging from 4 to 58 months. The higher-chlorinated PCBs were more persistent in the sludge-amended farmland. The concentration of PCBs in the runoff samples was correlated with surface soil concentration. The PCBs were associated with the runoff sediments, and there were no measurable PCBs in the liquid portion of the runoff. It was concluded that there was no PCBs translocation into either corn grain or corn stover samples.

Plant uptake by vapors and particulate

Lower-chlorinated biphenyls, i.e., mono, di, tri, and tetra congeners, have high vapor pressure and volatilize rapidly from the soil. The vapor phase uptake by the shoots may be an important route of their entry to the plant. Higher-chlorinated PCB congeners have low vapor pressure, are adsorbed by the various parts of the plants by soil particulate, and generally accumulate in the epidermal region of the stem and leaves.

Beall and Nash (1971) developed a method to differentiate between uptake of a pesticide through the plant vascular system in relation to the vapor phase movement. Following this technique, Fries and Marrow (1981) used ^{14}C individual isomers (2,2',5-trichlorobiphenyl, 2,2',5,5'-tetrachlorobiphenyl, and 2,2',4,5,5'-pentachlorobiphenyl) to study the uptake in soybean plants. Detectable residues were found only on plants grown with surface-treated soil, with most of the residues confined to the lower leaves. They attributed this uptake to vapor absorption. Uptake of PCBs by rice plants through vapors absorption was demonstrated by Ogiso et al. (1975). These investigators reported that considerable amounts of PCBs were detected in plants placed over a PCB solution in a closed box, showing that PCBs evaporated from the water surface and was absorbed by the rice plants through the atmosphere.

The presence of PCBs in a diverse flora was reported by Klein and Weisberger (1976). The plant samples were taken from two different areas: one from an industrial area containing high levels of PCBs and the other from a remote area in European Alps. The plant samples from the highway and remote area contained 0.09 to 0.95 ppm and 0.05 to 0.15 ppm of PCBs, respectively. The authors reported that PCBs were transported by air to remote areas and different plant species showed variable amounts of PCBs.

Iwata et al. (1976), in studies of uptake of Aroclor 1254 by carrots, found that 97% of the Aroclor taken up from the soil was retained by the carrot peels. The presence of Aroclor components was also reported in plant foliage and was attributed primarily to adsorption of contaminated dust. An accumulation of airborne PCBs by plant foliage has also been reported by Buckley (1982).

The loss of ^{14}C-labeled Aroclor 1254 from soybean leaves has been investigated by Weber and Mrozek (1979). Only 6.7% of applied PCBs were present in the plants after 12 d of application. It appeared that most of the ^{14}C-PCB applied to the soybean leaf evaporated or remained at the treated site.

Bush et al. (1986a) studied the transport of 42 PCB congeners by purple loosestrife (*Lythrum salicaria*). The contaminated site was beside the Hudson River at Albany, NY and the control site was 2 mi away on wasteland. By transplanting and translocating the plants between the sites, systemic uptake from ambient air was also measured. The dominant route of uptake by the plant was via the roots. At high ambient concentrations (140 ng/m^3), PCBs were absorbed from the air by the plants. At low ambient concentrations (8 ng/m^3), mono- and dichlorinated congeners were emitted by the plants.

Sheppard et al. (1991) investigated the mobility and uptake of 2,2',5,5'-tetrachlorobiphenyl (^{14}C-PCB) in soil with high and low organic contents in undisturbed soil cores. They reported that the mobility of ^{14}C-PCB depends on the profile distribution and amount of soil organic matter. There was evidence of atmospheric transfer of PCBs from the soil to plants. They concluded that uptake by the plant was attributed to vapor sorption.

Kylin et al. (1994), in their environmental monitoring studies of PCBs, found that PCBs were accumulated in the epicuticular wax of conifer needles. High concentrations of PCBs, with a profile toward low molecular species, were found. This was attributed to uptake from the atmosphere.

Metabolism of PCBs in plants

Metabolism of PCBs in plants has been studied and reviewed (Moza et al. 1973; Buckley 1987; Pal et al. 1980). Metabolism of PCBs in plants is very slow and essentially limited to mono-, di-, and trichloroisomers. Very little information is available in the literature regarding the uptake and metabolism of highly chlorinated compounds, i.e., tetra to deca, in plants.

The metabolism that has been detected in plants consists of hydroxylation of one or both rings, followed by the formation of adducts. Sugar adducts of hydroxylated PCB products may be a factor in the translocation of low-chlorinated congeners. However, the exact nature of the adduct and its formation in the plant are not known. The proposed pathway of 2,2'-dichlorophenyl metabolism in a marsh plant, *Veronica beccabunga*, has been outlined by Moza et al. (1973). The dichlorophenyl was hydroxylated to corresponding phenols. Moza et al. (1979) also studied the uptake of 2,4',5-trichlorobiphenyl ^{14}C and 2,2',4,4',6-pentachlorobiphenyl ^{14}C in carrots and sugar beets. Metabolism of pentachlorobiphenyl was insignificant. The trichlorobiphenyl was metabolized to phenolic products. The metabolic conversion of PCBs to phenols or their methylated products in animals is well known (Hutzinger et al. 1974; Goto et al. 1974; Greb et al. 1975; Safe et al. 1974, 1975; Sundström and Jansson 1976).

Response of plants to PCB contamination

McFarland and Clarke (1989) described an interesting feature of the toxicity of PCB congeners. Thirty-six PCB congeners were classified into four groups, depending upon their potential toxicity and frequency of occurrence in the environmental samples. Toxic potential was evidenced by type and specificity of mammalian microsomal mixed function oxidase (MFO) induction.

Plants respond very quickly to environmental changes and are known as chemical sensors which can detect the presence of pollutants in the air. When plants are exposed to a contaminated environment, even for a short period, acute injury occurs. Tissues are killed, causing necrosis, chlorosis, bronzing, and premature senescence. Typical symptoms have been described by many researchers and are reviewed for many plant–pollutant interactions (Reinert 1975; Ormrod et al. 1976; Heck et al. 1977).

Different plant species respond to PCB concentrations, depending upon their threshold values for tolerance and their detoxifying mechanisms. Limited information is available on the sensitivity of aquatic plants, particularly algae, to PCBs. In a series of studies on the effect of PCBs on algae, Mosser et al. (1972) found that high concentrations affect photosynthetic rates and cause distortion of chloroplasts and membranes. PCBs are relatively nontoxic to *Chlorella* species (Urey et al. 1976). Mahanty and Fineran (1976) studied the effect of PCBs in the aquatic plant *Spirodela oligorrhiza*. The plants became chlorotic, the chloroplasts were enlarged, and the thylakoids and stroma became irregular.

The chloroplasts of higher plants are assumed to be susceptible to PCBs, but there is very little evidence to support this view, except for the work of Sinclair et al. (1977) on isolated spinach chloroplasts, which demonstrated that oxygen evolution was inhibited by Aroclor 1221. At high concentrations, PCBs inhibited growth in higher plants. Weber and Mrozek (1979) demonstrated that at higher concentrations of PCBs, soybeans (*Glycine max* L.) were more inhibited than fescue (*F. arundi-*

naceae), while corn (*Zea mays*) and sorghum (*Sorghum bicolor*) were not affected by soil concentrations as high as 1000 ppm (Strek et al. 1981).

Objective

From the review of the literature, it is quite evident that the uptake of PCBs in plants remains a contradictory phenomenon. Part of the reason for this lies in the fact that most plant uptake studies have been conducted with uncharacterized or partially characterized commercial formulations, thus obscuring any correlation which might exist between the physicochemical properties of individual congeners and their observed concentration in plant tissues. Most of the findings indicate that PCBs remain trapped in the lipid-rich parts of the plant and do not enter into the plant transport system. However, higher-chlorinated PCB congeners (tetra to octa) are more persistent and may reach the end of the food chain with significant buildup in oily material and lipophilic substances.

A systematic study was undertaken to overcome the limitations of available literature and to produce a comprehensive database on the possible uptake and metabolism of PCBs in plants (Puri et al. 1990; Ye et al. 1992). It was perceived that the experimental approach would allow us to describe the correlation between uptake/metabolism and structure of PCBs. The salient features of the approach are described in Figure 1.

Experimental design

Selection of plant species

Hordeum vulgare (barley) and *Lycopersicon esculentum* (tomato) were selected for this study. The species were selected on the basis of monocot and dicot edible crops. These species can also be grown easily and rapidly in the laboratory. Further, monocot and dicot plants carry different genotypic characteristics which lead to different metabolic pathways. Both species were erect plants to minimize contamination with the soil.

Preparation of contaminated soil

To ensure the uniform distribution of PCBs in the soil, a PCB mixture containing Aroclors 1221, 1242, and 1260 was dissolved in hexane and mixed with clean sand. The sand was then mixed with soil in a rotary tumbler.

Prior to adding the Aroclor mixture, the sand was sieved with a #40-mesh screen, cleaned by washing with distilled water, followed by an acetone rinse, and then allowed to dry at room temperature. To

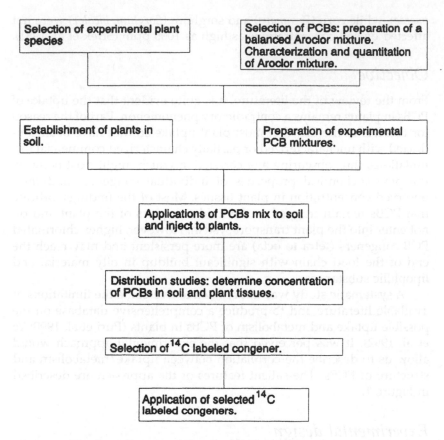

Figure 1   The flow diagram of the experimental approach.

ensure that it was free of organochlorine residues, the clean sand was
soxhlet extracted with *n*-hexane for 12 h, and the extracts were concentrated and injected into a gas chromatograph with an electron capture
detector.

Aliquots of clean, dry sand were spiked with the Aroclor mixture.
A total of 4.5 kg of sand containing 30 g each of Aroclors 1221, 1242,
and 1260 was obtained.

The soil for the study was obtained from the Sinclair Farm of the
University of Missouri. Approximately 1 ton of clay loam top soil (0 to
15 mm depth) was excavated and air dried at room temperature for 30
d. The physical characteristics of the soil are shown in Table 2. The air-
dried soil was sieved to remove stone and soil aggregates larger than
2 mm. A 60-kg portion was mixed with 1.5 kg of the PCB-spiked sand
in a rotary tumbler for 30 min. Ten-gram aliquots were taken during
the mixing and analyzed to determine the homogeneity of the contamination.

Table 2 Physical Characteristics of
the Soil

	Mean[a]
pH	6.50
Organic matter (%)	4.55
Sand (%)	21.32
Silt (%)	50.51
Clay (%)	28.20
Calcium (ppm)	185.25
Magnesium (ppm)	27.40
Potassium (ppm)	20.86
Zinc (ppm)	7.13
Iron (ppm)	70.49
Manganese (ppm)	65.64
Copper (ppm)	0.87
Sodium (ppm)	7.19

[a] Average of ten soil sample test results.

Establishment of plants in contaminated soil

Ceramic glazed pots (19 × 14 cm) were used. One variety of barley (Staptoe) and one of tomato (Early Red Rock) were used. The contaminated soil was placed in the pots containing a layer of clean gravel at the bottom. The soil was covered with a layer (3 cm) of vermiculite. Plant seeds were placed on the vermiculite and covered with a second layer (2 cm) of vermiculite. The pots were placed in a plant growth room under white artificial light, and the temperature was kept about 18°C. A 12/12 h light/dark cycle was maintained throughout the study. Nutrients and water were applied periodically to maintain optimal growing conditions. Control plants were grown in the same manner, except that clean, uncontaminated soil was used.

Direct introduction of PCBs into plants

Nonradiolabeled studies

To study the translocation of PCBs, a preparation of Aroclor mixture in dimethyl sulfoxide (DMSO) was injected into the tomato plants. These plants were grown in uncontaminated soil. After planting 10-cm tall seedlings, the pots were placed in an isolated plant growth room. After the plants had attained a height of approximately 20 cm, PCBs in DMSO were injected into the cortical region of the main lower stem 4 cm above the soil at 5 and 10 μg per plant. Control plants were injected with DMSO only.

Radiolabeled studies

To determine the degree of movement and metabolism of PCB congeners in the plant vascular system, radiolabeled PCB congeners were injected directly into the plant in a manner similar to the nonlabeled PCBs. Only two ^{14}C-labeled congeners were used: 3,3',4,4'-tetrachlorobiphenyl (IUPAC# 77) and 3,3',4,4',5,5'-hexachlorobiphenyl (IUPAC# 169). A total of 0.5 µCi was injected into the cortical region of the main stem of each plant, approximately 5 cm above the soil level; the ^{14}C-PCBs were dissolved in DMSO for those experiments. All experiments were performed in duplicate. These studies were carried out in 55 × 50 × 40 cm containment chambers (Figure 2) which were fabricated at the Science Instrument Shop, University of Missouri, Columbia. The front and back walls of the chamber were made of glass. Each chamber had one outlet and two inlets for air circulation. The outlets of the chambers were connected to serial traps which consisted of three 250-ml flasks: the first flask contained 150 ml of 1N NaOH, the second flask contained 150 ml of H_2O, and the third flask was used for collection of overflow from the second flask. The outlet of the third flask was connected to 15 × 0.63 cm o.d. copper tubing, which was packed with a 6-cm layer of activated carbon and a 6-cm layer of silica gel. The copper tubing was connected to a vacuum pump (Model PV-200, Bell & Gossett Inc., Chicago, IL), and an airflow rate of 80 cm^3/min was maintained throughout the duration of the experiment. The temperature of the chambers was maintained at 18°C.

Sampling

Plants were allowed to grow until fruits appeared on the plants. After this period (4 months), the plants were cut 5 cm above the soil level. The plants were separated, washed with distilled deionized water, and stored in a freezer. One-centimeter core samples of the soil were also taken at the same time. These were divided into 5-mm sections and stored in a refrigerator at 4°C.

Plants injected with nonlabeled PCBs were allowed to grow until fruiting (approximately 2 months after PCB injections). After this period, the plants were cut above the soil level. The plant stems were sectioned into four parts (Figure 3). Part a included 4 cm of stem just above the soil. The 2-cm section b, located just above section a, included the point of PCB introduction. Section c included 4 cm of stem above section b. Section d was 1 cm of stem above section c. These stem parts were wrapped in clean aluminum foil and stored in a freezer until the time of analysis, approximately 10 d.

A similar approach was implemented for plants injected with labeled PCBs. Those plants were allowed to grow for 35 d after injection. After this period, the plants were cut above the soil level. The plants were sectioned into three parts. A 4-cm section a included the

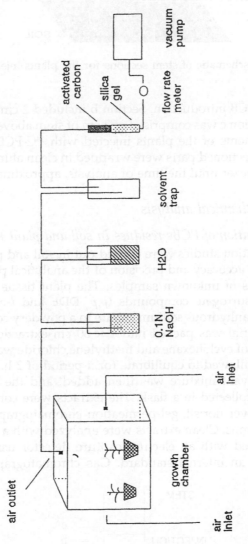

Figure 2 An overall scheme of the experimental setup for ^{14}C-PCB.

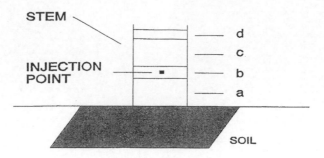

Figure 3 The schematic of stem sections for the plants injected with nonlabeled PCBs.

point of ^{14}C-PCB introduction. Section b included 2 cm of stem above section a. Section c was comprised of 2 cm of stem above section b. The sectioning scheme of the plants injected with ^{14}C-PCBs is shown in Figure 4. All sectioned parts were wrapped in clean aluminum foil and stored in a freezer until the time of analysis, approximately 10 d.

Methods of chemical analysis

Determination of PCBs residues in soil and plant tissues

Several validation studies were carried out in soil and plant tissues to determine the accuracy and precision of the analytical procedure prior to the analysis of unknown samples. The plant tissue samples were spiked with surrogate compounds (*p,p'* DDE and *t*-chlordane) and blended with anhydrous sodium sulfate to a powdery consistency. The blended material was packed into 4 × 60 cm extraction columns. A 50:50 mixture of cyclohexane and methylene chloride was added to the material and allowed to equilibrate for a period of 2 h. An additional aliquot of solvent mixture was then added, and the contents were drained and collected in a flask. The extracts were concentrated and fractionated over florisil, gel-permeation chromatography (GPC) and silica gel columns. Clean extracts were analyzed with a gas chromatograph equipped with an electron capture detector using pentachlorobenzene as an internal standard. Gas chromatograph separations

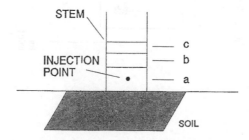

Figure 4 The schematic of stem sections for the plants injected with ^{14}C-PCBs.

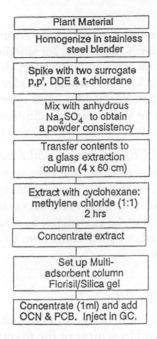

Figure 5 contents:

- Plant Material
- Homogenize in stainless steel blender
- Spike with two surrogate p,p′, DDE & t-chlordane
- Mix with anhydrous Na$_2$SO$_4$ to obtain a powder consistency
- Transfer contents to a glass extraction column (4 x 60 cm)
- Extract with cyclohexane: methylene chloride (1:1) 2 hrs
- Concentrate extract
- Set up Multi-adsorbent column Florisil/Silica gel
- Concentrate (1ml) and add OCN & PCB. Inject in GC.

Figure 5 Extraction and cleanup procedure for determination of the PCB concentration in plants.

were obtained with a 0.25 mm i.d. × 60 m fused silica capillary DB 5 column containing 95% methyl polysiloxane and 5% phenyl polysiloxane phase. Hydrogen was used as the carrier gas. The gas chromatograph oven was temperature programmed for 80 ~ 170°C at 10° per minute, followed by a slow ramp of 3° per minute to 280°. A flow scheme of the extraction and cleanup procedure is shown in Figure 5. The extraction and cleanup methods for both labeled and nonlabeled plant tissues and soil samples were similar to the one described earlier.

Extensive validation experiments and characterization of PCB congeners and Aroclor mixture were carried out with a two-dimensional reaction chromatographic system described by Duinker et al. (1988) and Duebelbeis et al. (1989). A standard, five-point calibration curve was run prior to analysis of unknown. The response calibration was made according to the procedure outlined in U.S. Environmental Protection Agency (EPA) Method 608 for chlorinated pesticides (Millar et al. 1984;Webb and McCall 1973). Individual PCB congeners for confirmation were obtained from the Canadian National Research Council, Halifax, N.S., Canada and Ultra Scientific, North Kingston, RI. The Aroclors used (1221, 1242, and 1260) were provided by the Department of Agricultural Chemistry, Oregon State University, Corvallis. The radiolabeled PCB congener 77 was obtained from Sigma Aldrich, St. Louis, MO, and congener 169 was acquired from Joint Research Center of European Communities, Ispra VA, Italy.

Determination of ^{14}C-PCBs

The extraction and cleanup scheme for determination of residual ^{14}C-labeled PCBs was similar to one used for unlabeled PCBs. The radioactivity in the clean extracts was monitored by diluting the extract with 10 ml of scintillation cocktail (a mixture of xylene, 2,5-diphenloxazale, polyoxyethylene octylphenol, *bis*-MSB, and oleic acid; Fisher Scientific Company), and counting with a liquid scintillation counter (Tri-Carb Liquid Scintillation Analyzer 1600CA, Packard Inc., Meriden, CT).

Determination of PCBs residues in soil cores

The analysis of the soil cores from the pots were carried out after the plants were harvested. At the end of the growth experiment (1 year), soil cores were taken from the pots and sliced into five sections comprising 0.5, 1.5, 2.5, 3.5, and 4.5 cm thickness, respectively. Each section was separately analyzed for PCB congeners to determine the loss of PCBs in the soil profile.

Results and discussion

Limited studies are available on the uptake and metabolism of PCBs. Among the available findings, most of the investigations indicate that PCBs do not translocate in the symplast or apoplast system of the plants. Higher-chlorinated PCB congeners limit their entry only up to the cuticle of the epidermal layers of the stems and leaves and do not enter into the cortical, phloem, or xylem portion of the plants. Findings on the plant uptake of mono- to trichlorinated congeners have been cited in the literature as discussed earlier, but lack evidence on their further translocation, transportation, and metabolism in the plants.

Studies in our laboratory have shown that measurable concentrations of PCBs can be detected in all parts of the plants grown in contaminated soil. In the tomato plants, concentrations were found to be highest in the leaves and lowest in the fruits (Table 3). These concentrations ranged from 8 ppb (ng/g) for 2,4',5-trichlorobiphenyl to 16 ppt (pg/g) for 2,2',3,3',4,4',5,6-octachlorobiphenyl and 2,2',3,3',4,5,5',6,6'-nonachlorobiphenyl. The prominent PCB congeners found in the plant tissues were 31, 44, 52, 60, 103, 101, 86, 87, 154, 153, 151, 138, 159, and 170. The concentration of PCBs was higher in leaves when compared to that of stem and fruits, as shown in Table 3. Chromatographic profiles of the PCB residue in exposed barley leaves and stems are shown in Figures 6 and 7. The concentration of PCB congeners in barley leaves ranged from 18 (ng/g) ppb for 2,4',5-trichlorobiphenyl to 99 ppt (pg/g) for 2,2',3,3',4,4',5,6-octachlorobiphenyl. The prominent PCB congeners found were 18, 31, 52, 44, 60, 101, 86, 87, 154, 153, 141, 138, 180, and 201, as shown in Table 4. The concentrations in the stems were lower than the concentrations in the leaves. The concentration of PCBs in plant tissues was found to be directly proportioned to the surface area

Table 3 Concentrations of PCB Congeners in Tomato
Plants Grown in Contaminated Soil

PCB congener IUPAC no.	Concentration (ppt)			
	Stems	Leaves	Flowers	Fruits
28,31	1320	7850	4840	345
52	254	1740	979	—[a]
44	150	1070	619	47
40,103	1220	2000	1250	178
60	108	592	390	23
101	136	750	477	28
86	33	183	117	—
87	153	919	612	37
154	115	722	508	28
151	46	291	198	10
118	67	638	454	14
153	133	633	411	—
141	33	185	141	—
138	124	573	438	24
129	14	27	27	—
159	53	168	67	—
183	28	95	76	—
185	—	19	19	—
156	12	53	41	—
180	28	97	80	—
170	30	99	80	—
201	27	67	54	—
203,196	20	62	51	—
208,195	—	16	16	—
194	—	39	24	—

[a] —: The concentration was lower than the detection limit (5 ppt). Standard deviation varies from 0.01 to 0.018.

of the plant organs. The correlation between available surface and PCB concentration in barley and leaves and stems was 735 and 74.6 pg/cm², respectively; in tomato leaves and stems, it was 306 and 54 pg/cm², respectively. PCB congeners profile in both the species also showed some different congeners; congeners 40, 103, 170, 203, and 208 were not found in barley plants.

Analysis of control plants grown in the vicinity of contaminated plants also showed appreciable amounts of PCB residues (Table 5). The chromatographic profile of this contamination was essentially the same as that seen in the exposed plants. These results point toward a vapor or particulate-mediated transfer of PCBs. Further analyses revealed that the concentrations of the lower-chlorinated PCB congeners were higher in both exposed and control plants; an inverse correlation between the number of chlorine substitutions and the concentration

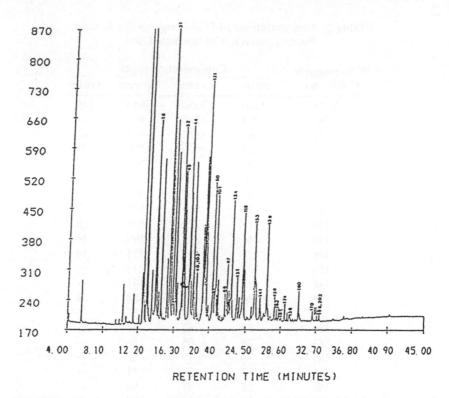

RETENTION TIME (MINUTES)

Figure 6 Chromatogram of PCB residues extracted from barley leaves of plants grown in contaminated soil.

was observed (Figure 8). Because the vapor pressure of PCB congeners decreases with the increased number of chlorine substitutions, a vapor-borne contamination should show a correlation between contaminant concentration and vapor pressure. Such a correlation was observed (Figure 9). A correlation coefficient of 0.991 was obtained. Indirect evidence for the lack of translocation of PCBs was obtained by analyzing the roots of the control plants; in these, no measurable PCB contamination was detected (Figure 10). Furthermore, the bark of the stem was removed, and the stem without the bark was analyzed. No detectable amount of congeners was found in the stem, whereas the analysis of the bark showed the same pattern of contamination. These results strongly indicate that there is no significant translocation of PCBs in plants and the contamination was restricted to bark only.

The results of stem section (tomato plant) analysis are summarized in Figure 11. The analyses show that even after 55 d, ≥95% of the total PCBs introduced could be accounted for within 6 cm of the point of introduction. Approximately 70% could be accounted for within the first 2 cm. The analysis of PCB residues in the different stem sections revealed an interesting trend. The PCB residue profile obtained in

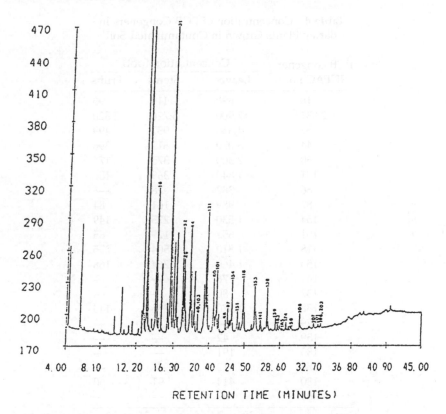

RETENTION TIME (MINUTES)

Figure 7 Chromatogram of PCB residues extracted from barley stems.

section b (the section where PCBs were introduced) resembled the profile of the original PCB mixture, but the PCB patterns in sections a and c—e above and below section b were quite distinct. The lower section contained a higher concentration of lower-chlorinated congeners. The upward mobility of PCB congeners can be related to the K_{ow} values of these molecules, and PCBs with a lower number of chlorines would be expected to move more readily. However, beyond this point, PCBs were not detected in the stem.

A similar trend was observed in experiments with labeled PCBs. The activity was largely confined within 4 cm of the point of introduction (Figure 12). Mobility of congener 77 was higher than the mobility of congener 169 (Figure 13). Greater than 95% of the activity remained in the form of the introduced congeners, showing little or no metabolism of these congeners in the plant system.

To observe the loss due to volatilization of PCB congeners, the top of the contaminated soil was covered with vermiculite. The analytical results showed that the vermiculite contained both lower- and higher-chlorinated PCBs. The chromatogram of PCB residue in vermiculite is shown in Figure 14. This indicated that the transfer of PCBs was due

Table 4 Concentration of PCB Congeners in
Barley Plants Grown in Contaminated Soil

| PCB congener | Concentration (ppt) | | |
IUPAC no.	Leaves	Stems	Fruits
18	690	119	96
28,31	17,900	5,720	2,820
52	4,350	932	499
44	3,070	612	396
60	2,300	375	174
101	1,840	369	186
86	382	78	—[a]
87	989	161	84
154	1,530	275	149
151	652	129	65
118	1,810	504	275
153	1,460	264	168
141	276	—	—
137	80	—	—
138	1,370	236	140
129	64	—	—
159	42	—	—
183	191	—	—
156	12	—	—
180	414	91	50
201	99	—	—

[a] —: The concentration is lower than detection limit (5
ppt). Standard deviation varies from 0.01 to 0.15.

to both volatilization and percolation from soil to vermiculite. The
analysis of soil and soil core before and after plants were grown showed
that there were significant losses (70%) from the surface soil by volatilization since photodegradation was prevented by vermiculite layer
which covered the top of the contaminated soil. Aerobic microbial
degradation of the lower-chlorinated biphenyls occurs at a faster rate
than the higher-chlorinated biphenyls (Hankin and Sawhney 1984;
Furukawa 1986), so the loss of PCBs at the surface soil could be partially
due to the microbial activities. The loss from the subsurface was com-
paratively less than that from the surface and that too could be attributed to microbial activities. Losses from 0 to approximately 70% were
found during 1 year after application. The chromatograms of the contaminated soil are shown in Figure 15. The concentrations of PCB
congeners in soil and soil cores are given in Table 6.

The percent loss of the top section was 15% larger than that of the
other four sections in Figure 15. For example, the relationship between
the depth of soil and percent loss of 31, 49, 52, 101, 103, 151, and 183

Table 5 Concentration of PCB Congeners in Control Barley and Tomato Plants Placed in the Vicinity of Contaminated Soil

PCB congener IUPAC no.	Concentration in barley (ppt)		Concentration in tomato (ppt)	
	Leaves	Stems	Leaves	Stems
18	461	70	—[a]	—
28,31	4350	985	2900	564
52	939	239	431	—
44	645	181	539	76
60	366	54	239	—
101	504	119	388	60
86	238	—	80	—
87	219	168	32	—
154	602	112	384	65
151	135	20	100	—
118	544	148	86	74
153	581	106	296	9
141	125	12	64	—
137	21	12	—	—
138	718	118	361	58
129	28	19	—	—
159	16	5	12	—
183	76	—	45	—
185	16	—	—	—
156	9	—	9	—
180	305	28	139	—
191	129	15	—	—
170,190	100	—	65	—
201	74	—	31	—

[a] —: The concentration is lower than the detection limit (5 ppt). Standard deviation varies from 0.01 to 0.16.

PCB congeners is shown in Figure 16. The loss is inversely proportional to the degree of chlorination in the biphenyl ring.

The results obtained clearly demonstrate a lack of active transport or metabolism of PCBs in tomato and barley. However, a good correlation between the vapor pressure of PCBs and their concentration in plant tissues was observed. The accumulation of PCBs in these plants was attributed to vapor absorption.

Conclusions

The results from the present investigation indicate that (1) measurable PCB concentrations were found in all plant parts; (2) the largest concentration was found in leaves; (3) control plants grown in the vicinity of contaminated plants contained detectable PCB residues, suggesting

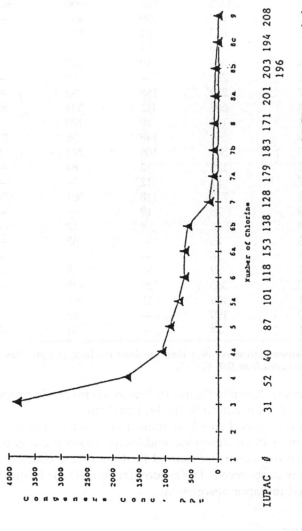

Figure 8 Graph showing correlation between the concentration of PCB congeners and the number of chlorine substitutions (tomato leaves).

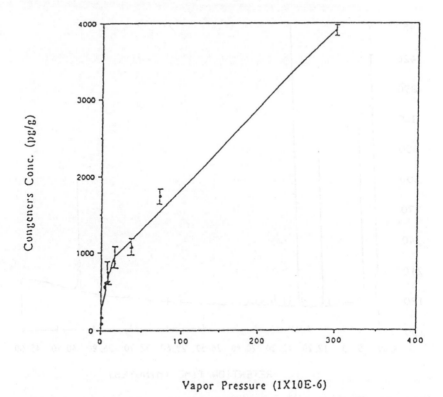

Figure 9 Graph showing correlation between the concentration of PCB congeners and their vapor pressure (tomato leaves).

that vapor absorption is an important process; (4) little translocation of PCBs occurs in stems of plants injected with radiolabeled PCBs; (5) the concentrations of PCB congeners in plants vary inversely with the number of chlorine substitutions in the congeners; (6) PCB uptake and metabolism did not differ between monocotyledon and dicotyledon plants; (7) greater than 95% of the PCBs remained in the epidermis of stem and leaves, showing little or no metabolism of these congeners in the plant system; and (8) analyses of soil core samples showed that the PCB loss was confined to the top 5 mm of the soil.

In view of this, it is concluded that plants uptake PCBs to a very limited extent from their environment and do not modify the extractable congeners appreciably. Higher-chlorinated PCBs are stored in the lipid contents of the plant parts, i.e., in the cuticles of the leaves, stems, or roots, and do not translocate through the symplast and apoplast systems of the plants. Plant uptake of airborne chemicals using plant cuticle–water partitioning and molecular connectivity has been shown by Sabijic et al. (1990). Plants act as chemical sensors to detect the presence of pollutants in the environment and are a salutary source for trapping lipophilic xenobiotics from the environment. Plants can be

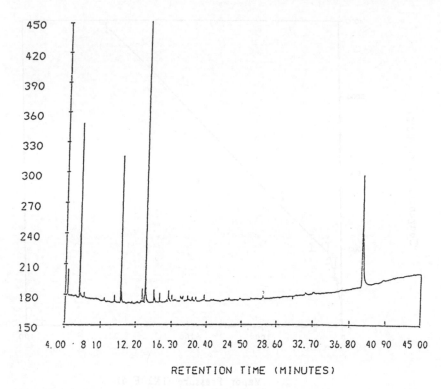

RETENTION TIME (MINUTES)

Figure 10 Chromatogram of root extract. Plants were grown in uncontaminated soil.

used as models to assess the fate and global distribution of contaminants in the food chain.

Acknowledgments

This study was supported in part by a grant from the U.S. Environmental Protection Agency, Environmental Research Laboratory, Corvallis, OR. However, this report has not been subjected to the agency's peer and administrative review and therefore may not necessarily reflect the views of the agency, and no official endorsement should be inferred. Thanks are due to Crystal Frey of ETSRC for her continuous encouragement and suggestions.

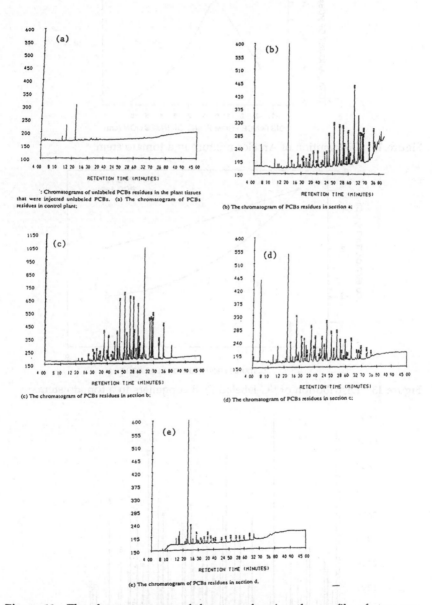

: Chromatograms of unlabeled PCBs residues in the plant tissues that were injected unlabeled PCBs. (a) The chromatogram of PCBs residues in control plant;

(b) The chromatogram of PCBs residues in section a;

(c) The chromatogram of PCBs residues in section b;

(d) The chromatogram of PCBs residues in section c;

(e) The chromatogram of PCBs residues in section d.

Figure 11 The chromatograms of the stem showing the profile of stem sections.

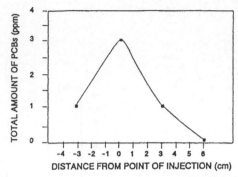

Figure 12   Distribution of Aroclor mixture in a tomato stem.

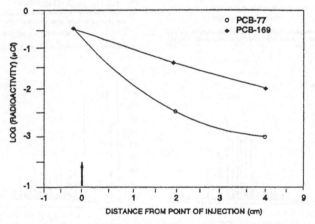

Figure 13   Distribution of ^{14}C-labeled PCB congeners in a tomato stem.

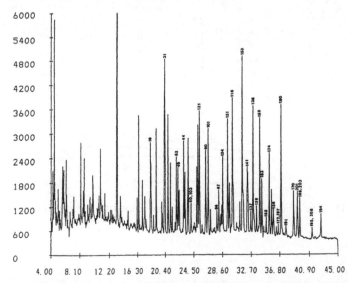

Figure 14   The chromatogram of PCB residues in vermiculite placed on top of contaminated soil.

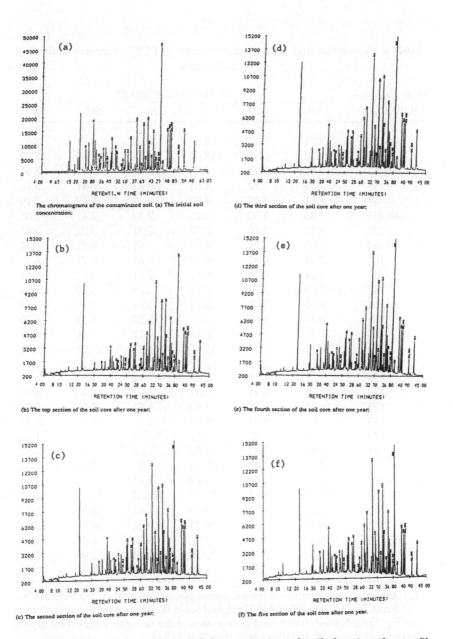

Figure 15 The chromatograms of the contaminated soil showing the profile of core sections.

Table 6   The Initial and Final Concentrations of PCB Congeners in the Soil
and Soil Cores

PCB congener	Concentration (ppm)					
IUPAC no.	Ic[a]	Top	Sec. 2	Sec. 3	Sec.4	Sec. 5
18	4.71	1.42	2.51	3.32	3.61	3.62
28,31	29.32	10.01	15.61	18.82	19.51	17.92
52	9.831	5.12	8.13	9.04	9.41	8.61
49	7.14	3.10	4.61	5.22	5.41	5.10
44	8.31	3.20	4.82	5.23	5.42	5.10
40,103	2.43	1.50	2.23	2.42	2.41	2.31
121	8.12	2.10	2.72	2.91	3.02	2.61
60	11.03	3.21	4.22	4.51	4.81	4.31
101	7.41	4.82	6.23	6.61	6.71	6.2
86	1.24	0.51	0.62	0.71	0.72	0.61
87	3.75	3.73	3.74	3.75	3.73	3.74
154	6.51	2.92	3.71	3.91	4.10	3.62
151	6.51	3.81	4.91	5.22	5.31	4.90
118	21.10	5.61	6.81	4.22	7.31	3.50
153	18.51	11.22	13.80	14.61	15.21	13.92
141	4.31	2.20	2.72	2.91	3.10	2.71
137	1.21	1.10	1.19	1.31	1.21	1.20
138	14.22	6.91	8.41	8.71	9.10	8.10
159	14.10	1.10	1.42	1.51	1.51	1.43
183	7.01	4.62	5.71	6.10	6.21	5.62
128	1.10	1.01	1.11	1.12	1.12	1.10
185	1.81	1.21	1.52	1.62	1.61	1.52
156	2.10	2.13	2.11	2.13	2.13	2.12
173,200	1.60	1.41	1.62	1.61	1.62	1.62
180	22.62	9.91	12.10	12.82	12.91	11.61
170	8.62	5.32	6.43	6.61	6.91	6.12
201	10.31	8.82	10.31	10.42	10.41	10.31
203,196	8.10	4.90	5.90	6.30	6.22	5.61
195,208	2.90	2.20	2.70	2.91	2.90	2.60
194	5.60	3.80	4.50	4.81	4.81	4.32

[a] Ic is the initial concentration of the soil before plants were grown; Top (0.5 cm), Sec.
2 (1.5 cm), Sec. 3 (2.5 cm), Sec. 4 (3.5 cm), and Sec. 5 (4.5 cm) are the concentrations of
the top, second, third, fourth, and fifth section of the soil after plants were grown,
respectively. Standard deviation varies from 0.01 to 0.14.

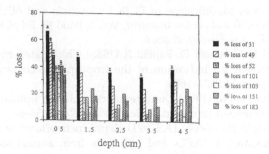

Figure 16 Relationship between the depth of soil and percent loss of PCB congeners.

References

Babish JG, Stoewsand GS, Furr AK, Parkinson TF, Bache CA, Gutenmann WH, Wszolek PC, Lisk DJ (1979) Elemental and polychlorinated biphenyl content of tissues and intestinal aryl hydrocarbon hydroxylase activity of guinea pigs fed cabbage grown on municipal sludge. *J Agric Food Chem* 27:399–402.

Beall ML, Nash RG (1971) Organochlorine insecticide residue in soybean plant tops: root uptake vs. vapor sorption. *Agron J* 63:460–464.

Buckley EH (1982) Accumulation of airborne polychlorinated biphenyls in foliage. *Science* 216:520–522.

Buckley EH (1987) PCBs in the atmosphere and their accumulation in foliage and crops. In *Recent Advances in Phytochem (Phytochemical Effect of Environmental Compounds)*, Vol. 21, Saunders JA, Kosa K, Channing L, Conn EE, Eds., Plenum Publishing, New York, pp. 175–201.

Bush B, Shane LA, Wilson LR, Barnard EL, Barnes D (1986a) Uptake of polychlorinated congeners by purple loosestrife (*Lythrum salicaria*) on the banks of the Hudson River. *Arch Environ Contam Toxicol* 15:285–290.

Bush B, Benett AH, Snow JT (1986b) Polychlorobiphenyl congeners, p,p'-DDE, and sperm function in humans. *Arch Environ Contam Toxicol* 15:333–341.

Cairns T, Doose GM, Froberg JE, Jacobson RA, Siegmund EG (1986) Analytical chemistry of PCBs. In *PCBs and The Environment*, Vol. 1, Waid JS, Ed., CRC Press, Boca Raton, FL, pp. 1–45.

Duebelbeis DO, Kapila S, Clevenger T, Yanders AF, Manahan SE (1989) A two-dimensional reaction gas chromatographic system for isomer-specific determination of polychlorinated biphenyls. *Chemosphere* 18:101–108.

Duinker JC, Schulz DE, Petrick G (1988) Multidimensional gas chromatography with electron capture detection for the determination of toxic congeners in polychlorinated biphenyl mixtures. *Anal Chem* 60:478–482.

Erickson MD (1986) *Analytical Chemistry of PCBs*, Butterworth Publishers, Boston, MA.

Fries GF, Marrow GS (1981) Chlorobiphenyl movement from soil to soybean plants. *J Agric Food Chem* 29:757–759.

Furukawa K (1986) Modification of PCBs by bacteria and other microorganisms. In *PCBs and the Environment*, Vol. 2, Waid JS, Ed., CRC Press, Boca Raton, FL, pp. 89–180, chap. 6.

Gaggi C, Bacci E, Calamari D, Fanelli R (1985) Chlorinated hydrocarbons in plant foliage: an indication of the tropospheric contamination level. *Chemosphere* 14:1673–1686.

Gan DR, Berthouex PM (1994) Disappearance and crop uptake of PCBs from sludge-amended farmland. *Water Environ Res* 66:54–69.

Golub MS, Donald JM, Reyes JA (1991) Reproductive toxicity of commercial PCB mixtures: LOAELs and NOAELs from animal studies. *Environ Health Perspect* 94:245–253.

Gosselin B, Naylor LM, Mondy NI (1986) Uptake of PCBs by potatoes grown on sludge-amended soils. *Am Potato J* 63:563–565.

Goto M, Sugiura K, Hattori M, Miyagawa T, Okamura M (1974) Metabolism of 2,3,5,6-tetrachlorobiphenyl-^{14}C and 2,3,4,5,6-pentachlorobiphenyl-^{14}C in the rat. *Chemosphere* 3:233–238.

Greb W, Klein W, Coulston F, Golberg L, Korte F (1975) Beiträge Lur, Ökologischen Chemie LXXXIV: Metabolism of Lower Polychlorinated Biphenyl-14C in the Rhesus Monkeys. *Bull Environ Contam Toxicol* 13:471–476.

Hankin L, Sawhney BL (1984) Microbial degradation of PCBs in soil. *Soil Sci* 137:401.

Heck WW, Heagle AS, Cowling EB (1977) Air pollution: impact on plants. In Proc. 32nd Meeting Soil Conservation Soc. of Am., pp. 193-203.

Hooper SW, Pettigrew CA, Sayler GS (1990) Ecological fate, effects and prospects for the elimination of environmental polychlorinated biphenyls (PCBs). *Environ Toxicol Chem* 9:655–667.

Hutzinger O, Jamieson WD, Safe S, Paulmann L, Ammon R (1974) Identification of metabolic dechlorination of highly chlorinated biphenyl in rabbit. *Nature (London)* 252:698–699.

Hutzinger O, Safe S, Litko V (1992) *The Chemistry of PCBs*, Lewis Publishers, Chelsea, MI.

Iwata U, Gunther FA (1976) Translocation of the polychlorinated biphenyl Aroclor 1254 from soil into carrots under field conditions. *Arch Environ Contam Toxicol* 4:44–59.

Jacobson JL, Humphrey HEB, Jacobson SW, Schantz SL, Mullin MD, Welch R (1989) Determinants of polychlorinated biphenyl (PCBs), polybrominated biphenyl (PBBs) and dichlorodiphenyl trichloroethane (DDT) levels in the sera of young children. *Am J Public Health* 79:1401–1404.

Jensen S (1966) Report of a new chemical hazard. *New Sci* 32:612.

Klein W, Weisberger I (1976) PCBs and environmental contamination. *Environ Qual Saf* 5:237.

Kylin H, Grimvall E, Östman C (1994) Environmental monitoring of polychlorinated biphenyls using pine needles as passive samplers. *Environ Sci Technol* 28:1320–1324.

Lawrence J, Tosine HM (1977) Polychlorinated biphenyl concentrations in sewage and sludges of some waste treatment plants in Southern Ontario. *Bull Environ Contam Toxicol* 17:49–56.

Mahanty HK, Fineran BA (1976) Effects of a polychlorinated biphenyl (Aroclor 1242) on ultrastructure of frond cells in aquatic plant *Spirodela oligorrhiza*. *NZ J Bot* 14:13.

McFarland VA, Clarke JU (1989) Environmental occurrence, abundance and potential toxicity of polychlorinated biphenyl congeners: considerations for a congener-specific analysis. *Environ Health Perspect* 81:225–239.

Millar JD, Thomas RE, Shattenber HJ (1984) *EPA Method Study 18, Method 608: Organochlorine Pesticides and PCBs,* EPA-600/4-84-061, U.S. Environmental Protection Agency/EMSL, Research Triangle Park, NC.

Mosser JL, Fisher NS, Teng TC, Wurster CF (1972) Polychlorinated biphenyl: toxicity to certain phytoplankters. *Science* 175:191–192.

Moza P, Weisgerber I, Klein W, Korte F (1973) Distribution and metabolism of carbon-14-labeled 2,2'-dichlorobiphenyl in the higher marsh plant *Veronica beccabunga. Chemosphere* 5:217–222.

Moza P, Weisgerber I, Klein W (1976) Fate of 2,2'-dichlorobiphenyl-¹⁴C in carrots, sugar beets and soil under outdoor conditions. *J Agric Food Chem* 24:881–885.

Moza P, Scheunert I, Klein W, Korte F (1979) Studies of 2,4',5-trichlorobiphenyl-¹⁴C and 2,2',4,4',6 pentachlorobiphenyl-¹⁴C in carrots, sugar beets and soil. *J Agric Food Chem* 27:1120–1124.

Nadeau RJ, Allen HL, Prince GR (1982) Hazard Assessment and Criteria Development Methodology Applied at PCB Incidents, Environ. Response Team, U.S. EPA, Edison, NJ, Hazar. Mater. Spills Conf. Proc., pp. 438–449, Ludwigson J, Ed., Govt. Inst., Inc., Rockville, MD; *Chem Abstr* 97:60243.

Nakanishi H, Shiyouichi N, Masao T (1975) Translocation of polychlorinated biphenyl (PCBs) in rice plant. *Shiga-Kenritsu Eisei Kenkyusho Ho* 11:38–40; *Chem Abstr* 91:187516S.

O'Connor GA, Kiehl D, Eiceman GA, Ryan JA (1990) Plant uptake of sludge-borne PCBs. *J Environ Qual* 19:113–118.

Ogiso M, Toyota I, Ido Y, Inagaki I (1975) Behavior of ¹⁴C-PCB in flooded soils. *Aichi-ken Nogyo, Shikengo Kenkyo, Hokoku* 7:77–81; *Chem Abstr* 85:117537.

Ormrod DP, Adedipe NO, Ballantyne DJ (1976) Air pollution injury to horticultural plants: a review. *Hortic Abstr* 46:241–246.

Pal D, Weber JB, Overcash MR (1980) Fate of polychlorinated biphenyls (PCBs) in soil-plant systems. *Residue Rev* 74:45–98.

Puri RK, Kapila S, Lower WR, Yanders AF (1990) Uptake of PCBs in *Hordeum vulgare* (barley) and *Lycopersicon esculentum* (tomato). Presented at Dioxin '90 Conference, Bayreuth/F.R.G., Germany.

Qiuping Y, Puri RK, Kapila S, Yanders AF (1992) Studies on the transport and transformation of PCBs in plants. *Chemosphere* 9:182–191.

Reinert RA (1975) Monitoring, detecting and effects of air pollutants on horticulture crops: sensitivity of genera and species. *Hortic Sci* 10:7–12.

Risebrough RW, Ricche P, Peakall DB, Herman SG, Kirven MN (1966) Polychlorinated biphenyls in the global ecosystem. *Nature* 220:1098.

Sabijic A, Güsten H, Schönherr J Riederer M (1990) Modeling plant uptake of airborne organic chemicals. I. Plant cuticle/water partitioning and molecular connectivity. *Environ Sci Technol* 24:1321–1326.

Safe S (1990) Polychlorinated biphenyl (PCBs), dibenzo-p-dioxins (PCDDs), dibenzofurans (PCDFs) and related compounds: environmental and mechanistic considerations which support the development of toxic equivalency factors (TEFs). *Crit Rev Toxicol* 22:51–88.

Safe S, Hutzinger O, Ecobichon D (1974) Identification of 4-chloro-4' hydroxybiphenyl and 4-4'-dichloro-3-hydroxybiphenyl as metabolites of 4-chloro and 4,4' dichlorobiphenyl fed to rats. *Experentia* 30:720–721.

Safe S, Platonow N, Hutzinger O (1975) Metabolism of chlorobiphenyls in the goat and cow. *J Agric Food Chem* 23:259–261.

Safe S, Safe L, Mullin M (1985) Polychlorinated biphenyls: congener-specific analysis of a commercial mixture and a human milk extract. *J Agric Food Chem* 33:24–28.

Sawhney BL (1986) Chemistry and properties of PCBs in relation to environmental effects. In *PCBs and the Environment*, Vol. 1, Waid JS, Ed., CRC Press, Boca Raton, FL, pp. 47–61.

Sawhney BL, Hankin L (1984) Plant contamination by PCBs from amended soils. *J Food Protec* 47:232–236.

Sheppard MI, Sheppard SC, Amiro BD (1991) Mobility and plant uptake of inorganic carbon-14 and C-14 labeled PCB in soils of high and low retention. *Health Phys* 61:481–492.

Sinclair J, Garland S, Arnason T, Hope P, Granville M (1977) Polychlorinated biphenyls and their effects on photosynthesis and respiration. *Can J Bot* 55:2679–2684.

Strek HJ, Weber JB, Shea PJ, Mrozek E, Overcash MR (1981) Reduction of polychlorinated biphenyl toxicity and uptake of carbon-14 activity by plants through the use of activated carbon. *J Agric Food Chem* 29:288–293.

Sundström G, Jansson B (1976) The metabolism of chlorobiphenyls — a review. *Chemosphere* 5:267–298.

Suzuki M, Aizawa N, Okano G, Takahashi T (1977) Translocation of polychlorobiphenyls in soil into plants: a study by a method of culture of soybean sprouts. *Arch Environ Contam Toxicol* 5:343–352.

Tilson HA, Jacobson JL, Rogan WJ (1990) Polychlorinated biphenyls and the developing nervous system: cross-species comparison. *Neurol Toxicol* 12:239–248.

Urey JC, Kricher JC, Boylan JM (1976) Bioconcentration of four pure PCB isomers by *Chlorella pyrenoidosa*. *Bull Environ Contam Toxicol* 16:81–85.

Wallnöfer PR, Engelhardt G, Safe S, Hutzinger O (1973) Microbial hydroxylation of 4-chlorobiphenyl and 4,4'-dichlorobipheyl. *Chemosphere* 2:69–72.

Wallnöfer PR, Koniger M, Engelhardt G (1975) The behavior of xenobiotic chlorinated hydrocarbon (HCB & PCBs) in cultivated plants and soil. *Z Pflanzenkr Pflanzenschutz* 82:91.

Walsh GE, Hollister TA, Forester J (1974) Translocation of four organochlorine compounds by red mangrove (*Rhizophora mangle* L.) seedlings. *Bull Environ Contam Toxicol* 12:129–134.

Webb RG, McCall AC (1973) Quantitative PCB standards for electron capture gas chromatography. *J Chromatogr Sci* 11:366.

Weber JB, Mrozek E Jr. (1979) Polychlorinated biphenyls: phytotoxicity, absorption and translocation by plants and inactivation by activated carbon. *Bull Environ Contam Toxicol* 23:412–417.

Wegmann MA, Daniel RC, Hani H, Iannone A (1987) Toxic organic substances in sewage sludges: a case study of soil plant transfer. *Toxicol Environ Chem* 14:287–296.

Winston SG, Gerstner HB (1978) *Polychlorinated Biphenyl, Polybrominated Biphenyls and Their Contamination — A Literature Compilation, 1965–1977*, ORNL/TIRC-78/2, National Library of Medicine, Washington, D.C.

chapter seventeen

Selection of phytotoxicity tests for use in ecological risk assessments

Lawrence A. Kapustka

Background

Ecological risk assessments (EcoRAs) evaluate the potential for adverse ecological effects of agents to occur. Though most societal interests lie with protection of humans and wildlife, plants are important dietary sources of some contaminants. Plants are also critical components of wildlife habitat. As the base source of organic energy for ecological systems, plants are essential for all life. The role of plants in soil development, stabilization, and nutrient cycling are also important.

Most human health and wildlife risk assessments rely primarily on toxicity and exposure data from laboratory studies. Relatively few plant species have been used to generate these data. As a consequence, many questions arise regarding relevance to ecological situations. Similar questions confront risk assessors regarding phytotoxicity. What is the ecological relevance of the standard test species? What is the basis for measuring different endpoints? Are different endpoints just measuring the same phenomenon? How does one interpret the results, especially if data on multiple endpoints were collected? None of these are new questions. Indeed, since Galileo established *inference* as the fundamental basis of scientific study (Gingerich 1980), similar queries have defined experimental biology.

Environmental toxicology developed during the late 1960s and early 1970s to meet new regulatory demands. Bioassays, however, had been used effectively for decades to determine nutritional requirements, pesticide efficacy, and disease responses. Agricultural experiments have been conducted in fields and laboratories since the late 1800s. The specialization of environmental toxicology merely grew from the basic (physiology and ecology) and applied (agronomy and horticulture) sciences.

Research on the effects of metals on plants has a long history. Work in the last century (Liebig 1840) and through the most of this century (Epstein 1972) focused on determining plant requirements for optimum growth of crops. This involved determining the list of essential elements, the form of uptake in plants, threshold levels for sufficiency, and toxic levels. In the 1960s, ecological interests turned to nutrient cycling dynamics and pollution concerns (Brown et al. 1980; Edmonds 1982; Howell et al. 1975; Risser et al. 1981). In the late 1960s and 1970s, a large effort was undertaken to evaluate the effects of municipal sewage sludge on crops, pastures, and other managed lands (Page et al. 1983; U.S. EPA 1985a–e). As the concepts of bioaccumulation and biomagnification were being developed to describe ecological consequences of isotopes from nuclear testing and DDT, a substantial effort was directed toward evaluating food chain transfers of metals from soil, sediment, or water into plants, and, from there, tracking movement into the various trophic levels (Woodwell 1967; ICF 1989; Page et al. 1983). Given the varied nature of the objectives for these studies, it is not surprising that study designs, measurements, and reporting of critical parameters lack consistency. Though the literature has thousands of technically valuable references containing several hundred thousand data records, the large number of environmental variables (soil type, parent material, pH, organic matter, plant species, associated microbial species, etc.) continue to restrict our ability to synthesize and interpret the full range of information.

The U.S. Environmental Protection Agency (EPA) sponsored studies to evaluate phytotoxicity and food chain transfer potential of metals in sludge (Page et al. 1983; U.S. EPA 1985a–e), and others have described potential ecological effects of metals in soils (ICF 1989). Review articles and books have condensed the vast quantity of information into manageable units (Alloway 1974; Kabata-Pendias and Pendias 1992). None of the efforts has synthesized the data adequately for site-specific EcoRAs or generalized hazard rankings. The vast number of critical variables that influence uptake and phytotoxicity of metals correlate poorly with environmental concentration, though the data most often available to the risk assessor are analyte concentrations.

Approach to phytotoxicity

Hazard evaluation of contaminated environmental samples, toxic chemicals, and pesticides has relied on a few standardized tests (U.S. EPA 1991; Holst 1986a–e). The linkage between inherent hazard, based on limited laboratory determinations of toxicity, and ecological risk often poses unacceptable levels of uncertainty. The tools used to acquire the primary data for EcoRAs (U.S. EPA 1992) come from classical sampling methods and standardized laboratory toxicity tests (Kapustka and Reporter 1993). The integration of toxicity and exposure is constrained by the power (or lack thereof) of the primary data (Kapustka 1996; Kapustka et al. 1995). The principal deficiencies of standardized tests *vis-à-vis* EcoRAs relate to exposure conditions, relevant endpoints, interspecies extrapolation, and lab-to-field extrapolation. Fundamental changes that broaden the types and scope of "standardized" tests and expand the suite of measurement endpoints are needed. The next generation of standardized tests must be developed in light of risk assessment requirements and expectations if they are to be effective.

Short-term plant toxicity tests originally were developed from simple measurements used in plant physiology and weed science (Cushman and Meyer 1990). The tests have been adopted to evaluate single chemical and mixed chemical effects. More recently, they have been used to evaluate soil contamination. They are used to test soils brought to the laboratory for ecological assessment of terrestrial waste (Garten and Frank 1984; Linder et al. 1990; Parkhurst et al. 1989; Lee et al. 1982), as well as for comparative hazard ranking among chemicals.

The seed germination assay, often promoted as testing a sensitive and critical stage in the plant lifecycle, is relatively insensitive to many toxic substances. The insensitivity results from two factors: (1) many chemicals may not be taken into the seed, and (2) the embryonic plant derives its nutritional requirements internally from the seed storage materials and is effectively isolated from the environment. Finally, from an ecological perspective, seed germination is relatively unimportant for perennial plant species. Even for nondomesticated annuals, extremely low percentages of seed germination are typical (Baskin and Baskin 1985).

Nevertheless, EcoRAs rely extensively on standard toxicity data. In simplified form, the EcoRA considers the inherent toxicity of individual chemicals on ecological resources, evaluates the likelihood of exposure applicable to specified pathways in site-specific situations or proscribed scenarios, and combines the toxicity and exposure information to quantify the risk to ecological resources. Under recent guidelines, the EcoRA focus has been expanded to consider nonchemical stressors (e.g., physical disturbance, temperature, drought, or herbivory).

Incorporating standard toxicity information into EcoRAs has been difficult. In an ideal situation, toxicity and exposure data would be available for the taxa and endpoints of concern. Typically, neither toxicity nor exposure data are available for the taxa or the endpoints of concern.

Standardized methods

The literature contains a mix of standard phytotoxicity laboratory tests, laboratory/greenhouse experiments, field experiments, and field surveys. Since the objectives of these methods vary, the information rarely can be applied directly in an EcoRA. Critical information might not be reported; alternatively, the information may be too detailed to readily allow comparison among different sources (see Fletcher et al. 1985; Royce et al. 1984). Expert judgment and computer-aided analysis often are required before data can be incorporated into EcoRAs. Tests discussed in this section are the seed germination test, the root elongation test, the early seedling growth assay, and the lifecycle bioassays.

Terrestrial plant tests

The most widely used acute phytotoxicity tests involving vascular plants are the seed germination test (a direct exposure method) and the root elongation test (typically performed with eluates). The seed germination test has been used extensively since standardized protocols were introduced (U.S. EPA 1985f; U.S. FDA 1987; Holst and Ellanger 1982). Presorted seed lots are exposed to test chemicals in a soil matrix. Site soil or test chemicals are mixed with control soils in a logarithmic series. Germination is measured 5 d after initiating the test. The effective concentration of the test soil to give a 50% decrease of seed germination is used for determination of the EC_{50}. This test is considered a direct soil toxicity test. Species commonly used are chosen to cover four to five types of plants. Alfalfa, beet, clover, corn, cucumber, lettuce, foxtail millet, mustard, oats, perennial ryegrass, pinto bean, soybean, sorghum, radish, and wheat have been reported most often. A modification of the seed germination tests was developed for field use (Nwosu et al. 1991). On-site containers were kept under a canopy and shaded from the sun and rain. Test performance was evaluated against companion laboratory tests. Biologically reasonable differences were obtained between field and laboratory protocols with cucumber, lettuce, and red clover, but not with wheat. The on-site version of the seed germination test requires special attention to ensure that quality control criteria are met. The principle advantage of the test is the reduction of shipment and handling effort and the accompanying costs.

The root elongation test was developed as an indirect toxicity test. Roots are exposed to water extracts and soluble test soil constituents potentially toxic to the growing roots. After incubation in a chamber controlled for temperature and moisture, root length is measured. The EC_{50} of the test group is calculated as the concentration of the extract that inhibits root length of test samples by half that of the control samples. Preference seems to have been given to lettuce as a test species (Kramer 1983; Krstich and Schwartz 1990; Weinstein et al. 1990; Linder et al. 1990). Other species that have different root morphology, development patterns, and carbon allocation patterns may provide more ecologically relevant data for risk assessments.

The early seedling growth assay overcomes deficiencies of the seed germination and root elongation tests. The duration of the test provides for exposure well into the autotrophic stage of plant development. Exposure occurs in soil (either synthetic soil mix or environmental sample), which provides a better approximation of field conditions than the root elongation test.

Lifecycle bioassays are used to assess sublethal responses of plants to toxic chemicals. Exposure may be either acute or chronic. The endpoints used to quantify the effects of toxic chemicals include morphological and phenological measurements which can be accomplished easily in a greenhouse, growth chamber, or field conditions. This system also allows examination of the roots for morphological impact. Two plant-related genera have been used in developing rapid lifecycle tests: *Arabidopsis* (Ratsch et al. 1986) and *Brassica* (Shimabuku et al. 1991). *Arabidopsis* is well characterized physiologically and genetically and is ideally suited for laboratory assays. Technical impediments arise from the prostrate growth habit and tiny seed size. The small seeds virtually preclude measures of any parameter involving seed counts (e.g., percentage germination, reproductive success). The rapid cycling _Brassica_ have been developed by the Crucifer Genetics Cooperative of the University of Wisconsin, Madison. This group of plants is gaining popularity as a model system, especially with molecular biologists and geneticists. The advantages of _Brassica_ compared to _Arabidopsis_ include their upright growing habit and large seed size. Relatively large variations in many growth parameters may limit the utility of some potential endpoints. However, the short lifecycle permits up to ten generations in a year. This offers a good opportunity to investigate nonlethal effects of considerable ecological import (e.g., reproductive potential). Legitimate questions regarding representativeness of these mustards as surrogates for other plants continue to slow acceptance of the lifecycle bioassays.

Aquatic macrophyte tests

Evaluation of wetland plants can be achieved best with aquatic macrophytes tests. Duckweed has been used to characterize single toxicant dose-response relationships (Walbridge 1977; Wang 1986a; Weinberger and Caux 1985). For some effluents, duckweed was more sensitive than daphnia or fish for determining effluent toxicity (Taraldsen and Norberg-King 1990). Bioassay endpoints include reduction of frond production, reduction of root length, biomass, ^{14}C uptake, total Kjeildahl nitrogen, and chlorophyll. Reduction of chlorophyll pigments can be more sensitive than frond production (Wang 1986b). The ease of culture and bioassay methods have favored the use of duckweed to evaluate contaminants in water and saturated soils.

Nevertheless, it is better to use rooted aquatic plants to evaluate sediment toxicity. _Hydrilla verticillata_ (hydrilla), a common aquatic angiosperm in the southeastern United States, is easy to culture and handle, is tolerant of a broad range of environmental conditions, and has a fast growth rate (Hinman 1989; Byl and Klaine 1990). The most reproducible and toxicant-related endpoints are new root growth and peroxides activity. Uptake and translocation of chemicals to shoots may be an important route of chemical mobility in the environment. Since it is an exotic species, however, it must not be used in the field for _in situ_ bioassays. Other plants that have similar growth and culture characteristics to hydrilla include _Elodea canadensis, Myriophyllum spicatum_, and _Potomogeton pectinatus_.

The Waterways Experiment Station of the U.S. Army Corps of Engineers has developed a method to quantify the uptake of heavy metals by marsh plants (APHA 1976; U.S. Army 1987; Folsom et al. 1981, 1988; Lee et al. 1978, 1982, 1983). The method was developed to evaluate the suitability of dredged material for disposal on uplands and for construction of wetlands. Sediments to be tested are homogenized, air dried, and placed in containers. Specimens of selected species are planted into the sediment and allowed to reach maximum standing stock (normal duration — 90 d) under favorable growth conditions in a greenhouse. The above-ground material is then harvested and analyzed for selected metals. The biomass of the harvested material is measured also. The test can be adapted to assess toxicity in a reducing environment by maintaining saturated conditions.

Terrestrial test species

Typically, the range of species incorporated in phytotoxicity tests is limited to agronomic plants. Of course, there have been many ecological studies over the years that relate plant performance or community development to soil metals, most notably those dealing with serpentine soils (Whittaker 1954; Kruckeberg 1954; Tadros 1957; McMillan 1956). However, summaries developed for phytotoxicity purposes seldom rely on these ecological investigations. The general *de facto* criteria which guided the development of tests emphasized availability and test performance features that favored agronomic species (Kapustka 1996). Therefore, relatively little has been done with what might be viewed as more ecologically relevant species. A synopsis of test species as presented in formal guidance and standards is presented in the following sections. Thirty-one plant taxa are explicitly identified in test guidelines and standard test procedures (Table 1). Many additional plant taxa were reported in phytotoxicity literature (Table 2).

Regulatory guidance

Under the Federal Insecticide Fungicide and Rodenticide Act (FIFRA) guidelines (Holst 1986a,b), ten species belonging to eight families are listed for toxicity testing (see Table 1). Plant testing under the Toxic Substance Control Act (TSCA), for premanufacturing notices (PMNs), has relied generally on algal assays, especially *Selenastrum capricornutum* growth. Benenati (1990) reported that only 17% (155 of 12,403) of PMNs had plant testing information submitted. Of these, only four used terrestrial plants (presumably lettuce) in an early seedling growth assay. One test was to evaluate uptake of a chemical, and one test used duckweed (*Lemna minor*). Proposals for routine use (seed germination/root elongation and early seedling growth) and less routine use (plant lifecycle) testing as part of TSCA Section 4 were withdrawn (Benenati 1990) from further evaluation for various political reasons. The U.S. Food and Drug Administration (U.S. FDA 1987) has relied on plant tests similar to those for FIFRA (see Table 1). International guidance (OECD 1984) relies on agronomic species, but has a broader selection of plants as compared to U.S. guidance.

Table 1 List of Plant Species Identified in Regulatory Documents and in Standard Test Procedures

Family	Species	Common name	FIFRA	TSCA	FDA	OECD	AWWA	ESG	ASTM
Chenopodiaceae	*Atriplex patula*	Seaside greens						☑	
Compositae	*Lactuca sativa*	Lettuce	☑		☑	☑	☑	☑	
Cruciferae	*Brassica alba*	Mustard				☑		☑	
	B. campestris var. chinensis	Chinese cabbage				☑		☑	
	B. napus	Rape				☑		☑	
	B. oleracea	Cabbage	☑	☑	☑			☑	
	B. rapa	Turnip				☑		☑	
	Lepidium sativum	Cress				☑		☑	
	Raphanus sativus	Radish				☑			
	Rorippa nasturtium-aquaticum	Watercress					☑		
Cucurbitaceae	*Cucumis sativus*	Cucumber	☑	☑	☑			☑	
Leguminosae	*Glycine max*	Soybean	☑	☑	☑	☑		☑	
	Phaseolus aureus	Mungbean						☑	
	P. vulgaris	Bean			☑			☑	
	Trifolium ornithopodioides	Fenugreek				☑		☑	
	T. pratense	Red clover				☑		☑	
	Vicia sativa	Vetch				☑		☑	

Family	Species	Common name
Liliaceae	*Allium cepa*	Onion
Nymphaeceae	*Nelumbo lutea*	American lotus
Poaceae	*Avena sativa*	Oat
	Echinochloa crusgalli	Japanese millet
	Leersia oryzoides	Rice cutgrass
	Lolium perenne	Perennial ryegrass
	Oryza sativa	Rice
	Sorghum bicolor	Sorghum
	Spartina alterniflora	Smooth cordgrass
	Triticum aestivum	Wheat
	Zea mays	Corn
	Zizania aquatica	Wild rice
Solonaceae	*Lycopersicon esculentum*	Tomato
Umbelliferae	*Daucus carota*	Carrot

Table 2   Partial Listing of Plant Taxa Studied for Toxicity Effects

Species	Common name	Ref.[a]	Species	Common name	Ref.[a]
Agrostis alba	Red top	2	*Lemna minor*	Duckweed	3
Agrostis sp.	Bentgrass	2	*Lespedeza* sp.	Lespedeza	2
Apocynum sp.	Milkweed	3	*Lolium perenne*	Perennial rye	2, 3
Arabidopsis thaliana	Mouse-ear-cress	3	*Lotus corniculatus*	Birdsfoot trefoil	2
Arachis hypogaea	Peanut	2	*Ludwigia natans*	Floating loosestrife	3
Avena sativa	Oats	1, 2, 3	*Lupinus* sp.	Lupine	2
Beta vulgaris	Beets	1, 2, 3	*Lycopersicon esculentum*	Tomato	1, 2
B. vulgaris	Chard	2	*Medacago sativa*	Alfalfa	1, 2, 3
B. vulgaris	Sugarbeet	1, 3	*Melilotus alba*	White sweet clover	2, 3
Brassica campestris	Kale	1, 2	*Melilotus officinale*	Yellow sweetclover	2
B. nigra	Mustard	1, 2, 3	*Musa paradisiaca*	Banana	1
B. oleracea	Broccoli	2	*Nicotiana tabaccum*	Tobacco	1
B. oleracea	Cauliflower	2	*Oryza sativa*	Rice	1
B. rapa	Turnip	2	*Panicum miliaceum*	Millet	3
Bromus	Smooth bromegrass	2	*P. virgatum*	Switchgrass	2
Bromus japonicus	Japanese bromegrass	2	*Phaseolus* sp.	Beans	1, 3
Cenchrus ciliaris	Buffelgrass	2	*P. vulgaris*	Pinto beans	3
Chrysanthemum sp.	Chrysanthemum	1	*Phleum pratense*	Timothy grass	1, 2
Citrus sinesnsis	Orange	1	*Pinus taida*	Loblolly pine	3
Cucumis sativa	Cucumber	2, 3	*Pistia statiotes*	Water lettuce	3

Species	Common name	
Cyperus esculentus	Yellow nutsedge	4
Dactylis glomerata	Orchardgrass	2
Daucas carota	Carrot	1, 2
Echinochloa crusgalli	Barnyard grass	5
Elodea densa	Elodea	3
Eragrostis curvula	Weeping lovegrass	2
Eragrostis lehmanniana	Lehman lovegrass	2
Erysimum capitatum	Wall flower	1
Fagopyrum esculentum	Buckwheat	1
Festuca arundinacea	Tall fescue	2, 3
F. pratensis	Meadow fescue	1
F. rubra	Red fescue	2
Fragaria sp.	Strawberry	1
Gladiolus sp.	Gladiolia	1
Glycine max	Soybean	2, 3
Gossypium	Cotton	1
Helianthus annuus	Sunflower	1
Hordeum vulgare	Barley	1, 3
Lactuca sativa	Lettuce	1, 2, 3
Pisum sativum	Pea	1
Poa pratense	Kentucky bluegrass	2
Raphanus sativus	Radish	2, 3
Rubus sp.	Raspberry	1
Setaria italica	Foxtail millet	3
Solanum tuberosum	Potato	1, 3
Sorghum bicolor	Sudangrass; sorghum	1, 2, 3
Spartina alterniflora	Cordgrass	5
Spinacia oleracea	Spinach	1, 2
Spirea alba	Meadowsweet	1
S. alba	Meadowsweet	1
Tagetes sp.	Marigold	2
Thalassia testidinum	Seagrass	3
Tradescantia paludosa	Spiderwort	3
Trifolium pratense	Clover	3
Triticum aestivum	Wheat	1, 3
Vicia faba	Broad bean	3
Vicia sp.	Vetch	2
Zea mays	Corn	1, 3

[a] 1 = ICF 1989; 2 = Page et al. 1983; 3 = Wang et al. 1990; 4 = Folsom and Price 1991; 5 = Walsh et al. 1991.

Guidance developed under the Comprehensive Environmental Response Compensation and Liability Act (CERCLA — also known as Superfund) offers limited direction with respect to plant testing. General methods recommended for the Remedial Investigation Baseline Risk Assessment portion of work listed by name only the seed germination and root elongation assays (Greene et al. 1988; Linder et al. 1990). Only lettuce (*Lactuca sativa*) is listed as the standard species of the test, although "other (taxa) can be used." The U.S. Department of Interior, in developing rules for Natural Resource Damage Assessment (CFR 1990a), referred to "economically important plant species."

Published standards

Relatively few standard methods on phytotoxicity have been published. The American Society for Testing and Materials (ASTM 1994) has only one terrestrial plant test standard (early seedling growth) at this time. The EPA has published test guidelines under FIFRA (Holst 1986a,b), TSCA (CFR 1990b), and CERCLA (Greene et al. 1988). The U.S. FDA (1987), American Public Health Association (APHA 1992), and OECD (1984) have published plant test standards and guidelines. The state of Washington Department of Ecology (WA-DOE) has produced a streamlined plant vigor test that uses lettuce growth as the endpoint (Norton and Stinson 1993).

General literature

Nearly a hundred plant taxa (Table 2) have been used routinely to study phytotoxicity. Fletcher et al. (1988) reported 1569 plant species from 682 genera in 147 families in the records included in an early version of PHYTOTOX. However, 42% of the records referred to only 20 species.

The formal phytotoxicity literature has remained limited to a narrow suite of taxa by virtue of self-perpetuating and restrictive recommendations that strongly encourage the use of "standard test species." Litigation also promotes continued use of the standard test species. Though valuable to build a reference base for comparative risk analysis of chemicals, reliance on so few taxa remains an impediment to developing more relevant EcoRAs. In the ASTM Standard for Early Seedling Growth (ASTM 1994), paragraph 1.4 explicitly authorizes modifying the test procedure to make the test more relevant. Even though the list of plant taxa includes 24 species from 8 families, the intent of the committee was to promote wider taxonomic selection as experience with other species became available. The list of species contained in the broader phytotoxicity literature certainly suggests room for expanding the range of test species where better ecological relevance warrants doing so.

Literature in ecology, horticulture, forestry, physiology, and tissue culture provides further evidence that many other plant species can be grown in controlled settings in ways that permit direct toxicological characterization. However, since the groups of people likely to examine the results may be unfamiliar with the growing requirements, phenology, and nominal development of the so-called nonstandard species, extra care and, therefore, extra costs are required for such tests. Justification for incurring the extra costs must be developed in the larger context of risk communications, risk management, and long-term benefits.

Endpoints

A measurement endpoint in phytotoxicity is an attribute of plants which can be scored either qualitatively or quantitatively. Measurement endpoints should have definable levels of precision and accuracy; relate to important physiological, morphological, or ecological features; and exhibit a graded response to one or more agents. These become the raw data used to evaluate the plant performance in a given test. Quantitative data generally are the most valuable; however, several attributes, especially signs of morbidity, can be scored subjectively and yield useful information.

Regulatory guidance and published standards

Standardized or regulatory test methods identify primary endpoints to be scored. Generally, these methods encourage collection of additional data such as observations of wilting, chlorosis, or other descriptions of plant health (Table 3).

General literature

Much of the primary phytotoxicity information appears in agronomic and physiology literature which preceded standardization of measurement endpoints (Table 4). It is unfortunate that the bulk of this information developed in research institutes has been ignored in the apparent effort to simplify standardized testing procedures.

Rationale

Improvements in the relevance of EcoRAs and hazard ranking could be realized if testing were expanded to nonstandard species and measurement of a wider array of toxicity endpoints. Below-ground measures of growth, as well as physiological and morphological biomarkers, have largely been ignored in the standard test procedures. If these nonstandard measurement endpoints were coupled with the more tra-

Table 3 Measurement Endpoints in Published Standards

Method	Primary endpoint(s)	Secondary endpoint(s)
Seed germination	Percentage germination	General observations
Root elongation	Length of longest root	General observations
Plant vigor	Shoot height Shoot mass	General observations
Early seedling growth	Percentage survival Shoot height Shoot mass Root length Root mass	General observations
Aquatic macrophytes	Percentage germination Percentage survival Seedling weight	General observations
WA-DOE	Percentage germination Shoot height Shoot mass	General observations
On-site seed germination	Percentage germination	General observations Shoot height Root length
Brassica lifecycle	Survival Growth Fruit set Seed set	General observations Shoot height, stem diameter, internode length, foliar length and width, branching morphology (number of axillary branches)
Arabidopsis lifecycle	Survival Growth Fruit set Seed set	General observations

ditional endpoints of germination, survival, and above-ground growth, the results would greatly enhance the utility of phytotoxicity data. Justifiably, there may be resistance from the regulated community against expanding the scope of regulatory testing. However, offering the opportunity for discretionary use of more relevant tests will lead to better information, wiser decisions, and long-term benefits.

Percentage germination

Germination is defined as the events associated with the reinitiation of embryo growth in a mature seed of higher plants (angiosperm or gymnosperm). Several excellent treatments of seed germination have been produced over the years (see Mayer and Poljakoff-Mayber [1989] for physiological characterization, Baskin and Baskin [1985] for ecological considerations, and various journals such as the _Journal of Seed Technology_ and the _United Nations Food and Agricultural Organization Seed_

Table 4 Partial List of Measurement Endpoints Reported from Nonstandardized Tests

Vegetative endpoints	Reproductive endpoints	Physiological biomarkers
Germination percentage	Fruit set	Chlorophyll content
Germination rate	Reproductive failure (no ears formed)	Chlorophyll fluorescence
"Marked damage"	Seed set	Chlorosis
Mortality	Yield	Peroxidase
"Poor condition"		Residue concentration
Premature death		
Reduced growth		
Reduced height		
Reduced vigor		
Root knotting		
Root weight		
Shoot weight		
Twig die-back		

Review). Many physiological processes, ranging from hormonal induction, derepression of genes, translation of preformed RNA, activation of proteins, cell expansion, and, in some species, cell division, occur after seeds have imbibed a sufficient quantity of water. Germination is considered complete when the elongating embryo penetrates the seed coat. As tests became standardized, the operational definition of germination was emergence of the embryo through the surface of the soil. As it is possible for seeds to germinate (i.e., elongate through the seed coat) but die before growing through the soil surface, some refer to these tests as seedling emergence tests.

The nature of plant seeds, except for some such as cottonwood (*Populus*) and willow (*Salix*), effectively insulates the embryo from environmental stressors. Seeds have evolved to afford protection to the embryo during periods of harsh (otherwise lethal) conditions, yet permit resumption of growth as conditions become favorable. These protective features in general make the seed germination test relatively insensitive to chemical insult. Nevertheless, to the extent that inhibition of germination/emergence has been demonstrated for many chemicals, the test has utility even though it is not particularly sensitive.

The ecological significance of seed germination varies for different species and for different ecological systems. Many perennials propagate vegetatively to the extent that germination is relatively insignificant once the species has colonized an area. Many nondomesticated species exhibit varying degrees and stages of seed dormancy that serve to limit the number of seeds that germinate at any given time. Multiple dormancy factors may provide short "windows of opportunity" for germination, with only a few percent of the viable seeds germinating.

Therefore, the ecological significance of germination must be interpreted in the context of the species of interest.

Percentage survival

The endpoint in a germination/emergence test, in most cases, is equivalent to the percentage germination. For longer tests, survival (or the complement, mortality) has greater meaning. Failure to germinate, at least in many plant species, may simply mean that the plants have not broken dormancy or have reverted to dormancy during the test period (a feature of considerable likelihood with nondomesticated species). As a species survival mechanism, this is not unfavorable. Once germination has occurred, the young plant cannot revert to dormancy. Exposure to toxic materials leading to death of the emerging embryo of test organisms indicates potentially serious adverse ecological consequences for similar species and similar exposures.

Growth

Growth can be defined in several different ways (e.g., changes in mass, volume, height, length, or cell number). In plants, new cells are formed in meristems and cambia. Meristems are located at the tips (apices) of stems and roots; in the internodal zones of grasses; and the margins, tips, or basal portions of leaves. As cell division occurs, the meristem is reconstituted and new tissues are formed, resulting in increased length of stems, increased length of roots, or expanse of leaves. Such growth is referred to as primary growth. Cambial cell division likewise reconstitutes the cambium plus adds cells that increase the girth of stems or roots.

Newly formed plant cells consist of a cell wall, cell membranes, and cytoplasm with organelles and a nucleus. Enlargement of the plant cell is due largely to expansion of the vacuole, which may occupy 90% or more of the volume of a mature cell. The vacuole consists mainly of water. During periods of rapid growth, cells increase in volume and wet weight with virtually no increase in organic matter (i.e., minimal increase in oven dry weight). Young plants can achieve relative growth rates of 20% or more during short periods under experimental conditions (Ingestad 1981, 1982).

Initially, the young plant derives its nutrient requirements from seed reserves. As growth proceeds and as roots develop, essential nutrients are acquired through the roots. In growth tests extending more than 2 weeks, adding nutrients to the soil medium may be warranted. However, if suitable reference soils are employed (and depending on the specific questions being addressed), it may be better to have smaller plants rather than compromise the interpretation of results through the addition of nutrients.

Shoot height.   An increase in shoot height is a convenient and relatively sensitive parameter of seedling growth. Various ways of measuring height are equally valid. These include the height of the canopy above the soil surface, the height of the central (primary) stem, and the "stretched" length of the tip of the highest reaching leaf. Regardless of the specific method of determining shoot height, the parameter is most sensitive during the early stages of seedling growth. As plants begin to approach reproductive maturing, the rate of growth slows so that differences among treatments may become less pronounced.

The ecological significance of shoot growth comes from two related factors. First, rapid early growth provides an initial advantage to the seedling with access to light. Second, as the stem grows, it can support more leaves or larger leaves that, in turn, enable greater photosynthetic capacity. Overall, this provides a better opportunity for increased primary production and all the attendant ecological consequences for ecological systems.

Shoot mass.   An increase in mass is an additional measure of growth. Most plant physiology studies report wet weight. Most plant ecology studies report oven dry weight. The different conventions undoubtedly arose from differences in experiences (lab/greenhouse plants of a common species exhibit fairly constant wet to dry weight ratios; field collected plants from many taxa in different seasons can exhibit wildly fluctuating ratios). Phytotoxicity literature often reports both wet weight and oven dry weight.

Mass is a better measure of growth for multiply branched shoots (e.g., many broadleaf species) because it incorporates all of the aboveground tissue, whereas height measures only the tallest part of the plant. Even for single-stemmed plants, mass may be preferable as a measure of growth. One clear example is that shoot height growth is limited in high light intensities, but stem thickness, lateral branching, and leaf growth are increased. Maximum information regarding shoot growth is obtained if both height and mass are measured.

Root length.   Root growth typically begins prior to shoot growth. In addition to anchoring the young seedling, the root begins to acquire water and nutrients needed to support the plant. Initially, root growth occurs with minimal branching in most species. This provides a relatively easy endpoint to measure. The root elongation assay capitalizes on this growth pattern to provide a rather sensitive measure of phytotoxic effects. The test method facilitates measurement by virtue of growing the root in a hydroponic or aeroponic condition. However, measurement of plant roots during the first week or two postemergence can be achieved with reasonable precision and repeatability even if the test plants are grown in soil. Heavy clays pose difficulties, especially if toxic materials are available. In such cases, roots may become weak and brittle, causing some difficulty with removal of the roots from the soil.

Nutrient conditions can have considerable influence on the growth pattern of roots (Fitter 1985). Generally, nutrient-deficient conditions will lead to long, unbranched root systems described as a herringbone pattern. Nutrient-sufficient conditions promote formation of lateral roots, resulting in a diffusely branched system. A root system passing through pockets of nutrients in soil may switch from herringbone (nutrient deficient) to highly branched (nutrient sufficient) and back to herringbone (another nutrient-deficient zone).

Root mass. Similar to the description for shoot mass, the measurement of root mass takes on more importance as branching occurs. Diffuse root systems may have most of the root biomass in secondary and tertiary root systems so that root length alone may not represent the full picture of plant growth response. The most difficult technical limitation in measuring root mass is the removal of soil particles and organic debris embedded among the roots. Vigorous washing results in loss of root tissue, whereas incomplete washing leaves large quantities of soil trapped among the roots. Care must be taken to minimize these two sources of error, as well as document the effectiveness of the root washing procedures. Estimates of the quantity of roots lost in washing and the residual quantities of soil add to the value of these data. As most soil contaminants reach plants through the roots, the roots typically exhibit the greatest response. Exceptions to this generalization would be compounds that exert toxicity to photosynthetic systems or stomatal control.

Total mass. Total plant mass provides the most definitive indication of an adverse plant response to toxic materials (Table 5). However, given the additional insights to phytotoxicity from separate measures of shoot mass and root mass, it is best to record the endpoints separately.

Table 5   Theoretical Phytotoxicity Response Patterns

	Shoot	Root	Total
Case 1	No change	No change	No change
Case 2	No change	Reduction	Reduction
Case 3	Reduction	No change	Reduction
Case 4	Reduction	Reduction	Reduction

Other permutations occur in which shoot or root growth are increased. Indeed, hormesis responses are commonly observed at low concentrations of many chemicals (Shirazi et al. 1992, 1994). There is no consensus as to whether stimulation should be considered toxic.

Shoot to root ratios. Plant physiology and ecology literature have provided valuable insights into the allocation of photosynthate among different plant parts. For a given species under favorable growth conditions, allocation patterns are relatively consistent. Developmental stages are associated with shifts of resource allocation. For example, many species initially direct more reserves toward root growth than shoot growth. Fogel (1985) reported allocation of photosynthate to roots ranges from 40 to 85%. As fruit set begins, resources shift toward growing the seeds and accessory fruit tissues. In crops, most of the gains in yield experienced in the past several decades have been due to successful shifting of energy from root growth to grain production. Nondomesticated species vary widely in the proportions allocated to shoots, roots, and reproductive tissues. Microbial associations also exert dramatic influence on the allocation patterns (Paul and Kucey 1981). Kapustka et al. (1985) determined that bacteria associated with roots cause shifts in net primary production ranging from 40 to 370% of controls. Allocation to roots accounted for most of the variation.

Photosynthesis

Photosynthesis is the most distinguishing feature of plants. Through a series of complex physical and chemical transformations, plants capture photosynthetically active radiation (approximately the same range as visible light) and convert a portion of the captured energy into sugar. It is this process that fuels everything else that occurs in ecological systems. The positive relationship between photosynthesis and productivity is intuitively obvious, though difficult to demonstrate (Zelitch 1982). Not until Christy and Porter (1982) measured total canopy photosynthesis was it possible to verify this intuitive relationship. Linkage to environmental stress is equally difficult due to the many annual, seasonal, and diurnal variations compounded by differences among species.

Faced with such problems, many are quick to dismiss photosynthetic analysis as impractical for toxicity testing. Much of the concern, and perhaps confusion over this issue, comes from our general persuasion in ecology to emphasize differences among species. Ecologically, plants exhibit wide differences in photosynthetic efficiency, rates of carbon assimilation, adaptation to light conditions, temperature, salinity, diurnal period, etc., and there are various alternative photosynthetic systems (i.e., C3, C4, CAM) adapted to different environmental conditions. However, photosynthesis is among the best-understood biological process. All key features of the photosynthetic apparatus have been characterized, including sequencing of the genes and the regulatory steps related to overall functioning of the system. In the process of characterizing the biophysics and biochemistry, many toxins (metabolic blocks) were used to verify models and chart the flow of electrons in photosynthesis. Consequently, a rich literature going back 60 years provides diagnostic insights to phytotoxic effects that involve photosynthesis.

Fluorescence. Fluorometric analysis of photosynthesis has been used to detect the genetic, biochemical, and physiological condition of plants (Jones and Winchell 1984; Miles et al. 1972; Miles 1980, 1990; Kramer et al. 1987; McClendon and Fukshansky 1990; Richardson et al. 1990; Walker 1988). The fundamental information regarding plant fluorescence dating to the 1930s was summarized by Franck and Loomis (1949). Plant fluorescence is better understood than most other biological methods used to evaluate environmental effects. Toxicological data, specific to photosynthetic systems, have been collected on hundreds of chemicals and several plant species over the past five decades.

The basis of this bioassay is the chlorophyll molecule, which serves as an intrinsic fluorescent probe of the performance and capacity of photosynthesis. Under normal conditions, 97% of the light energy absorbed by chlorophyll is converted to biochemical forms of energy in photosynthesis. Stress conditions can reduce the rate of photosynthesis, disturb the pigment–protein apparatus, or block the light-driven photosynthetic electron transport in the chloroplast. This results in 6 to 10% loss of absorbed light energy via chlorophyll fluorescence (the Kautsky Effect). Light-induced chlorophyll fluorescence from dark-adapted leaves can be recorded with portable, sensitive instruments using intact leaves. This nondestructive method monitors the physiological well-being of the plant. Any stress including disease, nutritional stress, water, temperature, radiation, and chemical stress can be recorded quickly and accurately. Chlorophyll fluorescence in intact native plants can be used to assess toxicity in the field or in a laboratory bioassay (Kapustka 1993).

Carbon fixation and oxygen evolution. Uptake of CO_2 or O_2 evolution are familiar biochemical techniques for studying the effects of chemicals on photosynthesis (Schafer and Bjorkman 1989; McFarlane et al. 1987a,b; McFarlane and Pfleeger 1987). Sophisticated methods of analyzing photosynthetic conditions are available (Nobel 1983; Larcher 1980). Portable units can be used to measure the "instantaneous" rates of net CO_2 uptake. Modifications of the basic method also permit full canopy measurements (Christy and Porter 1982). There are many technical considerations that require skilled personnel to design the study and ensure reliability of the resulting data. If the proper precautions are taken, however, excellent comparative data can be obtained to assess the impact of stress imposed by hazardous materials on the photosynthetic process. Relatively modest changes in protocols allow measurement of respiratory rates of nonphotosynthetic tissues or darkened photosynthetic tissues.

Isotope discrimination can also be used to assess long-term ecological conditions. The biophysical and biochemical features of leaves impose resistance to the incorporation of CO_2 (Farquhar et al. 1982; Hattersley 1982; O'Leary 1981). As a consequence of this resistance, plants discriminate among isotopes. This discrimination is confirmed by a comparison of the natural abundance of ^{13}C and ^{12}C to the abundance found in plants. Furthermore, the alternative photosynthetic pathways among plants exhibit differing levels of discrimination. Basically, any factor that affects the resistance of CO_2 influx enhances the discrimination. Thus, stressors that affect stomatal opening can be expected to alter the isotope discrimination. Peterson and Frye (1987) provide an excellent discussion of the processes of isotope discrimination and illustrate their uses for ecosystem analyses through several case studies.

Reproduction

Measures of yield are central to agronomic monitoring programs. Yet for some reason, phytotoxicity work has largely ignored the many ecologically relevant endpoints in various phases of reproduction. This appears to have been driven by the desire to promote simpler tests such as seed germination and root elongation. Though some extra costs would be involved in most tests using reproductive endpoints, the benefits gained in terms of better linkage to ecological assessment endpoints would often justify the costs.

A host of developmental events must happen correctly to shift plant growth from vegetative to reproductive structures. These range from production of flowers, pollen dispersal, and pollination through fruit or seed set, and, finally, the maturation of the fruit and seeds. Ecologically, this is a very important set of events both for the plant and for wildlife.

Morbidity

Many toxic chemicals, especially growth regulators, can affect organ and tissue development. Auxin-like chemicals often promote excessive apical elongation which may lead to curling or sagging of the stem. Chlorosis, premature senescence, rapid desiccation, and death of leaf tissues occur in response to some metals and selected organics. Deformed flowers and fruits, precocial senescence of fruits, and aborted seed set can be related to toxic conditions. Death of apical meristems in roots by lead results in blackened, stubby roots. Each of these symptoms can be early indicators of eventual death of the plant. At sublethal concentrations of toxins, these symptoms may be expressed, but the plant may recover. Seldom will these morphological biomarkers be sufficient in and of themselves because there are often several agents that could result in similar morphological features. Nevertheless, when evident, they should be noted. The manner of notation can range from a narrative description to a scalar that reflects increasing levels of effects.

Tissue concentration

Much of the phytotoxicity literature, especially on metal toxicity, reports plant tissue concentrations either in the leaves, shoots, fruits, or occasionally the roots. Mechanistically, the concentration of a toxic material is relevant only as it relates to the site of action. Therefore, it seems reasonable to measure the tissue concentration as a supplement to the other endpoints. In practical terms, there is relatively little value to this information, except in very carefully controlled research experiments.

First, it is impractical to measure the tissue concentration of most organics. Organic molecules are rapidly metabolized and become sequestered in cell walls and are insoluble, or they are modified slightly such that the analytical measures need to account for all metabolic transformations (McFarlane et al. 1990). Short of using a properly radiolabeled parent compound, measuring organics in plants is a wasted effort.

Metals can be measured more easily. Though similar difficulties arise if one is attempting to speciate metal or metalloid moieties, elemental concentrations can be determined with relatively high levels of precision and accuracy. Even so, one needs to distinguish between essential and nonessential elements. Most plants exhibit some capacity to regulate the concentration of essential elements. Therefore, even at toxic environmental concentrations, accumulation may be only slightly elevated in the plant tissues. Establishing statistical correlation between plant concentration and medium concentration can be frustrating. Better success is possible with nonessential metals, yet several pitfalls are likely. At high metal concentrations, roots are commonly the first part of the plant to be impaired. Consequently, with disruption of root functions, uptake of metals slows such that accumulation of metal in the plant (roots, shoots, or total) may be less than at subtoxic levels.

Analytical approaches

Data management, analysis, and interpretation are critical to the success of any phytotoxicity study. The general approaches must be anticipated at the design stage of the study to ensure that statistical assumptions are met and that proper collection of data can occur. Though most of the standardized tests were developed with an eye toward determining threshold toxicity values for single measurement endpoints, the tests are used most efficiently in hypotheses testing with multiple endpoints. Most standardized tests are remarkably vague regarding analytical methods. This can be interpreted to mean there is little agreement on the matter. Alternatively, given the varied uses of the tests, it is perhaps wise to leave considerable discretion to the investigator to design the test features and analyses to fit the questions at hand.

Virtually all of the standardized tests of terrestrial toxicity were modeled after aquatic tests. Aqueous environmental samples and single chemical additions are easily manipulated to produce various test concentrations. Soil or sediment samples taken from field situations present a number of nontrivial problems with respect to establishing dilutions series. Among the most difficult to address are heterogeneity of texture, differences among horizons, water content, and nutrient conditions. There is no universal way to solve the problems. Therefore, it is extremely important to characterize each assumption made in the study design. Also, proper quality assurance programming and quality control steps need to be incorporated.

Most standardized tests use replicated dilutions series (usually decade level or base two). Few descriptions speak directly to hypotheses testing of 100% environmental samples compared to reference samples. Yet this is a prominent use of terrestrial plant test data in survey or screening level risk assessments.

For quality assurance purposes, most tests require negative controls, many recommend use of a positive control, and a few recommend using spike additions of putative toxic materials. To the extent that any of these are built into the test design, they can influence the statistical model used to evaluate the results.

Effect level estimates

In cases where a dilution series is used, various conventions have been adopted to develop point estimates. Despite their widespread use, there are serious deficiencies and technical limitations embodied in these practices.

NOAEC and LOAEC

Regulatory practices have placed considerable weight on the no observable adverse effects concentrations (NOAEC). Though the concept of a NOAEC initially seems intuitively correct, it is fraught with difficulties, the least of which is that the NOAEC violates the basic premises of scientific method embodied in inference testing. Scientific experimentation relies on testing null hypotheses. Chapman et al. (1996) pointed out that the NOAEC is a poor construct for describing toxicity. Oris and others (California EPA 1995) suggested that the use of the NOAEC was the only situation where an experiment that fails to reject the null hypothesis establishes "proof" of a concept. Moreover, the statistics used to identify a NOAEC or the lowest observable effects concentration (LOAEC) fail to use most of the data collected in the study. Valuable information regarding slope of the response, as well as uncertainty associated with the point estimates, is lost. Finally, the presentation of a NOAEC or LOAEC is typically not usable in the context of risk assessment. With the widespread availability of powerful computers and sophisticated software to handle complex data, there is little justification for the extensive reliance on the simple point estimates. Their use should be discouraged in favor of more robust probabilistic statements that feed directly into risk assessments.

Median effects estimates

Various methods to estimate EC_{50} or EC_{xx} response concentrations have been used. These range from simple graphical methods to several regressions techniques. For ideal response profiles, the techniques are fairly reliable. Unfortunately, few tests are considerate enough to present ideal response profiles. Plants frequently exhibit classical stimulation patterns at low stress levels.

The traditional use of the median effects estimates has been to reduce the information to a point estimate without including the associated confidence interval. The EC_{50} has often been used for comparisons because it is the least responsive to the slope of the response and also has the smallest confidence interval. Despite widespread use, the estimates are not easily translated into estimates of toxicological risk.

An alternative approach to overcome some of these problems is to express toxicity test results as a probability of achieving an observed discrete effect level for a range of hypothesized effects (Erickson and McDonald 1995). For example, the data can be used to describe a concentration at which there is a 90% probability of seeing a 20% reduction of growth, a 95% probability of seeing a 30% reduction of growth, etc. This still would not translate directly to ecological effects, but it clearly would address risk estimates at the toxicity level.

Hypotheses testing

Though few of the standardized test methods speak directly to hypotheses testing, most of the phytotoxicity work related to site assessment or injury assessment is required to do so. By far, the most difficult task is to identify appropriate reference soil that can be used for comparisons of plant performance under the test conditions. Although artificial or synthetic soil mixtures can be used as a comparative standard, it is a virtual requirement to use reference site soil for the principle comparisons.

Dilutions series may be used either with artificial soil as the diluent or preferably mixing reference soil with the test soil; however, it is equally valid to use only 100% site soils (i.e., no dilution series). Ideally, several site soil samples can be compared to several reference soil samples. Parametric or nonparametric statistical methods can be used to test null hypotheses established in the study design.

Single vs. multiple endpoints

The various endpoints available in plant toxicity testing represent a wide range of observations and measurements (quantal, continuously distributed, and qualitative). Many widely used approaches are available to handle such data and are made readily accessible in software programs.

Multiple endpoints present a greater complication. Multivariate techniques (e.g., direct gradient analysis, ordination, and classification) have been used extensively in ecological research across ecological systems and agents (stressors). Many reference texts describe and evaluate the various techniques (Capen 1981; Gauch 1982; Gilbert 1987; Orloci et al. 1979). Many software packages offer ready access to both parametric and nonparametric analysis techniques. Phytotoxic studies typically yield data that fail homogeneity and normality tests. Though various methods to transform the data are available, it is often practical to use nonparametric statistical models to evaluate the data.

Ordering data into broad classes has been used in plant community ecology (Whittaker 1975; Greig-Smith 1983). Basic community and species taxonomic units are established from comparisons of multiple attributes; many of which may be more or less autocorrelated. Ranking toxicity responses into broad categories (e.g., quartiles or pentiles) can be effective in acknowledging precision limitations of a given test. It also is an effective method to reduce large quantities of data into meaningful comparisons that can be readily communicated to a general audience. Once such data are reduced to the unitless values of the class designation, it is also possible to accumulate rank scores of endpoints or of various taxa to obtain a synthesis of all the information (Phillips et al. 1994; Kapustka et al. 1996).

Summary and recommendations

The use of phytotoxicity as a critical part of EcoRAs has been limited by poor understanding of the field. Despite its strong technical foundation in physiology and ecology, phytotoxicity has been slow to advance. Serious impediments to a broader application of information come from excessive reliance on a few standardized protocols which have served to restrict the range of taxa examined and the endpoints recorded. If greater reliance were placed on scientific merit and less reliance placed on regulatory and legal precedent, risk assessments could be made more relevant. Potential improvements could be realized if testing was expanded to nonstandard species and measurement of a wider array of toxicity endpoints. The scientific integrity of tests that involve nonstandard species can be enhanced with relatively little extra effort. For nonstandard plant species, investigators should

1. Describe the germination or developmental stages (embryogensis in tissue culture or shoot and root initiation from cuttings) for the species, noting times and culture conditions required.
2. Describe the nominal performance standards (e.g., X% germinate in Y days) for the species and for the lot to be used in the test.
3. Characterize the statistical variability for each potential endpoint under negative control conditions. Particularly with slow growing plants or seeds with low germination rates, there may be unacceptable limits of precision available in the tests. Careful consideration of test performance requirements can be based on this information.
4. Evaluate the effect of a positive control (e.g., boric acid) for each endpoint.
5. Also, if there is relatively little information about the nonstandard species, it is advisable to use one or more standard test species in companion tests. The advantages of including some standard test species are that comparisons of responses among the nonstandard and standard species and comparisons of the standard species performance with other agents can be made.

In the current litigious atmosphere, the added costs to include more obviously appropriate or ecologically relevant species in the tests will likely be less than the cost of the lengthy arguments associated with justifying the use of standard species. This will be particularly true if nonconfrontational negotiations among regulators, managers, and other stakeholders are used to establish a strategic framework that emphasizes ecological relevance over past precedent.

The key to any successful study is the proper framing of the questions to be addressed, (i.e., the assessment endpoints). Care should be given to ensure the measurement endpoints are those best suited to answer the specific questions of a study. The choice of measurement endpoints involves technical issues of precision and accuracy as well as issues such as time and cost. As with the selection of test species, efforts should be made to include a broader array of measurement endpoints than that which is often used. Below-ground measures of growth and biomarkers (physiological and morphological) largely have been ignored in the standard test procedures. If coupled with the more traditional endpoints of germination, survival, and above-ground growth, the phytotoxicity test database would be greatly improved.

References

Alloway B J, (1974) *Heavy Metals in Soils,* John Wiley & Sons, New York.
APHA (1976) *Standard Methods for the Examination of Water and Wastewater,* 14th ed., Method No. 209E, American Public Health Association, Washington, D.C., pp. 95–96.

APHA, AWWA, and WEF (1992) Aquatic plants. In *Standard Methods for the Examination of Water and Wastewater,* 18th ed., No. 8220, American Public Health Association, Washington, D.C., pp. 8.42–8.45.

ASTM (1994) *Standard Practice for Conducting Early Seedling Growth Tests [E 1598 — 94],* American Society for Testing and Materials, Philadelphia, PA.

Baskin JM, Baskin CC (1985) The annual dormancy cycle in buried weed seeds: a continuum. *Bioscience* 35:492–498.

Benenati F (1990) Plants — keystone to environmental risk assessment. In *Plants for Toxicity Assessment,* Wang W, Gorsuch JW, Lower WR, Eds., ASTM STP 1091, American Society for Testing and Materials, Philadelphia, PA, pp. 5–13.

Brown J, Miller PC, Tieszen LL, Bunnell FL (1980) *An Arctic Ecosystem,* US/IBP Synthesis Series #12, Hutchinson Ross Publishing Company, Stroudsburg, PA.

Byl TD, Klaine SJ (1990) Peroxidase activity as an indicator of sublethal stress in the aquatic plant *Hydrilla verticillata* (Royle) in plants for toxicity assessment. In *Plants for Toxicity Assessment: Second Volume,* Gorsuch JW, Lower WR, Lewis MA, Wang W, Eds., ASTM STP 1115, American Society for Testing and Materials, Philadelphia, PA, pp. 101–106.

California EPA (1995) Ecotoxicological Risk Assessment Workshop Series. Ecotoxicology Unit, Office of Environmental Health Hazard Assessment, California Environmental Protection Agency, Sacramento, CA.

Capen DE, Ed. (1981) The use of multivariate statistics in studies of wildlife habitat. USDA Forest Service, General Technical Report RM-87, Rocky Mountain Forest and Range Experiment Station, Fort Collins, CO.

CFR (Code of Federal Regulations) (1990a) 43 CFR 11 (10-1-90 Edition). Natural resource damage assessments.

CFR (Code of Federal Regulations) (1990b) 40 CFR 1 (7-1-90 Edition). §797.2750 Seed germination/root elongation toxicity test.

Chapman PM, Caldwell RS, Chapman PF (1996) A warning: NOECs are inappropriate for regulatory use. *Environ Toxicol Chem* 15:77–79.

Christy LA, Porter CA (1982) Canopy photosynthesis and yield in soybean. In *Photosynthesis — Applications to Food and Agriculture,* Govindjee, Ed., Academic Press, New York, pp. 499–511.

Cushman R, Meyer RD (1990) Improving the utilization of non-traditional agricultural products through coordination during the registration process. *Commun Soil Sci Plant Anal* 21:1531–1540.

Edmonds RL (1982) *Analysis of Coniferous Forest Ecosystems in the Western United States,* US/IBP Synthesis Series #14, Hutchinson Ross Publishing Company, Stroudsburg, PA.

Epstein E (1972) *Mineral Nutrition of Plants: Principles and Perspectives,* Wiley, New York.

Erickson WP, McDonald LL (1995) Tests for bioequivalence of control media and test media in studies of toxicity. *Environ Toxicol Chem* 14:1247–1256.

Farquhar GD, O'Leary MH, Berry JA (1982) On the relationship between carbon isotope discrimination and the intercellular carbon dioxide concentration in leaves. *Aust J Plant Physiol* 9:121–137.

Fitter AH (1985) Functional significance of root morphology and root system architecture. In *Ecological Interactions in Soils: Plants, Microbes, and Animals,* Fitter AH, Atkinson D, Read DJ, Usher MB, Eds., British Ecological Society Special Publication No. 4, Blackwell Scientific Publications, London pp. 87–106.

Fletcher JS, Muhitch MJ, Vann DR, McFarlane JC, Benenati FE (1985) PHYTO-TOX database evaluation of surrogate plant species recommended by the U.S. Environmental Protection Agency and the Organization for Economic Cooperation and Development. *Environ Toxicol Chem* 4:523–532.

Fletcher JS, Johnson FL, McFarlane JC (1988) Database assessment of phyto-toxicity data published on terrestrial vascular plants. *Environ Toxicol Chem* 7:615–622.

Fogel R (1985) Roots as primary producers in below-ground ecosystems. In *Ecological Interactions in Soils: Plants, Microbes, and Animals,* Fitter AH, Atkinson D, Read DJ, Usher MB, Eds., British Ecological Society Special Publication No. 4, Blackwell Scientific Publications, London, pp. 23–36.

Folsom BL, Price RA (1991) A plant bioassay for assessing plant uptake of contaminants from freshwater soils or dredged material. In *Plants for Toxicity Assessment: Second Volume,* Gorsuch JW, Lower WR, Lewis MA, Wang W, Eds., ASTM STP 1115, American Society for Testing and Materials, Philadelphia, PA, pp. 172–177.

Folsom BL Jr., Lee CR, Bates DJ (1981) Influence of Disposal Environment on Availability and Plant Uptake of Heavy Metals in Dredged Material. Technical Report EL-81-12, U.S. Army Engineer Waterways Experiment Station, Vicksburg, MS.

Folsom BL Jr., Skogerboe JG, Palermo MR, Simmers JW, Pranger SA, Shafer RA (1988) Synthesis of the Results of the Field Verification Program Upland Disposal Alternative. Technical Report D-88-7, U.S. Army Engineer Waterways Experiment Station, Vicksburg, MS.

Franck J, Loomis W, Eds. (1949) *Photosynthesis in Plants,* American Society Plant Physiologist, Iowa State College Press, Ames.

Garten CT, Frank ML (1984) *Comparison of Toxicity to Terrestrial Plants with Algal Growth Inhibition by Herbicides,* ORLN/TM-9177, Oak Ridge National Laboratory, Oak Ridge, TN.

Gauch HG Jr. (1982) *Multivariate Snalysis in Community Ecology,* Cambridge University Press, Cambridge, England.

Gilbert RO (1987) *Statistical Methods for Environmental Pollution Monitoring,* Van Nostrand Reinhold, New York.

Gingerich O (1980) The Galileo Affair. *Sci Am* 1980:133–143.

Greene JC, Bartels CL, Warren-Hicks WJ, Parkhurst BR, Linder GL, Peterson SA, Miller WE (1988) *Protocols for Short-Term Toxicity Screening of Hazardous Waste Sites,* EPA/600/3-88/029, U.S. Environmental Protection Agency, Corvallis, OR.

Greig-Smith P (1983) *Quantitative Plant Ecology,* 3rd ed., University of California Press, Berkeley.

Hattersley PW (1982) Delta ^{13}C values of C_4 types in grasses. *Aust J Plant Physiol* 9:139–154.

Hinman ML (1989) Utility of rooted aquatic vascular plants for aquatic sediment hazard evaluations. Dissertation. Memphis State University, Memphis, TN, 31 pp.

Holst RW (1986a) Hazard Evaluation Division Standard Procedure Non-Target Plants: Seed Germination/Seedling Emergence — Tiers 1 and 2. EPA 5430/9-86-132, Office of Pesticides and Toxic Substances, U.S. Environmental Protection Agency, Washington, D.C.

Holst RW (1986b) Hazard Evaluation Division Standard Procedure Non-Target Plants: Vegetative Vigor — Tiers 1 and 2. EPA 5430/9-86-133, Office of Pesticides and Toxic Substances, U.S. Environmental Protection Agency, Washington, D.C.

Holst RW (1986c) Hazard Evaluation Division Standard Procedure Non-Target Plants: Growth and Reproduction of Aquatic Plants — Tiers 1 and 2. EPA 5430/9-86-134, Office of Pesticides and Toxic Substances, U.S. Environmental Protection Agency, Washington, D.C.

Holst RW (1986d) Hazard Evaluation Division Standard Procedure Non-Target Plants: Terrestrial Field Testing — Tier 3. EPA 5430/9-86-135, Office of Pesticides and Toxic Substances, U.S. Environmental Protection Agency, Washington, D.C.

Holst RW (1986e) Hazard Evaluation Division Standard Procedure Non-Target Plants: Aquatic Field Testing — Tier 3. EPA 5430/9-86-136, Office of Pesticides and Toxic Substances, U.S. Environmental Protection Agency, Washington, D.C.

Holst RW, Ellanger TC (1982) Pesticide Assessment Guidelines. Subdivision J. Hazard Evaluation: Non-Target Plants. EPA-540/9-82-020, Office of Pesticides and Toxic Substances, U.S. Environmental Protection Agency, Washington, D.C.

Howell FG, Gentry JB, Smith MH (1975) *Mineral Cycling in Southeastern Ecosystems,* U.S. Energy Research and Development Administration, NTIS, Springfield, VA.

ICF, Inc. (1989) Scoping study of the effects of soil contamination on terrestrial biota. Prepared for Office of Toxic Substance, U.S. Environmental Protection Agency, Washington, D.C.

Ingestad T (1981) Nutrition and growth of birch and grey alder seedlings in low conductivity solutions and at varied relative rates of nutrient addition. *Physiol Plant* 52:454–466.

Ingestad T (1982) Relative addition rate and external concentration; driving variables used in plant nutrition research. *Plant Cell Environ* 5:443–453.

Jones TW, Winchell L (1984) Uptake and photosynthetic inhibition by atrazine and its degradation products on four species of submerged vascular plants. *J Environ Qual* 13:250–300.

Kabata-Pendias A, Pendias H (1992) *Trace Elements in Soils and Plants,* 2nd ed., CRC Press, Boca Raton, FL.

Kapustka LA (1993) Chlorophyll fluorescence: its status and future as a non-intrusive assay of plant stress. In *Environmental Toxicology and Risk Assessment: Second Volume,* Gorsuch JW, Dwyer FJ, Ingersoll CB, T La Point TW, Eds., ASTM STP 1216, American Society for Testing and Materials, Philadelphia, PA, pp. 123–133.

Kapustka LA (1996) Plant ecotoxicology: the design and evaluation of plant performance in risk assessments and forensic ecology. In *Environmental Toxicology and Risk Assessment: Fourth Volume,* La Point TW, Price FT, Little EE, Eds., ASTM STP 1262, American Society for Testing and Materials, Philadelphia, PA, pp. 110–121.

Kapustka LA, Reporter M (1993) Terrestrial primary producers. In *Handbook of Ecotoxicology,* Calow P, Ed., Blackwell Press, Oxford, chap. 14, pp. 278–297.

Kapustka LA, Arnold PT, Lattimore PT (1985) Interactive responses of associative diazotrophs from a Nebraska Sand Hills grassland. In *Ecological Interactions in Soils: Plants, Microbes, and Animals*, Fitter AH, Atkinson D, Read DJ, Usher MB, Eds., British Ecological Society Special Publication No. 4, Blackwell Scientific Publications, London, pp. 149–158.

Kapustka LA, Lipton J, Galbraith H, Cacela D, LeJeune K (1995) Metal and arsenic impacts to soils, vegetation communities, and wildlife habitat in southwest Montana uplands contaminated by smelter emissions: II. Laboratory phytotoxicity studies. *Environ Toxicol Chem* 14:1095–1912.

Kapustka LA, Williams BA, Fairbrother A (1996) Evaluating risk predictions at population and community levels in pesticide registration — hypotheses to be tested. *Envrion Toxicol Chem* 15:427–431.

Kramer D (1983) Genetically determined adaptations in roots to nutritional stress: correlation of structure and function (iron deficiency, soil salinity). *Dev Plant Soil Sci* 8:33–39.

Kramer D, Adawi O, Morse PD, Crofts AR (1987) A portable double-flash spectrophotometer for measuring the kinetics of electron transport components in intact leaves. *Progr Photosynth Res* 2:665–668.

Krstich MA, Schwartz OJ (1990) Characterization of xenobiotic uptake utilizing an isolated root uptake test (IRUT) and a whole plant uptake test (WPUT). In *Plants for Toxicity Assessment*, Wang W, Gorsuch J, Lower W, Eds., ASTM STP 1091, American Society for Testing and Materials, Philadelphia, PA, pp. 87–96.

Kruckeberg QR (1954) The ecology of serpentine soils. III. Plant species in relation to serpentine soils. *Ecology* 35:267–274.

Larcher W (1980) *Physiological Plant Ecology*, 2nd ed., Springer-Verlag, Berlin.

Lee CR, Folsom BL Jr., Bates DJ (1978) Prediction of Heavy Metal Uptake by Marsh Plants Based on Chemical Extraction of Heavy Metals from Dredged Material. Technical Report D-78-5, U.S. Army Engineer Waterways Experiment Station, Vicksburg, MS.

Lee CR, Folsom BL Jr., Engler RM (1982) Availability and plant uptake of heavy metals from contaminated dredged material placed in flooded and upland disposal environments, *Environ Int* 7:65–71.

Lee CR, Folsom BL Jr., Bates DJ (1983) Prediction of plant uptake of toxic metals using a modified DTPA soil extractant. *Sci Total Environ* 28:191–202.

Liebig J (1840) *Chemistry in Its Agriculture and Physiology*, Taylor and Walton, London.

Linder G, Greene JC, Ratsch H, Nwosu J, Smith S, Wilborn D (1990) Seed germination and root elongation toxicity tests in hazardous waste site evaluation: methods development and applications. In *Plants for Toxicity Assessment*, Wang W, Gorsuch J, Lower W, Eds., ASTM STP 1091, American Society for Testing and Materials, Philadelphia, PA, pp. 177–187.

Mayer AM, Poljakoff-Mayber A (1989) *The Germination of Seeds*, 4th ed., Pergamon Press, New York.

McFarlane C, Nolt C, Wickliff C, Pfleeger T, Shimabuku R, McDowell M (1987a) The uptake, distribution and metabolism of four organic chemicals by soybean plants and barley roots. *Environ Toxicol Chem* 6:847–856.

McFarlane JC, Pfleeger T (1987) Plant exposure chambers for study of toxic chemical-plant interactions. *J Environ Qual* 16:361–371.

McFarlane JC, Pfleeger T, Fletcher J (1990) Effect, uptake, and disposition of nitrobenzene in several terrestrial plants. *Environ Toxicol Chem* 9:513–520.

McFarlane JC, Pfleeger T, Fletcher J (1987b) Transpiration effect on the uptake and distribution of bromacil, nitrobenzene, and phenol in soybean plants. *J Environ Qual* 16:372–376.

McClendon JH, Fukshansky L (1990) On the interpretation of absorption spectra of leaves. II. The non-absorbed ray of the sieve effect and the mean optical pathlength in the remainder of the leaf. *Photochem Photobiol* 51:211–216.

McMillan C (1956) The edaphic restriction of *Cupressus* and *Pinus* in the Coast Ranges of central California. *Ecol Monogr* 26:177–1738.

Miles CD, Brandle JR, Daniel DJ, Chu-Der O, Schnare PD, Uhlik DJ (1972) Inhibition of Photosystem II in isolated chloroplasts by lead. *Plant Physiol* 49:820–825.

Miles D (1980) Mutants of higher plants: maize. *Methods Enzymol* 69:3–23.

Miles D (1990) The role of chlorophyll fluorescence as a bioassay for assessment of toxicity in plants. In *Plants for Toxicity Assessment,* Wang W, Gorsuch JW, Lower WR, Eds., ASTM STP 1091, American Society for Testing and Materials, Philadelphia, PA, pp. 297–307.

Nobel PS (1983) *Biophysical Plant Physiology and Ecology,* Freeman Press, San Francisco, CA.

Norton D, Stinson M (1993) Soil Bioassay Pilot Study: Evaluation of Screening Level Bioassays for Use in Soil Toxicity Assessments at Hazardous Waste Sites Under the Model Toxics Control Act. Washington State Department of Ecology, Olympia, WA.

Nwosu JU, Ratsch HC, Kapustka LA (1991) A protocol for on-site seed germination test. In *Plants for Toxicity Assessment: Second Volume,* Gorsuch JW, Lower WR, Lewis MA, Wang W, Eds., ASTM STP 1115, American Society for Testing and Materials, Philadelphia, PA, pp. 333–340.

O'Leary MH (1981) Carbon isotope fractionation in plants. *Phytochemistry* 20:553–567.

Organization for Economic Cooperation and Development (OECD) (1984) *OECD Guideline for Testing of Chemicals. Guideline 208 'Terrestrial Plants Growth Test,'* OECD, Paris, 6 pp.

Orloci L, Rao CR, Stiteler WM (1979) *Multivariate Methods in Ecological Work,* International Cooperative Publishing House, Fairland, MD.

Page AL, Gleason TL, Smith JE, Iskandar IK, Sommers LE (1983) Utilization of municipal wastewater and sludge on land. Proceedings of the 1983 Workshop. University of California, Riverside, CA.

Parkhurst BR, Linder GL, McBee K, Bitton G, Dutka BJ, Hendricks CW (1989) Toxicity tests. In *Ecological Assessments of Hazardous Waste Sites: A Field and Laboratory Reference,* Parkhurst BR, Baker SS, Eds., EPA 600/3-89/013, U.S. Environmental Protection Agency, Corvallis, OR, pp. 6.1–6.66.

Paul EA, Kucey R (1981) Carbon flow in microbial associations. *Science* 213:473–476.

Peterson BJ, Frye B (1987) Stable isotopes in ecosystem studies. *Annu Rev Ecol Syst* 18:293–320.

Phillips CT, Checkai RT, Chester NA, Wentsel RS, Major MA, Amos JC, Simini M (1994) Toxicity testing of soil samples from Joliet Army Ammunition Plant, IL. Edgewood Research, Development, and Engineering Center, U.S. Army ERDEC-TR-137, Aberdeen Proving Ground, MD.

Ratsch HC, Johndro DJ, McFarlane JC (1986) Growth inhibition and morphological effects of several chemicals in *Arabidopsis thaliana* (L.) Heynh. *Environ Toxicol Chem* 5:55–60.

Richardson CJ, Sasek TW, DiGiulio RT (1990) Use of physiological and biochemical markers for assessing air pollution stress in trees. In *Plants for Toxicity Assessment*, Wang W, Gorsuch J, Lower W, Eds., ASTM STP 1091, American Society for Testing and Materials, Philadelphia, PA, pp. 143–155.

Risser PG, Birney EC, Blocker HD, May SW, Parton WJ, Wiens JA (1981) *The True Prairie Ecosystem*, US/IBP Synthesis Series #16, Hucthinson Ross Publishing Co., Stroudsburg, PA.

Royce CL, Fletcher JS, Risser PG, McFarlane JC, Benenati FE (1984) PHYTOTOX: a database dealing with the effect of organic chemicals on terrestrial vascular plants. *J Chem Inf Comput Sci* 24:7–10.

Schafer C, Bjorkman O (1989) Relationship between efficiency of photosynthetic energy conversion and chlorophyll fluorescence quenching in upland cotton (*Gossypium hirsutum* L.). *Planta* 178:367–376.

Shimabuku RA, Ratsch HC, Wise CM, Nwosu JU, Kapustka LA (1991) *Methodology for a New Plant Life-Cycle Bioassay Featuring Rapid Cycling Brassica*, Vol. 2, Gorsuch JW, Lower WR, St. John KR, Eds., ASTM STP 1115, American Society for Testing and Materials, Philadelphia, PA, pp. 365–375.

Shirazi MA, Ratsch HC, Peniston BE (1992) The distribution of relative error of toxicity of herbicides and metals to *Arabidopsis*. *Environ Toxicol Chem* 11:237–243.

Shirazi MA, Robideaux ML, Kapustka LA, Wagner JJ, Reporter MC (1994) Cell growth in plant cultures: an interpretation of the influence of initial weight in cadmium and copper toxicity tests. *Arch Environ Contam Toxicol* 27:331–337.

Tadros TM (1957) Evidence of the presence of an edaphobiotic factor in the problem of serpentine tolerance. *Ecology* 38:14–23.

Taraldsen JE, Norberg-King T (1990) New method for determining effluent toxicity using duckweed (*Lemna minor*). *Environ Toxicol Chem* 9:761–767.

U.S. Army Environmental Laboratory (1987) *Disposal Alternatives for PCB-Contaminated Sediments from Indiana Harbor, Indiana*, Vol. 11, U.S. Army Engineer District, Chicago, IL.

U.S. Environmental Protection Agency (1985a) Ambient water quality criteria for lead. EPA 440/5-84-027, U.S. Environmental Protection Agency, Washington, D.C.

U.S. Environmental Protection Agency (1985b) Environmental profiles and hazard indices for constituents of municipal sludge: cadmium. Office of Water Regulations and Standards, U.S. Environmental Protection Agency, Washington, D.C.

U.S. Environmental Protection Agency (1985c) Environmental profiles and hazard indices for constituents of municipal sludge: copper. Office of Water Regulations and Standards, U.S. Environmental Protection Agency, Washington, D.C.

U.S. Environmental Protection Agency (1985d) Environmental profiles and hazard indices for constituents of municipal sludge: lead. Office of Water Regulations and Standards, U.S. Environmental Protection Agency, Washington, D.C.

U.S. Environmental Protection Agency (1985e) Environmental profiles and hazard indices for constituents of municipal sludge: zinc. Office of Water Regulations and Standards, U.S. Environmental Protection Agency, Washington, D.C.

U.S. Environmental Protection Agency (1985f) Toxic substance control act test guidelines: environmental effects testing guidelines. 40 CFR Part 797, *Fed Reg* 50(188):39389.

U.S. Environmental Protection Agency (1991) *Plant Tier Testing: A Workshop to Evaluate Nontarget Plant Testing in Subdivision J Pesticide Guidelines.* EPA/600/9-91/041, U.S. Environmental Protection Agency, Washington, D.C.

U.S. Environmental Protection Agency (1992) *Framework for Ecological Risk Assessment. Risk Assessment Forum.* EPA/630/R-92/001, U.S. Environmental Protection Agency, Washington, D.C.

U.S. Food and Drug Administration (1987) Seed germination and root elongation. In *Environmental Assessment Technical Handbook 4.06,* Center for Food Safety and Applied Nutrition, Center for Veterinary Medicine, Washington, D.C.

Walbridge CT (1977) A flow-through testing procedure with duckweed (*Lemna minor* L.). EPA600/3-77-108, U.S. Environmental Protection Agency, Duluth, MN.

Walker D (1988) Measurement of O_2 and chlorophyll fluorescence. In *Techniques in Bioproductivity and Photosynthesis: Second Edition,* Coombs J, Hall D, Long S, Scurlock J, Eds., Pergamon Press, Oxford, England, pp. 95–106.

Walsh GE, Weber DE, Simon TL, Brashers LK, Moore JC (1991) Use of marsh plants for toxicity testing of water and sediment. In *Plants for Toxicity Assessment: Second Volume,* Gorsuch JW, Lower WR, Lewis MA, Wang W, Eds., ASTM STP 1115, American Society for Testing and Materials, Philadelphia, PA, pp. 341–354.

Wang W (1986a) Toxicity tests of aquatic pollutants by using common duckweed. *Environ Pollut* 11:1–14.

Wang W (1986b) The effect of river water on phytotoxicity of Ba, Cd and Cr. *Environ Pollut* 33(b):193–204.

Wang W, Gorsuch JW, Lower WR, Eds. (1990) *Plants for Toxicity Assessment,* ASTM STP 1091, American Society for Testing and Materials, Philadelphia, PA.

Weinberger P, Caux PY (1985) Effects of the solvent carrier Dowanol on some growth parameters of the aquatic angiosperm *Lemna minor* L. *Can Tech Rep Fish Aquat Sci* 1368:265–286.

Weinstein LH, Laurence JA, Mandl RH, Walti K (1990) Use of native and
    cultivated plants as bioindicators and biomonitors of pollution damage.
    In *Plants for Toxicity Assessment*, Wang W, Gorsuch J, Lower W, Eds.,
    ASTM STP 1091, American Society for Testing and Materials, Philadel-
    phia, PA, pp. 117–126.
Whittaker RH (1954) The ecology of serpentine soils. I. Introduction. *Ecology*
    35:258–259.
Whittaker RH (1975) *Communities and Ecosystems*, 2nd ed., Macmillan Press,
    New York.
Woodwell GM (1967) Toxic substances and ecological cycles. *Sci Am*
    216(March):24.
Zelitch I (1982) The close relationship between net photosynthesis and crop
    yield. *Bioscience* 32:796–802.

Index

A

Abiotic influences, 420
Above-ground organs, 400
Acclimation, of plants to UV-B, 5, 9, 15
Accumulation factor (AF), 380
Acenaphthylene, 470, 474
Acetyltransferase, 115
O-Acetyltransferase (OAT), 115, 116, 118
Acid
    activities, 292
    -mine drainage, 152, 162
    rain, 152
Action spectroscopy, 12
Action spectrum, 7, 13
Active oxygen
    damage to plant biomolecules via, 2
    detoxifixation, 20
Aerial contamination, 435, 438
Aerosols, 434
AF, see Accumulation factor
AFDM, see Ash-free dry mass
AFS, see Apparent free space
Agent standards, 134
Agricultural pesticides, 161
Air pollution
    bryophyte population and, 350
    germination of plants and, 352
    measuring, 351
    response of plants to, 351
Air-to-foliar transfer, 439
Alfalfa, 86, 519
Algae, 59, 141
    artificial substrata, 186
    autecological indices, 189, 191
    bioassays, 186
    biomass, 186, 187
    blue-green, 161
    community indices, 189, 193
    counting methods, 194
    diversity indices, 194

ecological importance, 179
freshwater, 144
functional indices, 196, 198
geographical distribution, 180
limitations as indicators, 183
multivariate analysis, 189, 193,
    196
pollution tolerance, 189
reference condition, 182
response to nutrients, 180, 183, 196,
    198
response to toxic chemicals, 180, 181,
    183
sensitivity to pollution, 180
sensitivity to stress, 189
similarity indices, 195
species richness, 193, 194
taxonomic analysis, 181
taxonomic indicators, 187, 188
toxic freshwater, 158
toxic marine, 158
types of indicators, 186
Algal biomass, 186, 287
Algal blooms, 156, 157
    blue-green, 157
    freshwater, 156
Algal production, 179, 183, 197
Algal toxicity, 159
Algicides, 160, 164
Alkaline phosphatase activities, 292
Allium test, for screening chemicals,
    307–333
    discussion, 323–329
        clastogens and aneugens, 325–326
        comparison of toxicity and
            genotoxicity, 323–325
        extracellular micronuclei, 328–329
        micronuclei correlated to other
            parameters, 327–328
        micronuclei as test parameter,
            326

D

E

T - #0222 - 101024 - C0 - 234/156/31 [33] - CB - 9781566700283 - Gloss Lamination